C0-BWQ-971

Using Computers to Solve Reservoir Engineering Problems

Second Edition

Using Computers to Solve Reservoir Engineering Problems

Second Edition

M. A. Nobles

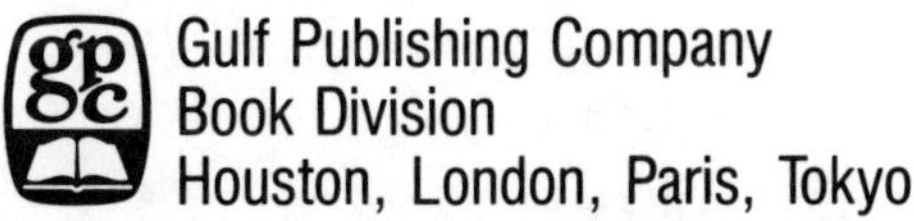

Gulf Publishing Company
Book Division
Houston, London, Paris, Tokyo

Using the Computer to Solve Petroleum Engineering Problems
Second Edition

Library of Congress Cataloging in Publication Data

Nobles, M.A.

Using computers to solve reservoir engineering problems.

Rev. ed. of: Using the computer to solve petroleum engineering problems. 1974.

Includes bibliographies and index.

1. Oil reservoir engineering—Data processing. 2. Microcomputers—Programming. 3. Basic (Computer program language) 4. FORTRAN (Computer program language) I. Nobles, M. A. Using the computer to solve petroleum engineering problems. II. Title.

TN871.N6 1984 622'.3382 84-3759
ISBN 0-87201-899-7

Contents

Preface

Since the first edition of this book was published in 1974, advances in solid-state physics have led to the mass production of programmable calculators and microcomputers. This new second edition is in response to the development of this technology, with an emphasis on microcomputers. It is now possible for a professional with a microcomputer to learn programming of scientific and engineering problems in a number of different languages, such as BASIC, Pascal, and FORTRAN. The principal advantage of a microcomputer over a large, central computer is that, over a given period of time, a person is able to write and test more programs; if there is a mistake in a program, the program can be rewritten and tested without anyone else being forced to deal with the mistake, nor does one have to wait for the program to be run on a large, central computer. Hours of time can be saved.

The basic content of this book remains the same; that is, mathematical and engineering material for writing many computer programs. It is written for engineers and scientists who have had a course in differential equations and a course in petroleum reservoir engineering. The mathematical and engineering material, while useful, is not comprehensive, since it is not the purpose of this book to cover in depth any topics as they are covered in mathematics and engineering courses. Note: if you are interested in machine language programming, read the appendix on number systems.

What is new in this second edition? Two chapters, one on BASIC programming and another on FORTRAN 77 programming, aimed at microcomputer use. These two chapters plus the documentation supplied by microcomputer manufacturers are enough to develop practical programming skills in BASIC and FORTRAN 77 languages. The other chapters help the reader learn to use microcomputers in the two languages and serve as reference material.

I used the Vic-20 by Commodore International, Ltd.; the TRS-80 Model III, TRS-80 Pocket Computer PC-2, and TRS-80 Extended BASIC Color Computer, all by Radio Shack; and the Apple II*e* by Apple Computer, Inc. The BASIC language programs included in Chapter 11 are written for the Apple II*e* and were tested on the same. All but one of the programs were tested on the Radio Shack PC-2 pocket microcomputer. Few changes were needed to alter a BASIC program written for the Apple II*e* so that it was suited for the other microcomputers.

This book is ideal for instructional purposes. With supporting material from selected readings, there is enough for a six semester-hour course. The programs in Chapters 10 and 11 are relatively simple and are graded, i.e., as one goes deeper into the chapters, the programs become more difficult. Each program was selected to demonstrate different computer skills.

The material that went into this book has been used at Mississippi State University for the past seven years. During the spring of 1983, the programs in Chapter 11 formed the basis for a petroleum engineering course in microcomputers. In addition to testing the programs on microcomputers, the programs were tested through terminals on the large UNIVAC computer at the computer center. Students registered in the course indicated they were pleased with the material.

The BASIC programs given in Chapter 11 are stored on floppy disks; also, most of the BASIC programs have been written in Apple FORTRAN and stored on floppy disks. Copies of these disks are available from the author.

M. A. Nobles
899 Southgate
Cookeville, TN. 38501

1 Electronic Digital Computers: Micro, Mini, Mainframe

Since the construction of the world's first electronic digital computer, the 30-ton ENIAC, built over a period from 1942 to 1945, there have been numerous technological advances in computers. These advances did not change the basic principles upon which electronic digital computers operate (i.e., the binary number system and Boolean algebra); instead, they brought about a reduction in their physical size while at the same time increasing computing capacity.

The first advance came in the early 1950s when the IBM 650 computers were manufactured. These machines contained compact vacuum tubes and a magnetic drum and marked the beginning of computer use in businesses, institutions of higher learning, and research laboratories. Machine languages were used by the early computers until the middle 1950s when symbolic languages were developed; however, symbolic languages were almost as unwieldy as machine languages and were rapidly supplanted by problem-oriented languages such as ALGOL, COBOL, FORTRAN, and BASIC. In the late 1950s solid-state computers with magnetic-core memories were produced. The IBM 1620 computer is an example of this type. Minicomputers such as the PDP-8 came on the market in the mid-1960s. These computers were very compact at that time (about the size of two desk drawers) and sold for about $20,000.

The development of the silicon chip (1967) led to the mass production of the programmable calculator, and then to the microcomputer (personal computer). Microcomputers now range in size from a handy pocket version (TRS-80 PC-2, Tandy Corp.) to a large typewriter and in price from less than $100 to $10,000. The cost depends on the amount of hardware built in or added to the basic microcomputer such as a printer interface, keyboard, memory, modem, etc.

The first electronic digital computers were produced for solving large engineering and scientific problems. Later it was discovered that computers were also suited for data processing, information retrieval, check writing, typesetting, and many other business-related tasks. All of these uses have greatly increased the demand for computers; in fact, the demand for computers designed for business is now greater than for engineering and scientific problem solving.

Microcomputer manufacturers have also added some features, such as color and sound, which makes them attractive to the whole family for entertainment. Sound effects such as explosions, music, etc., can be produced on a large number of microcomputers, and with some as many as eight or more colors can be produced on a color TV screen. Also, with some microcomputers it is possible to produce motion effects such as bouncing balls, flying objects, etc. With sound and animation, games can be programmed and produced on TV screens.

In summary, microcomputers can be used for entertainment, business, and solving engineering and scientific problems. This book concentrates on the latter application and is for learning and developing computer skills whether it be microcomputer, minicomputer, or mainframe.

Basic Components

Electronic digital computers have the following basic functional components: input, memory, control, arithmetic logical unit, and output. The functions of the basic components are:

1. The input receives information and instructions from an outside source such as a computer operator or an electronic device (another computer, etc.)
2. The memory stores the incoming information, incoming instructions, temporary information, temporary instructions, and final results.
3. The control unit converts the incoming instructions from the input for processing and in perfect synchronization activates the various sections of the computer for the processing.
4. The arithmetic logical unit (ALU) adds, subtracts, rearranges, and makes decisions about numbers.
5. The output transfers the processed information to an operator or an electronic device (a printer or another computer).

The ALU and the control form what is often called the nerve center of an electronic digital computer; the two taken together form the central processing unit (CPU) of the computer. The basic building block of microcomputers is the gate. The gate is similar to a switch, it is either 0 (off) or 1 (on). A silicon chip often contains thousands of gates, and the clusters of gates on the chip give the computer its many capabilities—adding, subtracting, counting, storing information, making decisions, etc.

So that the reader understands more fully how the functional components operate, a brief discussion of programming is presented. The discussion is based upon illustrations using a calculator and two obsolete computers, the LGP-30 and the IBM 1620. Also, six simple programs written in machine language are given for the LGP-30. For the majority of individuals using electronic digital computers, the use of machine language is a thing of the past; however, there is a very close analogy between programming in machine language for a computer and programming for a programmable calculator. Knowledge of a machine language aids understanding the organization of the functional components and operation of an electronic digital computer.

Programming

Programming is the planning of a method and/or procedure to follow for the solving of a problem. The type of program is controlled by the device used for obtaining the solution of the problem. The most common devices used are slide rules, desk calculators, electronic digital calculators, and analog calculators.

Computing devices which perform the operations of the procedure with discrete numbers are termed "digital computers." On the other hand, computing devices which deal with continuous quantities such as graphs, lengths, voltages, etc. are termed "analog computers." The common desk calculator is a "digital computer," and the slide rule is an "analog computer." An "electronic digital computer" is a digital computer which operates by electronic circuits; even the arithmetic operations are performed electronically—in other words, the computer is not in the strict sense a mechanical device. Electronic computers perform arithmetic operations in milliseconds and microseconds. Arithmetical operations are performed on the LGP-30 in milliseconds and on the IBM 1620 in microseconds.

The programming of a problem for an electronic digital computer is basically a process of translating from the language convenient to human individuals to the language convenient to the computer. Man understands the language in terms of words and mathematics; the language of the computer is in terms of coded form. For the basic machine language, the coding is essentially numerical in form except for symbols such as ; / *, etc.; for symbolic language and FORTRAN, the coding is of both numeric and alphabetic form plus special symbols such as ; / *, etc.

The steps carried out on a desk calculator during the addition of a column of numbers is actually a program for the desk calculator. The steps to follow for the addition of a column of three numbers on a desk calculator are as follows:

1. Depress the clear button—this causes zeros to appear in all the digit positions in the accumulator.
2. Depress the appropriate buttons on the keyboard for the first number.
3. Depress the add button—this causes the first number of the column of numbers to appear in the accumulator.
4. Depress the appropriate buttons on the keyboard for the second number of the column.
5. Depress the add button—this causes the second number to be added to the first number, and the total appears in the accumulator.
6. Depress the appropriate buttons on the keyboard for the third number of the column.
7. Depress the add button—this causes the third number to be added to the sum of the first two numbers; the total appears now in the accumulator.
8. Copy down on paper the value in the accumulator.

Programs can also be written for determining the difference of two numbers, the quotient of two numbers, and the product of two numbers.

The programming in basic machine language of an elecronic digital computer is somewhat similar to that of programming a desk calculator. Codes are used for the programming of electronic digital computers, and numbers are stored in designated locations of the memory. Another name for a step in a program is instruction. In other words, a program for the computer is composed of instructions in coded form. Once a program, with the appropriate data, is stored in the memory of the electronic digital computer and the computer has been given the appropriate signal (usually in the form of pressing down electrical contact buttons), the instructions of the program will be followed in logical order until a solution to the problem is obtained.

Memories and Registers

Royal Precision Electronic Digital Computer, LGP-30

Modern electronic digital computers usually are composed of several boxes or racks of mechanical and electronic equipment which are connected by electric cables. Five distinct functions are performed by the mechanical and electronic equipment. These functions are input, output, memory, arithmetic and control. Solving a problem on a desk calculator also requires these performance operations.

The memory of an electronic digital computer is equivalent, in the case of a desk calculator, to the paper which is required for the storing of initial data, intermediate data, and final results.

Arithmetic is performed in the accumulator. The LGP-30 computer has an accumulator which operates similarly to the accumulator in the desk calculator.

The control function is manual for the desk calculator, but it is automatic in a stored program for electronic digital computers. Control in a desk calculator requires the pressing of appropriate buttons (clear, add, subtract, divide, multiply, etc.). In electronic digital computers (stored program) the control function provides for the execution of the program either one step at a time by pressing a button for each step or all steps by pressing a button just once for the entire program.

Output from a desk calculator is performed by reading the result in the accumulator and copying it on paper. The electronic digital computer reads the results and prints it on paper with a printer, or it reads the results and punches the results in coded form on tape or cards.

Input to a desk calculator is performed by reading data from paper and punching the appropriate buttons on the keyboard. The electronic digital computer takes coded data and coded instructions from punched paper tape and punched cards. Then the computer stores, in machine language, the data and instructions in the memory of the computer.

There is, however, an advantage of the stored program computer over the desk calculator. The stored program LGP-30 computer can perform over 400 additions per second, whereas the desk calculator can perform approximately one addition per second.

The nerve center of the LGP-30 digital computer is the memory section. All instructions and data are stored in the memory. The memory also stores intermediate data, and all information must go through the memory or storage. The center memory section of the LGP-30 digital computer is the magnetic drum as sketched in Figure 1-1.

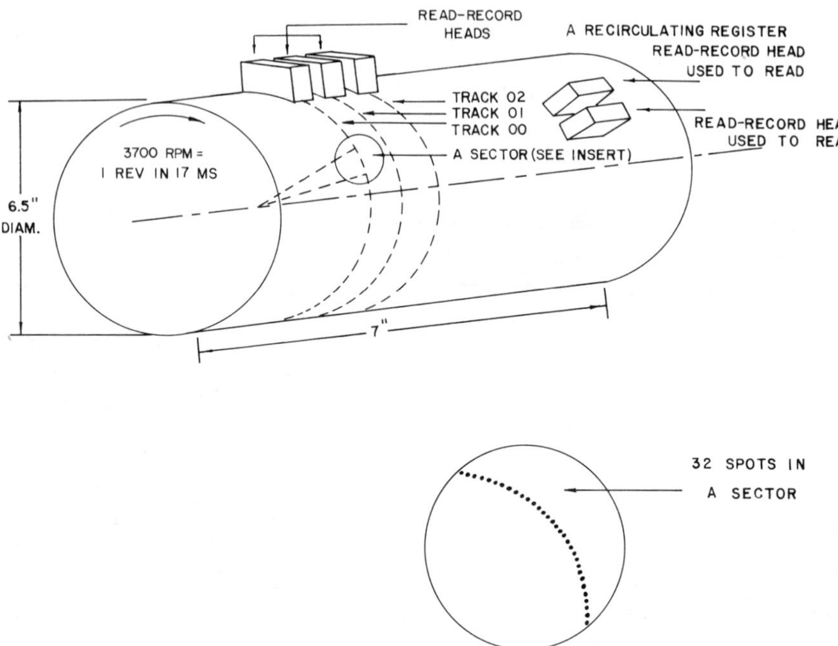

Figure 1-1. Drum memory.

The drum is a metal cylinder which is coated with a material which can be magnetized. It is 6.5 inches in diameter and 7 inches long, and it rotates at approximately 3,700 revolutions per minute.

In a frame around the drum are mounted 64 read-record heads. The read-record heads are for magnetizing portions of the drum and for detecting magnetized portions of the drum. The 64 read-record heads are spaced along the axis of the drum in order that spots in a circle around the drum can be read and recorded as the drum rotates. The number of circles is 64, and the circles are called "tracks." The use of the 64 tracks and 64 read-record heads enables any given portion of the drum to be available 64 times faster than if the memory consisted of a tape and one read-record head.

There are 64 groups of spots in each track. A group of spots is called a "sector," and there are 31 spots in a sector. Each group of 31 spots is separated from another group in a track by one spot which is called a spacer. Each of the 31 spots in a sector can be magnetized and demagnetized. The spacer is unaffected by recording, and it is never examined by reading.

A given location in memory is identified by its track and sector number. Tracks are given numbers from 00 to 63, and sectors are given numbers from 00 to 63. In designating a particular location, the number of the track is written first. With this method memory location will be numbered from 0000 to 6363—four digits are used in designating a memory location. As an example, the memory location for information in the position of track 24 and sector 02 is 2402. The 64 tracks and 64 sectors give $64 \times 64 = 4{,}096$ memory locations for storing information. Note that, although there are 4,096 memory locations, it is impossible to have a location number such as 2576 because 63 is the highest number for sectors.

A sector of a given track is accessible for reading or recording once each revolution as the drum is rotated. Since the time for each revolution of the drum is approximately 17 milliseconds, the access time is 17 milliseconds.

There are three tracks on the magnetic drum in addition to the 64 tracks for memory. These three tracks are for three recirculating registers; a recirculating register of one sector in length is on each track. The recirculating registers reduce the access time of a sector from approximately 17 milliseconds to approximately 0.26 milliseconds, which is equivalent to the time it takes a sector to pass under a read-record head. This is accomplished because, as the recirculating register passes under a read head, the information which is read is recorded back into the drum at a distance of one sector length from the read head.

One of the recirculating registers is the accumlator; the other registers are the instruction register and the counter register. Arithmetic operations are performed in the accumulator. The term "register" has been defined as the hardware for the temporary storing of one machine word until it is used. The computer finds the location of the instruction which is to be executed in the counter register. After the number of the location is found, this number is placed in the instruction register, and then the instruction is executed. At the same time the instruction is executed, 1 is added to the number in the counter register; this number then is the storage location of the next instruction.

The term "bit" is used for either designating a spot in a sector or for designating a "binary digit." Each spot can be in either of two states, unmagnetized or magnetized. An unmagnetized spot can be used to represent 0 and a magnetized spot can be used to represent 1.

Information is stored in memory locations in terms of magnetized and unmagnetized spots. The information stored in a sector (31 spots) is referred to as a word. Words can be either numbers or instructions. Each number word is represented by 30 magnitude bits, one sign bit and one spacer bit. Figure 1-2 indicates the manner in which a number word is represented in a sector.

A sign bit of 0 represents a positive number, and a sign bit of 1 represents a negative number.

Instructional words consist of an order part and an address part. The order part is represented by a letter, and the letter is represented by binary numbers. An order of add is represented by a, and a in binary form is 1110. The address part is represented in number form. Figure 1-3 indicates the manner in which an instructional word is represented in a sector.

In Figure 1-3 bit positions 12 through 15 in a word are used to represent the order, bit positions 18 through 23 are used to represent the track for the address, and bit positions 24 through 29 are used to represent the sector for the address. Six bit positions are allowed for the sectors. Therefore, 2^6 or 64 tracks and 2^6 or 64 sectors can be used for designating the address part of a word.

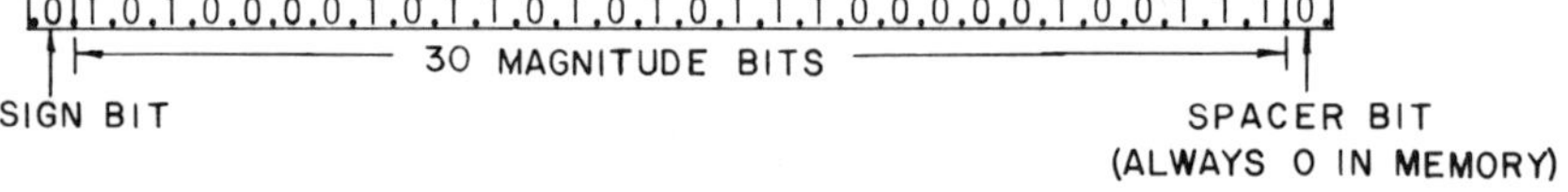

Figure 1-2. Bit representation of a sector of a LPG-30 magnetic drum.

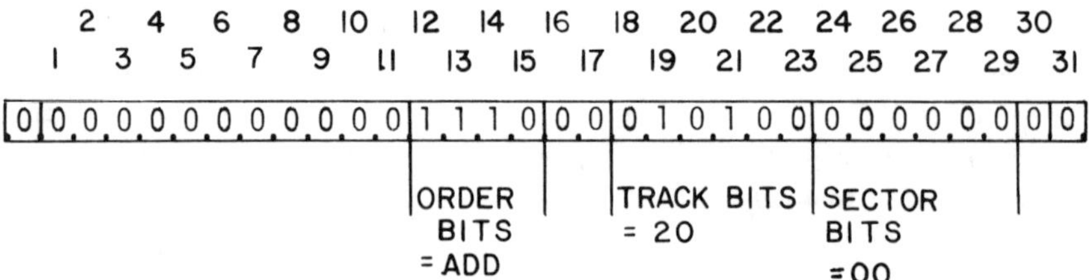

Figure 1-3. Word instructional word representation for a sector of a magnetic drum memory.

IBM 1620 Electronic Digital Computer

Similarly, the nerve center of the IBM 1620 digital computer is the memory section. Likewise, as for the LGP-30, all instructions and data are stored in the memory. The center memory section of the IBM 1620 digital computer is the magnetic core. The magnetic cores are tiny rings of ferromagnetic material which are only a few hundredths of an inch in diameter. These cores can be easily magnetized, and the magnetism is retained indefinitely unless it is deliberately changed. A large number of the cores are strung on a screen of wires to form a core plane. The cores are strung on the screen as illustrated in Figure 1-4.

Six cores are required to store a digit of information; of course, each core corresponds to one bit. Decimal digits are binary coded in the magnetic core storage of the IBM 1620 computer. A decimal digit is stored

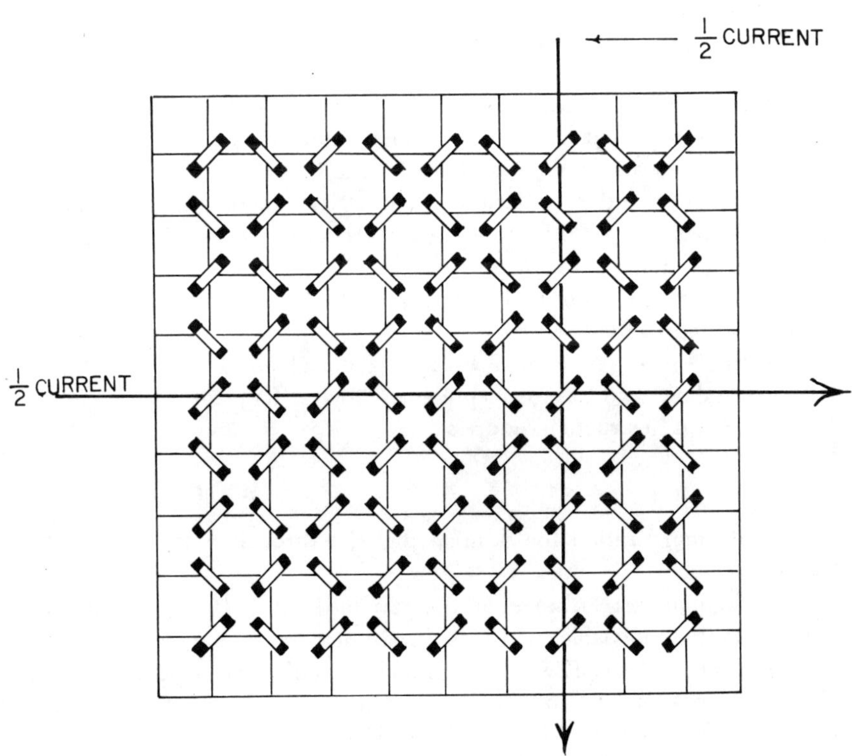

Figure 1-4. Core plane.

in bit positions. Each decimal digit requires six magnetic cores. The six bits are flag bit (F), check bit (C) and four numerical bits.

Both of the wires through each core are used for magnetizing the core; one-half the current which is required for the magnetization of the core is sent through each wire. By reversing the direction of the current through the two wires, the core is magnetized in the reverse direction. One direction of magnetization of the cores is taken arbitrarily as negative. A convention has been adopted so that when a core is magnetized positively the core is "on," and when the core is magnetized negatively the core is "off." For parity checking, each digit position within the 1620 computer must contain an odd number of coded digits, including a flag bit if there is one. Parity checking is a term which refers to the built-in method of checking the validity of the coded information. For creating the odd-bit number, a C-bit is automatically added to each digit when data enters the core storage. A digit position with an even number of bits causes the machine to signal a parity error. A plus zero is represented by a plus zero.

Depending upon the location of the flag bit and the operation which has been formed, the uses of the flag bit are as follows:

1. A data field is minus if the units digit has a flag bit. The absence of a flag bit represents a plus sign. The binary coded decimal representation of minus and plus 9 are F-8-1 and C-8-1.
2. The highest order (left-most) digit of a numerical data field is defined by a flag bit. A field represented as xxxxx where the dash is over the highest order digit is the field mark.
3. Carries in the add table are represented by a flag bit. A 7 with a carry is represented as 7.
4. A flag bit is represented by a flag bit alone.
5. An indirect address is represented by a flag bit over the units position of an instruction address.

The bit configuration for decimal digits 0 through 9 is represented in Figure 1-5.

The magnetic core storage of the standard IBM 1620 computer contains 20,000 addressable positions. Additional storage is available with the attachment of the IBM 1623. Models 1 and 2 of the IBM 1623 provide sufficient storage capacity for bringing, respectively, the total of the IBM 1620 computer to 40,000 and 60,000 positions.

The core storage in the IBM 1620 computer contains 12 core planes. There are 10,000 magnetic cores in each plane, giving a matrix of mag-

	DIGIT	C	F	8	4	2	1
	0	1	0	0	0	0	0
	1	0	0	0	0	0	1
POSITIVE SIGNS	2	0	0	0	0	1	0
	3	1	0	0	0	1	1
	4	0	0	0	1	0	0
	5	1	0	0	1	0	1
	6	1	0	0	1	1	0
	7	0	0	0	1	1	1
	8	0	0	1	0	0	0
	9	0	0	1	0	0	1
	0	0	1	0	0	0	0
	1	1	1	0	0	0	1
NEGATIVE SIGNS	2	1	1	0	0	1	0
	3	0	1	0	0	1	1
	4	1	1	0	1	0	0
	5	0	1	0	1	0	1
	6	0	1	0	1	1	0
	7	1	1	0	1	1	1
	8	1	1	1	0	0	0
	9	0	1	1	0	0	1

Figure 1-5. Bit configuration for decimal digits 0 through 9.

netic cores 100 by 100. The six top planes are the even-address planes, and the six lower planes are the odd-address planes. Figure 1-6 illustrates the arrangements of the 12 core planes.

Six planes (one bit from each plane) are required for representing one digit position. The core planes are labeled *C*, *F*, 8, 4, 2, 1. If the solid circles represent "on" and the open circles represent "off," Figure 1-6 represents the data shown in Figure 1-7.

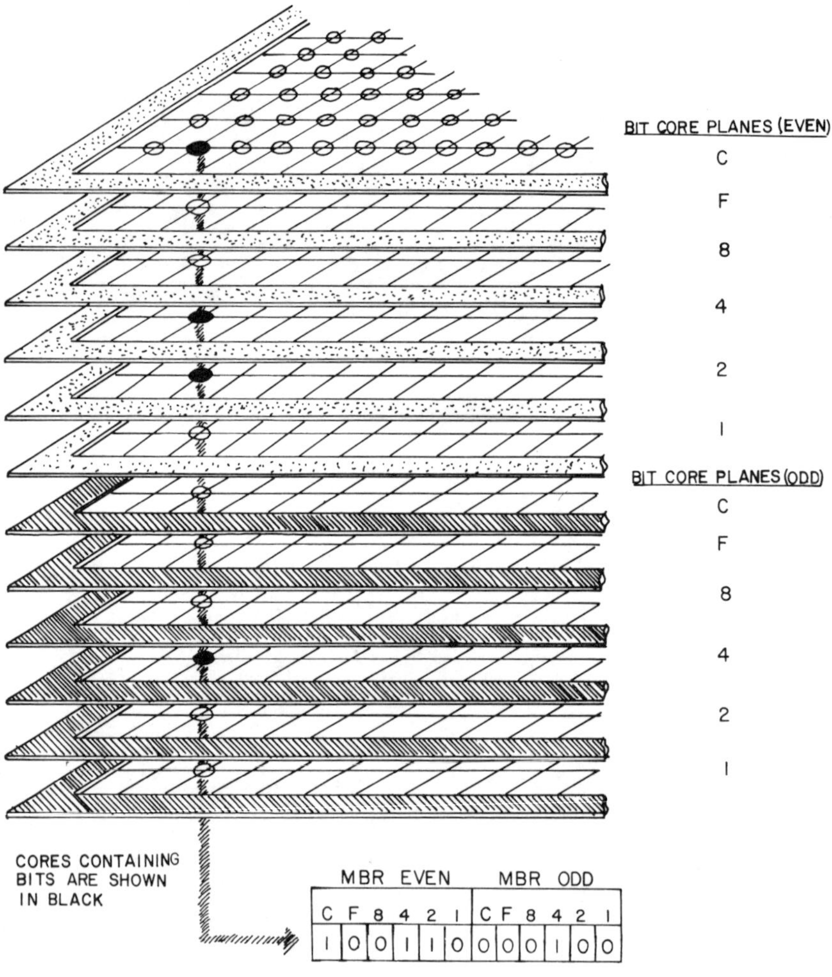

Figure 1-6. Arrangement of the 12 core planes.

MBR-EVEN						MBR-ODD					
C	F	8	4	2	1	C	F	8	4	2	1
1	0	0	0	1	1	0	1	0	1	0	1

Figure 1-7. Digit position representation by means of bits in planes.

This is + 3 stored in the even positions of the memory and − 5 stored in the odd position of the memory.

The 1620 machine has eight memory address registers:

1. IR-1: Instruction Address Register 1
2. IR-2: Instruction Address Register 2
3. OR-1: Operand Address Register 1
4. OR-2: Operand Address Register 2
5. OR-3: Operand Address Register 3
6. PR-1: Product Address Register 1
7. PR-2: Product Address Register 2
8. PR-3: Product Address Register 3

In addition to the above eight registers, there is a Memory Buffer Register (MBR). The memory buffer register is divided into two one-digit registers, MBR-Even and MBR-Odd. From the core storage, the even-address digits flow through the MBR-Even, and the odd-address digits flow through the MBR-Odd.

The eight Memory Address Register Storages (MARS) can contain five-digit addresses from 00000 to 19999; these registers address core storage through an additional five-digit Memory Address Register (MAR) which also comprises nonaddressable cores. The addressing of core storage (both instructions and data) is done through MAR. The function performed by the eight MARS and the MAR is determined by the operation code.

During a core storage readout cycle, all cores as indicated in Figure 1-6 by the vertical line (one core bit in each plane) are read out at the same time by the two-digit MBR; only "set" cores are read out as bits. The MBR receives digits entering or leaving core storage. The digits which leave the storage are regenerated through the MBR. The MBR is actually subdivided into two one-digit registers (MBR-Even and MBR-Odd or MBR-E and MBR-O). The odd-address digits flow through MBR-O and the even-address digits flow through MBR-E. During the entrance to core storage, digits are handled under the control of the units position of the MBR address. An even digit in the units position of the MAR address carries data to be selected from MBR-E. Since all core planes (12) are affected, a single core storage address affects the adjacent even storage position and the adjacent odd storage position. For example, if the digit at 00400 is addressed and programmed to be read from core storage, the digit at address 00401 is affected; if the digit at 00401 is addressed, the digit at 00400 is affected. The operation which is to be performed selects the digit which is to be performed. The digit which is addressed is moved to the one-digit Memory Data Register (MDR).

Each core position of the IBM 1620 computer is addressable; this is different from the LGP-30 where groups of 31 bits are addressable. The core storage positions are sequentially addressable from 00000 to 20000 for the basic unit of the IBM 1620 computer. For the basic 1620 unit of 20000 positions, there is wraparound storage because 00000 follows 20000 which is the highest-numbered address.

Data for the IBM 1620 computer can be in either numerical or alphanumeric mode whenever it is read or written. For given data, input and output should be in the same mode. A numerical character is represented by one decimal digit, but an alphanumeric character is represented by two decimal digits; these two decimal digits are called "zone digits" and "numerical digits." The alphanumeric codes are given in Figure 1-8.

The ring-core-type memory, as discussed in connection with the IBM 1620, has been used in minicomputers. Both the LGP-30 and the IBM 1620 used magnetic-type memories; microcomputers use chips containing thousands of gates for their storage memories and central processing units. Microcomputer memories for storing programs, data, files, etc. use magnetic memories in the form of cassette tapes and flat disks. Mainframe computers also use magnetic tapes and disks, but these are usually much larger than those used by microcomputers. The disks are similar to the rotating magnetic drum used by the LGP-30 in that words are stored in sectors of concentric tracks.

A reasonable-size microcomputer setup contains a printer and two or more disk drives. All microcomputers contain two types of storage memories: ROM and RAM. The ROM (read-only memory) contains permanent data which can be read only and not erased. The RAM (read-access memory) contains information which can be erased or modified. The BASIC compiler is stored in ROM memory. The unit of storage, one binary digit, is called a bit, and a group of eight bits which is used for storing a word is called a byte.

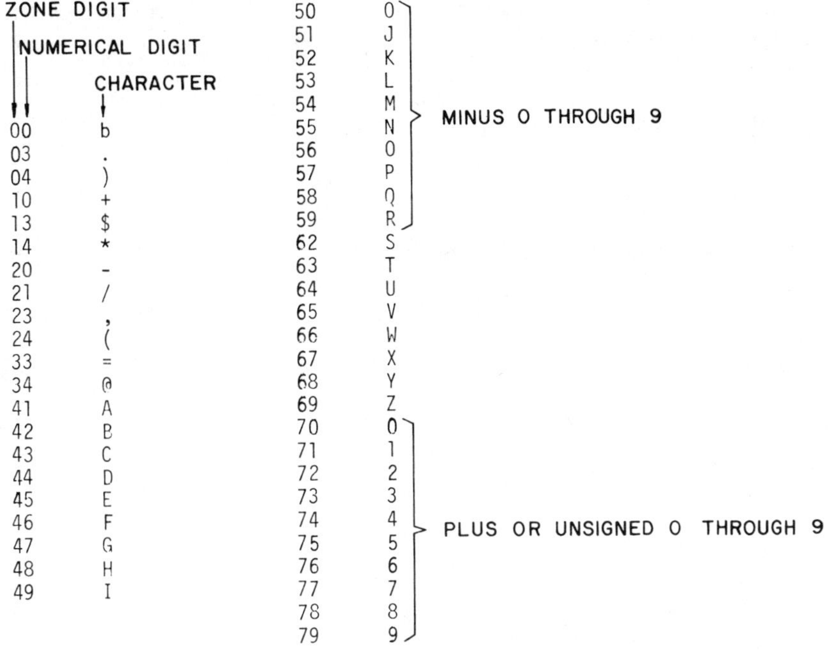

Figure 1-8. Alphanumeric codes.

Building Programs for the LGP-30

All instructions for the LGP-30 fall into five general categories; the function of the instruction determines the category.

1. Arithmetic
2. Internal data transmission
3. Logic (compare and branch)
4. Input/output
5. Program control

The LGP-30 has 16 instructions. The instructions are divided into categories as follows:

Arithmetic Instructions	**Code**
1. Add	a
2. Subtract	s
3. Multiply (most significant figures)	m
4. Multiple (least significant figures)	n
5. Divide	d

Internal Data Transmission	**Code**
6. Return address	r
7. Unconditional transfer	u
8. Store address	y
9. Bring from memory	b
10. Hold and store	h
11. Extract	e
12. Clear	c

Logic Instruction	**Code**
13. Test	t

Input-Ouput	**Code**
14. Input	i
15. Print	p

Program Control	**Code**
16. Stop	z

Instructions for the LGP-30 consist of an order part and an address part. For instance, b 3000 means bring the information which is stored in location 3000 and place this information in the accumulator. Table 1-1 is a list of the commands with brief discussions.

The address part of the instruction is denoted by M when is refers to a memory location.

Instructions for the LGP-30 computer are symbolic means (in computer language) of communication with the computer. These communication methods are used for telling the computer what to do. Each instruction

Table 1-1
Commands for the LGP-30 Computer

Instruction		Effect
b	M	**Bring.** Clear the accumulator and add the contents of M to it.
a	M	**Add.** Add the contents of M to the contents of the accumulator and retain the results in the accumulator.
s	M	**Subtract.** Subtract the contents of M from the contents of the accumulator and retain the result in the accumulator.
m	M	**Multiply.** Multiply the number in the accumulator by the number in memory location M and terminate the result at 30 binary places.
n	M	**Multiply.** Multiply the number in the accumulator by the number in M and retain the least significant half of the product.
c	M	**Clear.** Replace the contents of memory location, M, with the contents of the accumulator and replace the contents of the accumulator with zero.
y	M	**Store.** Replace the contents of the address portion of the word in location, M, with the contents of the address portion of the word in the accumulator.
r	M	**Return Address.** Add 1 to the address held in the counter register (C) and record in the address portion of the instruction in memory location M. The counter register normally holds the address of the next instruction to be executed.
e	M	**Extract.** Extract or logical product order—i.e., clear the contents of the accumulator to zero in the bit positions occupied by the zeros in M.
u	M	**Transfer Control.** Transfer control to M unconditionally—i.e., get the next instruction from M.
t	M	**Test or Conditional Transfer.** Transfer control to M only if the number in the accumulator is negative.
i	0	**Input.** Fill the accumulator from the Flexowriter.
d	M	**Divide.** Divide the contents in the accumulator by M and leave the results in the accumulator.
p	x	**Print.** Print a Flexowriter symbol. The symbol is denoted by the track number part of the address (x).
z	T	**Stop.** Contingent on five switch ($T_1, \ldots, T_5$) settings on the control panel.

consists of two parts—an order part and an address part. The instruction b 2000 means bring the contents (data) from location 2000 and store it in the accumulator. The instruction a 1000 means add the contents which are stored in memory location 1000 to the contents which are already in the accumulator.

A simple program for the addition of two numbers is illustrated in Example 1-1.

Example 1-1

Location	Instruction: Operation	Instruction: Address
0100	b	2000
0101	a	1000
0102	h	0500
0103	z	0000

This program indicates that one of the numbers is stored in memory location 2000, and the other number is stored in memory location 1000. The four instructions are stored in memory locations 0100, 0101 and 0102, 0103. The first instruction means to bring the number from memory location 2000 and store it in the accumulator and at the same time leave the number also in memory location 2000. The second instruction means add the number which is in memory location 1000 to the number which has been stored in the accumulator, and at the same time the number in memory location 1000 is not changed. At this stage, the sum of the two numbers is in the accumulator. The third instruction means store the contents (this time it is sum) of the accumulator in memory location 0500 and at the same time leave the contents of the accumulator unchanged. The last instruction simply means stop the computations by the computer. The computer, unless otherwise instructed, executes the instructions in consecutive order; the instructions are stored sequentially in the memory locations.

A program for computing five sets of two numbers or the solution to the equation, $x_i + y_i = z_i$, where i takes on values of 1 to 5, is illustrated in Example 1-2.

Example 1-2

Location	Instruction		
	Operation	Address	
0000	b	2000	
0001	a	1000	$x_1 + y_1 = z_1$
0002	h	0500	
0003	b	2001	
0004	a	1001	$x_2 + y_2 = z_2$
0005	h	0501	
0006	b	2002	
0007	a	1002	$x_3 + y_3 = z_3$
0008	h	0502	
0009	b	2003	
0010	a	1003	$x_4 + y_4 = z_4$
0011	h	0503	
0012	b	2004	
0013	a	1004	$x_5 + y_5 = z_5$
0014	h	0504	
0015	z	0000	Stop

In the program, the *x*'s were stored in memory locations 2000 to 2004, the *y*'s in memory locations 1000 to 1004 and the *z*'s in memory locations 0500 to 0504.

Observations of the program indicated that for each sum of two numbers, three instructions were required. If there had been 100 sets of numbers to be added, then 301 instructions would have been required. Large numbers of instructions make the computations by the LGP-30 more complicated than a desk calculator. Fortunately, the LGP-30 has the ability to modify instructions internally. The modification of the instructions is done by altering the addresses in the instructions. The addresses are altered by the addition or subtraction of numbers from the original addresses. An example of altering the first instruction of the second program is illustrated in Example 1-3.

Example 1-3

Location	Instruction Operation	Instruction Address
0025	b	0000
0026	a	0700
0027	h	0000

A 1 is stored in the memory location 0700 at the 29 bit position of this location. The first instruction takes the contents which were stored in memory location 0000 (the contents are *b* 2000) and places it in the accumulator. The second instruction causes a 1 to be added to *b* 2000 in the extreme right position; *b* 2000 now becomes *b* 2001. Now *b* 2001 is in the accumulator. The third instruction causes the contents of the accumulator to be placed back in memory location 000. The first instruction is now

0000 *b* 2001

The three steps for altering an instruction are generally referred to as "modifying the instruction."

By modifying instructions and employing a process called "looping," a program for solving the equation $x_i + y_i = z_i$ where $i = 1$ to 50 is illustrated in Example 1-4.

Example 1-4

Instruction No.	Location	Instruction Operation	Instruction Address	
1	0000	(b	2000)	Compute
2	0001	(a	1000)	
3	0002	(h	0500)	
4	0003	b	0000	Modify X
5	0004	a	0700	
6	0005	h	0000	
7	0006	b	0001	Modify Y
8	0007	a	0700	
9	0008	h	0001	
10	0009	b	0002	Modify Z
11	0010	a	0700	
12	1011	h	0002	

13	0012	b	0000	Test
14	0013	s	0701	
15	0014	t	0000	
16	0015	z	0000	Stop

The *X*s were stored in memory location 2000 to 2049.
The *Y*s were stored in memory location 1000 to 1049.
The *Z*s were stored in memory location 0500 to 0549.

One at bit position 29 was stored in memory location 0700, and *b* 2050 was stored in memory location 0701.

It is noted that the sums for 50 sets of two numbers can be calculated by the last program which utilizes modification of instructions and looping with only 16 instructions as compared to 151 instructions if the sums of the numbers were calculated by a program similar to Example 1-2.

Referring to the program for calculating 50 sets, the first three instructions are for calculating the sums. Instructions 4 through 12 are for modifying the data addresses of the first three instructions. Instructions 13 through 15 are for testing or decision making, and the last instruction, No. 16, stops the computer. At this stage, the program can be represented diagrammatically. Diagrammatical representations of programs are called "flow diagrams." The flow diagram for Example 1-4 is given in Figure 1-9.

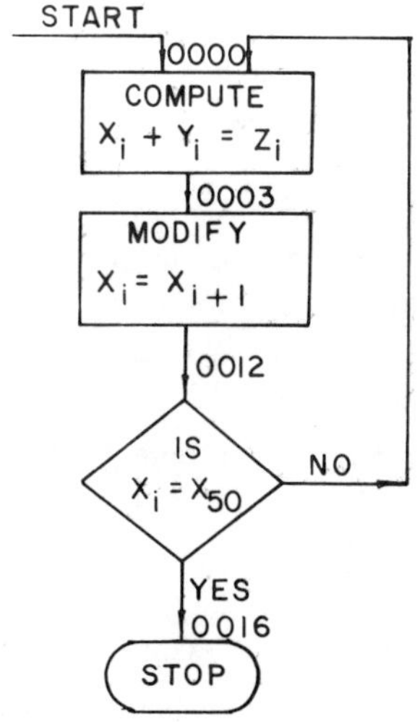

Figure 1-9. Flow diagram.

Instruction by instruction, the program for calculating the 50 sums for the equation $X_i + Y_i = Z_i$ is explained as follows:

1. The first instruction places X_i, which is stored in memory location 2000, in the accumulator.
2. The second instruction adds Y_i, which is stored in memory location 1000, to the X_i which is already in the accumulator; the sum of the two numbers is now in the accumulator.
3. Instruction 3 stores the sum of the two numbers, which is now in the accumulator, in memory location 0500; the sum remains in the accumulator until additional information is stored in the accumulator; at this time, the sum is cleared from the accumulator, and the additional information remains in the accumulator.
4. Instruction 4 takes the first instruction which is stored in memory location 1000 and place it (*b* 2000) in the accumulator.
5. Instruction 5 adds a 1, which is stored in location 0700 to the contents (*b* 2000) which is already stored in the accumulator. The contents in the accumulator now becomes *b* 2001.
6. Instruction 6 stores the contents of the accumulator back in the memory location of the first instruction, which is 0000. The first instruction now becomes 0000 *b* 2001.
7. Instruction 7 stores the contents of Instruction 2 which is stored in memory location 0001 in the accumulator.
8. Instruction 8 adds a 1, which is stored in memory location 0700 to the contents of the accumulator. The contents in the accumulator now becomes a 1001.
9. Instruction 9 stores the contents of the accumulator in the memory location of the second instruction. The second instruction now becomes 0001 *a* 1001.
10. Instruction 10 stores the contents of Instruction 3, which is stored in memory location 0002, in the accumulator.
11. Instruction 11 causes a 1, which is stored in memory location 0700, to be added to the contents of the accumulator. The contents of the accumulator now becomes *b* 0501.
12. Instruction 12 stores the contents (*b* 0501) of the accumulator in the memory location of the third instruction. The third instruction now becomes 0002 *b* 0501.
13. Instruction 13 brings the contents from the memory location of the first instruction and places it in the accumulator.
14. Instruction 14 causes the contents of memory location 0701 to be subtracted from the contents of the accumulator; *b* 2050 is stored in memory location 0701.

15. Instruction 15 is a logic instruction; if the contents of the accumulator are zero or positive, the next sequential instruction (No. 16) will be executed; if the contents of the accumulator are negative, the designated instruction (0000) will be executed.

16. Instruction 16 causes the computer to discontinue operations.

Summarizing briefly, the operations are as follows. The first three instructions perform the addition of two numbers, and the sum of these numbers is stored in a given memory location. The next nine instructions modify the data addresses of the first three instructions (computational instructions). The next three instructions are the testing or logic instructions. These three instructions cause the data addresses of the first three instructions to become modified for 50 times, and sums of two numbers are obtained for 50 times. At the end of the 50 calculations and storages of the 50 sums, the computer is instructed to execute the last instruction; this instruction causes the computer to stop.

From the discussion of the program, the meaning of the term "looping" is perhaps clear. Looping involves the modifications of data addresses of instructions and the use of these same instructions with the new data addresses for the determination of new information.

The parts of the program are (a) computation of sums, (b) modification of data addresses and (c) testing for the continuation of looping.

This program was sufficient for the calculation of 50 sums from 50 sets of two numbers. However, it is not sufficient at the end of the first 50 calculations for the calculations of 50 more sums from different data which may be stored in the original memory locations because all the data addresses in the instructions used for calculations have been altered. To calculate 50 sums from 50 new sets of data by the original program, all the data addresses in the calculating instructions must be set back to their original values. The procedure for altering the modified addresses to their original values is called "initializing." A counter is sometimes used for terminating a loop.

Example 1-5 is a revision of the previous program written with a counter and initializing steps.

Example 1-5

Instru. No.	Location	Instruction Operation	Instruction Address	
1	0000	(b	2000)	Compute
2	0001	(a	1000)	
3	0002	(h	0500)	

4	0003	b	0000	Modify *X*
5	0004	a	0700	
6	0005	h	0000	
7	0006	b	0001	Modify *Y*
8	0007	a	0700	
9	0008	h	0001	
10	0009	b	0002	Modify *Z*
11	0010	a	0700	
12	0011	h	0002	
13	0012	b	0702	Counter
14	0013	a	0700	
15	0014	h	0702	
16	0015	s	0703	
17	0016	t	0000	
18	0017	b	0704	Initiatize *X*
19	0018	h	0000	
20	0019	b	0705	Initiatize *Y*
21	0020	h	0001	
22	0021	b	0706	Initiatize *Z*
23	0022	h	0002	
24	0023	b	0707	Initiatize Counter
25	0024	h	0023	
26	0025	u	0000	

*X*s are stored in memory locations 2000 to 2049.
*Y*s are stored in memory locations 1000 to 1049.
*Z*s are stored in memory locations 0500 to 0549.
1 is stored at position 29 in memory location 0700.
Initially zeros are stored in all positions of memory location 0702.
50 is stored in memory location 0703, at bit positions 28 and 29.
b 2000 is stored in memory location 0704.
a 1000 is stored in memory location 0705.
h 0500 is stored in memory location 0706.
Zeros are stored in all positions of 0707.

The flow diagram for the program is given in Figure 1-10.

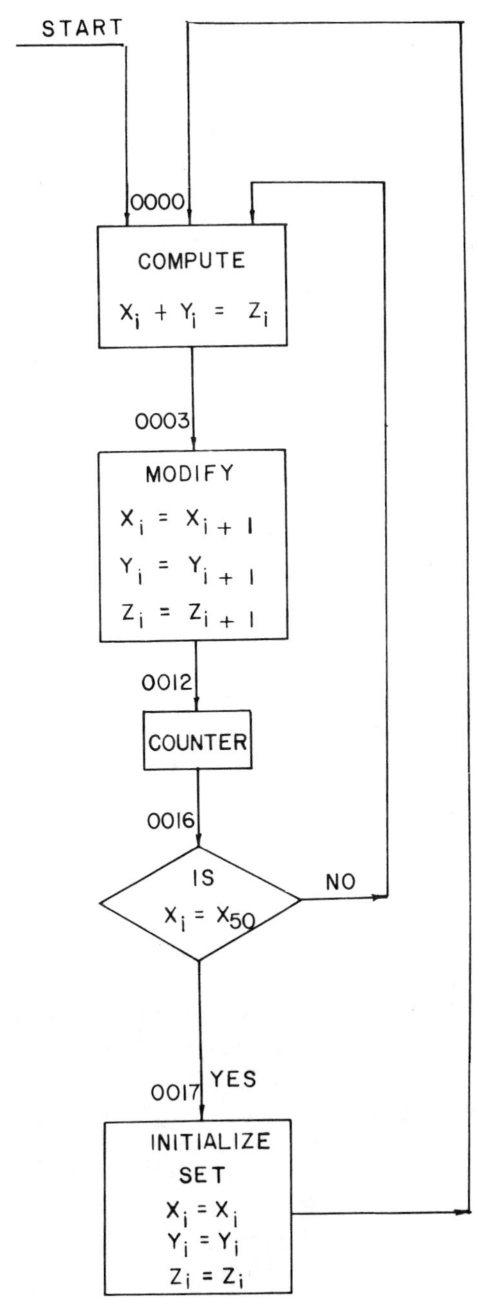

Figure 1-10. Flow diagram for computing the sums of two number sets by employing a counter.

The first 12 instructions of the last two programs are identical. The counter for the last program begins with the 13th instruction, and it takes in the next four instructions. The following explains the instructions of the counter.

1. Initially only zeros are stored in memory location 0702. Instruction 13 brings the contents, which is zero, from memory location 0702 and places it in the accumulator.
2. In memory location 0700, a 1 was stored. Instruction 14 causes a 1 to be added to the contents of the accumulator.
3. Instruction 15 stores the contents of the accumulator back in memory location 0702. Each time around the loop, the contents of the memory location is increased by 1.
4. Since the problem requires the computation of 50 sums of two numbers, the number 50 is stored in memory location 0703. Instruction 16 subtracts 50 from the number which is in the accumulator. The number in the accumulator is identical to the last number which has been stored in the memory location 0702.
5. Instruction 17 decides whether or not the number in memory location 0702 has reached 50; after Instruction 16 has been executed, the sign of the results in the accumulator will be negative until the number in memory location 0702 has reached 50. At this time, the results in the accumulator will be zero. Instruction 17 then directs the computer to return to the first instruction if the results in the accumulator have a negative sign. If the results in the accumulator are zero, the next instruction (No. 18) will be executed.

The operations of the counter can be summarized as follows: for each computation, a 1 is added to a memory location, and when the number in the memory location reaches the desired number of computations, the computer is directed to execute the next instruction of the program which is not included in the loop.

Instructions 18 through 25 are called "initializing instructions." After the completion of 50 computations, the data addresses of the computation instructions and the contents of the data address of the first instructions of the counter have increased by the number 50. Without storing a new program in the computer for calculating the sums of 50 additional sets of two numbers, the program in the computer can be used again by supplying new data in the original memory locations, if the data addresses of the first three instructions are set back to their original values and if the contents of the data address of the first instruction of the counter is set to zero.

The three data addresses are returned to their original numbers by instructions 18 through 23. The following explain these instructions.

1. During the initial storing of data and instructions in the computer, the first instruction was also stored in memory location 0704 (this instruction is *b* 2000) ; likewise, the second instruction was stored in memory location 0705, and the third instruction was stored in memory location 0706.
2. Instruction 18 places the contents of memory location 0704 in the accumulator; the contents of memory location 0704 is *b* 2000, and this is identical to the first instruction.
3. Instruction 19 stores from the accumulator *b* 2000 in memory location 0000; now the first instruction, which is stored in memory location 0000, has its initial value.
4. Instruction 20 places the contents of memory location 0705 in the accumulator. The contents in 0705 is *a* 1000.
5. Instruction 21 stores from the accumulator *a* 1000 in the memory location 0001 which is that of the second instruction.
6. Instruction 22 places the contents of memory location 0706 in the accumulator; the contents in 0706 is *h* 0500.
7. Instruction 23 stores from the accumulator *h* 0500 in memory location 0002 which is the addresss of the third instruction.

Now the contents of the data address of this first instruction of the counter must be returned to zero. This is accomplished by the following.

1. Zeros are stored in memory location 0707. Instruction 24 places the contents of the memory location in the accumulator.
2. Instruction 24 stores the contents of the accumulator in memory location 0702 which is the data address of the first instruction of the counter. Now the working storage of the counter has been set to zero.

After the initialization, the computer must return to the execution of the program again. Instruction 26 is an unconditional transfer instruction. This instruction directs the computer to return to memory location 0000 for the next instruction to be executed. This is the location of the first instruction of the program.

Perhaps the reader noted that each of the first three instructions were enclosed in brackets; this simply means that the addresses of these instructions would be modified during the execution of the program.

Figure 1-11 is a program for the solution of the equation $X_i + Y_i = Z_i$, where $i = 1$ to 15; the program is written on an LGP-30 coding sheet. Figure 1-12 is the data for the program, written on an LGP-30 data load sheet. The initializing steps are not included in the program.

LGP-30 CODING SHEET

PROGRAM INPUT CODES		STOP	LOCATION	INSTRUCTION OPERATION	INSTRUCTION ADDRESS	STOP	CONTENTS OF ADDRESS	NOTES
;.0.0.0	1.0.0.0	'						
/.0.0.0	1.0.0.0	'	X					
. . .	. . .		. .0.0	. .x.b	3.5.0.0	'	X_0	
. . .	. . .		. .0.1	. .x.a	3.5.1.5	'	Y_0	
. . .	. . .		. .0.2	. .x.h	3.5.3.0	'	Z_0	
. . .	. . .		. .0.3	. . .b	0.0.0.0	'	B3500	
. . .	. . .		. .0.4	. . .a	0.0.1.6	'	1 at 29	
. . .	. . .		. .0.5	. . .h	0.0.0.0	'	B3501	
. . .	. . .		. .0.6	. . .b	0.0.0.1	'	A3515	
. . .	. . .		. .0.7	. . .a	0.0.1.6	'	1 at 29	
. . .	. . .		. .0.8	. . .h	0.0.0.1	'	A3516	
. . .	. . .		. .0.9	. . .b	0.0.0.2	'	H3530	
. . .	. . .		. .1.0	. . .a	0.0.1.6	'	1 at 29	
. . .	. . .		. .1.1	. . .h	0.0.0.2	'	H3531	
. . .	. . .		. .1.2	. . .s	0.0.1.5	'	H3545	
. . .	. . .		. .1.3	. . .t	0.0.0.0	'		
. . .	. . .		. .1.4	. .x.z	0.0.0.0	'		
. . .	. . .		. .1.5	. .x.h	3.5.4.5	'		
,.0.0.0	0.0.0.1		. .1.6	. . .	. . .4	'		
. . .	. . .		. .1.7	. . .	. . .			
. . .	. . .		. .1.8	. . .	. . .			
. . .	. . .		. .1.9	. . .	. . .			
. . .	. . .		. .2.0	. . .	. . .			
. . .	. . .		. .2.1	. . .	. . .			
. . .	. . .		. .2.2	. . .	. . .			
. . .	. . .		. .2.3	. . .	. . .			
. . .	. . .		. .2.4	. . .	. . .			
. . .	. . .		. .2.5	. . .	. . .			
. . .	. . .		. .2.6	. . .	. . .			
. . .	. . .		. .2.7	. . .	. . .			
. . .	. . .		. .2.8	. . .	. . .			
. . .	. . .		. .2.9	. . .	. . .			
. . .	. . .		. .3.0	. . .	. . .			
. . .	. . .		. .3.1	. . .	. . .			

Figure 1-11. Program for solving $X_i + Y_i = Z_i$, where $i = 1$ to 15.

Upon observation of the program which is written on the LPG-30 coding sheet, several questions may have occurred to the reader. These questions, no doubt, concern codes or notations in the columns under the headings of "Program Input Codes" and "Stop" and the use of *X*s before some of the instructions.

1. The program input code ;000 1000 means start storing in location 1000.

2. The program input code 000 1000 means modify, during storage, by 1000 all data addresses and all instructional addresses in the program which are not preceded by an *X* before the operation code. The instructional address of which the operational codes are preceded by *X*s will not be modified during the storing of the program.

LGP-30 DATA LOAD SHEET

QUAN.	P	+	Q	LOCATION	STOP	+	NUMBER	STOP	CAR. RET.	
	2	+	2.0	3.5.0.0	'	+	0.0.5.2.1.4.1	'		
			.	. . .			0.0.4.2.5.6.7	'		
			.	. . .			0.0.3.3.3.3.2	'		0sPs9
			.	. . .			0.0.2.2.2.2.2	'		
			.	. . .			0.0.3.3.3.3.3	'		28sqs47
			.	. . .			0.0.4.4.4.4.4	'		
			.	. . .			0.0.5.5.5.5.5	'		0000sLocs5363
			.	. . .			0.0.4.1.1.1.1	'		
			.	. . .			0.0.1.2.3.4.5	'		
			.	. . .			0.0.6.7.8.9.0	'		
			.	. . .			0.0.1.6.7.5.4	'		
			.	. . .			0.0.4.3.2.3.1	'		
			.	. . .			0.0.6.7.7.4.2	'		
			.	. . .			0.0.7.7.7.7.1	'		
			.	. . .			0.0.1.1.1.1.1	'		
			.	3.5.1.5		-	0.0.6.6.6.6.6	'		
			.	. . .			0.0.2.2.2.2.2	'		
			.	. . .			0.0.3.3.3.3.3	'		
			.	. . .			0.0.4.2.4.6.7	'		
			.	. . .			0.0.5.2.3.4.6	'		
			.	. . .		-	0.0.6.4.3.2.1	'		
			.	. . .			0.0.5.4.3.2.1	'		
			.	. . .			0.0.7.4.8.7.2	'		
			.	. . .			0.0.3.4.3.4.5	'		
			.	. . .			0.0.6.7.8.8.8	'		
			.	. . .			0.0.4.2.4.2.1	'		
			.	. . .			0.0.6.6.7.7.8	'		
			.	. . .		-	0.0.2.4.6.4.4	'		
			.	. . .			0.0.3.3.4.2.1	'		
			.	. . .			0.0.8.9.7.4.7	'		
			.	. . .		-	0.0.0.0.0.0.0	'		
				. . .						
			.	. . .			. . . . . .			

Figure 1-12. Program data for solution of $X_i + Y_i = Z_i$.

3. The first two program input codes eliminate the necessity of assigning definite memory locations during the writing of a program. Also, with the use of the first input codes, a given program can be stored from time to time in different memory locations.

4. During the punching of a tape with a program, the individual instructions are separated one from another by means of apostrophes; this accounts for the apostrophes in the columns which are labeled "Stop."

5. The program input code demonstrates how hexadecimal notations are used for inserting a 1 at 29 in memory location 0016 + 1000 during the storing of the program in the computer. The program input code ,000 0001 means store one hexadecimal number beginning at sector location 31. The hexadecimal word is 4, which is equivalent to 0100 in binary. Figure 1-13 is a block diagram which explains how the 1 is placed at 29 in memory location 1016.

```
      2     4   6     8  10    12  14    16  18    20  22    24  26    28  30
    1   3     5   7     9  11    13  15    17  19    21  23    25  27    29  31
|0|0 0 0 , 0 0 0 0 , 0 0 0 0 , 0 0 0 0 , 0 0 0 0 , 0 0 0 0 , 0 0 0 0 , 0 1 0|0|
```

Figure 1-13. Block diagram illustrating how a one is placed in the 29th position.

The input from the Flexowriter to the computer and the sums which were calculated with the computer are given in Table 1-2.

The first four terms ,000 are the code for storing hexadecimal numbers. The second four terms 0001 indicate the number of hexadecimal terms. Then, the instruction address 0004 (the zeros are not required to be written in) is the hexadecimal number or word to be written in memory location 1016.

The general formula for storing hexadecimal is ,00000 N_1 N_2; the next N_1 N_2 decimal words are stored in hexadecimal form consecutively in the next N_1 N_2 storage locations.

Similarly, upon observation of the LPG-30 data load sheet, questions have perhaps arisen as to the functions of p and q. In the calculated data which is typed out in decimal, the number p denotes the number of digits following the decimal point which is in the seven digit field; p can have the range $0 \leqq p \leqq q$.

The LGP-30 is a fractional machine. Numbers are stored in binary; the binary point precedes the No. 1 bit position as shown in Figure 1-14.

For the binary number 10111, a shift of five places give .10111. The smallest number of places which could have been right and still have a fraction was six places. The scale factor q is used for keeping track of the shifts. If the number is small, it is represented as a negative q. In decimal, ¼ is equivalent to 0.01 in binary, and 0.01 in binary is ¼ at $q = -2$. In storing numbers, the q's should be large enough to avoid overflow of the computation and small enough to avoid truncation.

Numbers to be added must be of the same q's. Numbers to be subtracted must be of the same q's. The product of two numbers has a q which is equal to the sum of the q's of the two numbers. The quotient of two numbers has a q which is the difference of the q's of the two numbers. The limits of q's for given p's are given in Table 1-3.

The formation volume factor of a reservoir fluid is a function of the reservoir pressure. An approximate equation for expressing the relationship is

$$B = A_4P^4 + A_3P^3 + A_2P^2 + A_1P + A_0$$

or

$$B = \{[(A_4P + A_3)\ P + A_2)]\ P + A_1\}\ \ P + A_0$$

Table 1-2
Inputs and Sums Calculated with the LGP-30

```
';0006300'/0000000'r0740'u074u32'z0000'.0000000'

;006310';006310;00063100'/0000000'b3530'r1205'u1200'z1520'u0000';0002000/
002000

xb3500'xa3515'xh3530'b0000'a0016'h0000'b0001'a0016'h0001'b0002'a0016'h0002'
s0015't0000'xz0000'xh3545',0000001'4'.0000000'
;0006300'/0000000'r0740'u0732'z0000'.0000000';0006310'/0000000'b3530'r1205'
u.1200'z1520'u0000'7
;0006310'/0000000'b3530'r1205'u1200'z1520'u0000'
.0006300p
2+203500'52141'42567'33332'22222'33333'44444'55555'41111'12345'67890'
16754'43231'
677421'77771'11111'-0066666'22222'33333'42467'52346'-0064321'54321'74872'
34345'
67888'42421'66778'-0024644'33421'89747'-0000000'
'.0002000.0006310

00145.250                00647.890                00666.650
00646.890                00856.790               -00198.770
01098.760                01159.830                00466.900
01357.780                00591.750                01100.090
06527.770                01111.920                01008.580
```

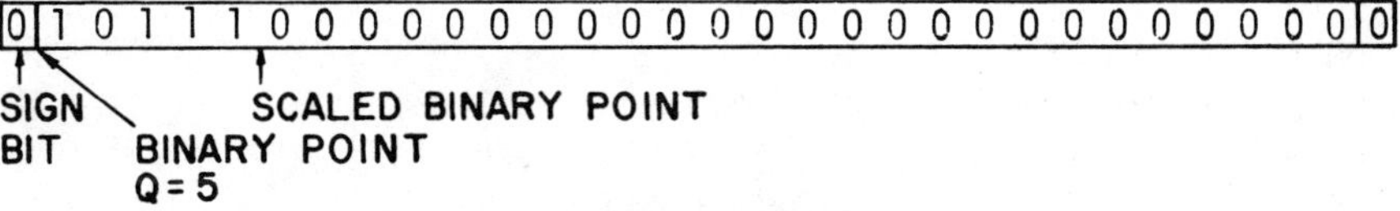

Figure 1-14. Storage of binary number 10111.

Table 1-3
p* versus *q

p	Max. q	Min. q
0	+47	+ 2
1	+43	− 2
2	+40	− 5
3	+37	− 8
4	+34	−11
5	+31	−14
6	+28	−17
7	+24	−21
8	+21	−25
9	+17	−28

A program for evaluating the polynomial in simplified form is given in Example 1-6.

Example 1-6

Location	Instruction of Address	Operand	Result or Notes
0000	b 1002	a 1005	Initial add instruction
0001	c 0005		
0002	h 1000	Zero	Initiating working storage
0003	b 1000	Working storage	
0004	m 1004	P	
0005	a (1005)	A_n	Intermediate and final
0006	h 1000	Working storage	results
0007	b 0005	a(1005 + n)	
0008	a 1001	1 at 29	
0009	h 0005	a(1005 +n + 1)	
0010	s 1003	a(1010) Flag	
0011	t 0003		
0012	b 1000	Final result	
0013	r 3050		
0014	u 3000	Print routine	
0015	z 0000	Stop	

Storage

1000	Working storage (initially zero)
1001	1 at 29
1002	a 1005
1003	a 1010
1004	P
1005	A_4
1006	A_2
1007	A_3
1008	A_1
1009	A_0

In the program (Example 1-6) the first three instructions are used to initialize the program. The command *c* of the second instruction is new.

The first instruction of the program causes *a* 1005 to be placed in the accumulator. The second instruction *c* 0005 causes the contents of the accumulator to be stored in location 0005, and, at the same time, the accumulator is cleared to zero. The third instruction causes the working storage, 1000, to be cleared to zero.

The computation during the first cycle of the loop is as follows:

1. Instruction 4 places the contents of the working storage in the accumulator—the contents are zero.
2. Instruction 5 multiplies the contents in the accumulator by the value of *x*—the product is zero for the first cycle.
3. Instruction 6 adds the contents of storage location 1005 to the accumulator—this is the value of A_4.
4. Instruction 7 stores the contents of the accumulator in the working storage location.
5. The next three steps alter the address number by 1 of the instruction stored in memory location 0005. This alteration gives the location of the next *a* value. Instruction 8 places a 1005 in the accumulator.
6. Instruction 9 adds 1 to *a* 1005 and makes the contents of the accumulator equal to *a* 1006.
7. Instruction 10 stores the contents of the accumulator in memory location 0005, and Instruction 6 now becomes *a* 1006.
8. Instruction 11 subtracts the contents which are stored in location 1003 from the contents of the accumulator; the contents in the accumulator are *a* 1006 and the contents stored in location 1003 are *a* 1010.

9. Instruction 12 tests the contents of the accumulator to determine whether the contents are either negative or zero. For the first cycle of the loop, if the contents of the accumulator are negative, the computer will return to Instruction 4 to begin the next cycle of the loop.
10. After five cycles in the loop, the contents for Instruction 11 in the accumulator will be zero. At this time, Instruction 12 will be performed. Instructions 13, 14 and 15 are used for causing the program to go into a subroutine which will print out the final answer and then return to the original program. Instruction 16 is used for stopping the computer.

At this stage, the computer is ready for a new set of data. With the addition of a new set of data, the computer is ready to calculate a new value for *B*.

Subroutines are sets of instructions used to perform standard tasks in a number of different programs. These tasks are of a general nature. Examples of subroutines are input data, output data, square root, arctangents, sine, tangent, etc. A given subroutine may be used several times in a given program. If a given subroutine is used several times in a given program, some method of going from the program into the subroutines and out again into the program at the same section should be provided. In this way, the subroutine can be used in several parts of the program. Usually, three instructions are required for going into and out of a subroutine. Instruction 13 causes the contents (computed formation volume value factor) to be placed in the accumulator. The subroutine for printing out this value is stored in locations 3000 to 3050. Instruction 14 causes the address part of the subroutine instruction which was stored in 3050 to become 0015; the last instruction of the program is stored in location 0015. The last instruction of the subroutine is now *u* 0015. After the value of the formation volume factor is punched out, the last instruction of the subroutine causes the computer to return to the original program. The computer will return from the subroutine to the last instruction (No. 16) of the program; this instruction (*z* 0000) causes the computer to stop.

The first cycle of the loop ends with Instruction 12. After this instruction is performed, the computer returns to Instruction 4. At the end of the first cycle A_4 is stored in location 1000. The explanation of the computer operations during the second cycle follows.

1. Instruction 4 causes the contents (A_4) of storage location 1000 to be placed in the accumulator.
2. Instruction 5 causes the contents of the accumulator to be multiplied by the value of *P*.

3. Instruction 6 causes the value of the second constant A_3 to be added to the contents of the accumulator. The accumulator now contains the value $(A_4P + A_3)$.

4. Instruction 7 causes the value of the accumulator to be stored in location 1000.

5. In a similar manner as discussed for the first cycle, Instructions 8, 9 and 10 cause the address portion of Instruction 6 to be modified so that the next value of the constant A will be used during the next cycle of the loop.

6. Instruction 11 causes *a* 1010 to be subtracted from *a* 1007; for this cycle the difference will be negative; therefore, Instruction 12 will cause the computer to return to Instruction 4 and start the third cycle of the loop.

Similarly, the computer will go through three more cycles. The computational operations during each cycle will consist of multiplying the computed results of the previous cycle by x and then adding the constant value of A to the result. At the end of the fifth cycle, the test, Instruction 12, will cause the computer to proceed further with the program. This will include going into the subroutine and punching out the value of B (formation volume factor).

Building Programs for the IBM 1620—Symbolic Languages

As the data processing system developed, larger and more sophisticated computers were designed and fabricated, and the programs for these computers became greater in length and complexity. At this stage, it was realized that the writing of programs in machine language with the use of numerical and alphabetic notation on lengthy and complex programs was conducive to an increase in the number of clerical and logical errors. Also, the problem of correcting errors in a program became intensified.

To simplify the writings of programs in machine language, symbolic programming systems were developed. In the writing of symbolic programs, names characteristic of instructions or closely related symbols are substituted for numbers. In other words, data to be processed and addresses are usually referred to by names, abbreviations or other meaningful designations, and the operation codes are written in mnemonic form. The elimination of numbers for specifying data, addresses, and codes greatly facilitates the writing of programs and reduces errors. However, before a symbolic program can be used on a computer, it must be translated into machine language. Programs written in symbolic language are translated into machine language programs by the use of a machine language program called a "processor". The computer itself with the "processor" does the work of translating symbolic programs into machine language programs. Usually, the translation is a one-to-one translation; that is, one machine language instruction is produced for each symbolic instruction. During the translation of a symbolic program into a machine language program, data is assigned machine addresses, and the symbolic program can be written without the assignment of addresses for data.

During the development of symbolic programming, the macro-instruction concept was developed. A macro-instruction is a synthetic instruction which is used during the translation of a symbolic program to a machine language program for referring the processor to a "library" from which a sequence of machine language instructions are extracted and placed in the machine language program. The macro-instructions are used to save the programmer from coding repetitious sequences of instructions.

Symbolic languages were an intermediate step between machine languages and the development of such problem-oriented languages as FORTRAN and BASIC. However, they never became popular or used to any great extent; therefore, there is no further discussion here of symbolic language. Lengthy discussions of FORTRAN and BASIC are given in Chapters 2, 3, 10, and 11.

Input-Output Devices

Before an electronic digital computer can process a given program, there has to be some means of communication between the programmer and the computer. Mechanically, the methods of communication are by means of punched tape or punched cards and, in some cases, by means of typing in information with a special typewriter. In coded form, the program and data are usually punched on either tape or cards. Perhaps the most convenient and most common practice is to use punched cards. In a similar manner, information comes for the electronic digital computers in the form of punched tape, punched cards or typed paper. The method used for communication between the computer and the programmer is determined by the input/output devices which are available and which were selected for a given computer.

Most mainframe computer systems still use punched cards for input, but printers are used for output. The use of punched paper tapes is now, perhaps, a thing of the past for both input and output. Today, most input and output is done by remote terminals which are connected directly to the mainframe computer by cable or through telephone hook-ups. Most terminals have a keyboard, similar to that of a typewriter, a monitor (CRT), and a printer. Microcomputers use a keyboard and a CRT for input and output, and many microcomputer setups have added printers for making copies of both input and output information. Input to a microcomputer can be from a cassette tape or from a magnetic disk. A description of the punched paper tape method used by the LGP-30 for input and output follows; the discussion is for historical purposes only.

Royal McBee LGP-30

The basic input/output device for the Royal McBee LGP-30 is a Friden typewriter (Flexowriter) which has an attachment for punching a six-track tape. Each of the characters on the keyboard of the typewriter has a code for being punched in the tape. The arrangement of the tracks on the tape is given in Figure 1-15.

The tracks on the section are numbered as indicated in Figure 1-15. The top track is No. 6, the second one from the top is No. 1, and the bottom track is No. 5. The dotted line on the tape represents feed holes. The keyboard code for the typewriter is given in Table 1-4.

With the typewriter, simultaneously as a given program is punched on the tape, the program is also typed out on the paper. To reduce the time for reading a program in the LPG-30 computer from a tape, there is an input device which reads the punched tape photometrically.

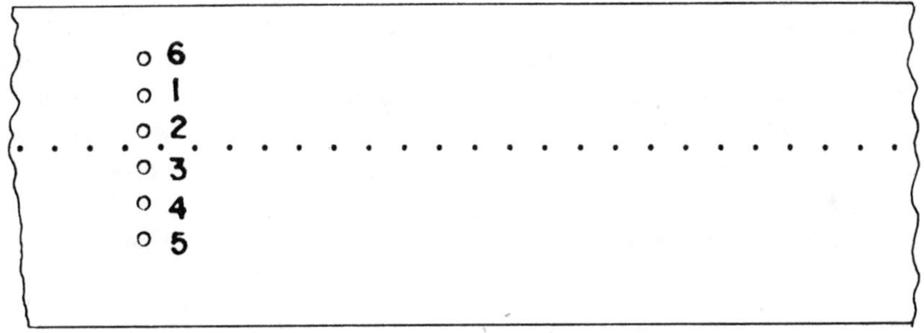

Figure 1-15. Punched tape section—six-track tape.

Table 1-4
LGP-30 Input/Output Tape Key Code

Numerical		Commands		Controls	
Track No.	123456		123456		123456
)0	000010	Zz	000001	Lower Case	000100
L1	000110	Bb	000101	Upper Case	001000
*2	001010	Yy	001001	Color Shift	001100
″3	001110	Rr	001101	Car Ret	010000
Δ4	010010	Ii	010001	Back Space	010100
%5	010110	Dd	010101	Tab	011000
$6	011010	Nn	011001	Cond Stop (′)	100000
π7	011110	Mm	011101	Start Read	000000
Σ8	100010	Pp	100001	Space	000011
(9	100110	Ee	100101	Delete	111111
Ff	101010	Uu	101001	Bell	111100
Gg	101110	Tt	101101		
Jj	110010	Hh	110001	**Signs**	
Kk	110110	Cc	110101		
Qq	111010	Aa	111001	= +	001011
Ww	111110	Ss	111101	— –	000111

Balance of Keyboard

Track No.	123456
:;	001111
?/	010011
].	010111
[,	011011
Vv	011111
Oo	100011
Xx	100111

IBM 1620 Computer

Communication between the programmer and the IBM 1620 computer can be with either punched tape or punched cards. Special hardware is required for either of the two methods of communication, depending upon the choice of the individuals who planned the particular IBM 1620 computer installation. The IBM 1622 Card Read-Punch is the input/output device used in the communication for punched cards. The IBM 1621 Paper Tape Reader and the IBM 1624 Paper Tape Punch are respectively the input/output devices used for the communication with punched tape. The typewriter on the console of the IBM 1620 computer is also an input output device which allows for the manual entry of information into and from the IBM 1620 computer.

For the punched card method, information is read into and read from the IBM 1620 computer by means of an IBM 1622 Card Read-Punch device. During input to the IBM 1620 computer, this device reads information from cards which have been punched by an IBM 26 Printing Card Punch device; however, during output, the IBM 1620 computer causes the IBM 1622 Card Read-Punch device to punch cards with the output information. The code for punching the characters into a card is illustrated in Figure 1-16.

Only one punched hole is required to represent a one decimal digit number. The one decimal digit number is coded by punching a hole in the row whose number corresponds to the one decimal digit. For instance, a 6 in the fourteenth column is represented by a hole punched in the sixth row of the fourteenth column. The punched card indicates that two punched holes in a given column are required to represent an alphabetical character in a given column, and the punched card indicates that from one to three punched holes are required to represent the special characters.

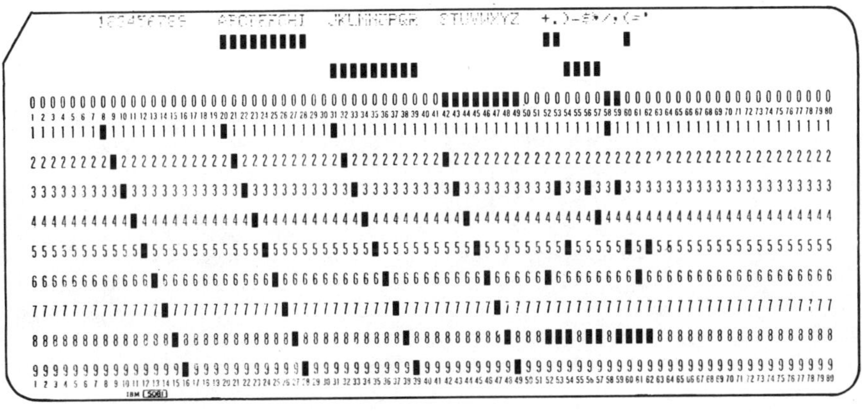

Figure 1-16. Paper card and codes.

Processing Computer Programs

Mainframe computers of central computer setups (sometimes called computer centers, computer laboratories, etc.) are operated by one or both of two methods. The two methods are sometimes called open-shop and closed-shop. With the open-shop method of operating a digital computer, the programmer processes the program, once it is prepared, on the computer. In other words, the programmer presses the electrical buttons, turns electrical switches, and performs the other physical manipulations required by the computer to process the prepared program. On the other hand, with the closed-shop method, the programmer gives the written program to a trained and skilled computer operator. This computer operator performs or supervises the physical manipulations required for processing the prepared program by the computer. After the computer program has been processed, the computed results are then given to the programmer. If the computer results are negative, the programmer will be required to correct, or attempt to correct, the original program before it is returned to the computer operator for a second processing.

The use of punched cards for communication with mainframe computers is being rapidly replaced or supplemented with terminals. Of course, the most common method of communication between a microcomputer and an operator is the keyboard.

Exercises

1. Royal McBee LGP-30 Computer
   a. Write a LGP-30 machine language program for determining the square root of a number by use of the Newton-Raphson iteration formula

$$y_i + 1 - y_i = \tfrac{1}{2}\left[\frac{a}{y_i} - y_i\right]$$

   where a is the number and y is the square root of the number.

   b. With the use of the program written for Exercise a, write a LGP-30 machine language program for determining the standard deviation for a sample of 50 numerical quantities by the equations

$$x = \sum_{i=1}^{n} \frac{x_i}{n}$$

$$S = \sqrt{\sum_{i=1}^{n} \frac{(x_i - \bar{x})^2}{n}}$$

where:

x_i = numbers.
n = sample size.
$\bar{x}$ = mean of the sample.
S = sample standard deviation.

c. Write a machine language program for finding the quotient of two numbers using the method of successive subtractions and shifts.

d. Write a machine language program for determining the average for 50 numerical quantities.

e. For fifty sets of data, write a machine language program for calculating hydraulic diffusivities.

$$\eta = \frac{k}{\phi c_\mu}$$

Use the division program written for Exercise 1c to obtain the quotients.

2. If you have a programmable pocket calculator, write a program for solving Example 1-5; compare the two programs.

Suggested Reading

1. Boraiko, A. A., "The Chip," *National Geographic*, pp. 421ff, 162, 1982.
2. Engineering Research Associates, Inc., supervised by C. B. Tompkins and J. H. Wakelin and edited by W. W. Stifler, Jr. *High-Speed Computing Devices*, New York: McGraw-Hill Book Co., 1950.
3. Leeson, D. N. and D. L. Dimitry. *Basic Programming Concepts and The IBM 1620 Computer*, New York: Holt Rinehart and Winston, Inc., 1962.
4. McCormick, E. M. *Digital Computer Primer*, New York: McGraw-Hill Book Co., 1959.
5. McCracken, D. D. *Digital Computer Programming*, New York: John Wiley & Sons, 1957.
6. McCracken, D. D. *A Guide to Fortran Programming*, New York: John Wiley & Sons, Inc., 1961.
7. Mims, F. M., III, *Understanding Digital Computers*, Fort Worth, Texas: Tandy Corporation, 1978.
8. Reference Manual A26-4500-2. *IBM 1620 Data Processing Systems*, New York: International Business Machines Corporation, 1961.
9. Reference Manual C26-5619-0. *IBM 1620 FORTRAN*, New York: International Business Machines Corporation, 1962.
10. General Information Manual F28-8074-1. *FORTRAN*, New York: International Business Machines Corporation, 1961.
11. IBM Systems Reference Library. *1620 IBM FORTRAN II Operator's Guide*, New York: International Business Machines Corporation, File No. 1620-25 Form C 26-5662-1, 1962.
12. IBM Systems Reference Library. *1620 IBM FORTRAN II Specifications*, New York: International Business Machines Corporation, File No. 1620-25 Form C 26-5602-2, 1962.

13. Osborne, A., *An Introduction to Microcomputers, Vol. 1,* Berkeley, California: Osborne/McGraw-Hill, 1980.
14. Richards, R. K. *Arithmetic Operations in Digital Computers,* New York: D. Van Nostrand Company, 1955.
15. Wrubel, M. H. *A Primer of Programming for Digital Computers,* New York: McGraw-Hill Book Company, 1959.
16. Bowlder, Henry J. and Roberta R. Smith. *Act III, An Algebraic Compiler for the LGP-30 Computer,* Los Angeles: Computer Users Organization No. H3-207, 1961.
17. *Royal Precision Electronic Computer LGP-30 Programming Manual,* Port Chester, New York: Royal McBee Corporation, 1957.
18. *RPC 4000 Compact Programming Manual—An Algebraic Compiler and Translator,* No. H-3-01-.0, New York: General Precision, Inc., 1963.
19. *RPC 4000 Purdue Floating Point Interpretive System (PINT),* No. H1-02.0, New York: Royal McBee Corporation, 1961.

2 Programming in FORTRAN

Substituting easy-to-remember mnemonics for numbers in the symbolic programming system relieved programmers of the tedious and error-breeding task of keeping track of numerical addresses. At this stage of program development, programmers still felt that too much time was being consumed for programming problems in sciences and engineering. Months were often spent programming a problem only to have a computer execute the program and produce the results in a few minutes. Also, during the development of symbolic programming systems, macro-instructions were developed. A macro-instruction is a pseudo-instruction which, during assembly, generates more than one machine language instruction. Now, a large number of macro-instructions are available for symbolic programming.

Following the development of a machine language program (called a processor), which would assemble a machine language instruction for each symbolic program statement and several machine language instructions for each macro-instruction, a machine language program (called a compiler) was developed for assembling one or many machine language instructions for each statement of a source program. In addition, the compiler was developed for taking source program statements written in algebraic-like language. This method of writing source statements further relieved programmers of the tedious task of keeping track of numerical addresses.

The repertoire of basic instructions for a given computer is determined by the design and construction of the computer; in other words, the repertoire of machine language instructions will vary from computer to computer.

It is also true that the repertoire of symbolic instructions is determined largely by the same factors which affect the repertoire of machine language instructions.

To alleviate further the process of preparing computer programs, computer programming systems were developed which were problem-oriented and could be used on many different computers with only minimum knowledge of their design and construction. Most problem-oriented languages require statements written in algebraic-like form, and a compiler which causes the computer to compile machine language instructions for the statements of the source program. The program of machine language instructions compiled from the source program is called the object program. Following the compilation of the object program, it is then executed on the computer, and this is the program which actually gives the much-wanted results.

Compilers for a given problem-oriented system will vary from computer to computer. The machine language instructions in a compiler are controlled by the design and construction of the computer. As an example, the source programs given in Chapter 10 were written in FORTRAN IV (a problem-oriented language) and processed on an IBM 360 (model 40) and a Xerox Sigma VI; the FORTRAN compilers were furnished by the manufacturer of the respective computer. The programmers writing the source programs in Chapter 10 had a very limited knowledge about the design and construction of either computer.

Many high-level languages have been written. The five most common high-level languages are:

1. FORTRAN IV, 77 (ANSI 66, 77)
2. COBOL
3. ALGOL
4. BASIC
5. Pascal

The most widely used problem-oriented language is FORTRAN; the name FORTRAN comes from the words *FORmula TRANslation.* The FORTRAN programming system was first developed by International Business Machine Corporation, and this corporation has stated that it is a system designed primarily for scientific and engineering computations. New versions of FORTRAN are continually being developed; today there are two predominant versions: FORTRAN IV and FORTRAN 77.

The name COBOL comes from the words *COmmon-Business-Oriented-Language.* This language was developed to provide a common set of procedures for defining and expressing business data-processing applications.

The name ALGOL comes from the words *ALGOrithmic* Language; this language, originally developed in Europe, uses a notation similar to everyday mathematical notations. The problem-oriented language BASIC was originally developed at Dartmouth College for the beginning programmer to learn and use; this language uses a series of numbered statements of common English words and familiar mathematical notations.

Since problem-oriented languages are easier to learn than machine languages (machine language and symbolic language), the present-day trend is for one to concentrate his efforts upon problem-oriented languages. Throughout the remainder of this text, problem-oriented languages will be stressed. FORTRAN IV and 77 are discussed in Chapters 2 and 10; BASIC is discussed in Chapters 3 and 11. Note: there are a number of FORTRAN IV and FORTRAN 77 statements which are identical.

FORTRAN IV and FORTRAN 77

A classification of FORTRAN statements is given in the following outline:

A. Arithmetic Statement

B. Control Statements
   1. GO TO
   2. IF
   3. DO
   4. CONTINUE
   5. PAUSE
   6. CALL EXIT
   7. STOP
   8. END

C. Subprogram Statements
   1. FUNCTION
   2. SUBROUTINE
   3. CALL
   4. RETURN

D. Input/Output Statements
   1. FORMAT
   2. READ
   3. WRITE
   4. PRINT

E. Specification Statements

1. DIMENSION
2. EQUIVALENCE
3. COMMON
4. DATA

Approximately half of the FORTRAN statements are needed for preparing FORTRAN source programs for the problems in reservoir engineering discussed in Chapter 10.

A FORTRAN programming system coding sheet is used for explaining the format for FORTRAN source programs (see Figure 2-1). The first field on the coding sheet, Columns 1 to 5, contains the numbers of the respective statements when the statements are numbered. If Column 1 of a given line contains a C, then this line is used for comments; information from this line is not processed. Column 6 of a given line is left blank if all the information on that line can be punched entirely on a card; otherwise, the column is used for noting a continuation, Columns 7 through 72 are used as the field for the FORTRAN statements, but the statement must not necessarily begin in Column 7; blanks are ignored in this field (7-72). Field Columns 73-80 are used for identification purposes; information in this field is not processed by the FORTRAN compiler.

In FORTRAN language, constants and variables can be expressed in one of two modes: fixed point, which is restricted to integers, and floating point, which is restricted to decimal notation. The restrictions on the two modes for expressing variables and constants in FORTRAN language can be summarized as in Table 2-1.

There are six basic arithmetic-like operations associated with the FORTRAN language. These operations are represented by specific symbols as given in Table 2-2.

The symbol = does not imply mathematical equality; it has the meaning "is assigned a value equivalent to"; that is, in the example P = R it means "compute the value of R, and P is then assigned the value equivalent to R." In Table 2-1 constants and variables of the modes are sometimes referred to as elements. The combination of arithmetic operations and elements give algebraic-like expressions which can be formulated in the notation of FORTRAN as shown in Table 2-3.

A FORTRAN expression has been defined as a constant, variable, or function—or any combinations of these that are separated by operation

IBM

FORTRAN Coding Form

GX28-7327-6 U/M 050**
Printed in U.S.A.

PROGRAM		PUNCHING INSTRUCTIONS	GRAPHIC	PAGE OF
PROGRAMMER	DATE		PUNCH	CARD ELECTRO NUMBER*

COMM.	STATEMENT NUMBER	CONT.	FORTRAN STATEMENT	IDENTIFICATION SEQUENCE
1	2 3 4 5	6	7 8 9 10 11 12 13 14 15 16 17 18 19 20 21 22 23 24 25 26 27 28 29 30 31 32 33 34 35 36 37 38 39 40 41 42 43 44 45 46 47 48 49 50 51 52 53 54 55 56 57 58 59 60 61 62 63 64 65 66 67 68 69 70 71 72	73 74 75 76 77 78 79 80

*A standard card form, IBM electro 888157, is available for punching statements from this form

**Number of forms per pad may vary slightly

Figure 2-1. IBM FORTRAN coding form GX28-7327-6U/M.

Table 2-1
Restrictions on the Two Modes for Expressing Variables and Constants

Mode	Constants	Variables
INTEGER	1. Must contain only digits and signs.	1. One to six characters in length.
	2. May be signed or unsigned; assumed positive if unsigned.	2. First character must be I, J, K, L, M, or N.
	3. Must be integers; no fractional values allowed.	3. Subsequent characters may be any mixture, in any order, of letters from the alphabet of digits.
	4. Examples: 1 +1 −1 0 3761 −3111 −4889	4. Examples: I IKE J JFK K NIXON L I1K2E3 M MEAN N J 36
FLOATING POINT	1. Must contain only digits, sign, and decimal point.	1. One to six characters in length.
	2. May be signed or unsigned; assumed positive if unsigned.	2. First character must be alphabetic, other then I, J, K, L, M, or N.
	3. Value may be integral or may have fractional parts.	3. Subsequent characters may be as described above.
	4. Examples: 1.0 +1.0 −1.0 0.0 37.61 3.111 3.333 67.6666	4. Examples: A SUM B XBAR A13 ALT ABE DIAM X ROOT Y ROOT1 P ROOT2 Q

Table 2-2
FORTRAN Language Symbols

Operation	Symbol	Example
Addition	+	P + R
Subtraction	−	P − R
Multiplication	*	P * R
Division	/	P / R
Exponentiation	**	P ** R (P^R)
Equality	=	P = R

Table 2-3
Algebraic Expressions

Algebraic Expression	FORTRAN Formulation
$3.1416\ R^2$	3.1416 * R * R
$A + BX + CX^2$	A + B*X + C*X**2

symbols, commas, and parentheses which form a meaningful expression. The rules for the construction of FORTRAN expressions are as follows:

1. Expressions may be composed of constants, variables, and functions which are written in either fixed-point or floating-point form, but not in both. Subscripts and exponents in a floating-point expression are the exceptions to this rule; subscripts and exponents can be fixed-point quantities.
2. Mixed modes are not permitted for expressions of constants, variables, and subscripted variables; either the mode is fixed point or floating point.
3. The name of the function determines the mode for the function, provided the ordinates are in the modes assumed in the definition of the function.
4. The mode of the expression is not affected by the exponentiation of the expression; however, a floating-point expression may be given a fixed-point exponentiation, but a fixed-point expression cannot be given a floating-point exponent.
5. The mode of an expression is not affected by a plus or minus sign which precedes the expression. For example, — R, + R, * R are all expressions of the same mode.

6. The mode of the expression is not affected by enclosing an expression in parentheses. For example, (R), ((R)) are expressions of the same mode.

7. Complex expressions can be formed by the connection of operation symbols, provided the items or the components of the complex expression are of the same mode and no two operation symbols appear in sequence; e.g., not R + − R, but R + (− R).

When not specific or doubt exists, the hierarchy of arithmetic operations in an expression as given in Table 2-4 can be used as a guide.

To eliminate ambiguity in expression, the use of parentheses for FORTRAN expressions is similar to the use of parentheses for algebraic expressions. Example 2-1 further explains the use of parentheses.

Example 2-1

Algebraic Expression	FORTRAN Expression
$\frac{P+R}{S}$	(P + R) /S
$y = ax^3 + bx^2 + cx + d$ or $y = ((ax + b)x + c)x + d$	Y = ((A* X + B) *X + C) *X + D
$\frac{P+R}{X-Y}$	(P + R) / (X − Y)

Table 2-4
Hierarchy of Arithmetic Operations in an Expression

Order	Symbol	Operation
1	**	Exponentiation
2	*	Multiplication
3	/	Division
4	+	Addition
5	−	Subtraction

In addition to the arithmetic operations, there are eight quasioperations which are used for forming FORTRAN expressions for evaluating arithmetic or algebraic expressions. The form and requirements of the quasioperations are given in Table 2-5.

Example 2-2 requires the quasifunctions. (A variable written to the left of = with an expression constitutes a statement.)

Example 2-2

$x = \frac{a + \sqrt{b}}{\ln c}$ X = (A + SQRT(B))/ALOG(C)

$P = \sinh(x)$ P = (EXP(X) − EXP(−X))/2.0

Expressions must not contain both modes of arithmetic. All quasifunctions are calculated in the floating-point mode. Exceptions will be the use of integers as subscripts and in the use of exponentiation when the power involved is an integer.

Table 2-5
Eight Quasioperations

Mathematical Function	FORTRAN Function	Requirement
Sine	SIN(X)	X in radians
Cosine	COS(X)	X in radians
Arc Tangent	ATAN(X)	Results in I or IV quadrant
Exponential ex	EXP(X)	
Natural Logarithm	ALOG(X)	X larger than zero
Logarithm to base 10	ALOG10(X)	X larger than zero
Square Root	SQRT(X)	
Absolute Value	ABS(X)	

Arithmetic Statements

The arithmetic statement is the most common statement; it resembles an algebraic formula. The general form of the arithmetic statement is

$r = s$

where r is a variable and s is an expression. The equal sign (as previously stated) does not take on the same meaning in FORTRAN as it does in mathematics—the equal sign signifies replacement rather than equivalence.

Previously, it was stated that mixed modes are not permitted in expressions except in the case of exponents and subscripts; however, the mode of the left-hand variable of a FORTRAN statement does not necessarily have to agree with the mode of the arithmetic expression of the FORTRAN statement. A variable on the left of the equality sign in fixed-point mode with an expression in floating-point mode will cause the result to be first computed in floating point and then truncated and converted to a fixed-point integer prior to being stored as the new value of the variable. On the other hand, a variable on the left of the equality sign in floating-point mode with an expression in fixed-point mode will cause the result to be computed in fixed point and then converted to floating point prior to being stored as the new value of the variable. Results calculated in fixed-point mode are truncated at the decimal point; for instance, 7 divided by 2 will be assigned the value of 3 by the computer instead of the true floating-point number 3.5. The different types of arithmetic statements are illustrated in Table 2-6.

Table 2-6
Types of Arithmetic Statements

Example	Meaning
S= R	Assign the equivalent value of R to S
N = R	Truncate R to an integer, convert to fixed point and assign the equivalent fixed point value of R to N
N = N + 1	Add 1 to N and the equivalent value of N is N + 1
P = 4 * R	Assign the equivalent value of 4R to P
P = N * R	Not permitted because the expression contains both fixed point and floating point variables
P = 3.0 * N	Not permitted because the expression is mixed
P = 3 * R	Not permitted because the expression is mixed
N = 7/2	Assign the value of 3 to N, not 3.5

Control Statements and End Statements

Similarly as for programs written in machine language and symbolic language, the statements of a program written in FORTRAN language are executed sequentially. Essentially, the control statements are used for controlling the order in which the FORTRAN statements are executed.

Unconditional GO TO Statement. The GO TO statement in FORTRAN is identical to unconditional transfer instructions of programs written in machine language. The general form of the unconditional GO TO statement is

GO TO n

where n is a number of a statement. The statement

GO TO 12

causes Statement 12 to be executed next.

Computed GO TO Statement. The general form of the computed GO TO statement is

GO TO $(N_1N_2 \ldots N_m)I$

where $N_1N_2 \ldots N_m$ are statement numbers, and I is a nonsubscripted fiixed-point variable. The value of I determines the number of the next statement to be executed; I takes on values which correspond to the subscripts of N, i.e., 1, 2, 3. . . . m. For the FORTRAN statement

GO TO (12, 42, 67, 8) I

if I is 2 at the time of execution, the next statement executed is Statement 42.

IF Statement. The general form of the IF statement is

IF $(r)\, n_1n_2n_3$

where r is an expression and n_1, n_2, and n_3 are FORTRAN statement numbers. The IF statement in the FORTRAN language corresponds very closely to the test or branch operations of machine language. The next statement to be executed is determined by the value of the expression r; if the computed value of r is less than zero, the next statement to be executed is Statement n_1. If the computed value of expression r is zero, the next statement to be executed will be Statement n_2; if the next value of the expression r is greater than zero, the next statement to be executed will be Statement n_3. For the example,

IF (N − KOUNT) 10, 4, 13

if the value of the expression (N − KOUNT) is less than zero, the next statement to be executed will be Statement 10; if the value of the expression (N − KOUNT) is equal to zero, the next statement to be executed will be Statement 4; and if the value of the expression (N − KOUNT) is greater than zero, the next statement to be executed will be Statement 13.

The logical IF FORTRAN statement permits the outcome of several comparisons to be combined into a single logical expression and used to effect branching within a FORTRAN program. The general form of the logical IF FORTRAN statement is

IF (B) R

where B is a logical expression and R is an executable FORTRAN statement other than another logical IF statement or a DO statement. During the operation of the logical IF statement, the expression B is evaluated. If the value of the expression B is true, then the statement R is executed; but if the value of the expression B is false, R is not executed and the control is passed directly to the next statement. The logical expression is formed by the use of relation operators. A list of relational operators is given in Table 2-7.

Table 2-7
Relational Operators for Logical IF Statements

Relational Operators	Meaning
.LT.	Less than
.LE.	Less than or equal
.EQ.	Equal to
.NE.	Not equal
.GT.	Greater than
.GE.	Greater than or equal to

By combining the logical operators (.AND., .OR., .NOT.) with relational operators, the logical IF statement uses are extended.

Examples 2-3 through 2-7 illustrate logical FORTRAN statements.

Example 2-3

```
IF (P(1) .LE. P(2)) GO TO 202
```

Example 2-4

```
IF (P(1).LT.P(2).AND.P(1).LE.P(3)) GO TO 101
```

2-13

Example 2-5

```
IF (X.LT.Y.AND.I.LE.J) GO TO 102
```

Example 2-6

```
IF (.NOT.(P.LT.R)) P=B
IF (P.GT.R) P=B
```

Example 2-7

```
IF (P(1).GT.P(2).OR.P(1).GE.P(3)) GO TO 202
```

The first logical IF statement (Example 2-3) causes Statement 202 to be executed if P1 is less than or equal to P2. If both conditions are met in the second logical IF statement (Example 2-4), then Statement 101 will be executed. For both conditions to be met means that P1 is less than P2 and P1 is less than or equal to P3; and if both conditions are not met, the following statement will be executed.

The logical IF statement in Example 2-5 illustrates the use of mixed arithmetic modes; mixed modes in this fashion are possible in logical IF statements. In Example 2-6, which illustrates the use of the logical operator .NOT., both FORTRAN statements have the same value: the last of the logical IF statements (Example 2-7) illustrates the use of the logical operator .OR. Briefly the statement says that if either P1 is greater than P2 or P1 is greater than or equal to P3, the next FORTRAN statement to be executed is Statement 202. If neither of the conditions have been met, the next statement in the program will be executed.

FORTRAN 77 contributed the useful IF-THEN-ELSE facility; the facility can be used to replace several IF statements. The general form of the facility is

```
IF (e) THEN
      .
      . Executable statements   (true IF)
      .
ELSE
      .
      . Executable statements   (false IF)
      .
END IF
```

where (e) is one of the relational operators (Table 2-7).

Another form of the IF-THEN-ELSE facility is

```
IF (e) THEN
      .
      . Executable statements   (true IF)
      .
ELSE IF (e) THEN
      .
      . Executable statements   (true IF)
      .
ELSE IF (e) THEN
      .
      . There can be other ELSE-IF statements   (true IF)
      .
ELSE
      .
      . Executable statements   Neither of the IF statements apply
      .
END IF
```

The IF-THEN-ELSE, IF-ELSE facility can be applied for an incremental change of a variable in given ranges as

```
IF (X.LE.2.0) THEN
      X=X+0.001
      GO TO 500
```

```
ELSE IF (X.LE.8.0) THEN
     X=X+0.01
     GO TO 500
ELSE
     GO TO 500
END IF
```

The previous partial program will cause X to be incremented in values of 0.001 between 0.0 and 2.0 and to be incremented in values of 0.01 between 2.0 and 8.0.

DO Statement. The general form of a DO statement is

$$\text{DO } n\ i = m_1, m_2 \quad \text{or} \quad \text{DO } n\ i = m_1, m_2, m_3$$

where n is a statement number, i is a nonsubscripted fixed-point variable, and m_1, m_2, m_3 each are either an unsigned fixed-point constant or a nonsubscripted fixed-point variable. If m_3 is omitted from the statement, it takes on the value of 1.

The DO statement causes all the statements which follow, up to and including the statement with the number n, to be executed repeatedly. For the first time during the execution of the statements, i will have the value of m_1; then, for each succeeding time, i will be increased by the value of m_3 until i becomes equal to m_2. The DO statement is illustrated by the partial program of Example 2-8.

Example 2-8

```
 .
 .
 .
13 DO 14 I = 1, 30
14  Z(I) = X(I) + Y(I)
15
 .
 .
 .
```

Perhaps the reader will recognize that the two statements of the partial program contain the essential statements for computing the sums of 30 sets of two numbers each or the essential statements for the solution of the equation.

$$Z_i = X_i + Y_i$$

for $i = 1$ to $i = 30$

In the example, suppose that the computer has advanced to Statement 13; then, Statement 14 will be executed with the value of $I = 1$. Thereafter, Statement 14 will be executed repeatedly for 29 more times; during each repeated execution of Statement 14, I will be increased in increments of 1. During the second pass of the loop, $I = 2$ and during the successive passes $I = 3, 4, 5 \ldots$ until $I = 30$. After the execution of Statement 14 with $I = 30$, the computer will transfer to Statement 15 for the next statement to be executed.

The set of statements up to and including statement n which follow the DO statement and which are executed successively and repeatedly is called the range of a DO statement. DO statements are used as one means of forming loops in a program. More than one DO statement is permitted within the range of a DO statement provided all of the latter DO statements are within the range of the first DO statement and there are no transfers into the range of any DO statement from outside its range. A set of DO statements which are within the range of a given DO statement is called a "nest of DO's" (see Figure 2-2). The following restrictions apply to the use of a nest of DO's: (a) the range of a second DO statement within the range of a first DO statement must not exceed the range of the first DO statement, (b) within the range of a DO statement, transfer of control statements as IF or GO TO are not permitted, (c) the range of a DO statement cannot end with a GO TO or IF statement, (d) neither of the statements CONTINUE, FORMAT or DIMENSION are permitted as the first statement in the range of a DO statement, and (e) a nest of DO's is executed from the innermost to the outermost.

CONTINUE Statement. The CONTINUE statement is a dummy statement which causes no instructions in the object program during the time the source program is being compiled. The general form of the CONTINUE statement is

```
CONTINUE
```

This statement is merely used as a dummy statement for an IF or GO TO statement. A table-type search program often requires the use of a CONTINUE statement. Example 2-9 illustrates the use of a CONTINUE statement.

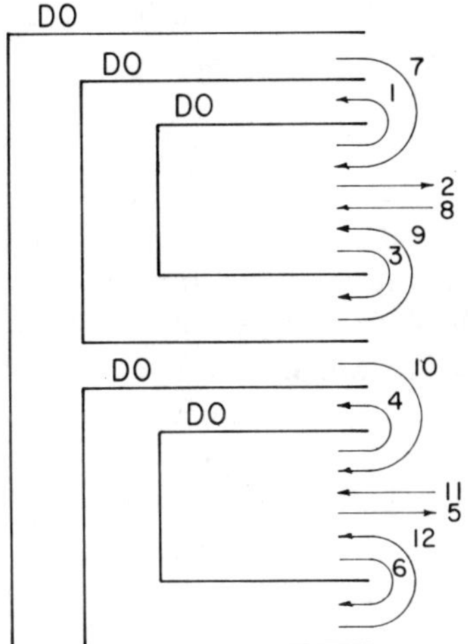

Figure 2-2. Nest of DO's: permissible transfer 1, 2, 3, 4, 5, 6; non-permissible transfers 7, 8, 9, 10, 11, 12.

Example 2-9

```
   .
   .
   .
14 DO 16 I = 1, 50
   IF (EIF – VALUE (I)) 16, 30, 16
16 CONTINUE
17 . . .
   .
   .
   .
```

The partial program of Example 2-9 causes a scan of a 50 entry-value table until it finds a value in the table which equals the value of the variable EIF; next the computer advances to Statement 30.

PAUSE Statement. The PAUSE statement halts the computer; to start the computer again, the START key is depressed. The general form of the PAUSE statement is

PAUSE

A PAUSE statement permits a temporary interruption of a program to check some intermediate result or take some other action. Most mainframe computers do not permit programmers to use this statement.

For computer systems which use monitor control for executing programs, the CALL EXIT statement returns the control of the computer to the monitor. Some compilers permit the use of the STOP statement in place of the CALL EXIT statement. The CALL EXIT statement does not actually stop the computer in its operation. The general form of the CALL EXIT statement is

CALL EXIT

STOP or STOP n Statements. The STOP or STOP n statements are used for terminating program execution; depressing the START key has no effect. The general forms of the statements are

STOP
STOP *n*

Upon execution of the statement, the message STOP is displayed on the console. For the statement with *n* being a five-digit or less number, the number *n* will be displayed by the console.

END Statement. The END statement is used for signaling to the compiler that the final statement of the FORTRAN source program has been received. The END statement is the last statement of a FORTRAN program, and every FORTRAN source program must have an END statement; otherwise, the FORTRAN source program will not be compiled. The general form of the END statement is

END

Functions and Subprogram Statements

Frequently, during the preparation of a FORTRAN source program, the programmer will find that identical kinds of computations, with different data, recur throughout the program. And sometimes the identical kinds of computations will recur in a number of different source programs. To eliminate the repetitious preparations of partial programs (better known as subprograms) for the identical kinds of computations, FORTRAN provides: (1) additional library functions, (2) arithmetic statement functions, and (3) FORTRAN functions.

Additional Library Functions. During the discussion of the arithmetic operations, eight quasioperations were given and discussed. These operations are often referred to as "closed" library functions, and these functions are provided by the IBM Corporation. The programmer can prepare other functions and add them to the library containing the closed functions; these functions will be used in the same manner as the closed functions.

Additional library functions consist of the two parts, name and argument. The argument follows the name, and it is enclosed in parentheses. The name consists of one to six alphabetic characters; special characters are not permitted. An example of an additional library function is

```
EIF (X)
```

The functions are called or used by the program in the same manner as the "closed" library functions. The name serves for calling the function. After the calling of the function, a single numerical quantity, using the argument, is then computed. A given library function produces only one value.

Arithmetic Statement Functions. The arithmetic statement functions are single FORTRAN arithmetic statements, and these functions apply only to the program in which they appear. The coding and naming of these functions are in the same manner as the library functions; that is, they consist of the two parts, name and argument. Similarly, the name is written first; and it is followed by the arguments, separated by commas, and enclosed in parentheses. The arguments must indicate whether it is fixed-point or floating-point mode. Examples of arithmetic statement functions are:

```
DIFF(H,U,F,C) = H(U*F*C)
MEAN(I,J,K) = ((I + J+ K)/3) + 5
```

As an arithmetic statement is needed in a program, it is called by using only the name and the arguments. For instance, the first example could be called in a program as illustrated in Example 2-10.

Example 2-10

```
DIFF(H,U,F,C) = H/(U*F*C)
.
.
.
READ 1,(A,B,D,E)
Q = SQRT  (DIFF(A,B,D,E)) + 42
```

The notations of the arguments in the calling expression and the expression of the arithmetical statement function are not required always to be identical; only the modes of the corresponding arguments are required to be identical.

FORTRAN Functions. The two FORTRAN functions are the FUNCTION subprogram and the SUBROUTINE subprogram. The two subprograms require the use of the four statements: (1) FUNCTION, (2) SUBROUTINE, (3) CALL, and (4) RETURN. The FORTRAN function which is a FUNCTION subprogram is always single-valued, but the SUBROUTINE subprogram may be multivalued. The FUNCTION subprogram is called by an expression which contains its name; a CALL statement is required for calling the SUBROUTINE subprogram.

FUNCTION Statement. In a FORTRAN function subprogram, the FUNCTION statement is always first. There are three parts to a FUNCTION statement; the parts are FUNCTION, name, and argument. The name is a symbolic name of a single-valued function. The arguments are separated by commas and are enclosed by parentheses, and one of the arguments must be nonsubscripted. The name of the FUNCTION consists of one to six characters, the first of which must be alphabetic. The following are examples of a FUNCTION statement:

```
FUNCTION ROOT (X,R,S,Y)
FUNCTION DARCY (H,U,R,X,A)
```

A FUNCTION subprogram may have several statements, whereas an arithmetic statement function can have only one statement. The name of the function must appear at least once on the left-hand side of an arithmetic statement. A partial program illustrating a FUNCTION subprogram is given in Example 2-11.

Example 2-11

```
FUNCTION DARCY (H,U,R,X,A)
.
.
.
DARCY = (H*U*X)/(A*R)
.
.
.
RETURN
END
```

A FUNCTION subprogram is called by using the name of the FUNCTION as a portion of an arithmetic expression. The partial program of Example 2-12 illustrates the calling of a FUNCTION subprogram for the FUNCTION statement, FUNCTION ROOT (X,I,K,Y).

Example 2-12

```
READ 5,J
.
.
.
Z =
N = . . .
M = . . .
R = . . .
Q = ROOT (Z,N,M,R) + 94.5
.
.
.
END
```

The arguments of the FUNCTION statement and the arguments of the arithmetic calling expression are not necessarily required to be identical in symbolic notation; however, the corresponding arguments of the FUNCTION statement and the arguments of the arithmetic expression must agree respectively in their modes. The arguments of the FUNCTION statement can be of mixed modes.

SUBROUTINE Statement. The general form of the SUBROUTINE statement is

SUBROUTINE Name $(a_1, a_2 \ldots a_n)$

The name is the symbolic name of a subprogram, and each argument is the name of a nonsubscripted variable. The name contains one to six alphabetic or numerical characters, the first of which is alphabetic.

The SUBROUTINE statement is used as the first statement of the subprogram. Examples of SUBROUTINE statements are as follows:

```
SUBROUTINE MANY (A,N,M,L,C)
SUBROUTINE MEAN (B,A,C, ROOT 1, ROOT 2)
```

The names of the arguments of SUBROUTINE statements are dummies which are replaced at the time of the execution by the true arguments supplied with the CALL statement. A SUBROUTINE subprogram can use one or more of its arguments and calculate more than one result. A partial program which illustrates a SUBROUTINE subprogram is shown in Example 2-13.

Example 2-13

```
SUBROUTINE ROOT (X,I,K,Y,R,N,P,L)
.
.
.
P = (X + Y) /2
N = K + I
R = X + Y
L = P*R
.
.
.
RETURN
END
```

The calling of the SUBROUTINE subprograms into the main program is illustrated by the partial program in Example 2-14.

Example 2-14

```
READ 21 (X,A,T)
.
.
.
X =
S =
L =
P =
CALL ROOT (X,I,L,P,R,N,Q,M)
AR = (P + 5.3) /R
.
.
.
END
```

CALL Statement. The CALL statement is used for calling SUBROUTINE subprograms; the statement transfers the control to the subprogram. The general form of the CALL statement is

CALL Name $(a_1,a_2,a_3 \ldots a_n)$

Name is the name of the SUBROUTINE subprogram, and the arguments a_1, a_2, a_3, . . . a_n take one of the following forms:

1. Fixed-point constant.
2. Floating-point constant.
3. Fixed-point variable, with or without subscripts.
4. Floating-point variable, with or without subscripts.
5. Arithmetic expressions.

Arguments of the CALL statement must agree in number, order, mode and array size with the corresponding arguments of the SUBROUTINE subprogram.

RETURN Statement. The RETURN statement is used for terminating the subprograms, FUNCTION and SUBROUTINE. It is the last statement of a subprogram which is executed. An example of the statement is

RETURN

Input/Output Statements

IBM computers have been fabricated with four input/output devices: (1) the card reader, (2) paper tape reader, (3) card punch, and (4) console typewriter. The input/output devices will vary from one computer installation to another. Selection of the input/output devices is determined at the time the computer setup is planned. Some computer installations may use punched tape for input and output, whereas other installations may use punched cards. Input/output FORTRAN statements are used for specifying the transmission of information during the execution of the object program, between the core storage and the input/output devices. There are seven input/output statements: READ, ACCEPT TAPE, ACCEPT, PUNCH, PUNCH TAPE, PRINT, and TYPE.

During the preparation of a FORTRAN program, the selection of the input/output statements is determined by the input/output devices of the given computer setup. The seven input/output statements are always used with a FORMAT statement. One or more input/output statements may be used with a given FORMAT statement. The FORMAT statement specifies how the information is arranged during the execution of the object program between the core storage and any of the input/output devices. FORMAT statements are not executable; they are only specification statements.

FORMAT Statements. The general form of the FORMAT statement is

n FORMAT $(S_1, S_2, S_3 \ldots S_n)$

where n is the assigned number of the FORMAT statement and S_1, S_2, $S_3 \ldots S_n)$ are the specifications. The symbol n is referred to by a previous FORTRAN statement such as a READ or WRITE statement. A list of the most common FORMAT specifications is given in Table 2-8. The special symbols—comma, apostrophe, and solidus—are not commonly classified as specifications; however, these symbols play an important part in a FORMAT statement. The following are examples of FORMAT statements:

```
12 FORMAT (I2,E12.4, F10.4)
14 FORMAT (F12.5,I6,E6.4)
```

Statements of the first example take a more general form as

12 FORMAT (*Iw,Ew.d,Fw.d*)

where *I, E,* and *F* are the control characters of the field; *w* is the width of the field; and *d* is the number of decimal positions which appear to the

Table 2-8
FORMAT Specifications

Number	FORMAT Specification	FORMAT Code	Use	External Representation	Field Width
1	I	rIw	Operation involving fixed point or integer variables	Integer ± xxxx	Number of significant variables + 1
2	E	rEw.d	Operations involving floating point or real variables with exponent	Decimal with exponent xx.xxxxE±xx	At least d + 7
3	F	rFw.d	Operations involving floating point or real variable with no exponent	Decimal ±xx.xxxx	At least d + 3
4	X	wX	Inserting blanks or ignoring columns	Skip field	Variable
5	H (Hollerith)	wH	Reading or writing alphanumeric		Variable
6	Comma	,	Separating fields or preventing ambiguity		None
7	Solidus	/	Terminating or beginning another line	Multi-cards or multi-lines	None
8	G	rGw.d	Operations involving real or complex variables, integers or logical data	Variable	Variable
9	Apostrophe	"	Reading or writing alphanumeric		Automatic adjusts
10	A	rAw	Input/output of alphanumeric information	Real or integral variables	Variable
11	T	Tw	Specifying columns to begin following FORMAT Code		None

right of the decimal point. The control character *I* will cause the computer to accept a variable in fixed-point form, but the control characters *E* and *F* will cause the computer to accept variables in floating-point form. Then, on output, the control character *E* will cause the output data to be in floating point, and the control character *F* will cause the output data to be fixed-decimal form.

The FORMAT statement of the first example has the assigned number 12, and, whenever this statement is used, it is referred to by this number.

The first specification of the example is I2 (Iw), meaning fixed point with a field width of two digits. The second specification is E12.4 (Ew.d) which means floating point with a field width of 12 digits, having four digits to the right of the decimal point. The third specification is F10.4 (Fw.d) which means floating point with a field width of 10 digits, having four digits to the right of the decimal point.

In the brief discussion on the FORMAT statement, only three FORMAT specifications were discussed to a limited extent. In the discussion which follows, eight FORMAT specifications are discussed along with three symbols: comma, apostrophe, and solidus. An extended discussion about the FORMAT statement is essential because the FORMAT statement (a non-executable statement) provides the computer with information about the input and output data. The design of the data on input cards (or their equivalents) is dictated by one or more FORMAT statements. With a computer system which types data from the computer system on a continuous sheet of paper, it is possible by means of appropriate FORMAT statements for the data from the computer to be listed in table form.

The FORMAT statements are used in conjunction with input and output statements such as READ and WRITE. The general form of the FORTRAN READ statement is

READ (*i*, *n*) list

where *i* is an unsigned fixed-point variable which refers to an input device, *n* refers to the assigned number of the associated FORMAT statement, and list refers to a sequence of one or more variable names separated by commas. And the general form of the FORTRAN WRITE statement is

WRITE (*i*, *n*) list

where *i* is an unsigned fixed-point variable which refers to an output device, n is the assigned number of the associated FORMAT statement, and list refers to a sequence of one or more variable names separated by commas. A few examples follow which illustrate the use of READ and WRITE statements used in conjunction with FORMAT statements.

Example 2-15

```
  READ (5,7) A,B,P,
7 FORMAT (2X,F7.2,2X,E12.4,3X,I6)
```

Unlike the READ statement in FORTRAN IV, the READ statement in FORTRAN 77 does not recognize the Hollerith FORMAT specification; however, the WRITE statement does.

The first number (5) in the parentheses of the READ statement in Example 2-15 indicates that the input device to the computer setup is a card reader. This number probably would be different for a different computer setup. The second number (7) refers to the assigned number of the associated FORMAT statement. The letters A,B,P refer to three data values. According to the FORMAT statement, A refers to a decimal number, B refers to an exponential number, and P refers to a fixed-point integer. The READ statement will cause three values of data to be read from a card and stored in the computer.

The terms in the parentheses of the FORMAT statement indicate the manner in which the data was punched in a card. The first 2X term indicates that Columns 1 and 2 of the data card were not punched. The term F7.2 indicates that the next seven columns in the card are punched with a decimal number such as ±xxx.xx; Column 3 is for the sign of the decimal number. If the decimal number were positive, it is not necessary to punch a plus sign; however, if the decimal number were negative, a minus sign must be punched in Column 3. The second 2X term in the FORMAT statement indicates that Columns 10-11 are blank.

The E12.4 term of the FORMAT statement indicates that an exponential number is punched in Columns 12 through 23; the exponential number is punched such as ±0x.xxxxE±yy. In the preparation of the card, the 12 columns are accounted for as follows:

Leading sign	1 column
Leading zero	1 column
Decimal point	1 column
5 Digits	5 columns
Character E	1 column
Sign of Exponent	1 column
Exponential Digit	2 columns
Total	12 columns

To punch a plus sign if the exponential number is positive is not essential; a blank in the sign column means the exponential number is positive. However, all negative signs for both sign of the exponential number and sign of the exponent must be punched.

The term 3X indicates that Columns 24 to 26 are not punched. The term I6 indicates that a five-digit fixed-point integer is punched in Columns 27 through 32. Column 27 is reserved for denoting the sign of the integer. Again, it is not essential to punch a positive sign, but a negative sign must be punched.

Example 2-16

```
     WRITE (6,101) X,Y,M
 101 FORMAT (5X,E15.6,5X,F8.3,5X,I7)
```

with computed values of

X = 685.3742
Y = —67.214
M = 842756

In Example 2-16 the first term (6) enclosed in the parentheses of the WRITE statement directs the computer to use device 6 (printer) for printing out information; the second term (101) enclosed in parentheses refers to the assigned number of the accompanying FORMAT statement. According to the specifications of the FORMAT statement, the WRITE statement will cause the computer to print out a line with Columns 1 through 5, 21 through 25 and 34 through 38 remaining blank and with Columns 6 through 20, 26 through 33 and 39 through 45 containing the values of X,Y,M. The values of X,Y,M will appear respectively on the printed line as 6.853742Eb03, —67.214 and 842756 (there is no period after the fixed-point number).

During the preparation of a computer program, it may be necessary to read into a computer or it may be necessary to print from the computer alphanumeric information in addition to numerical data. The FORMAT specifications, H, A, G, and the apostrophe are used for either reading in information or printing out information. Alphanumeric information is very useful for printing out headings for tabulated data. The specifications H, A, and apostrophe are often used for this purpose; however, if it is necessary to store alphanumeric information in a computer for comparative purposes, the A specification is used.

Alphanumeric information can be printed from the computer with a FORMAT statement and a WRITE statement, provided the information to be printed out is given in the FORMAT statement as illustrated in Example 2-17.

Example 2-17

```
   WRITE (6,51)
51 FORMAT (4X23HFORMATIONbVOLUMEbFACTOR)
```

where b represents a blank.

The two statements will cause the computer to print the 23 characters, FORMATION VOLUME FACTOR, on a line in Columns 5-27. The H specification requires a count of the number of characters which must be made and specified; also these characters must be punched in the FORMAT card.

In FORTRAN IV (but not in FORTRAN 77) the same effects of Example 2-17 could have been accomplished by using a READ statement and a card with the 23 characters punched in Columns 5 through 27 as illustrated in Example 2-18.

Example 2-18

```
   READ (5,51)
51 FORMAT (4X23H0123456789bbbbbABCDEFGH)
   WRITE (6,51)
```

The READ statement causes the computer to read a card with the 23 characters punched in Columns 5 through 27, according to FORMAT statement, 51, and store the 23 characters in the computer. Then the WRITE statement will cause the computer to print out the 23 characters in Columns 5 to 27. In using the READ statement with the WRITE statement, the 23 characters following the H specification in the FORMAT statement can be any 23 characters; these 23 characters must be supplied.

To eliminate the probability of errors in counting spaces, more recent versions of FORTRAN use two apostrophes to replace the H specification. If the characters in a FORTRAN statement are enclosed between two apostrophes, there is no need for counting the spaces; the computer will do the adjusting of spaces as illustrated in Example 2-19.

Example 2-19

```
     WRITE (6,72) I,J
  72 FORMAT (2X, 'MAY', 23,',',I5)
```

where *I* has a value of 1972. The FORMAT statement will cause the computer to print out MAY 23, 1972 in Columns 3 to 14. The first set of apostrophes causes the word MAY to be printed out and the second set of apostrophes causes a comma to be printed out.

Natural occurring apostrophes denoting possession, as John's, should be doubled (John''s) when the Hollerith specification is used.

The A specification is used for entering alphanumeric information into a computer which is to be processed. The processing may include: (1) moving of data from one place to another, (2) performing arithmetic operations upon data, and (3) sorting out alphanumeric information. The A specification permits the programmer to assign variable names to information.

The general form of the A specification is where *r* denotes the number of times the A specification is to be repeated and *w* denotes the width of the field. The width may be either greater than or equal to or less than the number of characters. If *h* is the number of characters, and *h* is less than *w*, then on output the *h* characters are printed out in the rightmost *w* columns; however, for input, if *h* is less than *w*, the rightmost *h* characters are stored. For output, if *h* is equal to or greater than *w*, the *w* leftmost characters in storage are printed out. For input, if *h* is equal to or greater than *w* the leftmost *w* characters are read from a card. Example 2-20a shows the punching of a card for the READ statement with its accompanying FORMAT statement; given the A alphanumeric data is bVOLUME.

Example 2-20a

```
      READ (5,102) D,E,F
  107 FORMAT (E7.0,A7,F5.0)
```

Solution: The card might be punched

bb2.5E2bVOLUMEbb2.7

where

D = bb2.5E2
E = bVOLUME
F = bb2.7

Either real or integer variables can be used for the storing of alphanumeric information.

Example 2-20b

Suggest the type of printout for the output statements

```
    WRITE (6,104) E
104 FORMAT (3X,A7)
```

Solution: if E has the same value as for Example 2-20a, then bVOLUME is printed out in Columns 4 through 10.

The G specification may be used for either input or output of integer, real or complex data. The general form of the G specification is

rGw.d

where *w* is the width of the field, *r* is the repetition number and *d* depends on the variables of the input or output. If *.d* is omitted, then the form of the G specification is equivalent to *r*L*w*. For input of real variables, *d* is the number of significant digits to the right of the decimal point; and for output, *d* is the number of significant variables to be transmitted.

A more powerful specification than the X specification for causing blanks to appear between the reading and writing of information is the T-specification. The T-specification specifies the column for the beginning of the following FORMAT specification. The general form of the T-specification is

T*w*

where *w* is the beginning column number for the information of the next FORMAT specification. The following illustrates the uses of the T-specification:

```
   READ (5,32) IP,FV
32 FORMAT (T3,I5,T10,F6.3)
```

where IP = 1225 and FV = 1.342.

The value 1225 is read from Columns 3 to 7 and the value 1.342 is read from Columns 10 to 15.

```
   WRITE (6,47) I,P
47 FORMAT (T4,I5,T10,A8)
```

where I has the value 1225 and P has the alphanumeric form PRESSURE, and both values for I and P have been stored in the computer. The two FORTRAN statements will cause the computer to print out the integer 1225 in Columns 4 to 8 and the alphanumeric characters PRESSURE in Columns 10 through 17. Specifications preceded by their corresponding T specifications may appear in any order. For instance, the two previous FORTRAN statements could have been written

```
   WRITE (6,47) I.P.
47 FORMAT (T10,A8,T4,I5)
```

and the same output record would have been obtained.

2-32

A solidus (/), sometimes referred to as a slash, is used in a FORMAT statement to cause a one-line record to terminate and another one-line record to begin, such as illustrated in Example 2-21a and 2-21b.

Example 2-21a

```
  WRITE (6,5) J,B
5 FORMAT (3X,I5/3X,F12.4)
```

The two FORTRAN statements in Example 2-21a will cause the value of J to be printed on one line in columns 4 through 8 and the value of *B* to be printed on another line in Columns 4 through 15. A slash may be used in a similar manner for input with a READ statement as in Example 2-21b.

Example 2-21b

```
    READ (5,5) J,B
5 FORMAT (3X,I5/3X, F12.4)
```

The two FORTRAN statements will cause an integer value to be read from Columns 4 to 8 of one card and a real value to be read from Columns 4 to 15 of another card.

Shorthand FORMAT statements can be used for input or output of successive I*w*, F*w*.*d*, E*w*.*d*, G*w* and A*w* fields. Example 2-22 illustrates the shorthand FORMAT statements for repetitive fields.

Example 2-22

Give equivalent shorthand FORMAT statements for the following FORMAT statements:

(a) 101 FORMAT (I4,E12.6,F6.4,F6.4,F6.4,A5)
(b) 101 FORMAT (I4,E12.5,I4,E12.5,I4,E12.5)
(c) 101 FORMAT (F10.4, F10.4, F10.4)
(d) 101 FORMAT (14,E12.5,F6.5,E12.5,F6.5,E12.5,F6.5,A4)

Solutions: The shorthand FORMAT statements are

(a) 101 FORMAT (I4,E12.6,3F6.4,A5)
(b) 101 FORMAT 3(I4,E12.5)
(c) 101 FORMAT (3F10.4)
(d) 101 FORMAT (I4,3(E12.5,F6.5),A4)

The following FORMAT statement is invalid because it has a repetitive group within another group:

57 FORMAT (3(A4,F6.2,2(I4,E12.5))).

Prior to the time of closed-shop computer systems, punched paper tapes were used for input and output; also, punched cards were used for output. Associated with the punched tapes and cards were the statements, ACCEPT TAPE, PUNCH TAPE, and TYPE; these statements are omitted because there is little need for them now.

Unformatted Input/Output Statements

Some computer installations permit unformatted READ and PRINT statements which can be used for short and simple FORTRAN programs. The general forms for the READ and PRINT statements are

READ *, list
PRINT *, list

where the * in the PRINT statement permits the computer to decide the output spacing of the values (constants, variables, or expressions) given in the list. If there is more than one value in the list, the values are separated by commas. If all the values in the printout will not fit on one line, more lines will be used until the printing of the entire list is complete.

The READ statement takes values from a data line and places them in memory cells. The first value from the data line goes into the first memory cell, the second value goes into the second memory cell, and so on. Examples of READ and PRINT statements are

```
READ *, DAR
READ *, X,Y,Z
PRINT *, "VALUE IS", B
PRINT *, X,Y,Z
```

Before using the unformatted input/output statements, one should consult a reference manual for a given setup or a local expert because the exact forms for READ and PRINT statements vary from one installation to another.

Specification Statements

Neither specification statements nor the FORMAT statement is executable. These statements supply information for increasing the efficiency of the object program. There are three specification statements: DIMENSION, EQUIVALENCE, and COMMON.

Many mathematical problems require the use of variables in the form of arrays or matrices. In this form, one variable which is subscripted can represent many quantities. FORTRAN permits the use of subscripted variables. In FORTRAN the subscripts to the variables are not written below the variable as in mathematics. The subscripts are separated by commas, enclosed in parentheses, and written on the same line as the variable. As many as three subscripts may be used in FORTRAN. Both floating-point and fixed-point variables can be subscripted; however, the variables used as the subscripts must be fixed point. An individual quantity of an array is called an element.

The following are valid subscripted variables:

```
A(I)
X(3)
P(5*J-2,K+2,L)
FORM(I,J,K)
```

The following are not valid subscripted variables:

A(I,)	Comma not needed
X(A)	A is not fixed point
ELE(I(3))	Unnecessary parentheses

DIMENSION Statement. The general form of a DIMENSION statement is

DIMENSION $v_1, v_2, \ldots v_n$

where each v represents a variable which is subscripted with 1, 2 or 3 integer constants. There may be any number of v's. The DIMENSION statement specifies to the computer the size of the arrays. The variables which appear in subscripted form in a program or subprogram must appear in a DIMENSION statement of the respective program or subprogram. A DIMENSION statement must precede the appearance in a program of a variable which is subscripted. Examples of DIMENSION statements are

```
DIMENSION X(10),Y(5,10),ZVAL(2,4,6)
DIMENSION A(10),J(8,8),SUMS(100)
DIMENSION C(20,20)
```

The first example will cause the computer to reserve storage space for 10 Xs, 50 Ys and 48 ZVALs. The third example will cause the computer to reserve storage space for a 20 by 20 array of elements; this gives a total of 400 Cs.

Floating-Point Arithmetic

Floating-point arithmetic was developed for simplifying lengthy and complex arithmetic calculations containing numbers with varying magnitudes. The reader has, no doubt, experienced difficulties in locating decimal points while solving mathematical or engineering problems.

In general, computers do not recognize decimal points in numerical quantities during the computational process. For instance, the product, 44184, results regardless of whether the factors are 84×526, 0.84×5.26, or 8.4×52.6, and so forth. The programmer has the responsibility of being cognizant of decimal point locations before and after calculations and in arranging the program accordingly. Even the lining up of numbers with varying magnitudes on desk calculators presents problems. Numbers in fixed-point mode, with varying magnitudes, are lined up on electronic digital computers by means of shifts (shift to the right or shift to the left so many digital positions) during the processing of a program. The lining up of digital positions becomes a very tedious operation during the solution of lengthy and complex problems.

In performing a series of arithmetical computations which involve one or more arithmetical operations of multiplication or division, the scientists and engineers often represent numerical quantities in terms of numbers expressed in powers of 10. With this representation, the location of the decimal point for the result of the computations is related to the sum obtained for the powers of 10. Subroutines have been developed for enabling electronic digital computers to process programs in which numerical quantities have been represented in terms of numbers expressed in powers of 10. These subroutines are known as "floating-point subroutines." IBM computer floating-point subroutines represent the data (whether very large or very small) as numbers between 0.1 . . . and 0.99 . . . times a power of 10. The powers of 10 can have values from —99 to +99. The floating-point arithmetic system expresses each quantity as an "n" digit number $(4 = n = 47)$ consisting of a 2-digit characteristic and an $(n - 2)$ digit fractional mantissa such as

$$\underbrace{x\ x\ x\ .\ .\ .x\ x\ x}_{m}\ \underbrace{x\ x}_{c}$$

where c represents the characteristic and m represents the mantissa. The representation for the original number is $m \times 10^c$.

The control characters which cause the computer to operate in terms of floating-point data are E and F. The E character causes the computer to work with data in floating-point form, and it causes the computed results to be transmitted from the computer in floating-point form. The F character causes the computer to work with the data in floating-point form, and it causes the computed results to be transmitted from the computer in decimal form. Of course, the fixed-point character I causes the computer to operate in terms of fixed-point data; this character also causes the computed data to be transmitted from the computer in fixed-point form.

EQUIVALENCE Statement. The general form of the EQUIVALENCE statement is as follows

EQUIVALENCE $(a_1, a_2, \text{---}), (b_1, b_2, \text{---}), \text{---}$

where $a_1, a_2, \text{---}, b_1, b_2, \text{---}, \text{---}$ are variable names which optionally are followed by a single unsigned fixed-point constant in parentheses. The EQUIVALENCE statement will cause the computer to assign two or more variable names, which are not used at the same time, to the same storage. The assigning of two or more variable names to the same storage space may be necessary in a lengthy program where there is a need for the conservation of storage spaces. With the statement, two or more variable names may be defined as meaning the same thing. For instance, in the preparation of a long program, the programmer may find that inadvertently several variable names should refer to the same address. Rather than changing all the names, which is a time-consuming task, an EQUIVALENCE statement is written. The following are examples of valid EQUIVALENCE statements.

```
EQUIVALENCE (X, Y(1), Z(5) )
EQUIVALENCE (A, B(6), C(8) ), (D, E,)
EQUIVALENCE (X, X5, XYZ7)
```

In the second example, the variables A, B(6), C(8) will all be assigned to the same storage space, and the variables D and E will be assigned to another storage space. All of the variables in the third example will be assigned to the same storage space.

COMMON Statement. The general form of the COMMON statement is

```
COMMON A, B, C, . . .
```

where A, B, C, . . . are the names of variables and nonsubscripted array names. The statement is often used with programs and subprograms where the same corresponding variables are named differently in the program and subprogram. The COMMON statement must appear in both the program and the subprogram as illustrated in the examples:

```
Main program:  COMMON A, B, J
Subprogram:  COMMON X, Y, K
```

In the example the variables A and X will be assigned the same memory location; likewise, the variables B and Y will be assigned the same memory location; and also the variables J and K will be assigned the same memory location. The COMMON statement provides a way to establish correspondence between variables in the main program and the subprogram.

Variables in COMMON statements are assigned storage area in the high end of core storage. The COMMON statement takes precedence over the EQUIVALENCE statement. If an array name appears in a COMMON statement, these names must also appear in a DIMENSION statement of the same program.

The DATA statement affords a programmer a convenient manner in which to set variables equal to some initial values; these initial values may be integer, real, logical, or alphanumeric. The DATA statement causes data to be placed in storage registers. The general form of the DATA statement is

DATA $v_1,v_2,v_3 \ldots v_n/d_1,d_2,d_3, \ldots d_n/,v_1,v_2,v_3, \ldots v_n/d_1,d_2,d_3, \ldots d_n/, \ldots /$

where $v_1,v_2,v_3, \ldots v_n$ refer to variables and $d_1,d_2,d_3, \ldots d_n$ refer respectively to the values of the variables. The use of the DATA statement is illustrated in Example 2-23.

Example 2-23

1. DATA NP, B/1225,1.232/,EI,D,C,U,EP/
1.23,0.234,0.0004,2.0,0.82/
2. DATA B/8HPRESSURE/
3. DATA B/5*1.23/
4. DATA L(1),L(2)/.TRUE.,F/

The first DATA statement causes the initial values of the real variables NP,B,EI,D,C,U,EP to be stored in storage registers; the initial values are 1225, 1.232, 1.23, 0.234, 0.0004, 2.0, 0.82, respectively.

The second data statement causes the initial alphanumeric datum for B to be stored in a storage register. The datum value is PRESSURE.

The third DATA statement illustrates a shorthand method of storing five identical values of 1.23 for B when B was given dimensions as containing five elements all equal to 1.23.

The fourth DATA statement causes logical data for variables L(1) and L(2) to be stored in storage registers; the logical values are TRUE and F.

Miscellaneous FORTRAN Statements

The accuracy of arithmetic computations on most computers is seven or eight significant digits. Frequently, the solution to a problem demands that the calculations be carried out to more than seven or eight significants; this is particlarly true in iterative processes where round off errors become appreciable. The precision of the calculation is limited by the allocation of storage for the numbers. With the DOUBLE PRECISION statement, it is possible to declare double the storage for each number used in the calculations, thereby doubling the digits in the number. The general form of the DOUBLE PRECISION statement is

DOUBLE PRECISION $n_1,n_2,n_3\cdots$

where n_1,n_2,n_3 . . . are real variables or function names which are in the program. The DOUBLE PRECISION statement is used for declaring only real variables. Integer, complex, and logical variables cannot be considered as double-precision variables. Example 2-24 illustrates the use of the DOUBLE PRECISION statement.

Example 2-24

Give a DOUBLE PRECISION statement with the variables Y,C,X,B and with at least three arithmetic expressions which explain the effects of the DOUBLE PRECISION statement.

DOUBLE PRECISION Y,C,X,B
a) Y = CX + B
b) W = CX + B + Y
c) X = (Y − B) /C + W

Commentary. The DOUBLE PRECISION statement declares that the real variables Y,C,X,B are double precision variables and that the use of these real variables in any fashion will be in double-precision.

The first arithmetic expression will be evaluated in double-precision, and the value of Y will be stored in double-precision.

The second arithmetic expression will be evaluated in double-precision; however, the result (value of W, most significant digits) will be stored in single-precision.

The third arithmetic expression will be evaluated with double-precision values for Y,B,C and single-precision value for W. And since X was declared to be in double-precision, the result of this expression will be stored in double-precision.

There is no double-precision specification for an integer variable; however, there are double-precision values for real variables. The DOUBLE

PRECISION FORMAT specifications for a real variable without exponentiation is

F*w*.*d*

where *w* has a value large enough for double-precision. The DOUBLE-PRECISION FORMAT specification for a real variable with exponentiation is the same as the E specification except E is replaced by D as

D*w*.*d*

where *w* is an integer which denotes the width of the field and *d* is an integer which denotes the number of digits to the right of the decimal point. The use of double-precision specifications is illustrated here with examples of input and output FORMAT statements:

```
   DOUBLE PRECISION X,Y,Z
   READ (5,37) X,Y
37 FORMAT (F16.5,D16.9)
   WRITE (6,42) Z
42 FORMAT (D20.8)
```

Commentary. The DOUBLE PRECISION statement declares that the real variables X,Y,Z are double-precision. The READ statement with its accompanying FORMAT statement causes the computer to read in X as a real value without an exponent. The WRITE statement with its accompanying FORMAT statement causes the values of Z to be printed out in double-precision; Z is printed out in exponential form. The FORTRAN functions listed in Table 2-5 are also given in double-precision. In writing the double-precision functions in a computer program, the letter D is placed in front of the function as DSIN(X), DEXP(X), DABS(X) etc.

Similarly, as for double-precision, a declarative statement can be written for declaring complex variables. Alphanumeric characters (maximum of six) are used to name the complex variable. The general form of the complex variable FORTRAN statement is

COMPLEX $n_1,n_2,n_3\cdots$

where $n_1,n_2,n_3,\cdots$ are complex variable names or functions which appear in the program. The complex variable names may be subscripted in the same manner as integer and real variables. The complex number for the complex variable is represented by two real numbers; one real number represents the real part of the complex number and the other real number represents the imaginary part of the complex number. The arithmetic IF statement will not permit the use of complex variables. The FORMAT specifications for complex quantities must consist of two specifications for each complex number.

Example 2-25

Given three complex variables,

$$P(2) = 25.42 - 15.26i$$
$$B = 11.2 + 3.6i$$
$$S = 22.64 + 37.37i$$

write FORTRAN statements for the input of P and B and output of S. Also, indicate how the card would be punched for input and how the output would appear when printed out.

Solution. The FORTRAN statements are

```
    COMPLEX P(2),B,S
    READ (5,101) P(2),B
101 FORMAT (2E12.4,F10.2,E12.2)
    WRITE (6,102) S
102 FORMAT (2X,2F10.2)
```

The input card would appear as

b0.25420Eb02-0.15260Eb02bbbbb11.20b0.36000Eb01

and the output printed line would appear as

bbbbbbb22.64bbbbb37.37

The following FORTRAN functions are available in complex variables: CABS, CSQRT, CEXP, CALOG, CSIN, CCOS.

A complex variable may appear in a DATA statement, and when a complex variable does appear in a DATA statement, it must have two constants, e.g.,

```
DATA C/5.23,-2.32/
```

FORTRAN permits the five arithmetic operations of addition, substraction, multiplication, division and exponentiation (exponentiation of complex variable to integer exponent only).

Debugging

An individual experienced in writing and processing computer programs fully understands the meaning of the word "perfection" because a computer will do only what it has been instructed to do. Even a very small mistake will either prevent it from working or giving correct results. In other words, the program must be free of errors, and no errors are tolerated in the processing of the program. A common name for the process of locating and correcting errors in a computer program is *debugging*. Common errors are:

1. Lack of conformity between input data and the program.
2. Computer calculating a value either too large or too small for which the computer was designed.
3. Improper specifications for output.
4. The instruction may not state what it was intended to.

The first three are examples of execution errors, and the fourth is an example of a logical error. Errors in a program may be the result of improper design; these errors may not be detected until the program has been processed on the computer. An example of this type of error is results which do not converge when it is known that the results should converge.

The debugging of a computer program may take hours or days on lengthy programs. To facilitate the process for locating errors, most compilers will detect compilation and execution errors. These compilers will give information about the nature of the errors and the location of the errors. There are special trace programs in addition to the compiler which can be used for locating errors in a program. The types and amount of programs for locating errors in a program depend upon the desires and nature of the computer center. The tracing programs also take up storage in a computer.

Apple FORTRAN

For those who have mastered problem solving on microcomputers with the BASIC language and are eager to explore problem solving with the FORTRAN language, some suggestions follow. In some instances one may become impatient with slow progress and want to put the blame on the authors of the Apple FORTRAN and Pascal books which are supplied with the packets. The Apple FORTRAN packet contains two disks and an instruction book. In developing FORTRAN skills for the Apple IIe, books supplied with the FORTRAN and Pascal packets are sufficient except for using the printer. Books for Apple FORTRAN study are *Apple Pascal: Operating Systems Reference Manual*, *Apple FORTRAN Language Reference Manual*, and *Apple FORTRAN*. The third title illustrates how to use the printer to get results from a FORTRAN program, but how to print out the text program is given vaguely in the second title.

To begin the study for working with FORTRAN, one should become very familiar with the first five chapters of the book *Apple Pascal: Operating Systems Reference Manual*, and Appendix A, Parts 1 and 2 of *Apple FORTRAN Language Reference Manual*. In addition, one should become very skilled in using disk drives; Apple FORTRAN works best with two or more disk drives. Disk skills to develop include: (1) listing files from disks, (2) transferring files between disks, (3) transferring files from the microcomputer to disks or vice versa, (4) copying disk files, (5) deleting files from a disk, and (5) linking files. Since the compiled coded form of Pascal programs is the same as the coded form of FORTRAN programs, one should know how to compile and run Pascal programs; these operations are similar for both language programs. Unfortunately, disks which come with the FORTRAN packet at this time are incomplete. To complete the FORTRAN disks, files have to be copied from disks which are supplied with the Pascal packet.

To obtain a printout of a FORTRAN text, the text has to be stored on a disk, then transferred directly from the disk to the printer. To obtain a printout from a compiled FORTRAN program, a file is opened (see *Apple FORTRAN*). Also, observe Example 2-26, which is a FORTRAN program (FORTRAN DEMO) for computing the squares of the integers 1 through 100. Both the FORTRAN text and the compiled FORTRAN code were filed under the names DEMO.TEXT and DEMO.CODE on disk FORT 2. To get the printout of the DEMO.TEXT, a transfer was made from disk FORT 2 to the printer. With disk FORT 2 in slot #4 and disk FORT 1 in slot #5, the screen displays

```
FILER: G, S, N, L, C, T, D, Q
```

T is pressed, after which a display of

```
TRANSFER ?
```

appears, then

```
DEMO.TEXT
```

is typed, followed by a display

TO WHERE ?

and

#6

is typed. The printer then prints out the FORTRAN program as it was originally typed into the Apple IIe.

Example 2-26. Printout of DEMO Text and Results

```
C FORTRAN DEMO
      DO 10 I=1,100
      X=I
      OPEN(6, FILE='PRINTER:')
10    WRITE(6,1)I,X,X*X
1     FORMAT (I6,3X,F7.2,3X,F8.2)
      END
```

```
     1      1.00       1.00
     2      2.00       4.00
     3      3.00       9.00
     4      4.00      16.00
     5      5.00      25.00
     6      6.00      36.00
     7      7.00      49.00
     8      8.00      64.00
     9      9.00      81.00
    10     10.00     100.00
    11     11.00     121.00
    12     12.00     144.00
    13     13.00     169.00
    14     14.00     196.00
    15     15.00     225.00
```

To get a printout of the executed program, the procedure is as follows, with FORT 2 in slot #4 and FORT 1 in slot #5 and the screen displaying

COMMAND: E, R, F, C, X, A, D

X is punched, after which the display

EXECUTE WHAT FILE ?

appears on the screen, then

DEMO.CODE

is typed, and the printer prints out the results from the FORTRAN program. The key statements for using the printer are

```
   OPEN (6,FILE='PRINTER:')
10 WRITE(6,1) I,X,X*X
```

Only the squares of the integers 1 through 15 are given in Example 2-26.

The book which comes in the FORTRAN packet lists the files which are to be transferred from the Pascal disks for configuring the FORTRAN disks. The book also explains in detail the procedure to follow in transferring files from one disk to another.

Exercises

1. Write FORTRAN variable notations for each of the following:

	Variable	*Symbol*
a.	Pressure	p
b.	Sum of pressures	Σp
c.	Sum of pressures squared	Σp^2
d.	Mean of a sample	$\overline{x}$
e.	Sample variance	s
f.	Subscripted variable	D_i
g.	Subscripted array	D_{ij}
h.	Sample size	η

2. Write FORTRAN arithmetic statements for each of the following:

   a. Give pressure the value of 2750.
   b. Give R the value of 10.73.
   c. Set I equal to 30.
   d. Change P from fixed point to floating point mode.
   e. The mean of a sample, $\overline{x} = \dfrac{\Sigma x}{n}$, using the variable names in (1).
   f. Increase (Σp) by the value of p.
   g. Increase (Σp^2) by the value of p^2.
   h. Increase (Σp^2) by the value of p_i.
   i. Increase (Σp^2) by p_i.

j. $Z = X \cdot Y$
k. $Z = X \cdot (-Y)$
l. $Z = X + 6Y$
m. $Z = A \cdot X/B \cdot Y$
n. $VAR = X^{I+2}$
o. $AVAR = X^{B+3.4}$
p. $COMR = \frac{X}{B} + Y \quad 3.6$
q. $SQR = \sqrt{COMR}$

3. Write a FORTRAN statement to read values for the following variables in the order listed, using FORMAT statement No. 147: barrels, permeability, viscosity, area, pressure, and length.

4. Write a FORTRAN statement to print the variables of (3) in the same order according to FORMAT statement No. 152.

5. If the general form of the values for the variables listed in (3) is:

barrels	xxx.
permeability	.xxx
viscosity	x.xxx
area	xxx.xx
pressure	xxxx.x
length	xxx.xx

a. Write a FORMAT statement No. 175 to describe the layout of the data on a single card punched without any blanks between each data field.
b. Write a FORMAT statement No. 191 to describe the printed layout of the same data printed with two blanks between each data field.

6. Write a FORTRAN program to calculate the sample mean. Assume that the values of X are punched one per card in the form xxx.xxx and that the value of N is punched on a separate card in the form xx.

$$\bar{x} = \sum_{i=1}^{N} x_i/N$$

Input: N = number of X_i's to read
X_i = one per card for N cards
Output: XBAR = mean of sample xxx.xxx
Note: Punch XBAR if sense switch 1 is on.
Punch: XBAR if sense switch 1 is off.

7. Using subscripted variable notation, write a FORTRAN program to find the smallest value from a random list of 100 variables.
Input: one-dimensional array of 100 variables
Output: smallest algebraic (not absolute) value of 100 variables.

8. Write a FORTRAN program for determining the quotient of two numbers by Equation (A-2) —reciprocal method.

Suggested Reading

1. Anderson, D. M., *Programming FORTRAN IV,* New York: Meredith Publishing Co., 1966.
2. Bauman, R., Felicianio, M., Bauer, F. L., and Samelson, K., *Introduction to ALGOL,* Englewood Cliffs: Prentice-Hall, Inc., 1964.
3. Dimitry, D. L. and Mott, Jr., T. H., *Introduction to FORTRAN IV Programming,* New York: Holt, Rinehart and Winston, Inc., 1966.
4. IBM Reference Library, *IBM System/360 FORTRAN IV Language,* File No. S36-25, Form C28-6515-7, New York: International Business Machines Corporation, 1968.
5. IBM Reference Library, *IBM System/360 Operating System Programmer's Guide to Debugging,* File No. S360-20, Form C28-6670-1, New York: International Business Machines Corporation, 1968.
6. IBM Reference Library, *IBM System/360 Operating System FORTRAN (G and H) Programmer's Guide,* File No. 1620-25, Form C38-6817-0, New York: International Business Machines Corporation, 1968.
7. IBM Reference Library, *IBM 1620 FORTRAN II,* File No. 1620-25, Form C26-5662-1, New York: International Business Machines Corporation, 1962.
8. IBM Reference Library, *IBM 1620 FORTRAN II Specifications,* File No. 1620-25, Form No. J26-5602-1, New York: International Business Machines Corporation, 1962.
9. *Introduction to Programming PDP-8 Family,* Maynard, Mass.: Digital Equipment Corporation, 1970.
10. Leeson, D. N. and Dimitry, D. L., *Basic Programming Concepts and the IBM 1620 Computer,* New York: Holt, Rinehart and Winston, Inc., 1962.
11. Murrill, P. W. and Smith, C. L., *FORTRAN IV Programming for Engineers and Scientists,* Scranton, PA: International Textbook Company, 1968.
12. *PDP-11 Handbook,* Maynard: Digital Equipment Corporation, 1970.
13. *PDP-8/e Small Computer Handbook,* Maynard: Digital Equipment Corporation, 1970.
14. *Programming Languages PDP-8 Family,* Maynard: Digital Equipment Corporation, 1970.
15. *Xerox ANS COBOL Sigma 5/6/7/9 Computers Language,* No. 90-15-00B, Xerox Reference Manual, El Segundo, Calif.: Xerox Data System, 1970.
16. *Xerox Basic Language and Operation Manual,* No. 90-15-46-D, Xerox Reference Manual, El Segundo: Xerox Data System, 1971.
17. *Xerox Extended FORTRAN IV Sigma 5/6/7 Computers,* No. 90-09-56-D, Xerox Reference Manual, El Segundo: Xerox Data System, 1970.
18. *Xerox SL-1 Sigma 5-6 Computers,* No. 90-16-76-B, Xerox Reference Manual, El Segundo: Xerox Data System, 1972.
19. *XDS Sigma 5/6 FORTRAN Debug Package (FDP) Operations* No. 90-16-77A, Xerox Manual, El Segundo: Xerox Data System, 1972.
20. Katzan, H., Jr., *FORTRAN 77,* New York: Van Nostrand Reinhold Co., 1978.
21. Lipschutz, S. and Poe, A., *Programming with FORTRAN,* Schaum's Series. New York: McGraw-Hill Book Co., 1978.
22. Silicon Valley Software, Inc., *Apple FORTRAN—Language Reference Manual,* Cupertino, Calif.: Apple Product #A2D0032 (030-0118-00) , Apple Computer, Inc., 1980.

23. Apple Computer, Inc., *Apple Pascal—Operating Systems Reference Manual,* Cupertino, Calif.: Apple Product #A2L0028 (030-0100-00) , 1980.
24. Blackwood, F. D. and Blackwood, G. H., *Apple FORTRAN,* Indianapolis: Howard W. Sams and Co., 1982.
25. Page, R., Didday, R., and Alpert, E., *FORTRAN 77 for Humans,* New York: West Publishing Co., 1983.
26. ANSI Task Group x 3.8.3, "Identity Code for Individuals," *The Office,* June 1970, p. 139.
27. Barlow, D. H., "Information Retrieval," *The Computer Bulletin* (British) , 16, 5, 1970, p. 250.
28. G.P.I. Computer Corporation, "Minicomputers Offer Improved Price/Performance for Dedicated Systems," *Computer Design,* 11, 7, 1972, p. 96ff.
29. Guard, J. R., Oglesby, F. C., Bennett, J. H., and Settle, L. G., "Semi-Automated Mathematics," *Journal of the Association for Computing Machinery,* 16, 1, 1969, pp. 49-62.
30. Lazak, D., "The Geometry of Program THRESH for Evaluating Bubble-Chamber Pictures," *Computer Reviews,* 10, 7, 1969, p. 305, (abstract) .
31. Lee, J. A., "The Formal Definition of BASIC Languages," *The Computer Journal* (British), 15, 2, p. 37ff, 1972.
32. Ness, D. N., Green, R. S., Martin, W. A., and Moulton, G. A., "Computer Education in a Graduate School of Management," *Communications of the ACT,* 13, 2, 1970, pp. 110-114.
33. Stamm, T. and Berkeley, E. E., "The Shooting of Presidential Candidate George C. Wallace: A Systems Analysis Discussion," *Computer and Automation,* 21, 7, 1972, p. 32ff.

3 Programming in BASIC

Most microcomputers have a BASIC interpreter stored in their ROM; the interpreter changes BASIC statements into machine language instructions one by one as the BASIC program is executed. The design of an interpreter is controlled by the make-up of the microcomputer, with the result that there are many versions of the BASIC language. Some micros, minis, and mainframes use compilers for processing BASIC programs. A compiler changes a whole BASIC program into a machine program before it is executed. If all computers used compilers for processing BASIC programs, it would be easier to standardize the various versions of the BASIC language. The execution of a computer program written in BASIC is faster with a compiler than with an interpreter; a compiler will increase the speed by a factor of 4 to 20.

The discussion on BASIC programming which follows is for the most part centered on Applesoft BASIC. Many similarities exist between the different versions of BASIC. If one learns a particular version of BASIC, he will have little difficulty in learning other versions. Throughout the discussion on Applesoft BASIC, brief comparisons with other versions are given. The discussion begins with a simple BASIC program using only three statements (LET, PRINT, END) and is followed by the use of the Apple computer as a calculator and an outline of Applesoft BASIC statements.

Engineers and scientists with little experience using computers should not have difficulty learning how to prepare BASIC programs for petroleum engineering problems—provided they study the BASIC manuals which are supplied with their given computer (this also includes running simple programs given in the BASIC manuals). Also, a reading of Chapter 11 at the same time should prove helpful. In this book only the portions of the BASIC language which apply to solving petroleum engineering problems are given; topics such as sound, graphics, color, and word processing have been omitted.

A Simple BASIC Program

Programming in BASIC is easy to learn. Example 3-1 illustrates the simplicity of the language; only three BASIC statements are used.

Example 3-1. Product of Two Numbers

```
]10 LET X=8
]20 LET Y=89
]30 LET Z=X*Y
]40 PRINT Z
]50 END
RUN
712
```

The statements are numbered sequentially in steps of 10 beginning just to the right of the closing square bracket; the position is considered to be the farthest left position on a line. The closing square bracket is called a prompt character; this particular character indicates the Apple computer is ready for programming in Applesoft BASIC. The first BASIC statement (line number 10) causes the computer to set the value of X equal to 8 and store an 8 in the memory position for X; likewise, the second BASIC statement (line number 20) causes the computer to set the value of Y equal to 89 and store 89 in the memory position for Y; and the third statement (line number 30) causes the computer to set the value of Z equal to the product of X times Y, which is 712, and store the product in the memory position for Z. The PRINT statement (line number 40) causes the computer to display the product (712) on the CRT of the computer. The END statement, which is the last statement of the program, causes the computer to cease the execution of the program.

If a calculator is needed during the development of a BASIC program, a microcomputer can function as a calculator by using the PRINT statement and the RETURN key; line numbers are not needed. Using a microcomputer as a calculator does not interfere with program instructions already in the computer. Example 3-2 illustrates how a microcomputer can be used as a calculator.

Example 3-2. Calculator Use of a Microcomputer

PRINT 7^2.37	RETURN KEY
100.665894	
PRINT (EXP(6) + 7)	RETURN KEY
410.428793	
PRINT LOG(6)/LOG(10)	RETURN KEY
.77815125	
PRINT ABS(3.241)	RETURN KEY
3.241	
PRINT INT (3.241)	RETURN KEY
3.	

The five calculations in the example are:

1. Raising 7 to the 2.37 power.
2. Raising e to the 6.0 power.
3. Calculating LOG (base 10) of 6.
4. Displaying the absolute value of 3.241.
5. Displaying the integer value of 3.241.

Applesoft BASIC

A classification of the Applesoft statements is given in the outline which follows. The list is not complete, but it is sufficient for programming a large number of petroleum engineering problems, including the problems in Chapter 11.

A. Arithmetic Statement
B. Control Statements
   1. NEW
   2. HOME
   3. STOP
   4. END
   5. CONTINUE
   6. GET
   7. FOR/NEXT
   8. IF/THEN
   9. GOTO

C. Subroutine Statements
   1. GOSUB
   2. RETURN
   3. DEFine FUnction
D. Input/Output Statements
   1. LET
   2. DATA/READ
   3. INPUT
   4. PRINT/TAB
   5. PR#
   6. LIST
   7. REM
E. Specification Statement
   1. DIMENSION

Example 3-1 illustrates the format of an Applesoft BASIC program as it appears on the CRT of an Apple computer. BASIC statements begin on a line at the position just to the right and next to the prompt. Each statement begins with a line number; the line number may be any number between 0 and 63999, including zero. BASIC statements are stored in the memory of a microcomputer in numerical ascending order of the line numbers; also, the statements are executed in that same order. The computer arranges the statements in numerical ascending order even if the statements are listed in a different order. Line number room, between the lines and before the first line number, should be left because new statements are often necessary during the writing of the program or when an existing program is modified.

Variables (other than string variables) are named by using one or two characters; the first character must be a letter but the second character can be either a letter or a number. Variable examples are A, D, D1, C2, MC, XC. Words such as MONEY, MONDAY, MORE, MOBY, MODEL all have the same variable name which is MO because only the first two letters are considered. Applesoft BASIC contains a number of words which cannot be used as variable names—the words are commands. A list of these words is in Table 3-1.

The basic arithmetic-like operations associated with BASIC are given in Table 3-2.

The symbol = does not necessarily imply mathematical equality; P = R has the normal meaning of equality and the meaning of compute the value

Table 3-1
Reserved Words in Applesoft*

&	GET	NEW	SAVE
	GOSUB	NEXT	SCALE=
ABS	GOTO	NORMAL	SCRN(
AND	GR	NOT	SGN
ASC		NOTRACE	SHLOAD
AT	HCOLOR=		SIN
ATN	HGR	ON	SPC(
	HGR2	ONERR	SPEED=
CALL	HIMEM:	OR	SQR
CHR$	HLIN		STEP
CLEAR	HOME	PDL	STOP
COLOR=	HPLOT	PEEK	STORE
CONT	HTAB	PLOT	STR$
COS		POKE	
	IF	POP	TAB(
DATA	IN#	POS	TAN
DEF	INPUT	PRINT	TEXT
DEL	INT	PR#	THEN
DIM	INVERSE		TO
DRAW		READ	TRACE
	LEFT$	RECALL	
END	LEN	REM	USR
EXP	LET	RESTORE	
	LIST	RESUME	VAL
FLASH	LOAD	RETURN	VLIN
FN	LOG	RIGHTS	VTAB
FOR	LOMEN	RND	
FRE		ROT=	WAIT
	MID$	RUN	
			XPLOT
			XDRAW

*Taken from **Applesoft BASIC Programming Manual,** Apple Computing Inc., 10260 Bandley Drive, Cupertino, California 95014.

Table 3-2
Arithmetic Symbols Used in Applesoft BASIC

Operation	Symbol	Example
Addition	+	P + R
Subtraction	—	P — R
Multiplication	*	P * R
Division	/	P / R
Exponentiation	^	P ^ R
Equal	=	P = R

of R and assign this value to P. Combinations of arithmetic operations and elements give algebraic-like expressions in the BASIC notation as shown in Table 3-3.

If there are several arithmetic operations in an algebraic expression, the operations are performed in the order of exponentiation, multiplication, division, addition, and subtraction. To clear or change the order of performing the operation, the new order is indicated with parentheses. When using parentheses, these rules apply:

1. Operations put inside parentheses are performed first: (7*8)*3 = 56 * 3 = 168.
2. Operations in parentheses within parentheses are performed first: (2 + (6 + 8 − 2)) − 6 is equal to (2 + 12) − 6 = 8.

Other arithmetic operators in BASIC expressions are as in Example 3-3.

Example 3-3. BASIC Expressions

Algebraic Expression	BASIC Expression
$\frac{P+R}{S}$	(P + R)/S
$y = ax^3 + bx^2 + cx + d$ $= ((ax + b)x + e)x + d$	Y = ((A*X + B)*X + C)*X + D
$\frac{P+R}{X-Y}$	(P + R)/(X − Y)

In addition to the arithmetic operations, there are several quasioperations which are used to form BASIC expressions for evaluating arithmetic or algebraic expressions. The forms and requirements of the quasioperations are given in Table 3-4.

Table 3-3
BASIC Algebraic-Like Expressions

Algebraic Expression	BASIC Formulation
$3.1416R^2$	3.1416*R*R or 3.1416*R^2
$A + BX + CX^2$	A + B*X + C*X^2

Example 3-2 illustrates the use of the BASIC functions ABS(X), EXP(X), INT(X), LOG(X), and SQR(X). The Apple microcomputer calculates only the natural logarithm of a number; to get the logarithm (base 10) of a number, the natural logarithm of the number is divided by the natural logarithm of 10.

In Applesoft BASIC, the microcomputer accepts real and integer quantities; the integer quantities may be positive or negative and fall within the range of —32,768 and 32,767. The real numbers may be between 10^{-38} and 10^{+38} with up to nine digits accuracy. When a number becomes greater than nine digits, it must be in scientific notation. Table 3-5 shows several different number types.

Table 3-4
Ten Quasioperations

Mathematical Function	BASIC Function	Requirement
Absolute value	ABS(X)	
Arc tangent	ATN(X)	
Cosine	COS(X)	X in radians
Cotangent	COT(X)	X in radians
Exponential	EXP(X)	
Integer value	INT(X)	
Natural logarithm	LOG(X)	X > 0
Sine	SIN(X)	X in radians
Square root	SQR(X)	X > 0
Tangent	TAN(X)	X in radians

Table 3-5
Number Types

Number	Type
876543	Integer or real
54.321	Real
—.72346E—03	Real (scientific)
—72.346E—05	Real (scientific)
5432.467	Real
54.32467E+02	Real

The last two numbers in Table 3-5 are equal—two ways for writing the same number. The terms E—03, E—05, E+02 are scientific notation and meaning 10 raised to the —3 power, 10 raised to the —05 power, and 10 raised to the second power; in other words, 5.342E+06 is read 5.342 times 10 raised to the sixth power. Other ways of writing this number are 5342000, 53.42E05, and .5342E07. The output and input forms of numbers are the same. Very small numbers between —3E—39 and 3E—39 are converted to zero by the Apple computer.

Control Statements and Commands

Control statements and commands are used for controlling the order in which BASIC statements are executed, how BASIC statements are executed, and preparation of the memory for the input of a BASIC program. All statements and commands must be typed in upper-case letters.

NEW Command. Typing the word NEW on the keyboard of an Apple computer and pressing the RETURN key causes the microcomputer to completely remove or erase the current BASIC statements from the memory (RAM). The NEW command is used before entering a new program into the computer; this may be followed by execution of a program and saving it on a disk.

HOME Command. Typing the word HOME and pressing the RETURN key causes the computer to remove all characters from the CRT. The HOME command is usually used prior to the input of a new BASIC program. The command also causes the cursor to appear in the upper left corner of the CRT. The cursor is the blinking rectangle which appears on the CRT and indicates the location of the next character to be typed. The HOME command can also be used as a part of a program for the clearing of the CRT. Example 3-4 is a BASIC program which, when executed, will cause the CRT to be cleared with only the answer 18 appearing in the upper left corner.

Example 3-4. HOME Statement

```
 5 HOME
10 LET X=3
20 LET Y=6
30 LET Z=X*Y
40 PRINT Z
50 END
```

STOP Statement. The STOP statement is used to cease execution of a program prematurely. As the execution of the program ceases, the line number of the statement at which the execution of the program ceases is printed out. STOP statements can be inserted throughout a program; the temporary halt is useful in the debugging of a program.

END Statement. In an Applesoft BASIC program the END statement is the last statement in the program, and it has the highest line number; no statement with a higher line number will be executed. The END statement causes the execution of a program to cease permanently, and the computer to return to the command level.

CONTINUE Command (CONT). The CONT command can be used for starting the execution of a program after a STOP statement. If a printer is used during the execution of a BASIC program containing a STOP statement, the printer statement must be given before giving the CONT command. Holding down the CONTROL key while striking the C key works in a fashion similar to the STOP command; this method has no control over the line number at which the execution of the program ceases.

The short program of Example 3-5 illustrates the use of the STOP and END statements and the CONT command.

Example 3-5. STOP and END Statements and CONT Command

```
10 LET X=3
20 LET Y=6
30 LET Z=X*Y
40 PRINT Z
50 STOP
60 LET W= (X^2)*Y
70 PRINT W
80 END
```

During the execution of the program, the number 18 appears on the CRT; then the phrase BREAK IN 50. After entering the command CONT and pressing the RETURN key, the number 54 appears and, finally, the blinking cursor appears, which indicates program execution is complete and the computer is to return to command level.

GET Statement. The GET statement causes the execution of a BASIC program to cease until a key (any key) is pushed. Example 3-6 illustrates the use of the GET statement.

Example 3-6. GET Statement

```
10 LET X = 56676
20 LET Y = 456
30 GET A$
40 LET Z = X*Y
50 PRINT Z
60 END
```

The execution of the program ceases at the GET statement until a key is punched from the keyboard, then the results are printed out (25845168), and the computer returns to the command level. The variable A$ associated with the GET statement is a string variable; it is a requirement of the GET statement.

The BASIC statements FOR/NEXT, IF/THEN, and GOTO are used for looping. Parallel with the study of these BASIC statements, the FORTRAN statements DO, IF, and GOTO in Chapter 2 should be noted; also, Figure 2-2 applies to BASIC languages.

FOR/NEXT Statements. The general form of these statements is

```
10 FOR I = i to n
   .
   .
   .
100 NEXT I
```

where I is a variable, i is the lower limit, and n is the upper limit of which 1 takes. The integer i usually has a value of 1, and n is the number of values calculated or the number of times something is done by the program and written on the lines between the FOR statement and the NEXT statement (see Example 3-7).

Example 3-7. FOR/NEXT Loop

```
10 X = 0
20 FOR I=1 TO 10
30 LET X = X+1
40 PRINT "  "X;
45 NEXT I
50 END
1  2  3  4  5  6  7  8  9  10
```

The statement on line 20 causes I to take on values of 1 to 10 in increments of 1 during each round of the loop. The statement on line 30 causes X to be increased by 1 each time around the loop. The statement on line 40 causes execution of the program to line 20 following each calculation until I becomes equal to 10.

IF/THEN Statements. An example of an IF/THEN statement is

```
10 IF X=6 THEN 200
```

The statement reads if X = 6 then go to statement 200, which means if X has the value of 6, the next statement to be executed is statement 200. If X is not equal to 6, the next statement to be executed is the statement with the next higher line number. Other rational line numbers are given in Table 3-6.

For IF/THEN statements containing any of the relational operators, if the statement is true, the number following THEN designates the line number of the next statement to be executed. If the statement is false, the statement with a line number just below the line number of the IF statement is executed. In addition to the six relational operators, there are three logical operators AND, OR, NOT. Examples of logical operators follow:

```
10 IF X=Y AND V=W THEN 150
20 IF X=Y NOT V=W THEN 150
```

The first example requires both equalities to be true before transferring to statement 150; the second example requires only one of the equalities to be

Table 3-6
Relational Operators for IF Statements

Relational Operators	Meaning
=	is equal to
<	is less than
>	is greater than
<= or =>	is less than or equal to
>= or =<	is greater than or equal to
<> or ><	is not equal to

true before transferring to statement 150. Other examples of relational operators are:

```
10 IF B+C>20 THEN LET A=6
20 IF B< >D THEN C=10
```

GOTO Statement. The GOTO statement causes the computer to switch the execution to the statement which has a line number designated by the GOTO statement such as:

```
40 GOTO 10
50 GOTO 200
```

The first statement causes statement 10 to be executed following statement 40. The second GOTO statement causes all the statements between 40 and 200 to be skipped and statement 200 to be executed next.

The IF/THEN statement combined with the GOTO statement is often used to perform repetitive calculations in loops (see Example 3-8).

Example 3-8. IF/THEN, GOTO Statements

```
10 X=0
20 LET X=X+1
30 PRINT "  " X;
40 IF X > 10 THEN 60
50 GOTO 20
60 END
RUN
1  2  3  4  5  6  7  8  9  10  11
```

Statement 10 initializes X and gives it a value of 0; statement 20 adds 1 to the value of X and sets this value equal to X; statement 30 prints out the values of X in a straight line; statement 40 tests to determine when the value of X becomes greater than 10 and, when the value becomes greater than 10, statement 60 is executed and the computer is returned to the command level. Statement 50 returns the execution of the program back to statement 20—this causes another loop for computing another value of X.

Subroutine Statements

During the process of writing a BASIC program, if there is reason to believe a section will be repeated several times throughout the program, it is possible to write this section only once and isolate this section so it can be called up as it is needed. This isolated section is called a subroutine. The BASIC statements for calling up the subroutine to the main program are GOSUB and RETURN.

GOSUB Statement. The GOSUB statement is a part of the main program. While executing a program, the computer upon the approach of the statement will transfer the execution to the subroutine. The beginning line number of the subroutine is designated to the right of the word GOSUB in this statement; at this stage the subroutine program will be executed.

RETURN Statement. The RETURN statement is the last statement in a subroutine. The statement causes the execution to be transferred back to the main program. Example 3-9 illustrates how a subroutine is used in a short program for rounding off the digits to the right of the decimal point.

Example 3-9. Significant Digits

```
10 LET X = 56.32
20 LET Y = 6.3267
30 LET Z = X*Y
40 GOSUB 1000
50 PRINT U
60 END
1000 LET W = (Z*10^3+.5)
1010 LET U = INT(W)/10^3
1020 RETURN
356.32
```

The short program caused the computer to multiply X times Y, which gave a value of 356.3197444. The product would have been in nine significant digits, but, using the subroutine, the number of digits was rounded to six digits including the decimal point; the Apple computer does not print out zeroes to the right of numbers in a decimal. The sequence for the execution of the program is: (1) the first three statements were executed in ascending order, (2) statement 40 caused the execution of the program to be transferred to the first statement (line number 1000) of the subprogram, (3) statements 1000 and 1010 of the subprogram were executed before the computer was transferred back to statement 50 of the main program by the RETURN statement, (4) following the execution of statement 50, the computer was returned to command level. Briefly, the GOSUB transfers out of the main program and the RETURN transfers out of the subprogram.

DEF FN Statement. Often, throughout a BASIC program, a certain form of calculation will occur again and again, such as the evaluation of an algebraic equation; if this is true it is not necessary to write the evaluation equation for each occurrence. In these instances the define function (DEF FN) statement is very useful. To write the DEF FN statement, a variable name with an argument is set equal to the equation or function:

```
10 DEF FN DA(X)=X^2+3X+4
20 DEF FN A(B)=SQR(B)+B+6
```

Then, throughout the program the functions can be referred to as:

```
40 LET Z=FN MM(Y)
```

or

```
80 LET Z=FN P(M)
```

and so on. The arguments are dummy variables and can be changed in the reference from the original.

Example 3-10. DEF FN Statement

```
10 DEF FN G(X) = INT(X*10^3+.5)/(10*3)
20 LET Y=34.675393
30 LET X=FN G(Y)
40 PRINT X
50 END
34.675
```

Examples 3-9 and 3-10 illustrate two methods for rounding off numbers to a designated number of significant figures. Define function statements cannot be used in every case to replace a subroutine. Subroutines are used when many statements are required. DEF FN statements are used when only one statement is required. Applesoft BASIC only recognizes one argument with the variable name; however, some versions of BASIC will recognize more than one argument with the variable name.

Input/Output Statements

INPUT statements are for placing information in the memory of a computer, and OUTPUT statements are for displaying information on the CRT and printing out information by means of a printer. The information can be words and/or data.

LET Statement. The LET statement causes the computer to store the value of the expression on the right side of the equal sign in the memory location which is reserved for the variable on the left side of the equal sign. For the statements

```
10 LET X=10
20 LET Y=SQR(34.6)
```

the first statement causes the computer to store 10 in the memory position which is reserved for X, and the second statement causes the computer to take the square root of 34.6 and store the result in the memory position which is reserved for Y. LET statements have been used in most of the Examples 3-1 through 3-10.

DATA/READ Statements. DATA and READ statements allow large volumes of fixed information (words, characters, or numbers) to be set up within a program The DATA statement contains the information which is transferred into variables by the READ statement during the execution of the program. DATA statements may be placed anywhere in a program; however, experience programmers usually place the DATA statements consecutively near the end of the program. The statements are read one by one, starting with the lowest number. Some rules to follow in writing DATA/READ statements are:

1. DATA items must correspond in order and type to the READ statement.
2. The number of elements in a data block must correspond to the number of variables in the READ statement—extra data items are ignored.
3. Items in a DATA statement must be separated with commas; a comma should not follow the last data item.
4. Variables and formulas are not permitted.

The use of DATA/READ statements are illustrated in the following example.

Example 3-11. DATA/READ Statements

```
5 I=0
10 DATA7,8
20 DATA5,6
30 DATA2,3
40 READ X,Y
50 LET Z=X+Y
60 PRINT Z
70 LET I=I+1
80 IF I=3 THEN 100
90 GOTO 40
100 END
15
11
5
```

DATA statements 10,20,30 each have a X and a Y value separated by a comma. The READ statement specifies the appropriate X,Y values for each time around the loop. There are three rounds in the loop. For the first time around, the X,Y values are 7,8; for the second time around, the values are 5,6; and for the third time around, the values are 2,3. Statement 70 causes the looping because it returns the computer to the READ statement following the calculation of each sum. Statements 5 and 70 form what is referred to as a counter. The counter determines the time around the loop at which all three sums have been computed and for this round the looping ceases. The counter works so that in statement 70, I is increased by 1 for each time around the loop; in statement 80 the value of I is checked to determine if its value is 3. If its value is 3, the execution of the program is switched to statement 100, which returns the computer to command level.

INPUT Statement. An INPUT statement allows a user to get data into a variable while a program is being executed. The INPUT statement causes the execution of the program to stop with a question mark on the CRT. Upon typing in a value and pressing RETURN, the execution of the program continues. If the INPUT statement is followed by a message inside quotes before the lists of variables, the message will be printed instead of the question mark. The message inside the quotes is followed by a semicolon before the variables whose values are to be typed. The variables must be separated by commas, but a comma should not be used following the last variable. Some INPUT statements are:

```
20 INPUT X,Y,Z
30 INPUT "THE CONSTANTS ARE     ";X,Y,Z
40 INPUT "WHAT IS THE RADIUS?     ":R
```

Example 3-12 is a short program containing an INPUT statement.

Example 3-12. INPUT Statement

```
10 INPUT "THE CONSTNATS ARE     ";X,Y,Z
20 LET W = X^2 + Y^2 + Z^2
30 PRINT W
40 END
RUN
THE CONSTANTS ARE 5,6,7
110
```

After typing RUN and pressing the RETURN key, the phrase "THE CONSTANTS ARE" appears on the CRT. Then the constants 5,6,7 (separated by commas) are typed and the RETURN key is pressed; after which 5,6,7 appears on the same line as the phrase. Finally, the sum of the squares is displayed on the line below the phrase.

PRINT Statement. The PRINT statement is one of the first statements an individual will learn, and it is probably the single statement which is the most often used. In the statement the word PRINT is followed by one or more of the following items (see Example 3-13):

1. Words inside quotes.
2. Variable names.
3. Functions.
4. Punctuation marks.

Example 3-13. PRINT Statements

```
10 PRINT "America the Beautiful."
20 PRINT "AMERICA!, AMERICA!"
25 PRINT "6 * 7 = 42"
30 LET X = 3
40 LET Y = 6
50 PRINT X^2 + Y^2
```

```
55 LET Z=7
60 PRINT X,Y,Z
70 PRINT X;Y;Z
80 PRINT X;TAB(8)Y;TAB(15)Z
90 PRINT LOG(56)/LOG(10), LOG(56)
100 END
America the Beautiful.
AMERICA!, AMERICA!
6 * 7 = 42
45
3               6               7
367
3      6      7
1.74818803      4.02535169
```

Statements 10, 20, and 25 are PRINT statements with phrases, words, and numbers within quotes. The computer causes the items within the quotes to be printed out just as they are given in the quotes. Both upper- and lower-case letters can be used within quotation marks; however, only upper-case letters can be used for other things in a BASIC program. Statements 50 and 90 indicate how a PRINT statement can be used to evaluate a function and cause the computer to print out the result. Statement 90 also indicates the set ups for calculating logarithm, base 10, of 56 and natural logarithm of 56. Statements 60,70 illustrate the use of commas and semicolons for separating variables in a print out. The commas cause the values for X,Y,Z, to be printed respectively in columns 1,17,33. Semicolons placed between variables cause their corresponding values to be printed with no space between the values. Statement 80 is a PRINT statement using the TAB function. The numbers in parentheses following the word TAB designate the column on the PRINT line for the beginning of the variable values. For instance, the value of Y will begin printing in column 8, and the value of Z will begin printing in column 15. In Applesoft BASIC the TAB scheme only works for 40 columns of printing (width of the screen). For printouts using all 80 columns, see Chapter 11.

PR#n Statement. The PR#n statement, where n designates the slot number of the printer card, tells the computer there is a printer attached to the card in the slot n and to send the results displayed on the CRT to the printer to be printed. For the Apple IIe computer, the printer card is usually placed in slot 1; therefore, n is 1 (PR#1).

LIST Statement. When a LIST statement is written into the computer program and there is also a PR#n statement in the program, the computer will print out the program prior to executing the program and printing out the results. If the word LIST is typed without a line number and used as a command, the computer will list the complete program on the CRT (often the program is too long for complete display; in this case only the latter portion will be displayed on the CRT). To list only a line at a time, example line 50, the command should be LIST 50. To list a section such as line 67 to line 145, the command should be LIST 67,145. The LIST commands given here are typed:

```
LIST
LIST 50
LIST 67,145 or LIST 67 — 145
LIST —70
LIST 20—
```

The LIST —70 command causes all statements with line numbers of 70 and below to be displayed, and the last LIST command causes all statements with line numbers greater than 20 to be displayed.

REM (REMark) Statement. REM statements are notes to whoever is reading a listing of the program. REM statements help in both reading and writing programs. Numerous examples of REM statements are given in the example programs of Chapter 11 and in Example 3-14.

Example 3-14. REM Statements

```
10 REM THE VARIABLE VALUES ARE
20 DATA 89,46
30 REM INFORMATION FROM DATA STATEMENTS
40 READ X,Y
50 THE SUM OF THE SQUARES OF X & Y IS
60 PRINT X^2 + Y^2
70 REM  THE NATURAL LOGARITHM OF X IS
80 PRINT LOG(X)
90 REM  THE LOG (BASE 10) OF Y IS
100 PRINT LOG(X) / LOG(10)
110 END
RUN
10037
4.48863637
1.94939001
```

In the REM program each executable statement is preceded by a REM statement which gives information about it. The REM statements have been used to provide program headings, identify important variables, distinguish the major logical segments of a program, etc.

Specification Statement

DIM (DIMension) Statement. The DIM statement is used for the input of large amounts of data, particularly when the data are in the form of tables or arrays. Subscripted variables are required for the DIM statement; subscripted variables can have one or more subscripts. The DIM statement is not required for single subscripted variables if a subscript is 10 or less; however, it is best to use DIM statements for all subscripted variables. Specific forms of single subscripted variables are:

A(10),AL(25),B(5),P(6)

Some general forms are:

A(I),AL(N),B(I),P(K)

The subscripts must be single integer constants, and 0 may be considered as a constant. The DIM statement causes the computer to reserve sufficient places in memory equal to the number of the pieces of data specified by the variable names. The data are placed in memory as they are needed for a calculation or as they are calculated. The statement DIM A(10) will cause the computer to reserve 11 spaces for the data of A(0) through A(10). The uses of the DIM statement and single subscripted variables are given in Example 3-15.

Example 3-15. Single Subscripted Variables

```
10 DIM A(4),B(4),C(4),D(4)
20 DATA 1,2,3,4
30 DATA 5,6,7,8
40 DATA 9,10,11,12
50 DATA 13,14,15,16
60 FOR I = 1 to 4
70 READ A(I),B(I),C(I),D(I)
80 LET A(I)=A(I)*A(I)
90 LET B(I)=B(I)*B(I)
```

```
100 LET C(I)=C(I)*C(I)
110 LET D(I)=D(I)*D(I)
120 PRINT A(I);TAB(5)B(I);TAB(10)C(I);TAB(15)D(I)
130 NEXT I
140 END
1    4    9    16
25   36   49   64
81   100  121  144
169  196  225  256
```

During the execution of the program, statement 10 caused the data to be stored in the reserved five spaces for each variable. The example also illustrates single looping. During the first time around the loop, the values for A(1),B(1),C(1),D(1) were computed and printed out; during the second time around, the values for A(2),B(2),C(2),D(2) were computed; and during times three and four, the values for the variables with subscripts 3 and 4 were computed and printed out.

The first element in a subscripted variable of the form A(I,J) refers to the number of rows in a two-dimensional array, and the second element refers to the number of the column in a two-dimensional array. For instance, the variable B(15,15) represents a 16 x 16 array—16 rows and 16 columns. The DIM B(15,15) statement causes the computer to reserve 16 x 16 or 256 spaces for data, which are either placed in the appropriate memory spaces as they are used or as they are computed during the execution of the program. Example 3-16 is a program for a 4 x 4 array which is designated by the A(4,4) subscripted variable.

Example 3-16. Double Subscripted Variables

```
10 DIM A(4,4),B(4,4),C(4,4)
20 DATA 1,2,3,4
30 DATA 5,6,7,8
40 DATA 9,19,11,12
50 DATA 13,14,15,16
60 FOR I=1 TO 4
70 FOR J=1 TO 4
80 READ A(I,J)
90 LET B(I,J) = A(I,J)^2
95 LET C(I,J) = INT (B(I,J))
100 NEXT J
110 NEXT I
```

```
120 FOR I = 1 TO 4
130 PRINT C(I,1);TAB(6)C(I,2);
    TAB(11)C(I,3);TAB(16)C(I,4)
140 NEXT I
150 END
1    4    9    16
25   36   49   64
81   100  121  144
169  196  225  256
```

Actually, there are 25 spaces reserved for data in each of the subscripted variables; only 16 spaces are utilized for each subscripted variable. The B and C subscripted variables are used to reserve spaces for the computed data. The B(4,4) variables cause the computer to reserve spaces for the squares of the A(4,4) data, and the C(4,4) variables cause the computer to reserve spaces for the integer values of the squares. There was a need for getting the integer values because some of the squares would have had values containing decimals. The decimals were caused by the computer working in binary numbers and printing out in decimal numbers. There are loops within a loop between statements 60 and 110 and a single loop between 120 and 140. To control the printout with designated spacings (TAB functions), the starting column values were specifically specified in PRINT statement number 130.

Miscellaneous Statements

Some of the more advanced versions of the BASIC language, particularly BASIC for mini or mainframe computers, include the MATrix statements; DEF FN statements for two or more variables in the arguments such as, DEF FN AN(X,Y,Z); LPRINT; and PRINT USING operations involving integer and double-precision data or real and double-precision data which will produce a double-precision result. A VAX 11-780 located at Tennessee Technological University, Cookeville, Tennessee was used for running the programs given in the examples which follow. An Apple IIe with a modem was used as the terminal.

MAT Statements. These statements greatly reduce the efforts of an individual for developing programs in which matrices or vectors are involved. MAT statements have been used for reducing the size of BASIC programs involving the solutions of simultaneous equations. A few MAT statements with their uses are:

1. 10 MAT READ A causes a set of values to be assigned to the appropriate elements of a matrix.

2. 20 MAT D=E causes each element of matrix D to be assigned to the corresponding element of matrix E.

3. 30 MAT D=E+F causes each element of matrix D to be assigned to the sum of the corresponding elements of matrix E and matrix F.

4. 40 MAT D=E—F causes each element of matrix D to be assigned the difference of the corresponding elements of matrix E and matrix F.

5. 50 MAT D=(K)*B causes each element of matrix D to be assigned the corresponding element of matrix B following its multiplication by K.

6. 60 MAT D=E*F causes the elements of matrix D to be replaced by an appropriate element following matrix multiplication of E*F.

7. 70 MAT PRINT B causes the elements of the data to be printed in matrix form.

8. 80 MAT D=INV(E) causes the computer to calculate the inverse of matrix E, and matrix D becomes the new matrix.

Examples 3-17 and 3-18 are two BASIC programs which illustrate the use of MAT READ, MAT PRINT, and MAT multiplication statements.

Example 3-17. MAT READ and MAT PRINT

```
10 DIM A(3,3)
20 MAT READ A
30 DATA 1,2,3,4,5,6,7,8,9
40 MAT PRINT A;
RUN
1  2  3
4  5  6
7  8  9
```

Example 3-18. MAT READ and MAT Multiplication

```
10 DIM A(3,2)
20 DIM B(2,4)
30 MAT READ A
40 DATA 2,2,2,2,2,2
50 MAT READ B
60 DATA 2,2,2,2,2,2,2,2
70 MAT C=A*B
80 MAT PRINT C;
90 END
RUN
8  8  8  8
8  8  8  8
8  8  8  8
```

DEF FN Statement (more than one argument). A program with a DEF FN statement with two arguments within parentheses is given in Example 3-19. From reading the Applesoft discussion on DEF FN it is believed the example is self-explanatory. Note the period between FN and SI(A,B).

Example 3-19. DEF.FN Statement, Two Variables in Parentheses

```
10 DEF FN.SI(A,B)=A*B
20 LET A=6
30 LET B=7
40 LET Z= FN.SI(A,B)
50 PRINT Z
60 END
RUN
42
```

PRINT USING Statement. A number of versions of the BASIC language include the PRINT USING Statement; Applesoft BASIC does not. This statement allows data items to be formatted when printed out. Examples 3-20 and 3-21 illustrate two uses of the PRINT USING statement. The first of the two examples illustrates the use of the statement for printing out a column of numbers with the decimal points aligned vertically and the same number of units to the right of the decimal points. Subroutines are required in Applesoft BASIC for vertically aligning the decimal points, but the zeroes are not added to the right of the farthest right number in the decimal.

Example 3-20. PRINT USING Statement

```
10 DATA A(4)
20 DATA 1234
30 DATA 123.4
40 DATA 12.34
50 DATA 1.234
60 FOR I=1 TO 4
70 READ A(I)
80 PRINT USING "####.####";A(I)
90 NEXT I
100 END
RUN
1234.0000
 123.4000
  12.3400
   1.2340
```

DOUBLE PRECISION Qualifier, Keyword. In scientific work and business there is often a need for a greater number of digits than the usual nine. By using a qualifier and a keyword, the accuracy can be extended to 16 digits. Some versions use the pound or number sign (#) such as A#, B#; # causes the computer to work and printout in 18 significant units. The VAX 11-780 uses /DOUBLE for doubling the number of digits. The qualifier and keyword are typed at the beginning before the program is entered (see Example 3-21).

Example 3-21. Double Precision Numbers

```
$ BASIS/DOUBLE
VAX-11 BASIC V2.0
  10 A=1
  20 B=6
  30 C=A/B
  40 PRINT USING "#.################",C
  50 END
  RUN
  0.1666666666666667
```

LPRINT Statement. Some computers use this statement to cause the printer to print information. The command is given in the program as the print statement. Applesoft requires entering PR#n, where n is the slot number containing the printer card.

Exercises

1. Using BASIC, repeat Exercise 2-1.
2. Using BASIC, repeat Exercise 2-2.
3. Using a microcomputer as a calculator, calculate the values of SQR(96), EXP(12), 7*25, 20^2.5.
4. Write a three- or four-line thank you note using the PRINT statement with quotation marks.
5. Using scientific notation for representing numbers, write a simple program for adding the sum of the squares of 54.32467E+04 and 674.5621E+07.
6. Referring to Example 3-16, write a similar program for the array of values:

36.82	84.3	220.24	467.32
274.929	32.7	886.67	784.94
328.724	67.9	992.82	34.67

Vertically align the decimal points in the columns and round off the results before printing one place to the right of the decimal point.

7. Rewrite Example 3-17 using explanation REM statements before each executable statement.
8. Determine the quotient of 85/34 by the reciprocal method given in Appendix A (Equation A-2).
9. Repeat Exercise 2-6 using BASIC.
10. Repeat Exercise 3-8 using DOUBLE precision methods.
11. Write a BASIC program for checking multiplications. Use an INPUT statement to supply the multiplicand, the multiplier, and the product. If the operator supplies the correct number for the answer, the word SMARTY will appear on the CRT; but if the operator supplies the incorrect answer, the word DUMMY will appear on the CRT.
12. Write a BASIC program using the DEF FN statement for the function:

    G (X) = .577 + X + X^2/2 + X^3/6

    when X = .3 and X = .6.
13. Obtain a book on BASIC for microcomputers, and prepare 10 programs as suggested by that book.

Suggested Reading

1. Apple Computer Inc., *The Apple Tutorial*, Cupertino, CA, Apple Computer Inc., 1981.
2. Blackwood, B. D. and Blackwood, G. H., *Applesoft Language*, Indianapolis, IN: Howard W. Sams and Co., 1981.
3. Commodore International, *Personal Computing on Vic-20*, Norristown, PA: Commodore International, Ltd., 1981.
4. Digital Equipment Corporation, *User's Manual—Vax-11 BASIC V 1.3*, Maynard, MA, 1981.
5. Dunn, S. and Morgan, V., *The Apple Personal Computer for Beginners*, Englewood Cliffs, NJ: Prentice Hall International, 1982.
6. Gottfried, B. S., *Theory and Problems of Programming with BASIC Including Microcomputer BASIC*, 2nd ed., New York, NY: McGraw-Hill Book Co., 1982.
7. Hennefeld, J., *Using BASIC—An Introduction to Computer Programming*, 2nd ed., Boston, MA: Prindle Weber and Schmidt, 1981.
8. Inman, D., Zamora, R., and Albrecht, B., *TRS-80 Advanced Level II BASIC*, New York, NY: John Wiley and Sons Inc., 1980.
9. Lee, J. A., "The Formal Definition of BASIC Languages," *The Computer Journal* (British), 15, 2, 37ff, 1972.
10. Murrill, P. W. and Smith, C. L., *BASIC Programming*, Scranton, PA: International Textbook Company, 1971.
11. Tandy Corporation, *TRS-80 Operation and BASIC Language Reference Manual*, Fort Worth, TX, 1980.
12. Waite, M. and Pardee, M., *BASIC Programming Primer*, Indianapolis, IN: Howard W. Sams and Co., Inc., 1982.

4 Selected Topics from Numerical Mathematical Analysis

Numerical mathematical analysis has an important role in petroleum engineering. This is particularly true since the advent of electronic digital computers. Petroleum engineers are interested in the production of actual solutions, rather than theoretical solutions, to their problems. This chapter on selected topics from numerical mathematical analysis is not an exhaustive treatment of the subject; however, the selection of topics is sufficient to give the petroleum engineer an appreciation of numerical mathematical analysis and the knowledge required for solving petroleum engineering problems.

Engineers often use numerical methods in the solutions of their engineering problems. Problems of petroleum engineering which usually require numerical solutions are material balance, pressure build-up potential distribution, multi-well and volumetric oil reserve estimates. This chapter is complete enough to enable the petroleum engineer to pursue additional information from textbooks on numerical analysis without outside assistance.

Solutions to engineering problems by numerical analysis are only approximate; however, the engineer often can control the accuracy of the computed results. Some of the errors commonly made during computations by numerical mathematical analysis are: (1) truncation errors, (2) roundoff errors, (3) human errors, (4) data errors, and (5) calculation errors. A Taylor series in general is an infinite series but for practical calculations a finite series must be used. If a Taylor series is terminated after a finite number of terms, the discarded terms represent a truncation error. Digital computers always work with numbers represented by a finite number of digits; thus the simple division ⅓ is subject to an error by the computer. The error due to rounding off the number of digits is called a round-off error.

Throughout this chapter, whenever possible, standard petroleum reservoir engineering symbols will be used. Also throughout this chapter, whenever possible, petroleum reservoir engineering problems will be used for the illustrations and examples.

Interpolation

In elementary mathematics, the engineer is taught a method for computing intermediate values of a function from a set of given or tabular values of that function. The process of computing the intermediate values is known as linear interpolation. Engineering problems often deal with functions whose analytical form is either totally unknown or else is such that the function cannot easily be subjected to the operations as required. In such cases, the petroleum engineer may wish to replace the given function with one which can be handled more easily. In the broad sense of the term, the operation of replacing the given function by a simpler function is known as polynomial approximation. The use of the polynomial for computing an intermediate value would constitute interpolation.

In 1885, the German mathematician Weierstrass[1] proved the two theorems: (1) a function which is continuous in a given region can be represented in that region by a polynomial to any degree of accuracy, and (2) a function which is continuous for a period of 2π can be represented by a finite trigonometric series to any degree of accuracy.

In this chapter methods are given for fitting a polynomial to a set of data. The methods are referred to as interpolation methods. Methods for fitting a trigonometric series to a set of data are given in the chapter on curve fitting.

Weierstrass' first theorem is explained with the use of Table 4-1; which contains the data for pressure and formation volume factors in a reservoir of crude oil:

[1]Scarborough, J. B., *Numerical Mathematical Analysis, 2nd Ed.*, Baltimore: The Johns Hopkins University Press, p. 52, 1950.

Table 4-1
Pressure-Formation Volume Factor Data

Pressure Psi (P)	Formation Volume Factor
100	1.059
500	1.145
900	1.197
1300	1.235
1700	1.268
2100	1.298
2500	1.327

Table 4-1 lists the values for formation volume factors corresponding to pressures in 400 psi increments. Now suppose it is desired to find a value for a formation volume factor corresponding to an intermediate pressure which is not listed in the table. What are the methods for finding the value of the formation volume factor? There are at least three methods: linear interpolation, curve drawing, and applying the theorem. A glance at the table indicates that the formation volume factor is not a linear function of pressure; therefore, values obtained by linear interpolation would be somewhat erroneous. If a curve were plotted with the data from the table it would have to be drawn large to give three significant digits accuracy. The most accurate method of determining values for formation volume factors at intermediate pressures would be to find a polynomial function which is suited to the data of the table. The use of the polynomial for determining the intermediate value would involve interpolation.

Four methods will be developed for finding polynomials which are suited to the data of the table. These polynomial formulas are known as interpolation formulas. The four interpolation formulas to be developed are know as:

1. Newton's Forward Interpolation Formula
2. Newton's Backward Interpolation Formula
3. Stirling's Interpolation Formula
4. Lagrange's Interpolation Formula

The first of the four formulas is used for determining values near the beginning of the table; the second formula is used for determining values near the end of the table; and the third formula is used for determining values near the center of the table. The fourth formula is used when it is not possible to obtain values of the independent variable at equi-distant intervals and it is used for any section of the table.

Newton's Forward Interpolation Formula

To find the polynomial which approximately represents the function near the beginning of the table, a diagonal difference table is developed. In terms of the function $B = f(P)$, if $B_0, B_1, B_2 \cdots B_n$ denote a set of values of the true function, then $(B_1 - B_0)$, $(B_2 - B_1)$, $(B_3 - B_2)$, . . . $(B_n - B_{n-1})$ are called the first differences of the function $f(P)$. By convention, these differences are denoted as ΔB_0, ΔB_1, $\Delta B_2 \ldots \Delta B_{n-1}$. In other words $\Delta B_0 = B_1 - B_0$, $\Delta B_1 = B_2 - B_1$, $\Delta B_2 = B_3 - B_2, \ldots, \Delta B_{n-1} = B_n - B_{n-1}$. The difference of two first differences is called the second difference as

$$\Delta^2 B_0 = (B_2 - B\beta) - (B_1 - B_0) = B_2 - 2B_1 + B_0$$
$$\Delta^2 B_1 = (B_3 - B_2) - (B_2 - B_1) = B_3 - 2B_2 + B_1 \text{ etc.}$$

Likewise, the third differences are the differences of the second differences. Then the nth differences are the differences of the $(n - 1)$th differences. Rewriting Table 4-1 in terms of alphabetic notation gives Table 4-2.

From the alphabetic notation in Table 4-2, a diagonal difference table is obtained (Table 4-3).

The numerical difference values for the data of Table 4-1 are given in Table 4-4.

The polynomial which suitably represents the data at the beginning of Table 4-3 is developed as follows. If $B = f(P)$ denotes the true function for the B_0, $B_1 \cdots B_n$ values which correspond respectively to the equi-incremented $P_0, P_1 \cdots P_n$ values, and if $B = \phi\ (P)$ denotes a polynominal of the nth degree which approximately represents the data of the table, then the

Table 4-2
Alphabetic Notation

Pressure	Formation Volume Factor
P_0	B_0
P_1	B_1
P_2	B_2
P_3	B_3
P_4	B_4
P_5	B_5
P_6	B_6

Table 4-3
Diagonal Difference

P	B	ΔB	Δ^2B	Δ^3B	Δ^4B	Δ^5B	Δ^6B
P_0	B_0						
		ΔB_0					
P_1	B_1		$\Delta^2 B_0$				
		ΔB_1		$\Delta^3 B_0$			
P_2	B_2		$\Delta^2 B_1$		$\Delta^4 B_0$		
		ΔB_2		$\Delta^3 B_1$		$\Delta^5 B_0$	
P_3	B_3		$\Delta^2 B_2$		$\Delta^4 B_1$		$\Delta^6 B_0$
		ΔB_3		$\Delta^3 B_2$		$\Delta^5 B_1$	
P_4	B_4		$\Delta^2 B_3$		$\Delta^4 B_2$		
		ΔB_4		$\Delta^3 B_3$			
P_5	B_5		$\Delta^2 B_4$				
		ΔB_5					
P_6	B_6						

Table 4-4
Numerical Difference Values

P	B	ΔB	Δ^2B	Δ^3B	Δ^4B	Δ^5B	Δ^6B
100	1.059						
		0.086					
500	1.145		−0.034				
		0.052		0.020			
900	1.197		−0.014		−0.011		
		0.038		0.009		0.004	
1300	1.235		−0.005		−0.007		−0.003
		0.033		0.002		0.007	
1700	1.268		−0.003		−0.000		
		0.030		0.002			
2100	1.298		−0.001				
		0.029					
2500	1.327						

polynomial which is suitable for the data of the table may be written in the form:

$$\begin{aligned}\phi(P) = {} & a_0 + a_1 (P - P_0) + a_2 (P - P_0)(P - P_1) \\ & + a_3 (P - P_0)(P - P_1)(P - P_2) + \ldots \\ & + a_n (P - P_0)(P - P_1)(P - P_2) \cdots (P - P_{n-1}) \qquad \text{4-1}\end{aligned}$$

By setting $\phi\ (P_0) = B_0, \phi\ (P_1) = B_1, \ldots \phi\ (P_n) = B_n$ the coefficients for $a_0, a_1, a_2, \cdots a_n$ are obtained. If the equi-increments of pressure ΔP are designated by h then $P - P_0 = h, P_2 - P_0 = 2h,$ etc., and

$$B_0 = a_0 \text{ or } a_0 = B_0$$

$$B_1 = a_0 + a_1 (P_1 - P_0) = B_0 + a_1 h$$

$$\text{or } a_1 = \frac{B_1 - B_0}{h} = \frac{\Delta B_0}{h}$$

$$B_2 = a_0 + a_1 (P_2 - P_0) + a_2 (P_2 - P_0)(P_2 - P_1)$$

$$= B_0 + \frac{B_1 - B_0}{h}(2h) + a_2 (2h)(h)$$

$$\text{or } a_2 = \frac{B_2 - 2B_1 + B_0}{2h^2}$$

$$\text{since } \Delta^2 B_0 = B_2 - 2B_1 + B_0$$

$$\text{then} \quad a_2 = \frac{\Delta^2 B}{2h^2}$$

$$B_3 = a_0 + a_1 (P_3 - P_0) + a_2 (P_3 - P_0)(P_3 - P_1) + a_3 (P_3 - P_0)(P_3 - P_1)(P_3 - P_2)$$

$$B_3 = B_0 + \frac{(B_1 - B_0)}{h}(3h) + \frac{(B_2 - 2B_1 + B_0)}{2h^2}(3h)(2h) + a_3 (3h)(2h)h$$

$$\text{or } a_3 = \frac{B_3 - 3B_2 + 3B_1 - B_0}{6\ h^3}$$

$$\text{and since } \Delta^3 B_0 = B_3 - 3B_2 + 3B_1 - B_0$$

$$\text{then } a_3 = \frac{\Delta^3 B_0}{3!h^3}$$

In a similar manner the nth coefficient is found to be

$$a_n = \frac{\Delta^n B_0}{n!h^n}$$

Substituting the values for coeffcients in Equation **4-1** gives

$$\phi\ (P) = B_0 + \Delta B_0 \left(\frac{P - P_0}{h}\right) + \frac{\Delta^2 B_0}{2!}\left(\frac{P - P_0}{h}\right)\left(\frac{P - P_1}{h}\right) + \frac{\Delta^3 B_0}{3!}\left(\frac{P - P_0}{h}\right)\left(\frac{P - P_1}{h}\right)\left(\frac{P - P_2}{h}\right)\cdots + \frac{\Delta^n B_0}{n!}\left(\frac{P - P_0}{h}\right)\left(\frac{P - P_1}{h}\right)\cdots\left(\frac{P - P_{n-1}}{h}\right) \quad \textbf{4-2}$$

$$\frac{P - P_0}{h} = u, \text{ or } u = P_0 + hu$$

If

$$\frac{P - P_0}{h} = u, \text{ or } u = P_0 + hu$$

and $P_1 = P_0 + h$, $P_2 = P_0 + 2h$, etc.

then $$\frac{P - P_1}{h} = \frac{P - (P_0 - h)}{h} = \frac{P - P_0}{h} - \frac{h}{h} = u - 1$$

$$\frac{P - P_2}{h} = \frac{P - (P_0 + 2h)}{h} = \frac{P - P_0}{h} - \frac{2h}{h} = u - 2$$

and

$$P - P_{n-1} = \frac{P - [P_0 + (n-1)\,h]}{h} = \frac{P - P_0}{h} - \frac{(n-1)\,h}{h}$$

$$= u - (n - 1) = u - n + 1$$

Now in terms of this u notation Equation **4-2** becomes

$$\phi\,(P) = B_0 + u\Delta B_0 + u\,\frac{(u-1)}{2!}\,\Delta^2 B_0 + \frac{u\,(u-1)\,(u-2)}{3!}\,\Delta^3 B_0 + \ldots + \frac{u\,(u-1)\,(u-2)\cdots(u-n+1)}{n!}\,\Delta^n B_0 \qquad \textbf{4-3}$$

The Use of Newton's forward interpolation formula for extrapolation is illustrated in Example 4-1.

Example 4-1. With the use of Equation **4-3** and Table 4-4 calculated the formation volume factor at 50 psi pressure.

Solution. The value of u is determined as

$$u = \frac{P - P_0}{h} = \frac{50 - 100}{400} = -\,0.125$$

Substituting the values for differences from Table 4-4 and using the calculated value for u in Equation **4-3** gives:

$$B = 1.059 + (-0.125)\,(0.086) + (1/2)\,(-0.125)\,(-1.125)\,(-0.034) + 1/6\,(-0.125)\,(-1.125)\,(-2.125)\,(0.022) + 1/24\,(-0.125)\,(-1.125)\,(-2.125)\,(-3.125)\,(-0.013) = 1.059 - 0.01075 - 0.002391 - 0.0009961 - 0.000428$$

$$B = 1.044863 \text{ or } 1.045$$

Newton's Backward Interpolation Formula

To find the polynomial which approximately represents the function near the end of the table, horizontal differences are used. The values of Table 4-3 are rewritten in the form of horizontal differences as indicated in Table 4-5.

The numerical values of Table 4-4 are rewritten in the horizontal difference form as indicated in Table 4-6.

To derive Newton's backward interpolation formula Equation **4-1** is rewritten in the following form:

$$\phi(P) = a_o + a_1 (P - P_n) + a_2 (P - P_n) (P - P_n) (P - P_{n-1}) + a_3(P - P_n) (P - P_{n-1}) (P - P_{n-2}) + \ldots + a_n (P - P_n) (P - P_{n-1}) \cdots (P - P_1) \qquad 4\text{-}4$$

Setting $\phi (P_n) = B_n, \phi (P_{n-1}) = B_{n-1}$, etc., the coefficients a_o, a_1, a_2, $\cdots a_n$ are determined as follows:

$$B_n = a_o$$
$$B_{n-1} = a_o + a_1 (P_{n-1} - P_n) = B_n + a_1 (-h)$$

$$\text{or } a_1 = \frac{B_n - B_{n-1}}{h} = \frac{\Delta_1 B_n}{h}$$

$$B_{n-2} = a_o + a_1 (P_{n-2} - P_n) + a_2 (P_{n-2} - P_n) (P_{n-2} - P_{n-1})$$
$$= B_n + \left(\frac{B_n - B_{n-1}}{h}\right)(-2h) + a_2 (-2h) (-h)$$

$$\text{or } a_2 = \frac{B_n - 2B_{n-1} + B_{n-2}}{2h^2} = \frac{\Delta_2 B_n}{2h^2}$$

Table 4-5
Horizontal Difference

P	B	Δ_1B	Δ_2B	Δ_3B	Δ_4B	Δ_5B	Δ_6B
P_0	B_0						
P_1	B_1	Δ_1B_1					
P_2	B_2	Δ_1B_2	Δ_2B_2				
P_2	B_3	Δ_1B_3	Δ_2B_3	Δ_3B_3			
P_3	B_4	Δ_1B_4	Δ_2B_4	Δ_3B_4	Δ_4B_4		
P_4	B_5	Δ_1B_5	Δ_2B_5	Δ_3B_5	Δ_4B_5	Δ_5B_5	
P_5	B_6	Δ_1B_6	Δ_2B_6	Δ_3B_6	Δ_4B_6	Δ_5B_6	Δ_6B_6

Table 4-6
Numerical Difference Values

P	B	Δ_1B	Δ_2B	Δ_3B	Δ_4B	Δ_5B	Δ_6B
100	1.059						
500	1.145	0.086					
900	1.197	0.052	−0.036				
1300	1.235	0.038	−0.014	0.022			
1700	1.268	0.033	−0.005	0.009	−0.013		
2100	1.298	0.030	−0.003	0.002	−0.007	0.006	
2500	1.327	0.029	−0.001	0.002	−0.000	0.007	−0.001

The other coefficients are determined in a similar manner. The general form of the coefficients is

$$a_n = \frac{\Delta_n B_n}{n!h^n}$$

With the coefficients Equation 4-4 takes the form:

$$\phi\,(P) = B_n + \frac{\Delta_1 B_n}{h}(P - P_n) + \frac{\Delta_2 B_n}{2!h^2}\,(P - P_n)\,(P - P_{n-1}) + \frac{\Delta_3 B_n}{3!h^3}$$
$$(P - P_n)\,(P - P_{n-1})\,(P - P_{n-2})\,\cdots + \frac{\Delta_n B_n}{n!h^n}$$
$$(P - P_n)\,(P - P_{n-1})\,\cdots\,(P - P_1). \qquad 4\text{-}5$$

By a change of variables Newton's backward interpolation formula, Equation 4-5 is changed to a simpler form as:

$$\phi\,(P) = P_n + \Delta_1 B_n \left(\frac{P - P_n}{h}\right) + \frac{\Delta_2 P_n}{2!}\left(\frac{P - P_n}{h}\right)\left(\frac{P - P_{n-1}}{h}\right)$$
$$+ \frac{\Delta_3 B_n}{3}\left(\frac{P - P_n}{h}\right)\left(\frac{P - P_{n-1}}{h}\right)\left(\frac{P - P_{n-2}}{h}\right) + \cdots + \frac{\Delta_n B_n}{n!}$$
$$\left(\frac{P - P_n}{h}\right)\left(\frac{P - P_{n-1}}{h}\right)\cdots\left(\frac{P - P_1}{h}\right) \qquad 4\text{-}6$$

Now letting

$$u = \frac{P - P_n}{h} \qquad \text{or} \qquad P = P_n + hu$$

and since $P_{n-1} = P_n - h, P_{n-2} = P_n - 2h$, etc.,

then

$$\frac{P - P_{n-1}}{h} = \frac{P - (P_n - h)}{h} = \frac{P - P_n + h}{h} = \frac{P - P_n}{h} + \frac{h}{h} = u + 1,$$

$$P - P_{n-2} = \frac{P - (P_n - 2h)}{h} = \frac{P - P_n}{h} + \frac{2h}{h} = u + 2$$

$$P - P_{n-3} = \frac{P - (P_n - 3h)}{h} = \frac{P - P_n}{h} + \frac{3h}{h} = u + 3$$

and so on. In terms of the u values Equation **4-6** now becomes

$$\phi(P) = B_n + u\Delta_1 B_n + \frac{u(u+1)}{2}\Delta_2 B_n + \frac{u(u+1)(u+2)}{3!}\Delta_3 B_n + \ldots + \frac{u(u+1)(u+2)\ldots(u+n-1)}{n!}\Delta_n B_n \qquad \textbf{4-7}$$

The use of Newton's backward interpolation formula for determining values near the end of a set of values is illustrated in Example 4-2.

Example 4-2. With the use of Equation **3-7** and Table 4-6 find the formation volume factor at 2300 psi pressure.

Solution: The value for u is determined as

$$u = \frac{P - P_n}{h} = \frac{2300 - 2500}{400} = -0.500$$

Using the u value and four difference values from Table 4-6 and substituting the values into Equation **4-7** gives

$$B = 1.327 + (-0.5)(0.029) + \frac{(-0.5)(0.5)(-0.001)}{2!}$$

$$+ \frac{(-0.5)(0.5)(1.5)(0.002)}{6} + \frac{(-5)(0.5)(1.5)(2.5)(0.000)}{24}$$

$$= 1.327 - 0.0145 + 0.000125 - 0.000125 + 0$$

or $B = 1.327 - 0.0145 = 1.3125$

Stirling's Interpolation Formula

Newton's two formulas are applicable for the majority of cases of interpolation; however, these formulas do not converge as rapidly as the central-difference interpolation formulas. The central-difference interpolation formulas are more suited for interpolating values near the middle of a tabulated set. Only one central-difference interpolation formula is developed here; this interpolation formula is known as Stirling's interpolation formula.

A diagonal difference table with special marking, as shown in Table 4-7, is required for developing Stirling's interpolation formula. In deriving the formula special consideration is given to differences lying as near as possible to the horizontal line through B_0; these quantities are underscored.

Beginning with B_o, Newton's forward interpolation formula is written as

$$B = B_o + u\Delta B_o + \frac{u(u-1)}{2}\Delta^2 B_o + \frac{u(u-1)(u-2)}{3!}\Delta^3 B_o + \frac{u(u-1)(u-2)(u-3)}{3}\Delta^4 B_o + \ldots \tag{4-8}$$

or

$$B = B_o + C_1\Delta B_o + C_2\Delta^2 B_o + C_3\Delta^3 B_o + C_4\Delta^4 B_o + \ldots \tag{4-9}$$

Table 4-7
Central Differences

P	B	ΔB	Δ²B	Δ³B	Δ⁴B	Δ⁵B	Δ⁶B
P_{-3}	B_{-3}						
		ΔB_{-3}					
P_{-2}	B_{-2}		$\Delta^2 B_{-3}$				
		ΔB_{-2}		$\Delta^3 B_{-3}$			
P_{-1}	B_{-1}		$\Delta^2 B_{-2}$		$\Delta^4 B_{-3}$		
		ΔB_{-1}		$\Delta^3 B_{-2}$		$\Delta^5 B_{-3}$	
P_0	B_0		$\Delta^2 B_{-1}$		$\Delta^4 B_{-2}$		$\Delta^6 B_{-3}$
		ΔB_0		$\Delta^3 B_{-1}$		$\Delta^5 B_{-2}$	
P_1	B_1		$\Delta^2 B_0$		$\Delta^4 B_{-1}$		
		ΔB_1		$\Delta^3 B_0$			
P_2	B_2		$\Delta^2 B_0$				
		ΔB_2					
P_3	B_3						

where the C's denote binomial coefficients.

Expressions for averaging the odd differences above the horizontal line through B_o are

$$\text{(a)}\quad m_1 = \frac{\Delta B_{-1} + \Delta B_o}{2}$$

$$\text{(b)}\quad m_3 = \frac{\Delta^3 B_{-2} + \Delta^3 B_{-1}}{2}$$

$$\text{(c)}\quad m_5 = \frac{\Delta^5 B_{-3} + \Delta B_{-2}}{2}$$

$$\text{(d)}\quad m_7 = \frac{\Delta^7 B_{-4} + \Delta^7 B_{-3}}{2}$$

The next steps are to eliminate the ΔB_0, $\Delta^2 B_0$, $\Delta^3 B_0$, etc., from Equation **4-9** by a process of elimination. The process begins with ΔB_o, $\Delta^2 B_o$, etc., and works diagonally upward to the right until the quantities in the horizontal line are reached.

By definition $\Delta^2 B_{-1} = \Delta B_o - \Delta B_{-1}$ or

(e) $\Delta B_o = \Delta^2 B_{-1} + \Delta B_{-1}$

From (a) $\Delta B_{-1} = 2m_1 - \Delta B_o$

(f) $\therefore \Delta B_o = m_1 + \frac{1}{2}\Delta^2 B_{-1}$

For the second difference term $\Delta^3 B_{-1} = \Delta^2 B_o - \Delta^2 B_{-1}$

(g) or $\Delta^2 B_o = \Delta^2 B_{-1} + \Delta^3 B_{-1}$ and by definition

(h) $\Delta^4 B_{-2} = \Delta^3 B_{-1} - \Delta^3 B_{-2}$

and from (b)

(i) $\Delta B_{-1} = 2m_3 + \Delta B_{-2}$

Substracting (h) from (i) gives

(j) $\Delta^3 B_{-1} = m_3 + \frac{1}{2}\Delta^4 B_{-2}$

Now substituting (j) into (g) gives

(k) $\Delta^2 B_o = \Delta^2 B_{-1} + m_3 + \frac{1}{2}\Delta^4 B_{-2}$

For the third difference term $\Delta^4 B_{-1} = \Delta^3 B_o - \Delta^3 B_{-1}$

(l) or $\Delta^3 B_o = \Delta^3 B_{-1} + \Delta^4 B_{-1}$

and from (j) $\Delta^3 B_o = m_3 + \frac{1}{2}\Delta^4 B_{-2} + \Delta^4 B_{-1}$.

But by difinition $\Delta^5 B_{-2} = \Delta^4 B_{-1} - \Delta^4 B_{-2}$

(m) or $\Delta^4 B_{-1} = \Delta^4 B_{-2} + \Delta^5 B_{-2}$ and likewise

(n) $\Delta^6 B_{-3} = \Delta^5 B_{-2} - \Delta^5 B_{-3}$ and from (c)

(o) $\Delta^5 B_{-2} = 2m_5 - \Delta^5 B_{-3}$

Subtracting (n) from (o) and solving for $\Delta^5 B_{-2}$ gives

(p) $\Delta^5 B_{-2} = m_5 + \frac{1}{2}\Delta^6 B_{-3}$.

By substituting (p) in (m) gives

(q) $\Delta^4 B_{-1} = \Delta^4 B_{-2} + m_3 + \frac{1}{2}\Delta^6 B_{-}^{3}\,\Delta^6 B_{-3}$.

Now substituting (q) in (l) gives the value for the third difference term.

(r) $\Delta^3 B_o = m_3 + {}^3/_2\, \Delta^4 B_{-2} + m_5 + \frac{1}{2}\, \Delta^6 B_{-3}$.

In a similar manner the fourth difference term was found to be

(s) $\Delta^4 B_o = \Delta^4 B_{-2} + 2m_5 + 2\Delta^6 B_{-3}$
$+ m_7 + \frac{1}{2}\, \Delta^8 B_{-4}$.

Now substituting (e), (h), (r), and (s) into Equation **4-9** gives

$$B = B_o + C_1\,(m_1 + \tfrac{1}{2}\,\Delta^2 B_{-1}) + C_2\,(m_3 + \Delta^2 B_{-1} + \tfrac{1}{2}\,\Delta^4 B_{-2})$$
$$+ C_3\,(m_3 + m_5 + \tfrac{1}{2}\,\Delta^4 B_{-2} + \tfrac{1}{2}\,\Delta^6 B_{-3})$$
$$+ C_4\,(2m_5 + m_7 + \Delta^4 B_{-2} + 2\Delta^6 B_{-3} + \tfrac{1}{2}\,\Delta^8 B_{-4}) \qquad \textbf{4-10}$$

or

$$B = B_o + C_1 m_1 + \left(\frac{C_1}{2} + C_2\right)\Delta^2 B_{-1} + (C_2 + C_3) m_3$$
$$+ \left(\frac{C_2}{2} + \frac{3C_3}{2} + C_4\right)\Delta^4 B_{-2} + \text{etc.} \qquad \textbf{4-11}$$

Replacing the C's and m's by their equivalences in Equation 3-11 gives

$$B = B_o + u\,\frac{\Delta B_{-1} + \Delta B_o}{2} + \left\{\frac{u}{2} + \left[\frac{u\,(u-1)}{2}\right]\right\}\Delta^2 B_{-1}$$
$$+ \left[\frac{u\,(u-1)}{2} + \frac{u\,(u-1)(u-2)}{6}\right]\left(\frac{\Delta^3 B_{-2} + \Delta^3 B_{-1}}{2}\right)$$
$$+ \left(\frac{u\,(u-1)}{4} + \frac{3u\,(u-1)(u-2)}{12} + \frac{u\,(u-1)(u-2)(u-3)}{\cdot}\right)$$
$$\Delta^4 B_{-2} + \ldots \qquad \textbf{4-12}$$

or

$$B = B_o + u\left(\frac{\Delta B_{-1} + \Delta B_o}{2}\right) + \frac{u^2}{2}\,\Delta^2 B_{-1} + \frac{u\,(u^2-1)}{3!}\,\frac{\Delta^3 B_{-2} + \Delta^3 B_{-1}}{2}$$
$$+ \frac{u^2\,(u^2-1)\Delta^4 B_{-2}}{4!} + \ldots \qquad \textbf{4-13}$$

By continuing the process to more terms, Stirling's central-difference interpolation formula is derived. The formula is:

$$B = B_o + u\,\frac{\Delta B_{-1} + \Delta B_o}{2} + \frac{u^2}{2}\,\Delta^2 B_{-1} + \frac{u\,(u^2-1)}{3!}\,\frac{\Delta^3 B_{-2} + \Delta^3 B_{-1}}{2}$$
$$+ \frac{u^2\,(u^2-1)}{4!}\,\Delta^4 B_{-2} + \frac{u\,(u^2-1^2)(u^2-2^2)}{5!}\,\frac{\Delta^5 B_{-3} + \Delta^5 B_{-2}}{2}$$

$$+\frac{u^2(u^2-1)(u^2-2^2)}{6!}\Delta^6 B_{-3}+\ldots$$

$$+\left[\frac{u(u^2-1^2)(u^2-2^2)(u^2-3^2)\ldots u^2-(n-1)^2}{(2n-1)!}\right]$$

$$\left(\frac{\Delta^{2n-1}B_{-n}+\Delta^{2n-1}B_{-(n-1)}}{2}\right)$$

$$+\frac{u^2(u^2-1)(u^2-2^2)(u^2-3^2)\ldots[u^2-(n-1)^2]}{(2n)!}\Delta^{2n}B_{-n} \qquad \textbf{4-14}$$

There are $(2n+1)$ terms in Eq. **4-14** and this equation coincides the true function at $(2n+1)$ points.

Lagrange's Interpolation Formula

The three interpolation formulas which have already been derived are for equi-spaced sets of values of functions. For sets of values which are not equi-spaced several interpolation formulas have been developed, one of which is the Lagrange interpolation formula. There follows a development of this formula in the (P, B) notation. The given function is replaced by an nth degree polynomial of the form:

$$\begin{aligned}\phi(P) &= a_0(P\text{-}P_1)(P\text{-}P_2)(P\text{-}P_3)\cdots(P\text{-}P_n)\\ &+ a_1(P\text{-}P_0)(P\text{-}P_2)(P\text{-}P_3)\cdots(P\text{-}P_n)\\ &+ a_2(P\text{-}P_0)(P\text{-}P_1)(P\text{-}P_3)\cdots(P\text{-}P_n)+\cdots\\ &+ a_n(P\text{-}P_0)(P\text{-}P_1)(P\text{-}P_3)\cdots(P\text{-}P_{n-1})\end{aligned} \qquad \textbf{4-15}$$

In Equation **4-15** there are $n+1$ terms with n factors in each term and $n+1$ constants. To find expressions for the constants the identities $\phi(P_0)=B_0$, $\phi(P_1)=B_1$, $\cdots$, $\phi(P_n)=B_n$ are formed. Now setting $P=P_0$ and $\phi(P_0)=B_0$ gives

$$B_0 = a_0(P_0-P_1)(P_0-P_2)\ldots(P_0-P_n)$$

or

$$a_0=\frac{B_0}{(P_0-P_1)(P_0-P_2)\ldots(P_0-P_n)};$$

also setting $P=P_1$, and $\phi(P_1)=B_1$ gives

$$B_1 = a_1 (P_1 - P_o)(P_1 - P_2) \ldots (P_1 - P_n)$$

or

$$a_1 = \frac{B_1}{(P_1 - P_o)(P_1 - P_2) \ldots (P_1 - P_n)}$$

In a similar manner the other constants for Equation **4-15** are determined, for instance

$$a_2 = \frac{B_2}{(P_2 - P_o)(P_2 - P_1)(P_2 - P_3) \ldots (P_2 - P_n)}$$

and

$$a_n = \frac{B_n}{(P_n - P_o)(P_n - P_1) \ldots (P_n - P_{n-1})}$$

Substituting expressions for the constants in Equation **4-15** gives

$$\begin{aligned}\phi(P) = {} & \frac{(P - P_1)(P - P_2) \ldots (P - P_n)}{(P_o - P_1)(P_o - P_2) \ldots (P_o - P_n)} B_o \\ & + \frac{(P - P_o)(P - P_2) \ldots (P - P_n)}{(P_1 - P_o)(P_1 - P_2) \ldots (P_1 - P_n)} B_1 \\ & + \frac{(P - P_o)(P - P_1)(P - P_3) \ldots (P - P_n)}{(P_2 - P_o)(P_2 - P_1)(P_2 - P_3) \ldots (P_2 - P_n)} B_2 \\ & + \ldots \\ & + \frac{(P - P_o)(P - P_1) \ldots (P - P_{n-1})}{(P_n - P_o)(P_n - P_1) \ldots (P_n - P_{n-1})} B_n \end{aligned} \qquad \textbf{4-16}$$

which is Lagrange's interpolation formula.

Notes Concerning Interpolation

The interpolation formulas were derived in terms of petroleum engineering symbols, using symbols P and B. If the P and B symbols were replaced with x and y, then the interpolation formulas would be in the notations of mathematicians. There are interpolation formulas other than the four which have been developed. If one is interested in pursuing the subject of interpolation further, several good references on the subject are given at the end of the chapter. In general, errors in tabular values increase with successive differences.

Differences and Derivatives

Frequently, the two terms, differences and derivatives, may become confused; then it becomes desirable to know the relations which exist between differences and derivatives. *Differences* refers to differences between two consecutive values of a function or differences between two consecutive differences of the same order; whereas *derivatives* refers to differences divided by increments raised to the power of the derivative as the increments approach zero. Perhaps an illustration for computing the differences of a polynomial and an illustration for determining the derivatives of Newton's forward interpolation formula will further explain the relationship between differences and derivatives.

The illustration for computing the differences of an nth order polynomial follows. If the nth order polynomial is

(a) $B = f(P) = aP^n + bP^{n-1} + cP^{n-2} + \ldots + kP + q$

then the expression $B + \Delta B$ is

(b) $B + \Delta B = a(P + h)^n + b(P + h)^{n-1} + c(P + h)^{n-2} + \cdots + k\,(P + h) + q$

where $h = \Delta P$.

Subtracting (a) from (b) gives

(c) $\Delta B = a[(P + h)^n - P^n] + b[(P + h)^{n-1} - P^{n-1}] + c[(P + h)^{n-2} - P^{n-2}] + \cdots + kh$. Expanding the terms $(P + h)^n$, $(P + h)^{n-1}$, $(P + h)^{n-2}$, $\cdots$, by the binomial theorem gives

$$
\begin{aligned}
\text{(d) } \Delta B = {} & a\Big[P^n + nhP^{n-1} + \frac{n(n-1)}{2}h^2P^{n-2} \\
& + \frac{n(n-1)(n-2)}{3!}h^3P^{n-3} + \ldots - P^n\Big] \\
& + b\Big[P^{n-1} + (n-1)hP^{n-2} + \frac{(n-1)(n-2)}{2}h^2P^{n-3} \\
& + \ldots - P^{n-1}\Big] \\
& + c\Big[P^{n-2} + (n-2)hP^{n-3} + \frac{(n-2)(n-3)}{2}h^2P^{n-4} \\
& + \ldots - P^{n-2}\Big] + \ldots + kh.
\end{aligned}
$$

or

$$(e)\ \Delta B = anhP^{n-1} + \left[ah^2 \frac{n(n-1)}{2} + bh(n-1)\right]P^{n-2}$$

$$+ \left[ah^3 \frac{n(n-1)(n-2)}{3!}\right.$$

$$\left. + bh^2 \frac{(n-1)(n-2)}{2} + ch(n-2)\right] P^{n-3} + \ldots$$

If $h = \Delta P$ is constant then the bracketed coefficient of P^{n-2}, P^{n-3}, etc. . . . are constants and can be replaced by the simple coefficients b^1, c^1, etc., and can be written as

$$(f)\ \Delta B = anhP^{n-1} + b'P^{n-2} + c'P^{n-3} + \cdots k'P + q'$$

The second difference is determined by considering Equations. (f) and (g) and following the same procedure as for the first difference.

(g) $\Delta B + \Delta(\Delta B) = anh\ (P + h)^{n-1} + b'\ (P + h)^{n-2} + c'\ (P + h)^{n-3} + \cdots + k'(P + h) + q'$. Then by subtracting (*f*) from (*g*), expanding the difference by the binomial theorem and simplifying gives

(h) $\Delta(\Delta B) = \Delta^2 B = an\ (n - 1)h^2\ P^{n-2} + b''P^{n-3} + c''P^{n-4} \ldots + k''P + q''$.

By continuing the procedure for each successive difference, the nth difference becomes

$$\Delta^n B = a[n(n-1)\ (n - 2 \ldots 1]\ h^n P^{n-n} = an!h^n P^o = ah^n n! \qquad \textbf{4-17}$$

It can be concluded from the illustration concerning the differences of an nth degree polynomial that the nth difference is a constant when values of the independent variable are taken in arithmetic progression, that is, at equal intervals apart. The converse is also true.

The illustration for determining the derivatives of Newton's forward interpolation formula follows:

Let

$$(a)\ u = \frac{P - P_o}{h} \text{ and } \frac{dB}{du}\frac{du}{dP} = \frac{1}{h}\frac{dB}{du} \text{ for } \frac{du}{dP} = \frac{1}{h}$$

Given Newton's forward interpolation formula

$$\text{(b)}\ B = B_o + u\Delta B_o + \frac{u(u-1)}{2!}\Delta^2 B_o + \frac{u(u-1)(u-2)}{3!}\Delta^3 B_o + \frac{u(u-1)(u-2)(u-3)}{4!}\Delta^4 B_o + \ldots$$

Expanding the coefficients of the differences which contain parentheses gives

$$\text{(c)}\ B = B_o + u\Delta B_o + \frac{u^2 - u}{2}\Delta^2 B_o + \frac{(u^3 - 3u^2 + 2u)}{3!}\Delta^3 B_o + \frac{(u^4 - 6u^3 + 11u^2 - 6u)}{4!}\Delta^4 B_o + \ldots$$

Taking the derivative of (c) with respect to P gives

$$\text{(d)}\ \frac{dB}{dP} = \frac{1}{h}\left[\Delta B_o + \frac{2u - 1}{2}\Delta^2 B_o + \frac{3u^2 - 6u + 2}{3!}\Delta^3 B_o + \frac{4u^3 + 18u^2 + 22u - 6}{4!}\Delta^4 B_o + \ldots\right]$$

The second derivative which is the derivative of (d) is

$$\text{(e)}\ \frac{d^2B}{dP^2} = \frac{1}{h^2}\left[\Delta^2 B_o + \frac{6u - 6}{3}\Delta^3 B_o\ \frac{12u^2 - 36u + 22}{4!}\Delta^4 B_o + \ldots\right]$$

The third derivative which is the derivative of (e) is

$$\text{(f)}\ \frac{d^3B}{dP^3} = \frac{1}{h^3}\left[\Delta^3 B_o + \frac{24u - 36}{4!}\Delta^4 B_o + \ldots\right]$$

The fourth derivative which is the derivative of (f) is

$$\text{(g)}\ \frac{d^4B}{dP^4} = \frac{1}{h^4}\left[\frac{24}{4!}\Delta^4 B_o + \ldots\right]$$

And the nth derivative is

$$\frac{d^nB}{dP^n} = \frac{1}{h^n}\Delta^n B_o \qquad \text{4-18}$$

which is constant.

Numerical Integration

To obtain a solution to a problem in petroleum engineering, numerical integration for a set of tabular data is often required. In the section on "Interpolation," methods for representing sets of tabular data in the form of interpolation formulas were presented. Logically this section suggests a method for numerical integration. The method is to represent the set of tabular data by an interpolation formula and then integrate the same between limits.

Since the petroleum engineer is familar with the use of graphical methods for obtaining the volume of a reservoir from isopachous data, the variables (V,A,H) are used. The formula for the volume of a reservoir is

$$V = \int_{H_o}^{H_o + nh} A\,dH \quad \text{or} \quad V = \int_{H_o}^{H_o + nh} f(H)\,dH \qquad \textbf{4-19}$$

where h represents an interval thickness, A represents area, H represents thickness, and V represents the volume.

If $H = H_o + hu$, then, $dH = hdu$ and in terms of Newton's forward interpolation formula Equation **4-19** is written as

$$\int_{H_o}^{H_o + nh} AdH = h \int_{0}^{n} \left[A_o + u\Delta A_o + \frac{u(u-1)}{2!} \Delta^2 A_o + \frac{u(u-1)(u-2)}{3!} \Delta^3 A_o + \frac{u(u-1)(u-2)(u-3)}{4!} \Delta^4 A_o + \ldots \right] du \qquad \textbf{4-20}$$

Multiplying the terms in the parentheses gives:

$$\int_{H_o}^{H_o + nh} AdH = h \int_{0}^{n} \left[A_o + u\Delta A_o + (u^2 - u) \frac{\Delta^2 A_o}{2} + (u^3 - 3u^2 + 2u) \frac{\Delta^3 A_o}{3!} + (u^4 - 6u^3 11u^2 - 6u) \frac{\Delta^4 A_o}{4!} + \ldots \right] du \qquad \textbf{4-21}$$

Intergrating Equation **4-21** gives

$$\int_{H_o}^{H_o + nh} AdH = h\left[uA_o + \frac{u^2}{2}\Delta A_o + \left(\frac{u^3}{3} - \frac{u^2}{2}\right)\frac{\Delta^2 A_o}{2} + \left(\frac{u^4}{4} - \frac{u^3}{3} + \frac{u^2}{2}\right)\frac{\Delta\, A_o}{3!}\right.$$

$$\left. + \left(\frac{u^5}{5} - \frac{3u^4}{2} + \frac{11u^3}{3} 3u^2\right)\frac{\Delta^4 A_o}{4!} + \ldots\right]\Bigg|_o^n \qquad \textbf{4-22}$$

or

$$\int_{H_o}^{H + nh} AdA = h\left[nA_o + \frac{n^2}{2}\Delta A_o + \left(\frac{n^3}{3} - \frac{n^2}{2}\right)\frac{\Delta^2 A_o}{2} + \left(\frac{n^4}{4} - n^3 + n^2\right)\frac{\Delta^2 A_o}{3!}\right.$$

$$\left. + \left(\frac{n^5}{5} - \frac{3n^4}{2} + \frac{11n^3}{3} - 3n^2\right)\frac{\Delta^4 A_o}{4!} + \ldots\right] \qquad \textbf{4-23}$$

The familiar Simpson's Rule for numerical integration is derived, by setting $n = 2$, integrating Equation **4-20**, and neglecting all differences above the second, as

$$\int_{H_o}^{H_o + 2h} AdH = h\int_0^2\left[A_o + u\Delta A_o + u(u-1)\frac{\Delta^2 A_o}{2}\right]du$$

$$= h\left[A_o u + \frac{u^2}{2}\Delta A_o + \left(\frac{u^3}{3} - \frac{u^2}{2}\right)\frac{\Delta^2 A^o}{2}\right]\Bigg|_0^2$$

$$= h\left[2A_o + 2\Delta A_o + \left(\frac{8}{3} - 2\right)\frac{\Delta^2 A_o}{2}\right] \qquad \textbf{4-24}$$

But since $\Delta A = A_1 - A_o$ and $\Delta^2 A_o = A_2 - 2A_1 - A_o$ then Equation **3-24** becomes

$$\int_{H_o}^{H + 2h} AdH = [2A_o + 2A_1 - 2A_o + 1/3\,(A_2 - 2A_1 + A_o)]$$

$$= \frac{h}{3}[A_o + 4A_1 - A_2] \qquad \textbf{4-25}$$

Similarly for the next two intervals from H_2 to $H_2 + 2h$

$$\int_{H_2}^{H_2 + 2h} AdH = \int_0^2 h\ A_2 + u\Delta A_2 + u(u-1)\frac{\Delta^2 A_2}{2}\,du$$

$$= \frac{h}{3}(A_2 + 4A_3 + A_4) \qquad \textbf{4-26}$$

and similarly for the third pair of intervals from H_4 to $H_4 + 2h$

$$\int_{H_4}^{H_4 + 2h} AdH = \int_0^2 h\left[A_4 + u\Delta A_4 + u(u-1)\frac{\Delta^2 A_4}{2}\right]du$$

$$= \frac{h}{3}\left[A_4 + 4A_5 + A_6\right] \qquad \textbf{4-27}$$

and so on. By adding expressions as Equations **4-25, 4-26** and **4-27** from H_o to H_n, where n is even, gives

$$\int_{Ho}^{Ho + nh} Adh = \frac{h}{3}\left[A_o + 4A_1 + 2A_2 + 4A_3 + 2A_4 + \ldots\right.$$

$$\left.2A_{n-2} + 4A_{n-1} + A_n\right] \qquad \textbf{4-28}$$

And Equation **4-28** is known as Simpson's Rule. This formula applies only when the interval of integration is divided into an even number of subintervals of width h.

Another integration formula known as Weddle's Rule is obtained by setting $n = 6$ in Equation **4-20** and neglecting all differences above the sixth as

$$\int_{H_o}^{H_o + 6h} AdH = h\left[6A_o + 18\Delta A_o + 27\Delta^2 A_o + 24\Delta^3 A_o + \frac{123}{10}\Delta^4 A_o\right.$$

$$\left.+ \frac{33}{10}\Delta^5 A_o + \frac{41}{140}\Delta^6 A_o\right] \qquad \textbf{4-29}$$

Since the coefficient $\left(\frac{41}{140}\right)$ of the last term differs by $\frac{1}{140}$ from $\frac{3}{10}$ it is replaced by $\frac{3}{10}$. With the changing of the last coefficient to $\frac{3}{10}$ and simplifying, Equation **4-29** becomes

$$\int_{H_o}^{H_o + 6h} AdH = \frac{3h}{10}\left[A_o + 5A_1 + A_2 + 6A_3 + A_4 + 5A_5 + A_6\right] \tag{4-30}$$

In a similar manner the integration for the next six intervals (H_6 to H_{12}) becomes

$$\int_{H_6}^{H_6 + 6h} AdH = \frac{3h}{10}\left[A_6 + 5A_7 + A_8 + 6A_9 + A_{10} + 5A_{11} + A_{12}\right] \tag{4-31}$$

Adding expressions such as Equations **4-30** and **4-31** for H_o to H_n where n is a multiple of 6 gives

$$\int_{H_o}^{H_o + nh} AdH = \frac{3h}{10}\Big[A_o + 5A_1 + A_2 + 6A_3 + A_4 + 5A_5 + 2A_6 + 5A_7$$
$$+ A_8 + 6A_9 + A_{10} + 5A_{11} + 2A_{12} + \ldots$$
$$+ 2A_{n-6} + 5A_{n-5} + A_{n-4} + 6A_{n-3} + A_{n-2} + 5A_{n-1} + A_n\Big] \tag{4-32}$$

Weddle's Rule is more accurate, in general, than Simpson's Rule; however, it requires at least seven consecutive values of the function.

The use of Simpson's Rule for calculating the bulk volume of a reservoir from isopachous data is illustrated in Example 4-3.

Example 4-3. Given isopachous map data in Table 4-8, calculate with the use of Simpson's Rule the bulk volume of an idealized reservoir for h equal to 10 feet.

Table 4-8
Isopachous Map Data

Productive Area	A_0	A_1	A_2	A_3	A_4	A_5	A_6
Area Acres	450	375	303	231	154	74	2

Solution.

$$V_b = \frac{h}{3}\left[A_o + 4A_1 + 2A_2 + 4A_3 + 2A_4 + 4A_5 + A_6\right]$$

$$= \frac{10}{3}\left[450 + 4 \times 375 + 2 \times 303 + 4 \times 231 + 2 \times 154 + 4 \times 74 + 2\right]$$

$$= \frac{10}{3}\left[450 + 1500 + 606 + 924 + 308 + 296 + 2\right]$$

$$= \frac{40860}{3} = 13{,}620 \text{ acre-ft.}$$

Simultaneous Linear Equations

A method is developed for solving problems in petroleum engineering which requires desk calculator solutions of simultaneous linear equations when the number of equations is greater than three and less than ten. The method can be readily programmed for microcomputers; however, computer programs are available for solving larger numbers of equations. The method is developed for three equations with three unknowns; however, the method will apply for solving n simultaneous equations with n unknowns. In the development of the method for solving the equations, let Equations (a), (b) and (c) represent the three equations:

(a) $A_1X + B_1Y + C_1Z = D_1$

(b) $A_2X + B_2Y + C_2Z = D_2$

(c) $A_3X + B_3Y + C_3Z = D_3$

Dividing (a) by A_1 gives

$$\text{(d)}\ X + \frac{B_1}{A_1} Y + \frac{C_1}{A_1} Z = \frac{D_1}{A_1}$$

Solving (d) for X gives

$$\text{(e)}\ X = \frac{D_1}{A_1} - \frac{B_1}{A_1} Y - \frac{C_1}{A_1} Z$$

Substituting (e) in (b) gives

$$\text{(f)}\ A_2\left(\frac{D_1}{A_1} - \frac{B_1Y}{A_1} - \frac{C_1Z}{A_1}\right) + B_2Y + C_2Z = D_2$$

or

$$\text{(g)}\ \left(B_2 - \frac{A_2B_1}{A_1}\right) Y + \left(C_2 - \frac{A_2C_1}{A_1}\right) Z = \left(D_2 - \frac{A_2D_1}{A_1}\right)$$

Similarly (c) reduces to

$$\text{(h)}\ \left(B_3 - \frac{A_3B_1}{A_1}\right) Y + \left(C_3 - \frac{A_3C_1}{A_1}\right) Z = \left(D_3 - \frac{A_3D_1}{A_1}\right)$$

Now setting

$$\text{(i)}\ B_1' = \frac{B_1}{A_1};\ C_1' = \frac{C_1}{A_1};\ D_1' = \frac{D_1}{A_1}$$

$$B_2' = \left(B_2 - \frac{A_2B_1}{A_1}\right);\ C_2' = \left(C_2 - \frac{A_2C_1}{A_1}\right);\ D_2' = \left(D_2 - \frac{A_2D_1}{A_1}\right)$$

$$B_3' = \left(B_3 - \frac{A_3B_1}{A_1}\right);\ C_3' = \left(C_3 - \frac{A_3C_1}{A_1}\right);\ D_3' = \left(D_3 - \frac{A_3D_1}{A_1}\right)$$

With the substitution of the identities (i) in (d), (g), and (h) the original equations reduce to

$$\text{(j)}\ X + B_1'Y + C_1'Z = D_1'$$

$$\text{(k)}\ O + B_2'Y + C_2'Z = D_2'$$

$$\text{(l)}\ O + B_3'Y + C_3'Z = D_3'$$

Dividing (k) by B_2' gives

$$\text{(m)}\quad Y + \frac{C_2'}{B_2'} Z = \frac{D_2'}{B_2'}$$

or

$$\text{(n)}\quad Y = \frac{D_2'}{B_2'} - \frac{C_2'}{B_2'} Z$$

Substituting (n) into (j) and (l) gives

$$\text{(o)}\quad X + B_1'\left(\frac{D_2'}{B_2'} - \frac{C_2'}{B_2'} - Z\right) + C_1'Z = D_1'$$

or

$$\text{(p)}\quad X + \left(C_1' - \frac{B_1'C_2'}{B_2'}\right)Z = D_1' - \frac{B_1'D_2'}{B_2'}$$

Similarly

$$\text{(q)}\quad B_3'\left(\frac{D_2'}{B_2'} - \frac{C_2'}{B_2'} Z\right) + C_3'Z = D_3'$$

$$\text{(r)}\quad \left(C_3' - \frac{B_3'C_2'}{B_2'}\right)Z = D_3' - \frac{B_3'D_2'}{B_2'}$$

The three original equations now reduce to

$$\text{(s)}\quad X + \left(C_1' - \frac{B_1'C_2'}{B_2'}\right)Z = D_1' - \frac{B_1'D_2'}{B_2'}$$

$$\text{(t)}\quad Y + \frac{C_2'}{B_2'} Z = \frac{D_2'}{B_2'}$$

$$\text{(u)}\quad \left(C_3' - \frac{B_3'C_2'}{B_2'}\right)Z = D_3' - \frac{B_3'D_2'}{B_2'}$$

Again setting

$$\text{(v)}\quad C_1'' = C_1' - \frac{B_1'C_2'}{B_2'};\ C_2'' = \frac{C_2'}{B_2'};\ D_1'' = D_1' - \frac{B_1'D_2'}{B_2'}$$

$$D_2'' = \frac{D_2'}{B_2'};\ C_3'' = C_3' - \frac{B_3'C_2'}{B_3'};\ D_3'' = D_3' - \frac{B_3'D_2'}{B_2'}$$

With the identities of (v), Equations (s), (t), and (u) become

(w) $X + C_1''Z = D_1''$

(x) $Y + C_2''Z = D_2''$

(y) $C_3''Z = D_3''$

Dividing Equation (Y) by C_3'' gives

(z) $Z = \dfrac{D_3''}{C_3''}$

and substituting (z) into (w) and (x) gives

(aa) $X + \left(C_1'' - \dfrac{D_3''}{C_3''}\right) = D_1''$

(bb) $Y + \left(C_2'' - \dfrac{D_3''}{C_3''}\right) = D_2''$

or

(cc) $X = D_1'' - C_1' \dfrac{D_3''}{C_3''}$

(dd) $Y = D_2'' - \dfrac{C_2''\, D_3''}{C_3''}$

(ee) $Z = \dfrac{D_3''}{C_3''}$

The final set of equations gives the values for the unknowns (x, y, z). Written in matrix form the method becomes

(ff) $$\begin{pmatrix} A_1 & B_1 & C_1 & D_1 \\ A_2 & B_2 & C_2 & D_2 \\ A_3 & B_3 & C_3 & D_3 \end{pmatrix}$$

(gg) $$\begin{pmatrix} A_1' & B_1' & C_1' & D_1' \\ 0 & B_2' & C_2' & D_2' \\ 0 & B_3' & C_3' & D_3' \end{pmatrix}$$

(hh) $$\begin{pmatrix} A_1'' & 0 & C_1'' & D_1'' \\ 0 & B_2'' & C_2'' & D_2'' \\ 0 & 0 & C_3'' & D_3'' \end{pmatrix}$$

(ii) $$\begin{pmatrix} 1 & 0 & 0 & D_1''' \\ 0 & 1 & 0 & D_2''' \\ 0 & 0 & 1 & D_3''' \end{pmatrix}$$

Generalizing the procedure for proceeding from one matrix to the next matrix is according to the following rules:

1. First, select a variable in a row to be retained.
2. Then, divide all elements of the row by the coefficient of the selected variable.
3. These quotients are the new elements for the corresponding row of the next matrix.
4. Next, select an element in one of the other rows to be evaluated. Call the old value H.
5. Move sideways in the old matrix to the column which contained the coefficients of Rule 1 and note its value. Call it K.
6. Go to the element in the new matrix corresponding to the H element of Rule 4. Then go up or down in this column to the values calculated by Rule 2. Call this value L.
7. Multiply K by L.
8. Subtract KL from H and post this value for the appropriate element in the new matrix.

The procedure is continued until a coefficient of an unknown has been selected for each element of the new matrix.

An illustration of the application of the rules to finding the solution for four simultaneous linear equations is given in Example 4-4.

Example 4-4. Apply the rules for determining the solution of simultaneous equations and find the solution for the four equations:

$$X + Y + Z + W = 5$$

$$X - Y + Z + W = 1$$

$$X + 2Y - Z + W = 1$$

$$2X + Y + Z + W = 6$$

Solution. By applying the rules, the following matrices were obtained:

$$\begin{pmatrix} 1 & 1 & 1 & 1 & 5 \\ 1 & -1 & 1 & 1 & 1 \\ 1 & 2 & -1 & 1 & 1 \\ 2 & 1 & 1 & 1 & 1 \end{pmatrix} \quad \text{(A)}$$

$$\begin{pmatrix} 1 & 1 & 1 & 1 & 5 \\ 0 & -2 & 0 & 0 & -4 \\ 0 & 1 & -2 & 0 & -4 \\ 0 & -1 & -1 & -1 & -4 \end{pmatrix} \quad \text{(B)}$$

$$\begin{pmatrix} 1 & 0 & 1 & 1 & 3 \\ 0 & 1 & 0 & 0 & 2 \\ 0 & 0 & -2 & 0 & -6 \\ 0 & 0 & -1 & -1 & -2 \end{pmatrix} \quad \text{(C)}$$

$$\begin{pmatrix} 1 & 0 & 0 & 1 & 0 \\ 0 & 1 & 0 & 0 & 2 \\ 0 & 0 & 1 & 0 & 3 \\ 0 & 0 & 0 & -1 & 1 \end{pmatrix} \quad \text{(D)}$$

$$\begin{pmatrix} 1 & 0 & 0 & 0 & 1 \\ 0 & 1 & 0 & 0 & 2 \\ 0 & 0 & 1 & 0 & 3 \\ 0 & 0 & 0 & 1 & -1 \end{pmatrix} \quad \text{(E)}$$

From the matrix (E) it is obvious that $X = 1$, $Y = 2$, $Z = 3$, and $W = -1$. Errors are eliminated if elements of a whole row are calculated at one time as indicated below:

Row 2 Matrix (B)

H	K	L	KL	$\triangle$
−1	1	1	1	−2
1	1	1	1	0
1	1	1	1	0
1	1	5	5	−4

Row 3 Matrix (B)

2	1	1	1	1
−1	1	1	1	−2
1	1	1	1	0
1	1	5	5	−4

Row 4 Matrix (B)

1	2	1	2	−1
1	2	1	2	−1
1	2	1	2	−1
6	2	5	10	−4

Row 1 Matrix (C)

H	*K*	*L*	*KL*	△
1	1	0	0	1
1	1	0	0	1
5	1	2	2	3

Row 3 Matrix (C)

−2	1	0	0	−2
0	1	0	0	0
−4	1	2	2	−6

Row 4 Matrix (C)

−1	−1	0	0	−1
−1	−1	0	0	−1
−4	−1	2	−2	−2

Row 1 Matrix (D)

H	*K*	*L*	*KL*	△
1	1	0	0	1
3	1	3	3	0

Row 2 Matrix (D)

$$\begin{matrix} 0 & 0 & 0 & 0 & 0 \\ 2 & 0 & 3 & 0 & 2 \end{matrix}$$

Row 4 Matrix (D)

$$\begin{matrix} -1 & -1 & 0 & 0 & -1 \\ -2 & -1 & 3 & -3 & +1 \end{matrix}$$

Row 1 Matrix (E)

$$\begin{matrix} 0 & 1 & -1 & -1 & +1 \end{matrix}$$

Row 2 Matrix (E)

$$\begin{matrix} 2 & 0 & -1 & 0 & 2 \end{matrix}$$

Row 3 Matrix (E)

$$\begin{matrix} 3 & 0 & -1 & 0 & 3 \end{matrix}$$

This method for solving simultaneous linear equations requires a large number of calculations. Consequently, when the number of equations is large, round-off errors are likely to show up in the values obtained for the variables unless a sufficient number of significant digits is used during the calculations.

Numerical Solutions of Ordinary Differential Equations

The simplest form of Muskat's Material Balance Equation is

$$\frac{dS_o}{dP} = \frac{\dfrac{S_o}{B_o B_g}\dfrac{dR_s}{dP} + \dfrac{S_o}{B_o}\dfrac{k_g}{k_o}\dfrac{\mu_o}{\mu_g}\dfrac{dB_o}{dP} + (1 - S_o - S_w)\dfrac{1}{B_g}\dfrac{dB_g}{dP}}{1 + \dfrac{k_g}{k_o}\dfrac{\mu_o}{\mu_g}}$$

Since all the terms on the right side of the equal sign are functions of either pressure or fluid saturation, the equation reduces to a first-order ordinary differential equation as

$$\frac{dS_o}{dP} = f\,(P, S) \qquad \textbf{3-34}$$

Five methods for solving ordinary differential equations of the first order are given. These methods are known as Euler's Method, Picard's Method, Approximating Polynomial Method, Milne's Method, and Runge-Kutta Method.

Relatively few differential equations can be solved analytically. However, there are methods for finding, with a high degree of accuracy, numerical solutions of any ordinary differential equation with given initial conditions. Beginning with the initial values, the solutions involve the calculations of values ahead for equal and small intervals, $\Delta P = h$, of P. Usually, each step is checked by some method before proceeding to the next step.

Euler's Method. Generally during the first course in reservoir engineering, the petroleum engineer is taught to solve Muskat's material balance equation by simplified form of Euler's Method for solving first order differential equations. While performing the calculations for a depletion-drive performance curve the values of the points on the curve are computed one at a time. Starting from initial condition, an incremental value of oil produced is computed for an incremental drop in pressure. Then, starting with this computed value of oil produced, the next incremental value of oil is computed for another drop in pressure. The process of computing incremental values of oil produced for incremental drops of pressure is continued until the pressure has dropped to that of atmospheric pressure.

The ordinary differential equation of the first order for consideration is written in symbolic form as

$$\frac{dS}{dP} = f\,(P,S) \qquad \textbf{4-34}$$

The integral of Equation **4-34** can be written as

$$S = F\,(P)$$

where S is a function of P. The graph for Equation **4-35** is given in Figure 4-1. The true solution to Equation **4-35** is represented by the smooth curve.

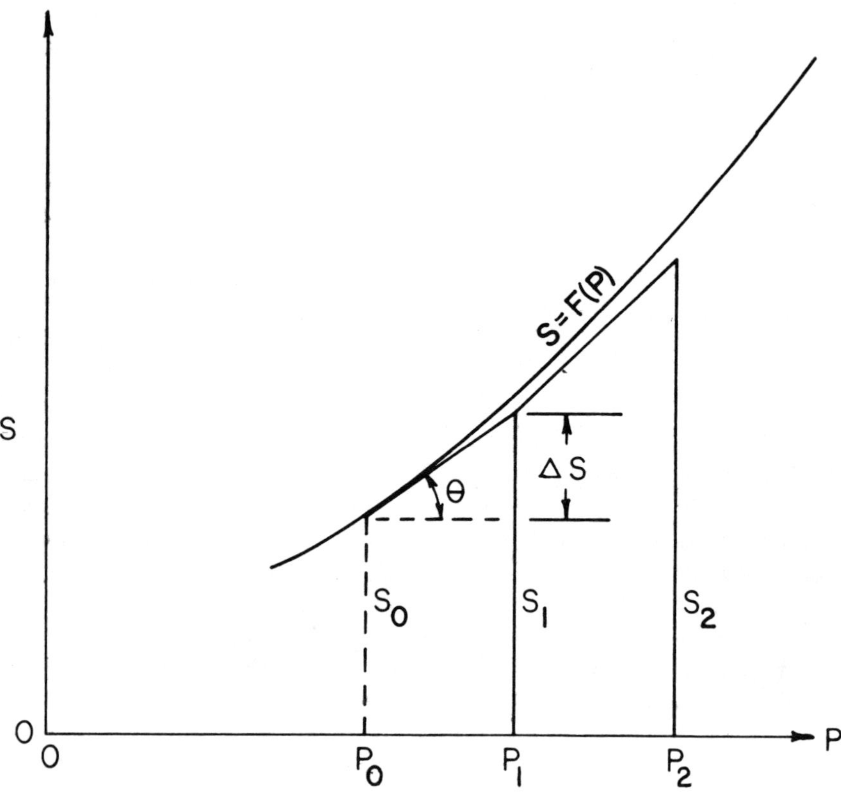

Figure 4-1. Approximate solutions of first order ordinary differential equation by Euler's method.

For a smooth curve, small sections are approximately straight; therefore,

$$\Delta S = \Delta P \tan\theta = \left(\frac{dS}{dP}\right) \Delta P$$

and

$$S_1 = S_o + \left(\frac{dS}{dP}\right)_0 \Delta P$$

or

$$S_1 = S_o + \left(\frac{dS}{dP}\right)_o h \qquad \textbf{4-36}$$

Since $\Delta P = h$, other values of S corresponding to $P_2 = P_1 + h, P_3 = P_2 + h$, etc., can be represented as

$$S_2 = S_1 + \left(\frac{dS}{dP}\right)_1 h \qquad \textbf{4-37}$$

$$S_3 = S_2 + \left(\frac{dS}{dP}\right)_2 h, \text{ etc.} \qquad \textbf{4-38}$$

Step by step in increments of $\Delta P = h$, values for points on the curve $S = F(P)$ are computed. First, a value for S_1 is computed by using the initial values for (P_o, S_o); next, a value for S_2 is computed by using the computed values of S_1; and this process is continued until a value for S_n is computed by using the computed value for S_{n-1}. In general the smaller the increment h the more accurate the computed values. But for very small values of h, sometimes the computed values for S are inaccurate. In order that more accurate values of S can be obtained Euler's Method has been modified.

The modification of Euler's Method takes into consideration the average of two slopes. A procedure for computing S values by the modified method follows. With the initial values P_o, S_o an approximate value for S_1 is computed from the relations

$$S_1^{(1)} = S_o + \left(\frac{dS}{dP}\right)_o h \qquad \textbf{4-39}$$

where

$$\left(\frac{dS}{dP}\right) = f(P_o, S_o)$$

Next this approximate value of S_1 is use for estimating the slope or

$\frac{dS}{dP}$ at S_1 is

$$\left(\frac{dS}{dP}\right)_1^{(1)} = f(P_1, S_1^{(1)}) \qquad \textbf{4-40}$$

and then an improved value of S_1 is computed by using the average value for the slopes at P_o and P_1 by the formula

$$S_1^{(2)} = S_o + \frac{\left(\frac{dS}{dP}\right)_o + \left(\frac{dS}{dP}\right)_1^{(1)}}{2} h \qquad \textbf{4-41}$$

The improved value of $S_1{}^{(2)}$ is now used for computing a second value of $\left(\frac{dS}{dP}\right)_1$ by the formula

$$\left(\frac{dS}{dP}\right)_1^{(2)} = f(P_1, S_1^{(2)}) \qquad \textbf{4-42}$$

and with the second approximate value of $\left(\frac{dS}{dP}\right)_1$ a second improved value for S_1 is calculated by the formula

$$S_1^{(2)} = S_0 + \frac{\left(\frac{dS}{dP}\right)_o + \left(\frac{dS}{dP}\right)_1^{(2)}}{2} h \qquad \textbf{4-43}$$

The process of computing values for S_1 from the average of two slopes is continued until there is no change in the value of S_1 to the required number of digits.

After a value of S_1 is found the next interval corresponds to $P_2 = P_1 + h$. Using P_1, S_1 as initial values, then a value for S_2 is computed in a similar manner as S_1 was computed. The process of computing values of S continues until S_n is computed.

The solution of a first-order ordinary differential equation by Euler's Method is illustrated with Example 4-5.

Example 4-5. Compute two values of S for the differential equation

$$\frac{dS}{dP} = (2P + S)$$

with the initial conditions $P_o = 0, S_o = 1$, by taking $h = 0.05$.

Solution: The initial slope at (P_0, S_0) is

$$\left(\frac{dS}{dP}\right)_o = 2P_o + S_o = 0 + 1 = 1$$

Using Equation **4-39** and the computed value of $\frac{dS}{dP_o}$ gives

$$S_1^{(1)} = S_o + \left(\frac{dS}{dP}\right)_o h = 1 + (0.05) = 1.05$$

The slope at $P_o + h$ or P_1 is

$$\left(\frac{dS}{dP}\right)^{(1)} = 2P_1 + S_1^{(1)}$$
$$= 2 \times 0.05 + 1.05 = 1.15$$

and

$$S_1^{(2)} = 1 + \left(\frac{1 + 1.15}{2}\right)(0.05)$$
$$= 1 + 0.05375 = 1.05375$$

Similarly

$$\left(\frac{dS}{dP}\right)_1^{(2)} = (2 \times 0.05 + 1.05375)$$
$$= (0.10 + 1.05375) = 1.15375$$

and

$$S_1^{(3)} = 1 + \left(\frac{1 + 1.15375}{2}\right) 0.05 = 1 + 0.053844 = 1.053844$$

and

$$S_1^{(4)} = 1 + \left(\frac{1 + 1.153844}{2}\right) 0.05 = 1 + 0.053846 = 1.053846$$

Therefore, by rounding off for six digits, $S_1{}^{(3)}$ and $S_1{}^{(4)}$ are equal. As a first approximation of the second value for S

$$\begin{aligned} S_2^{(1)} &= S_1 + \left(\frac{dS}{dP}\right)_1 h \\ &= 1.05384 + (2 \times 0.05 + 1.05384)\ 0.05 \\ &= 1.05384 + 1.15384 \times 0.05 \\ &= 1.05384 + 0.05796 = 1.11153 \end{aligned}$$

Now

$$\begin{aligned} \left(\frac{dS}{dP}\right)_2^{(1)} &= 2P_2 + S_2^{(1)} = 2 \times 0.10 + 1.11153 = 1.31153 \\ S_2^{(2)} &= 1.05384 + \left(\frac{1.15384 + 1.31153}{2}\right) 0.05 \\ &= 1.115474 \end{aligned}$$

and

$$\begin{aligned} \left(\frac{dS}{dP}\right)_2^{(2)} &= 2P_2 + S_2^{(2)} = 0.02 + 1.115474 = 1.315474 \\ S_2^{(3)} &= 1.05384 + \left(\frac{1.15384 + 1.315474}{2}\right) 0.05 \\ &= 1.05384 + 0.061733 = 1.115573 \end{aligned}$$

Therefore, by rounding off for five digits, $S_2{}^{(2)}$ and $S_2{}^{(3)}$ differ by one in the fourth place to the right of the decimal point.

Therefore by rounding off for for five digits, $S_2{}^{(2)}$ and $S_2{}^{(3)}$ differ by one in the fourth place to the right of the decimal point.

Picard's Method. For the equation

$$\frac{dS}{dP} = f(P, S) \tag{4-44}$$

an increment of S is

$$dS = f(P, S)\ dP \tag{4-45}$$

Integrating Equation **4-45** between corresponding limits of P and S gives

$$\int_{S_o}^{S_1} dS = \int_{P_o}^{P} f(P, S)\, dP = \int_{P_o}^{P} \left(\frac{dS}{dP}\right) dP \qquad \textbf{4-46}$$

or

$$S = S_o + \int_{P_o}^{P} f(P, S)\, dP = S_o + \int_{P_o}^{P} \left(\frac{dS}{dP}\right) dP \qquad \textbf{4-47}$$

An equation of the form

$$S = S_o + \int_{P_o}^{P} f(P, S)\, dP \qquad \textbf{4-48}$$

is known as an integral equation. Integral equations can be solved by a process of successive approximations for S; that is, if the integration can be performed in successive steps.

Picard's Method of solving differential equations such as Equation **4-44** uses successive approximations for S of the integrand as

$$S^{(1)} = S_o + \int_{P_o}^{P} f(P, S_o)\, dP \qquad \textbf{4-49}$$

Now the first approximation for S is substituted into the integrand of Equation **3-44,** which is

$$S^{(2)} = S_o + \int_{P_o}^{P} f(P, S^{(1)})\, dP \qquad \textbf{4-50}$$

The process of finding an approximate value for S and substituting the approximate value in the integrand of Equation **4-44** is continued as many times as may be desirable; the nth approximation for S is given by the equation

$$S^{(n)} = S_o + \int_{P_o}^{P} f(P, S^{(n-1)})\, dP \qquad \textbf{4-51}$$

Picard's Method is applied to a simple problem in Example 4-6.

Example 4-6. Find, by Picard's Method, the value of S at $P = 0.10$, and $P = 0.20$, for the differential equation

$$\frac{dS}{dP} = (2P + S)$$

if the initial conditions are $P_o = 0.0$ and $S_o = 1$

Solution.

$$S_1^{(1)} = 1 + \int_0^P \left(\frac{dS}{dP}\right) dP = 1 + \int_0^P (2P + 1)\, dP = P^2 + P + 1$$

and

$$S_1^{(2)} = 1 + \int_0^P (2P + P^2 + P + 1)\, dP$$

$$= 1 + \int_0^P (P^2 + 3P + 1)\, dP$$

$$= \frac{P^3}{3} + \frac{3}{2} P^2 + P + 1$$

and

$$S_1^{(3)} = 1 + \int_0^P \left(2P + \frac{P^3}{3} + \frac{3P^2}{2} + P + 1\right) dP$$

$$= 1 + \int_0^P \left(\frac{P^3}{3} + \frac{3P^2}{2} + 3P + 1\right) dP$$

$$= 1 + \frac{P^4}{12} + \frac{3P^3}{6} + \frac{3P^2}{2} + P$$

Substituting in numerical values

$$S = 1 + \frac{0.0001}{12} + \frac{0.003}{6} + \frac{0.03}{2} + 0.1$$
$$= 1 + 0.000008 + 0.0005 + 0.015 + 0.1$$
$$= 1.115508$$

The value for S obtained by Picard's Method checks to four decimal digits with the value obtained by Euler's Method.

For the second increment a value of S is calculated as

$$S_2^{(1)} = 1.1155 + \int_{0.1}^{P} (2P + 1.1155)\, dP$$

$$= (1.1155 + P^2 + 1.1155P) \Big|_{0.1}^{P}$$

$$= 1.1155 + P^2 + 1.1155P - 0.01 - 0.11155$$

$$= P^2 + 1.1155P + 0.99395$$

$$S_2^{(2)} = 1.1155 + \int_{0.1}^{P} (P^2 + 3.1155P + 0.99395)\, dP$$

$$= 1.1155 + \left(\frac{P^3}{3} + \frac{3.1155P^2}{2} + 0.99395P\right) \Big|_{0.1}^{P}$$

$$= \frac{P^3}{3} + \frac{3.1155P^2}{2} + 0.99395P + 0.9972$$

$$S_2^{(3)} = 1.1155 + \int_{0.1}^{P} \left(\frac{P^3}{3} + \frac{3.1155P^2}{2} + 2.99395P + 0.9972\right) dP$$

$$= 1.1155 + \left(\frac{P^4}{12} + \frac{3.1155P^3}{6} + \frac{2.99395}{2}P^2 + 0.9972P\right) \Big|_{0.1}^{P}$$

$$= \frac{P^4}{12} + \frac{3.1155P^3}{6} + \frac{2.99395}{2}P^2 + 0.9972P + 1.1155 - 0.000008$$
$$- 0.000519 - 0.01497 - 0.09972$$

or

$$S_2^{(3)} = \frac{P^4}{12} + \frac{3.1155P^3}{6} + \frac{2.99395}{2} P^2 + 0.9972P + 1.001337$$

At $P_2 = 0.2$ then

$$S_2^{(3)} = \frac{0.0016}{12} + \frac{0.024924}{6} + \frac{0.119758}{2} + 0.19944 + 1.001337$$

$$= 0.000133 + 0.004154 + 0.059879 + 0.1994 + 1.001337$$

$$S_2^{(3)} = 1.2649$$

Often the integral associated with the differential equations is so complex· that integration is difficult. Picard's Method is not used for differential equations of this kind.

Approximating Polynomial Method. Both Euler's and Picard's Methods for solving first order differential equations were of the form

$$S = S_o + \int_{P_o}^{P} f(P_1S)\, dP$$

or

$$\Delta S = \int_{P_o}^{P} \left(\frac{dS}{dP}\right) dP \qquad \textbf{4-52}$$

The approximating polynomial method uses the same forms. However, with this method $\frac{dS}{dP}$ is represented by an interpolation formula. Since, in solving differential equations, values ahead are usually computed. Newton's backward interpolation formula is used. Replacing $\frac{dS}{dP}$ by S' then Newton's backward interpolation formula becomes

$$S' = S_n' + u\Delta_1 S_n' + \frac{u(u+1)}{2}\Delta_2 S_n' + \frac{u(u+1)(u+1)}{6}\Delta_3 S_n'$$

$$+ \frac{u(u+1)(u+2)(u+3)}{24}\Delta_4 S_n'$$

or

$$= S_n' + \Delta_1 S_n' u + \frac{\Delta_2 S_n'}{2}(u^2 + u) + \frac{\Delta_3 S_n'}{6}(u^3 + 3u^2 + 2u)$$

$$+ \frac{\Delta_4 S_n'}{24}(u^4 + 6u^3 + 11u^2 + 6u) \qquad \textbf{4-53}$$

where

$$u = \frac{P - P_n}{h} \text{ or } P = P_n + hu$$

Substituting Equation 4-53 into Equation 4-52 gives

$$\Delta S = \int_{P_k}^{P_{k+1}} \Bigg[S_n' + \Delta_1 S_n' u + \frac{\Delta_2 S_n'}{2}(u^2 + u) + \frac{\Delta_3 S_n'}{6}(u^3 + 3u^2 + 2u) + \frac{\Delta_4 S_n'}{24}(u^4 + 6u^3 + 11u^2 + 6u) \Bigg] dP \qquad \text{4-54}$$

Since $P = P_n + hu$, $dP = hdu$. Substituting the value for dP in Equation 3-54 and changing limits accordingly, then

$$\Delta S = h\int_{u_k}^{u_{k+1}} \Bigg[S_n' + \Delta_1 S_n' u + \frac{\Delta_2 S_n'}{2}(u^2 + u) + \frac{\Delta_3 S_n'}{6}(u^3 + 3u^2 + 2u) + \frac{\Delta_4 S_n'}{24}(u^4 + 6u^3 + 11u^2 + 6u) \Bigg] du \qquad \text{4-55}$$

or

$$\Delta S = h\Bigg[S_n' u + \Delta_1 S_n' \frac{u^2}{2} + \frac{\Delta_2 S_n'}{2}\left(\frac{u^3}{3} + \frac{u^2}{3}\right) + \frac{\Delta_3 S_n'}{6}\left(\frac{u^4}{4} + u^3 + u^2\right) + \frac{\Delta_4 S_n'}{24}\left(\frac{u^5}{5} + \frac{3u^4}{2} + \frac{11u^3}{3} + 3u^2\right)\Bigg]\Bigg|_{u_k}^{u_{k+1}} \qquad \text{4-56}$$

The limits of P in terms of u are determined as

$$u_{k+1} = \frac{P_{n+1} - P_n}{h} = \frac{h}{h} = 1$$

$$u_k = \frac{P_n - P_n}{h} = 0$$

$$u_{k+2} = \frac{P_{n+1} - P_n}{h} = \frac{2h}{h} = 2, \text{ etc.}$$

Substituting in limits in Equation 4-56 and simplifying gives

$$\Delta S_n^{\ n+1} = h\left[S_n' + \frac{1}{2}\Delta S_n' + \frac{5}{12}\Delta_3 S_n' + \frac{3}{8}\Delta_2 S_n' + \frac{251}{720}\Delta_4 S_n'\right] \qquad \text{4-57}$$

Substituting other limits in Equation 4-56 and simplifying gives the following equations:

$$\Delta S_{n-1}^{\ n} = h\left[S_n' - \frac{1}{2}\Delta_1 S_n' - \frac{1}{12}\Delta_2 S_n' - \frac{1}{24}\Delta_3 S_n' - \frac{19}{720}\Delta_4 S_n'\right] \qquad \text{4-58}$$

$$\Delta S_{n-2}^{\ n-1} = h\left[S_n' - \frac{3}{2}\Delta_1 S_n' + \frac{5}{12}\Delta_2 S_n' + \frac{1}{24}\Delta_3 S_n' + \frac{11}{720}\Delta_4 S_n'\right] \qquad \text{4-59}$$

$$\Delta S_{n-3}^{\ n-2} = h\left[S_n' - \frac{5}{2}\Delta_1 S_n' + \frac{23}{12}\Delta_2 S_n' - \frac{3}{8}\Delta_3 S_n' - \frac{19}{720}\Delta_4 S_n'\right] \qquad \text{4-60}$$

$$\Delta S_{n-4}^{\ n-3} = h\left[S_n' - \frac{7}{2}\Delta_1 S_n' + \frac{53}{12}\Delta_2 S_n' - \frac{55}{24}\Delta_3 S_n' + \frac{251}{720}\Delta_4 S_n'\right] \qquad \text{4-61}$$

Replacing the coefficients $\frac{251}{720}$ and $\frac{19}{720}$ by $\frac{1}{3}$ and $\frac{1}{38}$ respectively, then adding Equations 4-57 and 4-58 gives

$$\Delta S_{n-1}^{\ n+1} = h\left[2S_n' + \frac{1}{3}\Delta_2 S_n' + \frac{1}{3}\Delta_3 S_n' + \frac{232}{720}\Delta_4 S_n'\right] \qquad \text{4-62}$$

Similarly the adding of Equations 4-58 and 4-59 gives

$$\Delta S_{n-2}^{\ n} = 2h\left[S_n' - \Delta_1 S_n' + \frac{1}{6}\Delta_2 S_n' - \frac{1}{180}\Delta_4 S_n'\right] \qquad \text{4-63}$$

Each of the Equations **4-57** to **4-63** has its place in computing values for a first-order differential equation. Equations **4-57** and **4-62** are the equations to use for computing values ahead. Equation **4-58** is used as a check on the value which was computed by either Equation **4-57** or Equation **4.62**. Equation **4-58** is also used to improve upon the value which was computed by either Equation **4-57** or **4-62**. The other equations, **4-59, 4-60,** and **4-61,** are used for the starting values of a solution.

The approximating polynomial method is applied in Example 4-7.

Example 4-7. Given the initial conditions and computed values of Example 4-5, compute for the differential equation

$$\frac{\partial S}{\partial P} = \left(2P + S\right)$$

two additional values by using the approximating polynomial method.

Solution. The difference table for the data of Example 4-5 is:

Before calculating new values, the values calculated by the Euler's Method are checked:

By Equation **4-59:**

$$\begin{aligned}\Delta S &= 0.05\left[1.3156 - \frac{3}{2}(0.1618) + \frac{5}{12}(0.0080)\right]\\ &= 0.05\left[1.3156 - 0.2427 + 0.0033\right]\\ &= 0.05 \times 1.0762 = 0.05381\end{aligned}$$

or

Table 4-9
Pressure-Saturation Data

P	S	ΔS	S′	$\Delta_1 S'$	$\Delta_2 S'$
0.0	1.00		1.0000		
0.05	1.0538	0.0538	1.1538	0.1538	
0.10	1.1156	0.0618	1.3156	0.1618	0.0080

$$S_1 = 1.0000 + 0.0538 = 1.0538$$

By Equation **4-58:**

$$\Delta S = 0.05 \left[1.3156 - \frac{1}{2} (0.1618) - \frac{1}{12} (0.0079) \right]$$

$$= 0.05 \, (1.3156 - 0.0809 - 0.0007)$$

$$= 0.05 \times 1.2342 = 0.06171$$

$$S_2 = 1.0538 + 0.06170 = 1.1155$$

With the newly computed values for S a new difference table is made in Table 4-10.

To find a value for S_3 Equation **4-57** is used as follows:

$$\Delta S = 0.05 \left[1.3155 + \frac{1}{2} (0.1617) + \frac{5}{12} (0.0079) \right]$$

$$= 0.05 \times 1.3997 = 0.07000$$

or

$$S_3 = 1.1155 + 0.0700 = 1.1855$$

Table 4-10
Computed Saturation Values

P	S	ΔS	S_1'	$\Delta_1 S'$	$\Delta_2 S'$	$\Delta_3 S'$
0.00	1.0000		1.000			
0.05	1.0538	0.0538	1.1538	0.1538		
0.10	1.1155	0.0617	1.3155	0.1617	0.0079	
0.15	1.1855	0.0700	1.4855	0.1700	0.0083	0.0004
0.20	1.2642	0.0787	1.6642	0.1787	0.0087	0.0004

Next Equation **4-58** is used to check the value obtained by Equation **3-57** as

$$S_3' = (2 \times 0.15 + 1.1855) = 1.4855$$

and

$$\Delta S_3 = 0.05\left[1.4855 - \frac{1}{2}(0.1700) - \frac{1}{12}(0.0083)\right]$$

$$= 0.05 \times 1.3998 = 0.0700$$

The computed values for ΔS check by both Equations **4-57** and **4-58.** Now the new computed value of S_3 is added to the second table.

Similarly the value for S_4 is computed by Equation **4-57:**

$$\Delta S = 0.05\left[1.4855 + \frac{1}{2}(0.1700) + \frac{5}{12}(0.0083)\right] = 0.0787$$

and

$$S_4 = 1.1855 + 0.0787 = 1.2642$$

Checking with Equation **4-58,**

$$S_4' = (2 \times 0.2 + 1.2642) = 1.6642$$

$$\Delta S = 0.05\left[1.6642 - \frac{1}{2}(0.1787) - \frac{1}{12}(0.0087)\right] = 0.0787$$

Also the computed values for S_4 check by two different equations; that is, to four decimal places.

Milne's Method. In starting a solution to a first-order ordinary differential equation the first few values are often determined by one of the three methods, Euler, Milne, or Runge-Kutta. The Runge-Kutta Method is explained in this section; the Milne Method which is explained in this article uses several Taylor series equations for its development. The series equations are derived as follows. The general form of the Taylor series equation is

$$f(P) = f(P_o) + f'(P_o)\,(P - P_o) + f''\,(P_o)\left(\frac{P - P_o}{2!}\right)^2 + \frac{f'''\,(P_o)}{3!}\,(P - P_o)^3 + \ldots \qquad 4\text{-}64$$

or

$$S = S_o + S_o'\,(P - P_o) + S''\left(\frac{P - P_o}{2!}\right)^2 + \frac{S'''}{3!}\,(P - P_o)^3 + \ldots \qquad 4\text{-}65$$

and for determining the value of S_1 Equation **4-65** becomes

$$S_1 = S_o + S_o'\,(P_1 - P_o) + \frac{S_o''}{2!}\,(P_1 - P_o)^2 + \frac{S_o'''}{3!}\,(P_1 - P_o)^3 + \ldots \qquad 4\text{-}66$$

For equal interval sets of data $P_1 - P_0 = h$, and by replacing $P_1 - P_0$ by h, Equation **4-66** becomes

$$S_1 = S_o + S_o'\,(h) + \frac{S_o''}{2}\,(h)^2 + \frac{S_o'''}{3!}\,(h)^3 + \ldots \qquad 4\text{-}67$$

Similarly

$$S_2 = S_o + S_o'\,(2h) + \frac{S_o''}{2}\,(2h)^2 + \frac{S_o'''}{3!}\,(2h)^3 + \ldots \qquad 4\text{-}68$$

$$S_{-1} = S_o - S_o'\,(h) + \frac{S_o''}{2!}\,(h)^2 - \frac{S_o'''}{3!}\,(h)^3 + \ldots \qquad 4\text{-}69$$

$$S_{-2} = S_o - S_o'\,(2h) + \frac{S_o''\,(2h)^2}{2!} - \frac{S_o'''\,(2h)^3}{3!} + \ldots \qquad 4\text{-}70$$

The use of a Taylor series equation for determining two starting values is illustrated in Example 4-8.

Example 4-8. Find the values of S_1 andS_2 by Equations **4-67** and **4-68** for the differential equation

$$\frac{dS}{dP} = (2P + S)$$

the initial condition $P = 0$, $S = 1$, and the increment $\Delta P = h = 0.05$.

Solution.

$$S_o' = (0 + 1) = 1\ ;$$

$$S_o'' = (2 + S') = (2 + 1) = 3\ ;$$

$$S_o''' = S'' = S^{IV} = 3$$

By Equation (4-67):

$$S_1 = 1 + 0.05 + 0.0038 + 0.000063 = 1.053863$$

By Equation (4-68):

$$S_2 = 1 + 0.10 + 0.015 + 0.0005 = 1.1155$$

Frequently higher derivatives of $S' = f(P,S)$ are difficult to find. In these cases the Taylor series equations cannot be used for starting the computations. A method for computing the first five values of differential equations, provided the second derivative of S can be found, was developed by W. E. Milne[2]. The method of Milne is developed as follows.

Representing the derivatives of Equations **4-67** and **4-69** as

$$S_1' = S_o' + S_o''h + \frac{S_o'''h^2}{2!} + \frac{S_o^{IV}h^3}{3!} + \frac{S_o^{V}h^4}{4!} + \ldots \qquad \textbf{4-71}$$

$$S_{-1}' = S_o' - S_o''h + \frac{S_o'''}{2!}h^2 - \frac{S_o^{IV}h^3}{3!} + \frac{S_o^{V}}{4!}h^4 + \ldots \qquad \textbf{4-72}$$

And upon adding Equations **4-71** and **4-72** gives

$$S_1' + S_{-1}' = 2S_o' + S_o'''h^2 + \frac{S_o^{V}}{12}h^4 + \ldots \qquad \textbf{4-73}$$

[2]American Mathematical Monthly, Vol 48 (1941), p. 52.

and subtracting Equation **4-72** from Equation **4-71** gives

$$S_1' - S_{-1}' = 2S_o''h + \frac{S_o^{IV}h^3}{3} + \ldots \qquad \textbf{4-74}$$

Solving Equation **4-73** for S_o''' and Equation **4-74** for S_o^{IV} and then substituting these values in Equations **4-67** and **3-69** gives

$$S_1 = S_o + \frac{h}{24}(S_{-1}' + 16S_o' + 7S_1') + \frac{S_o''h^2}{4} - \frac{S_o^V h^5}{180} \qquad \textbf{4-75}$$

$$S_{-1} = S_o - \frac{h}{24}(7S_{-1}' + 16S_o' + S_1') + \frac{S_o''h^2}{4} + \frac{S_o^V h^5}{180} \qquad \textbf{4-76}$$

Similarly the equations for S_2 and S_{-2} were obtained. These equations are:

$$S_2 = S_o + \frac{2h}{3}(5S_1' - S_o' - S_{-1}') - 2S_o''h^2 + \frac{7}{45}S_o^V h^2 \qquad \textbf{4-77}$$

$$S_{-2} = S_o - \frac{2h}{3}(5S_{-1}' - S_o' - S_1') - 2S_o''h^2 - \frac{7}{45}S_o^V h^5 \qquad \textbf{4-78}$$

Subtracting Equation **4-76** from Equation **4-75** gives

$$S_1 - S_{-1} = \frac{h}{3}(S_{-1}' + 4S_o' + S_1') - \frac{S_o^V h^5}{90} \qquad \textbf{4-79}$$

or

$$S_1 = S_{-1} + \frac{h}{3}(S_{-1}' + 4S_o' + S_1') - \frac{S_o^V h^5}{90} \qquad \textbf{4-80}$$

The general form for Equation **4-80** is

$$S_{n+1} = S_{n-1} + \frac{h}{3}(S_{n-1}' + 4S_n' + S_{n+1}') - \frac{S_o^V h^5}{90} \qquad \textbf{4-81}$$

The expression $\frac{h}{3}(S_{n-1}' + 4S_n' + S_{n+1}')$ is Simpson's Rule and it is an approximation for the definite integral

$$\int_{-h}^{h} \left(\frac{dS}{dP}\right) dP$$

which represents the increment from $P_n - h$ to $P_n + h$.

Runge-Kutta Method. The Runge-Kutta Method differs from the other methods which have been explained in the preceeding pages. This method is perhaps the easiest to use because it requires only a single initial point of the solution curve. With this method, increments of the function (or functions) are calculated by means of a definite set of equations.

The set of equations for the first increment is

$$k_1 = f(P_o, S_o)\, h$$

$$k_2 = f\left(P_o + \frac{h}{2}, S_o + \frac{k_1}{2}\right) h$$

$$k_3 = f\left(P_o + \frac{h}{2}, S_o + \frac{k_2}{2}\right) h$$

$$k_4 = f(P_o + h, S_o + k_3)\, h$$

$$\Delta S = \frac{1}{6}(k_1 + 2k_2 + 2k_3 + k_4)$$

and $P_1 = P_o + h; S_1 = S_o + \Delta S$ 4-82

And the set of equations for the second increment is

$$k_1 = f\,(P_1, S_1)\, h$$

$$k_2 = f\left(P_1 + \frac{h}{2}, S_1 + \frac{k_1}{2}\right) h$$

$$k_3 = f\left(P_1 + \frac{h}{2}, S_1 + \frac{k_2}{2}\right) h$$

$$k_4 = f\,(P_1 + h, S_1 + k_3)\, h$$

$$\Delta S = \frac{1}{6}(k_1 + 2k_2 + 2k_3 + k_4)$$

and $P_2 = P_1 + h; S_2 = S_3 + \Delta S$ 4-83

Similar sets of equations can be written for succeeding intervals.

The application of the two sets of Equations **4-82** and **4-83** is illustrated in Example 4-9.

Example 4-9. Find the values of S_1 and S_2 by the sets of Equations **4-82** and **4-83** for the differential equation

$$\frac{dS}{dP} = (2P + S),$$

with the initial conditions $P = 0$, $S = 1$, and the increment $\Delta P = h = 0.05$.

Solution. The value for S_1 is computed as follows:

$$k_1 = (0 + 1)\,0.05 = 0.05$$

$$k_2 = \left(0 + \frac{0.05}{2} + 1 + \frac{0.05}{2}\right) 0.05 = 0.0525$$

$$k_3 = \left(0 + \frac{0.05}{2} + 1 + \frac{0.0525}{2}\right) 0.05 = 0.053565$$

$$k_4 = (0 + 0.05 + 1 + 0.052565)\,0.05 = 0.055128$$

$$\Delta S = \frac{1}{6}\,(0.05 + 0.1050 + 0.10513$$

$$+ 0.055128) = \frac{1}{6}\,(0.0315258) = 0.052544$$

$$S_1 = 1.00000 + 0.05254 = 1.05254$$

The value for S_2 is computed as follows:

$$k_1 = (0.10 + 1.05254)\,0.05 = 0.05763$$

$$k_2 = \left(0.10 + \frac{0.05}{2} + 1.05254 + \frac{0.05763}{2}\right) 0.05 = 0.06027$$

$$k_3 = \left(0.10 + \frac{0.05}{2} + 1.05254 + \frac{0.06027}{2}\right) 0.05 = 0.06038$$

$$k_4 = (0.10 + 0.05 + 1.05254 + 0.06038)\,0.05 = 0.063146$$

$$\Delta S = \frac{1}{6}\,(0.5763 + 0.12054 + 0.12076 + 0.063146) = 0.06047$$

$$S_2 = 1.05254 + 0.06047 = 1.11301$$

Numerical Solutions of Partial Differential Equations

Problems concerning the flow of fluids in porous media require solutions to two partial differential equations. In this book only solutions to these two partial differential equations are considered. Problems concerning steady-state flow of fluids through porous media require solutions to the Laplace equation. The Laplace equation is

$$\nabla P = 0 \text{ or } \frac{\partial^2 P}{\partial x^2} + \frac{\partial^2 P}{\partial y^2} + \frac{\partial^2 P}{\partial z^2} = 0 \qquad \text{4-84}$$

Problems concerning unsteady-state flow through porous media require solutions to the hydraulic diffusivity equation. The hydraulic diffusivity equation is

$$\nabla^2 p = \frac{1}{\eta}\frac{\partial \rho}{\partial t} \text{ or } \frac{\partial^2 \rho}{\partial x^2} + \frac{\partial^2 \rho}{\partial y^2} + \frac{\partial^2 \rho}{\partial z^2} = \frac{1}{\eta}\frac{\partial \rho}{\partial t} \qquad \text{4-85}$$

For the two equations, P denotes pressure, x, y, z denote directions or coordinates, ρ denotes density, and η denotes hydraulic diffusivity $k/\phi c\mu$.

The developments of Equations **4-84** and **4-85** and specific applications of the two equations are given in subsequent chapters.

Laplace Equation. The equation for the pressure distribution of the one-dimensional Darcy equation is a solution to the Laplace equation in one dimension or the equation

$$\nabla^2 P = 0 \text{ or } \frac{\partial^2 P}{\partial x^2} = 0 \qquad \text{4-86}$$

One solution to Equation **4-86** takes the form

$$P = \frac{x}{L}(P_e - P_o)$$

where P denotes the pressure at any point in the x direction, L denotes overall length, P_e denotes the pressure at $x = L$, and P_o denotes the pressure at $x = 0$.

Numerical solutions to partial differential equations are often found from difference quotients. Difference quotients are similar to derivatives except the difference of the independent variable does not approach zero. For instance the expression

$$\text{(a)}\quad \frac{f(P+h) - f(P)}{h}$$

is a first difference quotient for a single variable and the expression

$$\text{(b)}\quad \frac{\dfrac{f(P+h) - f(P)}{h} - \dfrac{f(P) - f(P-h)}{h}}{h}$$

is a second difference quotient. Expression (a) is called a forward first-difference quotient. The expression

$$\text{(c)}\quad \frac{f\,(P) - f(P-h)}{h}$$

is called a backward first-difference quotient. A second-difference quotient is the difference of a forward first-difference quotient and a backward first-difference quotient divided by the difference of the independent variable.

With the explanation for difference quotients, and the coordinate system for one variable as given in Figure 4-2, the difference solution for Equation 4-86 is determined.

The forward first-difference quotient is

$$\left(\frac{\partial P}{\partial x}\right)_x = \frac{P(x+h) - P(x)}{h} \qquad \text{4-88}$$

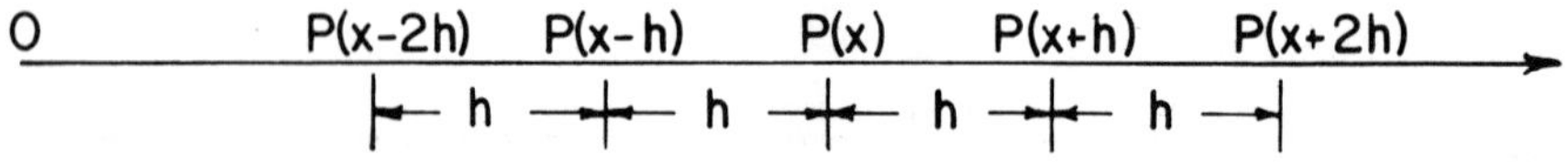

O P(x-2h) P(x-h) P(x) P(x+h) P(x+2h)

h h h h

Figure 4-2. One-dimensional coordinate system.

and the backward first-difference quotient is

$$\left(\frac{\partial P}{\partial x}\right)_{\bar{x}} = \frac{P(x) - P(x-h)}{h} \qquad \text{4-89}$$

where $P(x)$ denotes the value for the pressure at x, $P(x+h)$ denotes the pressure at $x+h$, $P(x-h)$ denotes the pressure at $x-h$ and $\Delta x = h$.

The second difference quotient is

$$\left(\frac{\partial^2 P}{\partial x^2}\right)_{\bar{x}x} = \frac{\dfrac{P(x+h) - P(x)}{h} - \dfrac{P(x) - P(x-h)}{h}}{h}$$

$$= \frac{P(x+h) + P(x-h) - 2P(x)}{h^2} \qquad \text{4-90}$$

It should be noted that $\dfrac{\partial P}{\partial x}$ approximately equals $\left(\dfrac{\partial P}{\partial x}\right)_x$ but for practical purposes, when h is taken small these, two expressions are considered equal. In a similar manner $\left(\dfrac{\partial^2 P}{\partial x^2}\right)_{\bar{x}x}$ is approximately equal to $\left(\dfrac{\partial^2 P}{\partial x^2}\right)$.

Setting Equation **4-90** equal to zero and solving for $P(x)$ gives the solution

$$P(x) = \frac{P(x+h) + P(x-h)}{2} \qquad \text{4-91}$$

An application of Equation **4-91** by the method of iteration is given in Example 4-10.

Example 4-10. For steady-state flow of oil through a sand-packed pipe which is 1000 ft. long, find the pressures at 250 , 500, and 750 ft. from the outlet, if the inlet pressure is 500 psig. and the outlet pressure is zero psig.

Solution. For steady-state flow a linear relationship exists between distance along the pipe and pressure; also it is obvious that the values for the pressures are 125 psig, 250 psig, and 375 psig. To find the pressures by Equation **4-91.** Table 4-11 is constructed and the pressures at the designated distances are estimated.

Table 4-11
Pressure Values Versus Distance from Outlet

No.	Distance from Outlet (feet) 0 (P_0)	250 (P_1)	500 (P_2)	750 (P_3)	1000 (P_e)
1.	0	120	260	380	500
2.	0	130.0	255.0	377.5	500
3.	0	127.5	252.5	376.3	500
4.	0	126.3	251.3	375.6	500
5.	0	125.7	250.7	375.4	500
6.	0	125.4	250.4	375.2	500
7.	0	125.2	250.2	375.1	550
8.	0	125.1	250.1	375.1	500
9.	0	125.0	250.0	375.0	500

The values for the pressures in the first row of the table were estimated. The values for the pressures in the other rows were computed from the equations:

$$P_1 = \frac{P_o + P_2}{2}, \; P_2 = \frac{P_1 + P_3}{2} \text{ and } P_3 = \frac{P_2 + P_e}{2}$$

The computational scheme for calculating the values for the table is:

1. P_1 of the second line was computed from P_o and P_2 of the first line.
2. P_2 of the second line was computed from the last computed value of P_1 in the second line and the value of P_3 in the first line.
3. P_3 of the second line was computed from the last computed value of P_2 in the second line and P_e of the first line. Of course P_0 and P_e remain constant.
4. P_1 of the third line was computed from the values of P_o and P_2 of the second line.
5. P_2 of the third line was computed from the value of P_1 in the third line and the value of P_3 in the second line.
6. P_3 of the third line was computed from the value of P_2 in the third line and the P_e value of the second line.
7. By a similar manner the values of P were computed for the remaining values of the table. The calculations were continued until the computed values for the P's approached their limits.

8. The digits to the right of the decimal points were omitted for values of P in the last line.

The development of the difference quotient for the two-dimensional Laplace equation

$$\frac{\partial^2 P}{\partial x^2} + \frac{\partial^2 P}{\partial y^2} = 0 \qquad \textbf{4-92}$$

follows with the aid of Figure 4-3.

The expressions such as $P(x,y,)$, $P(x+h,y)$, $P(x,y,+k)$, etc., denote pressures at nodal points of the mesh in Figure 4-3. The vertical parallel lines in the figure are spaced $\Delta x = h$ units apart and the horizontal lines are spaced $\Delta y = k$ units apart. The first forward difference quotient in the $x-$ direction is

$$\left(\frac{\partial P}{\partial x}\right)_x = \frac{P(x+h,y) - P(x)}{h} \qquad \textbf{4-93}$$

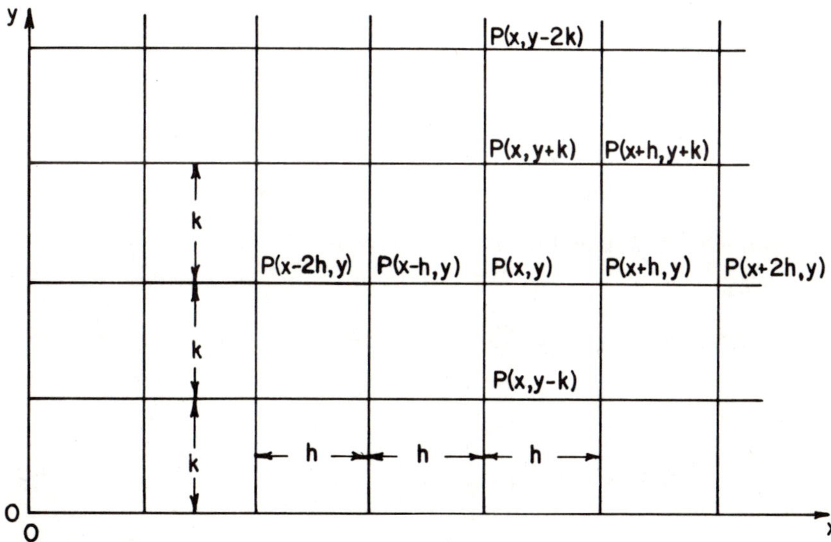

Figure 4-3. Two-dimensional coordinate system which illustrates pressures at mesh points (nodal points).

and the first backward difference in the x direction is

$$\left(\frac{\partial P}{\partial x}\right)_{\bar{x}} = \frac{P(x) - P(x - h, y)}{h} \tag{4-94}$$

Then the second difference quotient in the x direction is

$$\left(\frac{\partial^2 P}{\partial x^2}\right)_{x\bar{x}} = \frac{\dfrac{P(x + h, y) - P(x)}{h} - \dfrac{P(x) - P(x - h, y)}{h}}{h} \tag{4-95}$$

or

$$\left(\frac{\partial^2 P}{\partial x^2}\right)_{x\bar{x}} = \frac{P(x + h, y) + P(x - h, y) - 2P(x, y)}{h^2} \tag{4-96}$$

By a similar procedure, the second difference quotient in the y direction is

$$\left(\frac{\partial^2 P}{\partial y^2}\right)_{y\bar{y}} = \frac{P(x, y + k) + P(x, y - k) - 2P(x, y)}{k^2} \tag{4-97}$$

Replacing the second partial derivations of Equation **4-92** by their corresponding difference quotients, Equations **4-96** and **4-97,** gives

$$\frac{P(x + h, y) + P(x - h, y) + 2P(x, y)}{h^2} + \frac{P(x, y + k) + P(x, v - k) - 2P(x, y)}{k^2} = 0 \tag{4-98}$$

If h is set equal to k, then Equation **4-92** becomes

$$4P(x, y) = P(x + h, y) + P(x - h, y) + P(x, y + k) + P(x, y - k)$$

or

$$P(x, y) = \frac{P(x + h, y) + P(x - h, y) + P(x, y + k) + P(x, y - k)}{4} \tag{4-99}$$

which is the solution for the two-dimensional Laplace equation. Briefly, Equation **4-99** states that the pressure at any one point is the average of the four surrounding pressures taken in the x and y directions on lines which pass through the point in question.

The following example illustrates the iteration procedure for applying Equation 4-99 to a two-dimensional flow problem.

Example 4-11. For the following mesh, with values for the exterior pressures as so given, find the values for the pressures at the interior points.

Solution: The pressure points are numbered in the mesh. To begin the solution the starting values for the interior points are taken as the average of four surrounding values. Also the mesh is redrawn and the starting values for the interior points are calculated by the procedure which follows in Table 4-12.

The starting value for the center point No. (13) is the average for the sum of the value at the points Nos. (3), (15), (23), and (11) or

$$\frac{60 + 120 + 120 + 50}{4} = \frac{350}{4} = 87.5$$

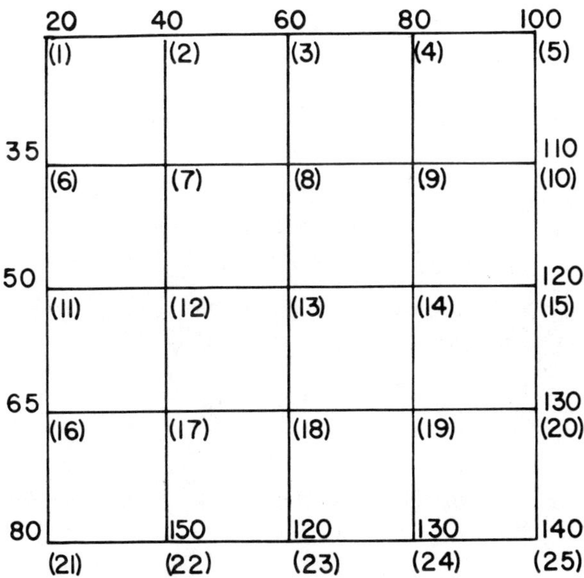

Figure 4-4. Two-dimensional mesh with exterior values.

Table 4-12
Computed Values for Interior Mesh Points

20	40	60	80	100
				110
35	54.38 54.38 54.38 54.35 54.39	73.44 73.44 73.34 73.31 73.26	91.88 91.88 91.83 91.79 91.77	
				120
50	69.07 69.06 68.88 68.84 68.80	87.5 87.19 87.06 86.97 86.92	104.07 103.99 103.85 103.80 103.78	
				130
65	84.34 84.06 83.98 83.93 83.91	102.19 102.03 101.88 101.83 101.81	116.88 116.51 116.43 116.41 116.40	
				140
80	100	120	130	

The starting value for point No. (7) is the average for the sum of the values at points Nos. (1), (3), (11), and (13) or

$$\frac{20 + 60 + 50 + 87.5}{4} = \frac{217.5}{4} = 54.38$$

The starting value for point No. (9) is the average for the sum of the values at points Nos. (3), (5), (13), and (15) or

$$\frac{60 + 100 + 87.5 + 120}{4} = \frac{367.5}{4} = 91.88$$

The starting value for point No. (8) is the average for the sum of the values at points Nos. (7), (3), (9), and (13) or

$$\frac{54.38 + 60 + 91.88 + 87.5}{4} = \frac{293.76}{4} = 73.44$$

The starting value for point No. (17) is the average for the sum of the values at points Nos. (11), (13), (21), and (23) or

$$\frac{50 + 87.5 + 80 + 120}{4} = \frac{337.5}{4} = 84.38$$

The starting value for point No. (19) is the average for the sum of the values at points Nos. (13), (15), (23), and (25) or

$$\frac{87.5 + 120 + 120 + 140}{4} = \frac{467.5}{4} = 116.88$$

The starting value for point No. (12) is the average for the sum of the values for points Nos. (11), (7), (13), and (17) or

$$\frac{50 + 54.38 + 87.5 + 84.38}{4} = \frac{276.26}{4} = 69.07$$

The starting value for point No. (14) is the average for the sum of the values for points Nos. (13), (9), (15), and (19) or

$$\frac{87.5 + 91.88 + 120 + 116.88}{4} = \frac{416.26}{4} = 104.07$$

And the starting value for point No. (18) is the average for the sum of the values at points Nos. (17), (13), (19), and (23) or

$$\frac{120 + 84.3 + 87.5 + 116.88}{4} = \frac{408.76}{4} = 102.19$$

These starting values for the interior points are used to begin the iteration process for computing the values of the interior points. To calculate a new value for an interior point, four surrounding values are used. The four surrounding values are those horizontally to the left, horizontally to the right, vertically above, and vertically below. The last computed value at a given point is used in the computation of a new value at the center point. For instance the first iterated value for point No. (7) is the average of the sum of the values at points Nos. (6), (2), (8), and (12) or

$$\frac{35 + 40 + 73.44 + 69.07}{4} = \frac{217.51}{4} = 54.37$$

The first iterated value for point No. (8) is the average for the sum of the values at points Nos. (7), (3), (9), and (13) or

$$\frac{54.38 + 60.00 + 91.88 + 87.5}{4} = \frac{293.76}{4} = 73.44$$

The first iteration value for point No. (9) is the average for the sum of the values at the points Nos. (8), (4), (10), and (14) or

$$\frac{73.44 + 80 + 110 + 104.07}{4} = \frac{367.51}{4} = 91.88$$

The first iteration value for point No. (12) is the average for the sum of the values at the points Nos. (11), (7), (13), and (17).

In a similar manner the first iteration values for points Nos. (13), (14), (17), (18), and (19) are computed. These computed values complete the computations for the first five interior points of the first cycle of the iteration.

And in a similar manner a new set of values are computed for the interior points. The cycles of the iteration process are continued until the last set of computed values for the interior points have approached their limits to as many digits to the right of the decimal point as is required for the desired accuracy. For this example, four cycles of iteration were required to give an accuracy of a maximum difference of 6 units in the second digit to the right of the decimal point for the last two sets of values of the interior points.

Hydraulic Diffusivity Equation. Solutions of the one-dimensional hydraulic diffusivity equation are required for problems concerning unsteady-state linear flow of oil through porous media. The one-dimensional hydraulic diffusivity is

$$\frac{\partial^2 p}{\partial x^2} = \frac{1}{\eta}\frac{\partial p}{\partial t} \qquad \textbf{4-100}$$

With the aid of the (x,t) coordinate as indicated in Figure 4-5 the difference quotient solution of Equation **4-100** is derived.

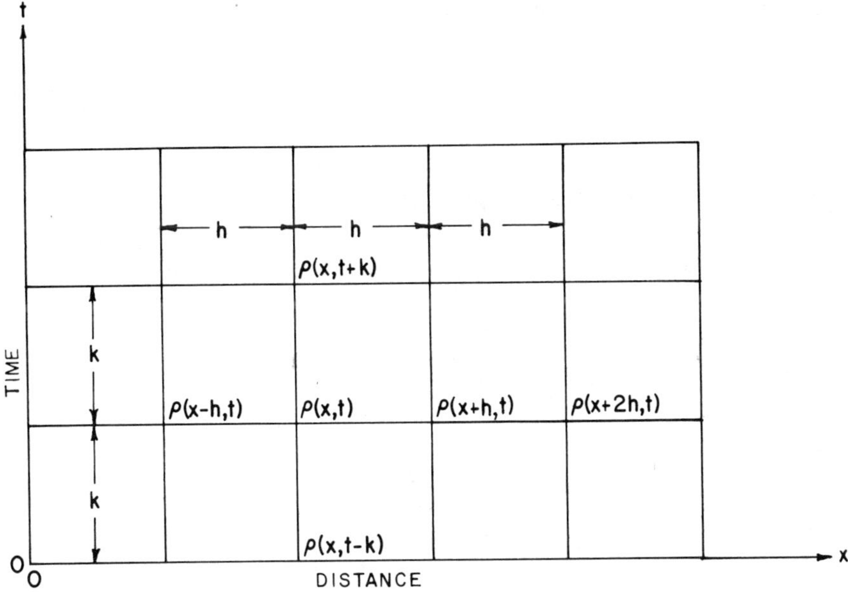

Figure 4-5. Distance-time mesh for one-dimensional hydraulic diffusivity equation.

The forward first-difference quotient in the x direction is

$$\left(\frac{\partial \rho}{\partial x}\right)_x = \frac{\rho(x + h, t) - \rho(x, t)}{h} \qquad \text{4-101}$$

The backward first-difference quotient in the x direction is

$$\left(\frac{\partial \rho}{\partial x}\right)_{\bar{x}} = \frac{\partial(x, t) - \rho(x - h, t)}{h} \qquad \text{4-102}$$

The second difference quotient in the x direction is

$$\left(\frac{\partial^2 \rho}{\partial x^2}\right)_{x\bar{x}} = \frac{\dfrac{\rho(x + h, t) - \rho(x, t)}{h} - \dfrac{\rho(x, t) - \rho(x - h, t)}{h}}{h}$$

or

$$\left(\frac{\partial^2 \rho}{\partial x^2}\right)_{x\bar{x}} = \frac{\rho(x + h, t) + \rho(x - h, t) - 2\rho(x, t)}{h^2} \qquad \text{4-103}$$

The forward first-difference quotient in the t direction is

$$\left(\frac{\partial \rho}{\partial t}\right)_t = \frac{\rho\,(x, t\ + k) - \rho\,(x, t)}{k} \quad \textbf{4-104}$$

Replacing the partial derivatives of Equation **4-85** in one dimension with their respective difference quotients gives

$$\frac{\rho\,(x + h, t) + \rho\,(x - h, t) - 2\rho\,(x, t)}{h^2} = \frac{1}{\eta}\,\frac{\rho\,(x, t + k) - \rho\,(x, t)}{k} \quad \textbf{4-105}$$

Solving for $\rho(x{,}t{+}k)$ in Eq. **4-105** gives

$$\rho\,(x, t + k) = \frac{\eta k}{h^2}\,[\rho\,(x + h, t) + \rho\,(x - h, t) - 2\rho\,(x, t)] + \rho\,(x, t) \quad \textbf{4-106}$$

If $\frac{\eta k}{h^2}$ is arbitrarily set equal to ½ then Equation **4-105** becomes

$$\rho\,(x, t + k) = \frac{\rho\,(x + h, t) + \rho\,(x - h, t)}{2} \quad \textbf{4-107}$$

which is a difference quotient solution of the one-dimensional hydraulic diffusivity equation. The value $\frac{\eta k}{h^2}$ must be ½ or less for the solution to converge.

The difference quotient solution to the one-dimensional diffusivity equation is applied in Example 4-12.

Example 4-12. A pipe of length 800 ft. is filled with sand of porosity, $\phi = 0.3$, permeability of $k = 0.2$, Darcy, contains an oil of viscosity $\mu = 3$ cps., compressibility of 3×10^{-5} atm^{-1}, and density of $\rho = 0.60$ at atmospheric conditions which is compressed to a density of 0.80. If a valve at one end of the pipe is opened to atmoshperic pressure, how long will it take for pressure waves to arrive at the closed end of the pipe?

Solution. The pipe is divided into 8 equal segments of 100 ft. ($\Delta x{=}h{=}100$) each. The increment of time $\Delta t{=}k$ now becomes

$$\Delta t = \frac{1}{2}\frac{\Delta x^2}{\eta}$$

$$= \frac{1}{2}\frac{0.3 \times 3 \times 3 \times 10^{-5}}{0.2}(100 \times 30.5)^2$$

$$= 627.8 \text{ sec or } 0.00727 \text{ days.}$$

Next a (x,t) mesh is constructed in Table 4-13.

Below the mesh the distance along the pipe is labeled in terms of feet from the open valve. The horizontal lines represent increments of time, $\Delta t = 0.00727$ days. The vertical lines indicate divisions of the pipe into increments along its length ($\Delta x = 100$ ft). The initial densities of the oil along the pipe are written above the first horizontal line at the corresponding nodal points. With the initial values for the densities of the oil, the densities of the oil at the various positions along the pipe, for the first increment of time, are computed. The computed densities of the oil are then written above the second line at the respective nodal points. A general formula for computing the densities of the oil for each increment of time is

$$\rho(x, t + \Delta t) = \frac{\rho(x + h, t) + \rho(x - h, t)}{2}$$

The density values at the end of the first increment of time are used for computing the increment values at the end of the second increment of

Table 4-13
Computed Values of Density

Time									
	0.600	0.640	0.680	0.728	0.772	0.787	0.797	0.799	0.800
	0.600	0.663	0.680	0.757	0.776	0.794	0.797	0.800	0.800
	0.600	0.663	0.725	0.757	0.788	0.794	0.800	0.800	0.800
	0.600	0.675	0.725	0.775	0.788	0.800	0.800	0.800	0.800
	0.600	0.675	0.750	0.775	0.800	0.800	0.800	0.800	0.800
	0.600	0.700	0.750	0.800	0.800	0.800	0.800	0.800	0.800
←Δt→	0.600	0.700	0.800	0.800	0.800	0.800	0.800	0.800	0.800
←Δt→	0.600	0.800	0.800	0.800	0.800	0.800	0.800	0.800	0.800
	P	100	200	300	400	500	600	70	800

Distance, feet

time. In a similar manner, the density values at the end of the $(n\text{-}1)$ increment of time are used for computing the density values at the end of the nth increment of time. The computed values for the densities are given at the respective nodal points of the (x,t) mesh. The tabulated values in the (x,t) mesh indicate that seven increments of time were required before pressure waves appeared at the closed end of the sand-packed pipe. Therefore the answer for the example is

$$7 \times 0.00727 = 0.05089 \text{ days.}$$

Solutions of the two dimensional hydraulic diffusivity equation are required for problems concerning unsteady-state two-directional flow of oil through a porous medium. The two-dimensional hydraulic diffusivity equation is

$$\frac{\partial^2 \rho}{\partial x^2} + \frac{\partial^2 \rho}{\partial y^2} = \frac{1}{\eta}\frac{\partial \rho}{\partial t} \qquad \textbf{4-108}$$

A three-dimensional mesh (x,y,t) is required for illustrating the conditions for developing a difference-quotient solution for Equation **4-108**. Since a mesh drawn in three dimensions is rather complicated, a figure containing a three-dimensional mesh will be omitted. Recalling the developments of previous difference quotients, it is not essential that the three dimensional mesh be given. From the developments of previous difference quotients, it is known that the second difference quotient in the x direction is

$$\left(\frac{\partial^2 \rho}{\partial x^2}\right)_{\bar{x}\bar{x}} = \frac{\rho\,(x+h,y,t) + \rho\,(x-h,y,t) - 2\rho\,(x,y,t)}{h^2} \qquad \textbf{4-109}$$

If the equal increments h are taken both in the x direction and the y direction, the second difference quotient in the y direction is

$$\left(\frac{\partial^2 \rho}{\partial y^2}\right)_{\bar{y}\bar{y}} = \frac{\rho\,(x,y+h,t) + \rho\,(x,y-h,t) - 2\rho\,(x,y,t)}{h^2} \qquad \textbf{4-110}$$

and the forward first-difference quotient in the t direction is

$$\left(\frac{\partial \rho}{\partial t}\right)_{t} = \frac{\rho\,(x,y,t+k) - \rho\,(x,y,t)}{k} \qquad \textbf{4-111}$$

Substituting the difference quotients of Equations **4-109, 4-110,** and **4-111** into Equation **4-108** gives

$$\frac{1}{\eta}\frac{\rho(x,y,t+k)-\rho(x,y,t)}{k}=\frac{\rho(x+h,y,t)+\rho(x-h,y,t)-2\rho(x,y,t)}{h^2}$$

$$+\frac{\rho(x,y+h,t)+\rho(x,y-h,t)-2\rho(x,y,t)}{h^2} \quad \textbf{4-112}$$

Now setting $\frac{\eta k}{h^2}=\frac{1}{4}$ or $\Delta t=\frac{h^2}{4\eta}$ and solving Equation **4-12** for $\rho(x,y,t+k)$ gives

$$\rho(x,y,t+k)$$

$$=\frac{\rho(x+h,y,t)+\rho(x-h,y,t)+\rho(x,y+h,t)+\rho(x,y-h,t)}{4} \quad \textbf{4-113}$$

which is the difference-quotient solution of the two-dimensional hydraulic diffusivity equation in terms of rectangular coordinates.

The mesh for the solution by Equation **4-113** is three dimensional, it can be represented by a series of (x,y) planes for recording the density values at the nodal point. A whole (x,y) plane is required for recording a set of density values for a given increment of time.

Exercises

1. Construct a diagonal difference table for the data:

P	100	300	500	700	900	1100	1300	1500	1700	1900	2100	2300	2500
R_s	94	194	264	320	369	412	450	486	520	553	586	618	650

2. With the data of Exercise 1 construct a horizontal difference table.

3. For the data of the table construct diagonal and horizontal difference tables.

x	0.1	0.2	0.3	0.4	0.5	0.6	0.7	0.8	0.9
y	0.1	0.8	2.7	6.4	12.5	21.6	34.3	51.2	72.9

4. Using the appropriate interpolation equations and the data of Exercise 1 find the gas solubilities of the oil (R_s) at pressures, 2400 psi, 1200 psi, and 50 psi.

5. With the isopachous data of the table find the bulk volume of the reservoir by means of Simpson's Rule if $h = 10$ ft.

Productive Area	A_0	A_1	A_2	A_3	A_4	A_5	A_6
Area Acres	800	750	612	452	312	152	10

6. With Weddle's Rule for numerical integration, find the value of the integral

$$\int_{1.0}^{9.0} \frac{z}{P_r}\, dP_r$$

for the data of the table:

P_r	1.0	2.0	3.0	4.0	5.0	6.0	7.0	8.0	9.0
z	0.705	0.585	0.550	0.628	0.715	0.806	0.915	0.992	1.075

7. Solve the first order ordinary differential equation

$$\frac{dS}{dP} = (P + S)$$

with the initial conditions $P=100$, $S=100$ for the limits $100 < P < 1000$ if $\Delta P = h = 100$.

a. By Euler's Method
b. By Picard's Method
c. By the approximating polynomial method
d. By Milne's Method
e. By Runge-Kutta Method

and compare the results by the different methods.

8. Determine the values for the interior point of the following mesh by using the iteration method for the difference quotient solution of the Laplace two-dimensional equation.

	50	250	300	400	325
	100				275
y	150				250
	200				250
	250	275	300	275	250
			x		

9. A pipe of length 800 ft., filled with sand of porosity $\phi = 0.25$, and permeability $k = 0.2$ darcy, contains an oil of viscosity $\mu_o =$ cps and compressibility $c_o = 1.5$ x 10^{-5} psi^{-1}. If the density distribution of the oil is as indicated in the (x,t) mesh, determine the time it will take before the density throughout the pipe will be 99% of the steady-state. The density at each end of the pipe is 0.80

	0.80								0.80
	0.80								0.80
	0.80								0.80
t	0.80								0.80
	0.80								0.80
	0.80								0.80
	0.80	0.75	0.70	0.65	0.60	0.65	0.70	0.75	0.80
	0	100	200	300	400	500	600	700	800
					X				

Note. Each of the problems of the previous exercises can be programmed for microcomputers and solved by the same.

10. Prepare a microcomputer (BASIC language) program for interpolating between the values given in Exercise 1. In the program utilize: (a) Newton's forward interpolation formula for computing corresponding values of R_s between $P=50$ and $P=900$, (b) Stirlings interpolation formula for computing corresponding values of R_s between $P=900$ and $P=1900$, and (c) Newton's backward interpolation formula for computing corresponding values of R_s between $P=1900$ and $P=2400$. Prepare the program so that the computer will select the appropriate formula to use.
11. Prepare a microcomputer (BASIC language) program for solving linear simultaneous equations for n equations and n unknowns. In preparing the computer program, use the method for solving linear simultaneous equations which is presented in this chapter.

Suggested Reading

1. Engineering Research Associates, Inc., supervised by C. B. Tomkins and J. H. Wakelin and edited by Stifler, W. W., Jr., *High-Speed Computing Devices,* New York: McGraw-Hill Book Company, 1950.
2. Forsythe, G. E., and Wasow, W. R., *Finite Difference Methods for Partial Differential Equations,* New York: John Wiley and Sons, Inc., 1960.
3. Grinter, L. E., *Numerical Methods of Analysis in Engineering,* New York: The Macmillan Company, 1949.
4. Hildebrand, F. B., *Methods of Applied Mathematics,* New York: Prentice-Hall, Inc., 1952.
5. Hildebrand, F. B., *Introduction to Numerical Analysis,* New York: McGraw-Hill Book Company, 1956.
6. Lipka, J., *Graphical and Mechanical Computation,* New York: John Wiley and Sons, Inc., 1918.
7. Milne, W. E., *Numerical Solutions of Differential Equations,* New York: John Wiley and Sons, Inc., 1953.
8. Ralston, A., and H. S. Herbert, *Mathematical Methods for Digital Computers,* New York: John Wiley and Sons, Inc., 1959.
9. Salvadori, M. G., and M. L. Baron, *Numerical Methods in Engineering,* New York: Prentice-Hall, Inc., 1952.
10. Scarborough, J. B., *Numerical Mathematical Analysis,* 2nd Ed., Baltimore: The Johns Hopkins University Press, 1950.
11. Shaw, F. S., *An Introduction to Relaxation Methods,* New York: Dover Publications, Inc., 1953.
12. Stanton, R. G., *Numerical Methods for Science and Engineering,* New York: Prentice-Hall, Inc., 1961.

5 Empirical Equations

The petroleum engineer collects data of petroleum production operations to use as a guide in searching for new knowledge and techniques or in analyzing and controlling current and future petroleum operations. Most frequently the data are collected in sets of values for two variables which indicate what is happening to one variable as the other variable changes. The values for a set of two variables can be plotted on coordinate paper to give a graphical picture of the relationships between the two variables. Next it may be desirable to obtain a mathematical equation for a curve which fits the plot of the data very closely. This mathematical equation is called an empirical equation, and the procedure followed in obtaining it is called curve fitting. An empirical equation can be used for interpolating between values of a tabular set of data. The empirical equation is used in the place of a difference table and an interpolation formula. The computer programmer often prefers to store data in the form of an empirical equation: the corresponding values are computed as they are needed. This method of storing data is usually possible and is much simpler than storing the data in tabular form and incorporating interpolation formulas into the computer program.

The procedure most often followed for obtaining an empirical equation for a set of data is: (1) to plot the data, (2) to determine the type of curve which most closely fits the data, and (3) to determine the constants in the equation for the curve. The type of curve which most closely fits the data is found in some cases by transforming the data so that a straight line is obtained for the plot of the data. After a straight line is obtained for a given set of data (whether it be from the original data or from transformed data), the constants for the empirical equation are determined by one of three methods. They are: (1) the method of selected points, (2) the method of averages, and (3) the method of least mean squares. The procedures to follow for each of the methods are demonstrated below with several examples.

Non-Periodic Curves

The simplest of the non-periodic curves are the straight line curves. Two straight line curves are discussed in plane analytical geometry; therefore, the petroleum engineer may already be familiar with them. One of the straight line curves passes through the origin, and the other curve cuts the y-axis. The first curve contains one constant and the second curve contains two constants.

The Straight Line, $q = K'p$

For viscous flow of a liquid through a sandstone sample mounted in a permeameter, Darcy's equation applies. The equation is

$$q = \frac{k}{\mu} A \frac{p}{\Delta L} \qquad \text{5-1}$$

where the outlet is atmospheric and p is the reading in gauge pressure (psig). If the terms $kA/\mu\Delta L$ are replaced by K', Equation **5-1** becomes

$$q = K'p \qquad \text{5-2}$$

Equation 5-2 for K' gives

$$K' = \frac{q}{p} \qquad \text{5-3}$$

which is the equation for determining the constant of an empirical equation by the method of selected points for a straight line curve which passes through the origin. The procedures for determining the value of K' of Equation **5-2** by the three methods (selected points, averages, and least squares) is illustrated with the data of Table 5-1.

Table 5-1
Pressure-Flowrate Data

p psig	q mls/sec	pq	p^2
1.0	1.117	1.117	1.0
2.0	2.412	4.824	4.0
3.0	3.564	10.692	9.0
4.0	4.812	19.248	16.0
5.0	6.046	30.230	25.0
6.0	7.111	42.666	36.0
7.0	8.334	58.338	49.0
8.0	9.624	76.992	64.0
9.0	10.781	97.029	81.0
10.0	11.980	119.800	100.0
Σ 55.0	65.781	460.936	385.0

To determine a value for K' by the *method of selected points* the (p,q) data of Table 5-1 are plotted on rectangular coordinates as shown in Figure 5-1 and a straight line is drawn through the points by eye. The points on the plot are marked at the intersection of two short straight lines, one horizontal and one vertical. The straight line passes through the point $p = 0$ and $q = 0$ and the point $p = 10.0$ and $q = 11.95$. Now K' is determined by substituting 10.0 for p and 11.95 for q in Equation **5-3** as

$$K' = \frac{11.95}{10.0} = 1.195$$

The empirical equation for the (p,q) data of Table 5-1 by the method of selected points is

$$q = 1.195p$$

The vertical distances between the plotted points and the straight line of Figure 5-1 are known as residuals. The residuals are the differences between the observed value for p and the value for p on the straight line. For the best straight line, the algebraic sum of the residuals is equal to zero as

$$\Sigma(q - K'p) = 0 \text{ or } \Sigma q = K'\Sigma p = 0$$

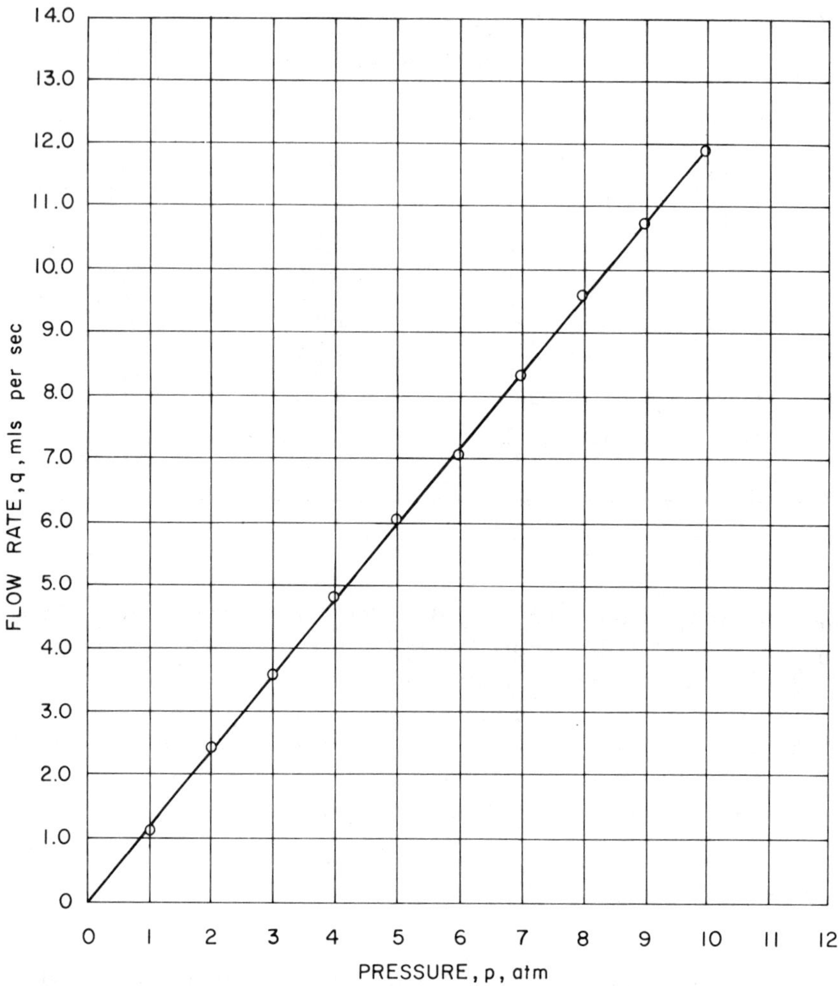

Figure 5-1. Viscous flow of a fluid through a sandstone; equation of the form $q = k'p$.

Solving Equation (5-4) for K' gives

$$K' = \frac{\Sigma q}{\Sigma p} \tag{5-5}$$

Equation 5-5 is the equation for computing the average value for K' by the *method of averages.*

Using the sums of the values of p and q of Table 5-1 gives

$$K' = \frac{\Sigma q}{\Sigma p} = \frac{65.781}{55.0} = 1.197$$

and the empirical equation by the method of averages is

$$q = 1.197p$$

Values of constants for straight line curves obtained by the method of averages are better than those obtained by the method of selected points; however, the best values for the constants are obtained by the *method of least squares.* With this method, the object is to find the K' value which minimizes the sum of the squares of the residuals.

$$\Sigma(q_i - K'p_i)^2 = \text{minimum} \qquad \textbf{5-6}$$

For Equation **5-6** to be true its derivative with respect to K' must equal zero as

$$\frac{d}{dK'}\Sigma(q - K'p_1)^2 = 0$$

or

$$\Sigma p\ (q - K'p) = 0$$

Solving for K' gives

$$K' = \frac{\Sigma pq}{\Sigma p^2} \qquad \textbf{5-8}$$

Equation **5-8** is the equation for computing the value of K' by the method of least squares.

Substituting the values from Table 5-1 for Σpq and Σp^2 gives

$$K' = \frac{460.936}{385.0} = 1.197$$

and the empirical equation obtained by the method of least squares is

$$q = 1.197p$$

The Straight Line $q = a + bp$

For linear flow through a porous core sample where the pressure is measured in absolute pressure (atm), Equation **5-1** becomes

$$q = \frac{kA}{\mu} \frac{(p - p_o)}{\Delta L}$$

or

$$q = \frac{kA}{\mu \Delta L} (p - p_o)$$

or

$$q = \frac{-kAp_o}{\mu \Delta L} + \frac{kAp}{\mu \Delta L} \qquad \textbf{5-9}$$

The symbol p_o is that for atmospheric pressure and the atmospheric pressure will remain practically constant throughout an experiment. Replacing

$-\dfrac{kAp_o}{\mu \Delta L}$ by a and $\dfrac{KA}{\mu \Delta L}$ by b gives

$$q = a + bp \qquad \textbf{5-10}$$

In Equation **5-10** the q intercept is $-p_o kA/\mu\Delta L$ or a, and the slope of the straight line is $kA/\mu\Delta L$.

The three methods for determining the values for the constants a and b are illustrated with the data from a permeability experiment of a sandstone core sample. The data of the experiment are given in Table 5-2. For the experiment, the efflux q from the permeameter was measured in mls/sec and the pressure p was measured in psia.

To determine the constants of Equation **5-10** by the *method of selected points* the values for p and q are plotted on rectangular coordinates as indicated in Figure 5-2, and by eye the best straight line is drawn through the plotted points. Then two points on the straight line of Figure. 5-2 are

Table 5-2
Permeability Data

psia	ml/sec	p^2	pq
15.7	2.577	246.49	40.459
16.7	3.872	278.89	64.662
17.7	5.024	313.29	88.925
18.7	6.272	349.69	117.286
19.7	7.506	388.09	147.682
20.7	8.571	428.49	177.420
21.7	9.794	470.89	212.530
22.7	11.084	515.29	251.607
23.7	12.241	561.69	290.112
24.7	13.440	610.09	331.968
Σ 202.0	80.381	4162.90	1722.561

selected. The (p,q) values of the points are $p = 14$, $q = 0.59$, and $p = 24$, $q = 12.65$. Next these values of (p,q) are substituted in Equation **5-10** which gives

$$12.65 = a + 24b$$

and

$$0.59 = a + 14b$$

Solving the two equations for a and b gives $a = -16.294$ and $b = 1.206$. The empirical equation is determined by substituting the values for a and b into Equation **5-10** which is

$$q = -16.294 + 1.206p$$

The equation for determining the constants of Equation **5-10** by the *method of averages* is

$$\sum_i (q_i - a - bp_i) = 0$$

$$\text{or } \sum_i q_i = na + b\sum_i p_i$$

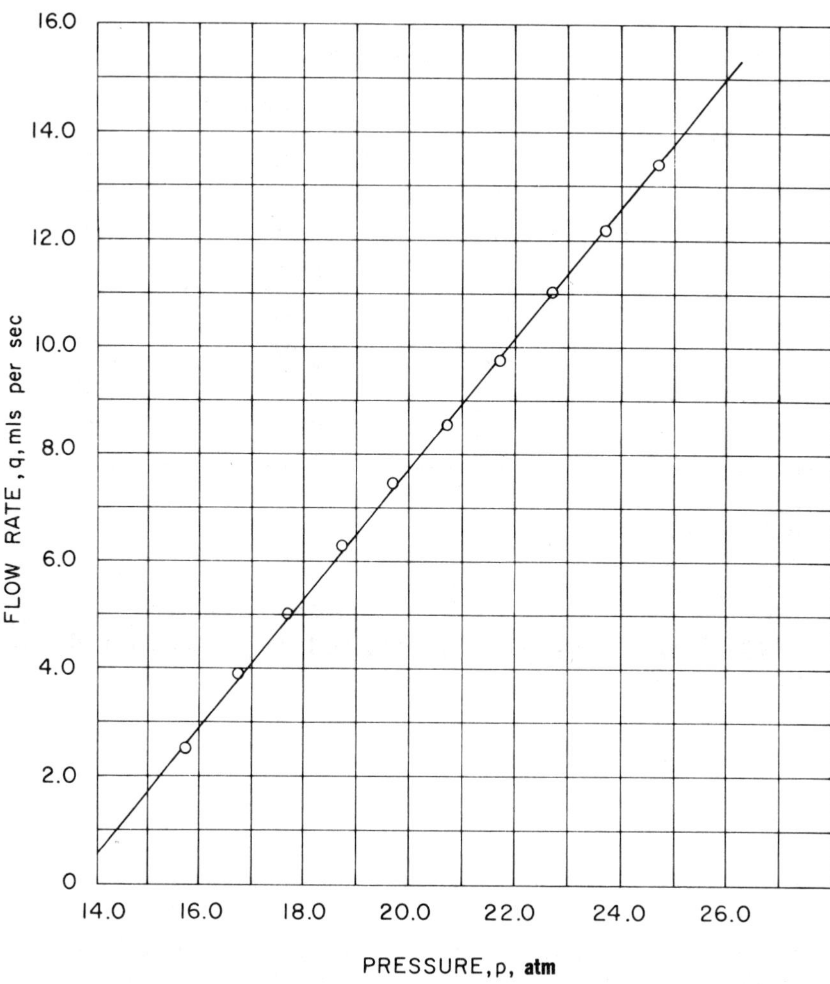

Figure 5-2. Viscous flow of a fluid through a sandstone; equation of the form $q = a + bp$.

Since there are two constants in Equation **5-11** the (p,q) data of Table 5-2 is divided into two equal groups and the sum of the residuals for each group is equal to zero. By substituting the sums for each (p,q) group into Equation 5-11, two equations are obtained:

$$25.251 = 5a + 88.5b$$

and

$$55.130 = 5a + 1135b$$

Solving the two equations gives

$$a = -16.11 \quad , \quad b = 1.196$$

and the empirical equation is

$$q = -16.11 + 1.196p$$

To determine the values for the constants by the *method of least squares* the sum of the squares of the residuals is a minimum, *i.e.*

$$\Sigma(q - a - bp)^2 = \text{minimum} \qquad \textbf{5-12}$$

and the partial of Equation 5-12 with respect to a and b must be zero; thus

$$\frac{\partial}{\partial a}\Sigma(q - a - bp)^2 = 0, \frac{\partial}{\partial b}\Sigma(q - a - bp)^2 = 0$$

or

$$\Sigma q = an + b\Sigma p \qquad \textbf{5-13}$$

and

$$\Sigma pq = a\Sigma p + b\Sigma p^2 \qquad \textbf{5-14}$$

From the (p,q) values in Table 5-2, the values pq, p^2, Σq, Σp, Σpq, and Σp^2 are computed and substituted into Equations **5-13** and **5-14** as

$$80.381 = 10a + 202.0b$$
$$1722.650 = 202a + 4162.9\,b$$

Solving the two equations for the constants gives

$$a = -16.182, \; b = 1.199$$

and the empirical equation obtained by this method is

$$q = -16.182 + 1.199p$$

Simple Parabolic and Hyperbolic Curves $p = ax^b$

A large number of sets of empirical data can be approximated with parabolic and hyperbolic curves. A number of these curves are given in Figure 5-3. The curves all have the same value for *a,* but the exponents are different for each curve. Curves are drawn for $a = 1.5$, and $b = -2, -1.5, -0.5, 0.5, 1.0$, and 2.0. All the parabolic curves pass through the points $(0,0)$ and $(1,a)$, and as one of the variables increases the other variable

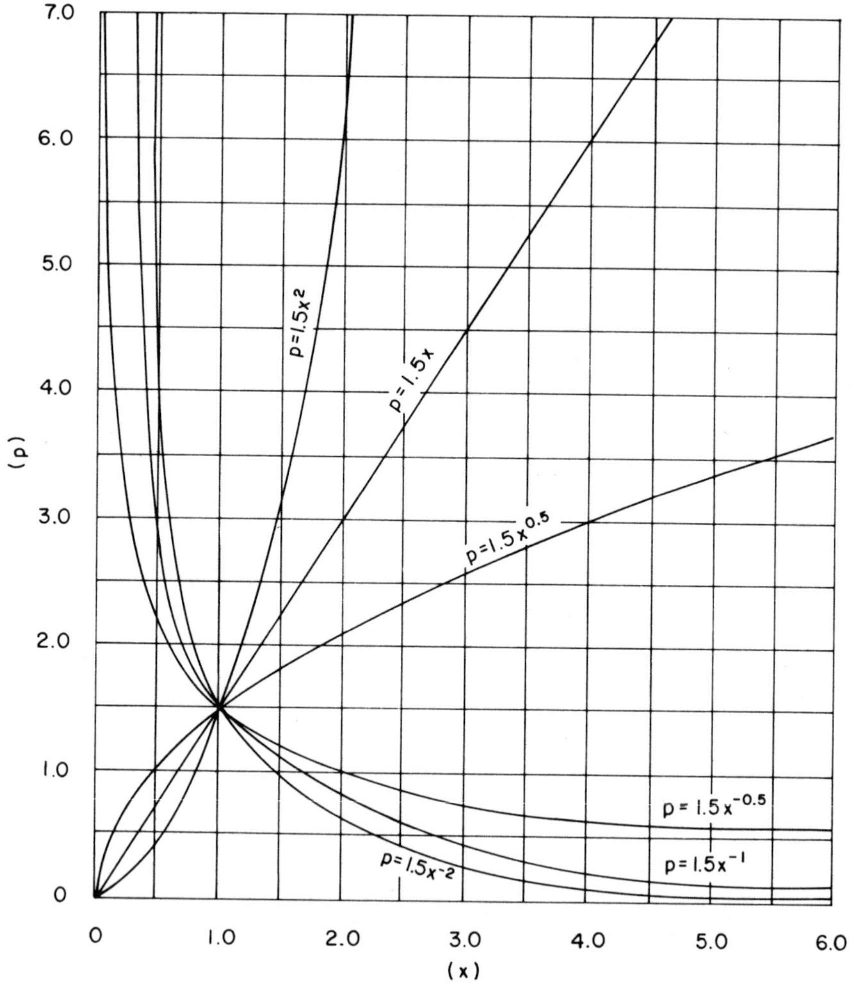

Figure 5-3. Simple parabolic and hyperbolic curves of the form $p = ax^b$.

Table 5-3
Gas Absorption Data

ml CO_2 gm Char	P mm	log x	log p	$(\log x)^2$	log x · log p
19.5	100	1.2900	2.0000	1.6641	2.5800
27.2	200	1.4346	2.3010	2.0581	3.3010
33.3	300	1.5224	2.4771	2.3177	3.7711
38.2	400	1.5821	2.6021	2.5030	4.1168
43.5	500	1.6385	2.6990	2.6847	4.4273
46.8	600	1.6702	2.7782	2.7896	4.6413
49.5	700	1.6946	2.8451	2.8717	4.8213
51.5	800	1.7118	2.9031	2.9303	4.9695
Σ		12.5442	20.6056	19.8192	32.6233

increases also. All the hyperbolic curves pass through the point $(1,a)$ and have the coordinate axes as asymtotes. For the parabolic curves, the exponents of x are positive; for the hyperbolic curves, the exponents are negative.

For a given weight of adsorbent with a given surface area the amount of a gas adsorbed depends on the partial pressure of the gas around the adsorbent. Adsorption of this type may often be represented by the empirical equation prepared by Freundlich[1]. The Freundlich equation is

$$p = ax^b \tag{5-15}$$

where p is the partial pressure of the gas in equilibrium with the solid and x is the amount of gas adsorbed per unit quantity of adsorbent. The data for the adsorption of CO_2 on charcoal is given in Table 5-3.

A plot of the (x,p) data of Table 5-3 is given in Figure 5-4. Since the curve of Figure 5-4 is not a straight line, the next step is to transform the (x,p) data of Table 5-3 so that its plot will give a straight line. Writing Equation **5-15** in logarithmic forms gives

$$\log p = \log a + b \log x \tag{5-16}$$

If $\log p = p_1'$, $\log a = a'$, and $\log x = x_1'$ then Equation **5-16** can be written as

[1]Bikerman, J. J. *Surface Chemistry: Theory and Applications.* New York: Academic Press, 1958.

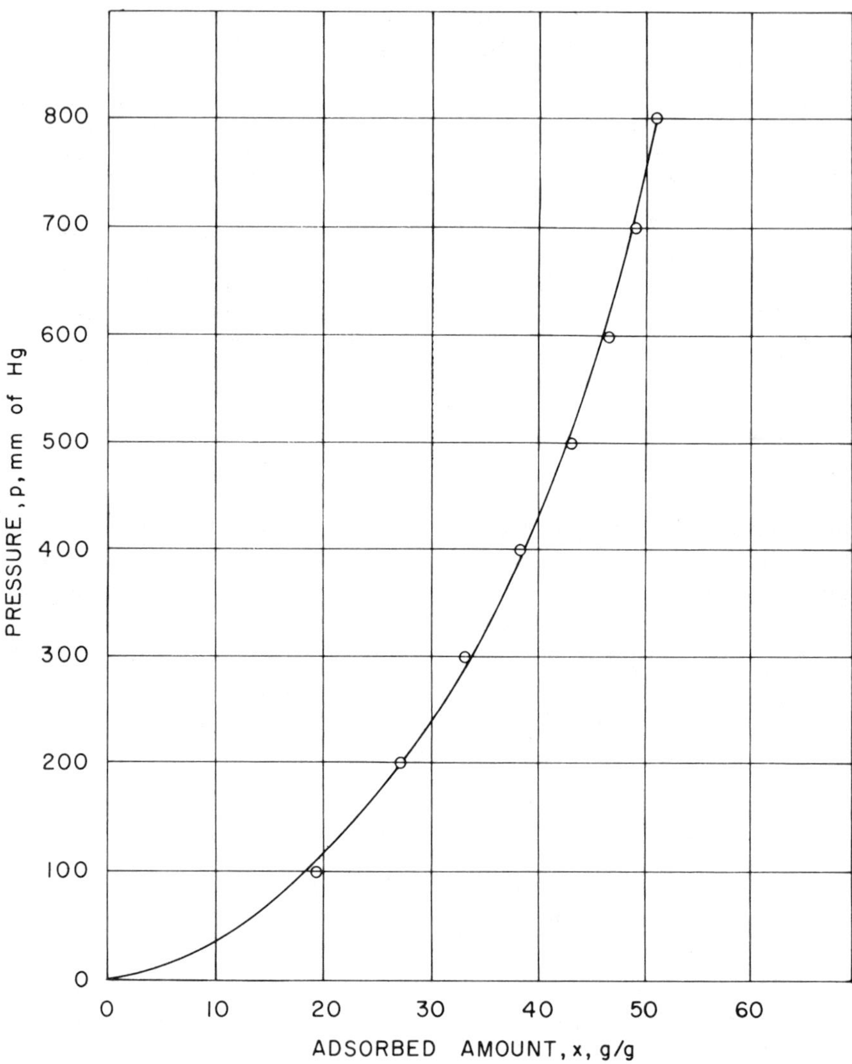

Figure 5-4. A plot of the adsorption of carbon dioxide on charcoal-equation of the form $p = ax^b$.

$$p' = a' + bx' \qquad \text{5-17}$$

But Equation **5-17** is the equation of straight line. Now a plot of (log x, log p) should yield a straight line with an intercept equal to log a and a slope equal to b. A (log x, log p) plot of the data of Table 5-3 is plotted in Figure 5-5 and the curve of the plotted points is a straight line.

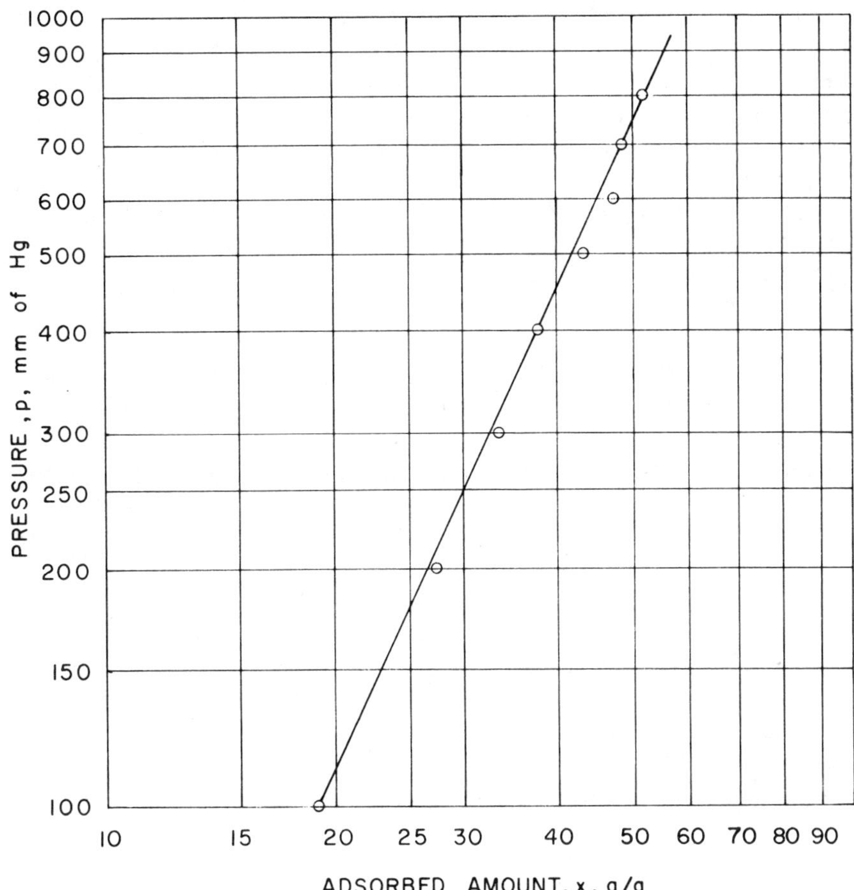

Figure 5-5. The log-log plot of the adsorption of carbon dioxide on charcoal-equation of the form $p = ax^b$.

To find the constants by the method of selected points, two points are selected on the curve of Figure 5-5. The values at these points are

$$\log x = \log 19.3 = 1.2856, \log p = \log 100 = 2.0000$$

and

$$\log x = \log 51.3 = 1.7101, \log p = \log 800 = 2.9031$$

Substituting the values of Equation **5-16** gives

$$2.000 = \log a + 1.2856\, b$$

and

$$2.9031 = \log a + 1.7101\, b$$

Subtracting the first equation from the second equation and solving for a and b gives

$$a = 0.1721$$
$$b = 2.1274$$

To find the values of the constants (a,b) by the method of averages, the data of the log x and log p columns of Table 5-3 is divided into two equal groups of four. The sums of the two groups are substituted into the equation

$$\Sigma \log p = n \log a + b\, \Sigma \log x \qquad \textbf{5-18}$$

to give the following two equations:

$$9.3802 = 4 \log a + 5.8291\, b$$

$$11.2254 = 4 \log a + 6.7151 b$$

Solving the two equations for a and b gives

$$a = 0.2001\ ;\ b = 2.0826$$

The empirical equation obtained by this method is

$$p = 0.2001 x^{2.0826}$$

Two equations for determining the values for the constants (a,b) by the method of least squares are obtained by differentiating Equation **5-19** with respect to log a and b.

$$\Sigma(\log p - \log a - b \log x) = \text{minimum} \qquad \textbf{5-19}$$

The derivatives of Equation 5-19 are

$$\frac{\partial}{\partial \log A} \Sigma\,(\log p - \log a - b \log x)^2 = 0$$

or

$$\Sigma \log p = n \log a + b \, \Sigma \log x \qquad \textbf{5-20}$$

and

$$\frac{\partial}{\partial b} \Sigma (\log p - \log a - b \log x)^2 = 0$$

or

$$\Sigma \log x \log p = \log a \, \Sigma \log x + b \, \Sigma (\log x)^2 \qquad \textbf{5-21}$$

Substituting the appropriate sums from Table 5-3 into Equations **5-20** and **5-21** gives the two equations

$$20.6056 = 8 \log a + 12.5442\, b$$
$$32.6233 = 12.5442 \log a + 19.8192\, b$$

Solving the two equations for the constants (a,b) gives

$$a = 0.2008, \; b = 2.0833$$

And the empirical equation obtained by the *method of least squares* is

$$p = 0.2008x^{2.0833}$$

The values for the constants obtained by the *method of averages* and by the *method of least squares* check to four significant digits; however, the values obtained by the *method of selected points* are close to the values obtained by the other two methods.

The empirical equation obtained by the *method of least squares* is checked for $x = 38.2$ as follows:

$$p = 0.2008\,(38.2)^{2.0833}$$

or

$$p = 395$$

The value ($p = 395$ mm) obtained for the pressure by the empirical equation compared favorably with the value ($p = 400$ mm) which is listed in Table 5-3.

Simple Exponential Curves $\gamma = ae^{bS}$

Frequently a set of data for a given petroleum production operation may be approximated by a simple exponential curve. Simple exponential curves have been used for approximating the relationships: (a) between density and pressure of an undersaturated reservoir oil; (b) between volume of an undersaturated reservoir oil and pressure; and (c) between the relative permeability ratios and fluid saturation. The equation which approximates the relationship between relative permeability ratios and fluid saturation is

$$\frac{k_g}{k_o} = ae^{bS}$$

or

$$\gamma = ae^{bS}$$

where

$$\gamma = \frac{k_g}{k_o} \qquad \text{5-22}$$

Since e^x may be represented by a series as

$$e^x = 1 + x + \frac{x^2}{2!} + \frac{x^3}{3!} + \ldots \frac{x^n}{n!}$$

then Equation 5-22 may be represented as

$$\gamma = a\left[1 + bS + \frac{(bS)^2}{2!} + \frac{(bS)^3}{3!} + \ldots \frac{(bS)^n}{n!}\right]$$

or

$$\gamma = a + abS + \frac{ab^2S^2}{2!} + \frac{ab^3S^3}{3!} + \ldots . \frac{ab^nS^n}{n!} \qquad \text{5-23}$$

The terms a, ab, $ab^2/2!$, etc. are constants; therefore, Equation 5-23 may be represented by the polynomial as

$$\gamma = a_0 + a_1S + a_2S^2 + a_3S^3 + \ldots a_nS^n \qquad \text{5-24}$$

where

$a_O = a_1, a_1 \doteq ab, a_2 = a^2b^2/2!$ etc.

A comparison of Equation 5-22 and Equation 5-23 indicates that simple exponential curves may be represented by a polynomial. The present discussion is concerned with treatments of empirical equations such as Equation 5-22. In a subsequent discussion empirical equations in the form of polynomials are treated.

Several simple exponential curves are drawn in Figure 5-6. The curves are of the form

$$\gamma = ae^{bS}$$

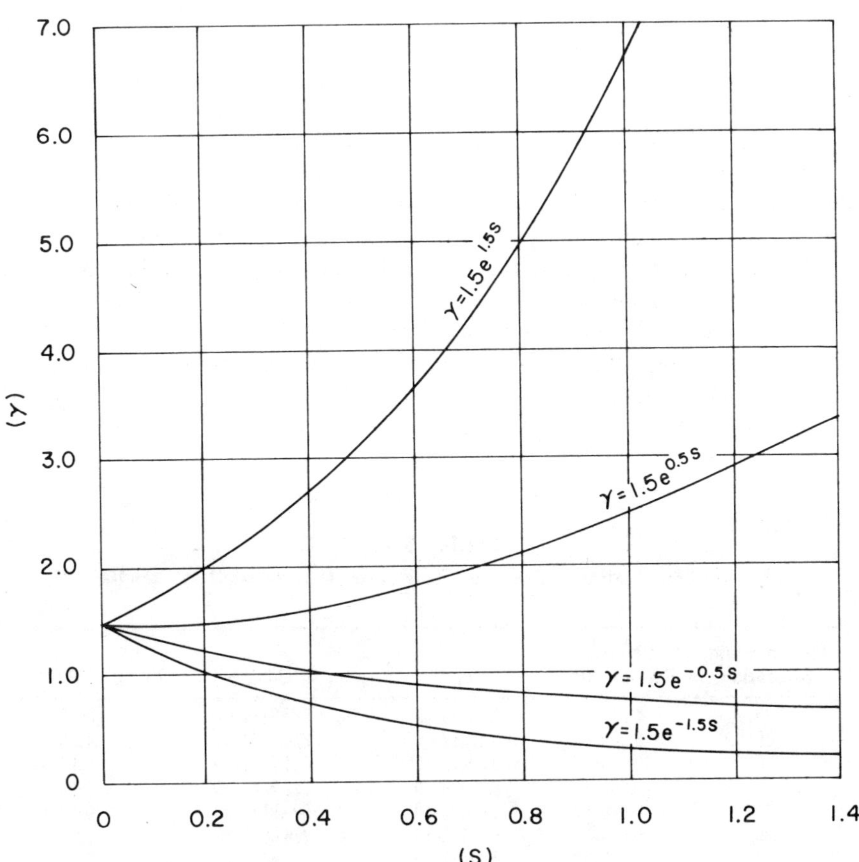

Figure 5-6. Simple exponential curves of the form $\gamma = ae^{bS}$.

where $a = 1.5$, and $b = -1.5, -0.5, 0.5$ and 1.5. All of the curves pass through the points (a, 0) and have the S-axis for asymptote.

Since the curves of Figure 5-6 are not straight lines, Equation **5-22** is transformed into a form which gives a straight line. Taking the logarithms of both numbers of Equation **5-22** gives.

$$\log \gamma = \log a + (b \log e) S$$

or

$$\ln \gamma = \ln a + bS \qquad \textbf{5-25}$$

If, in Equation **4-25,** $\ln \gamma = \gamma'$ and $\ln a = a'$; it then becomes

$$\gamma' = a' + bS \qquad \textbf{5-26}$$

and Equation **5-26** is an empirical equation of a straight line. Referring to Equation **5-24** a plot of $(S, \log_{10} \gamma)$ produces a straight line.

The data for relative permeability ratios and fluid saturations of a core sample are given in Table 5-4. The S,γ values of Table 5-4 are plotted in Figure 5-7 on ordinary coordinate paper. The curve through the plotted points deviates considerably from a straight line and indicates that the curve appears to be exponential.

The S,γ values of Table 5-4 are plotted on Figure 5-8 on semi-logarithmic coordinate paper. The curve through the plotted points approximates a straight line. Taking the S,γ values at two points on the curve of Figure 5-8 and substituting the values in Equation **5-25** gives the two equations

Table 5-4
Fluid Saturation Versus Relative Permeability Data

Percent Fluid Saturation	Relative Permeability	log γ	S log γ	S^2
60	0.540	9.73239–10	583.9434–600	3600
65	0.300	9.47712–10	616.0128–650	4225
7C	0.160	9.20412–10	644.2884–700	4900
75	0.092	8.96379–10	672.2843–750	5625
80	0.050	8.69897–10	695.9176–800	6400
85	0.027	8.43136–10	716.6656–850	7225
90	0·015	8.17609–10	735.8481–900	8100
Σ 525	1.184	62.68384–70	4664.9602–5250	40,075

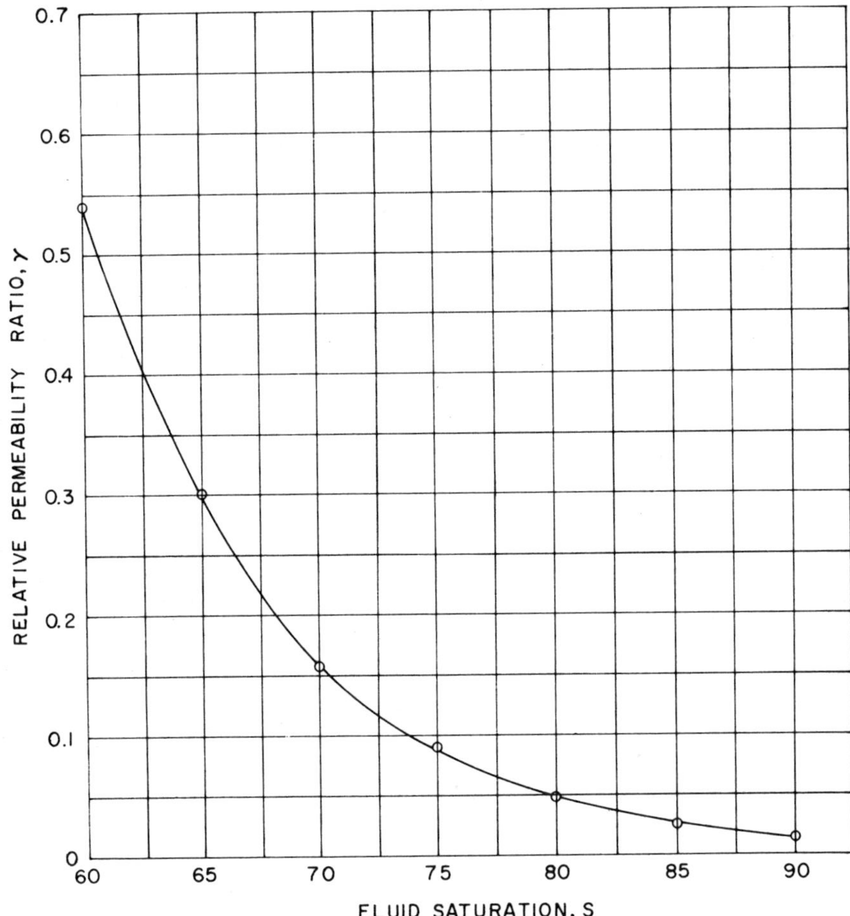

Figure 5-7. A plot of fluid saturation vs. relative permeability in rectangular coordinates.

$$\log 0.296 = \log a + (b \log e)\ 65$$
$$\log 0.0970 = \log a + (b \log e)\ 85$$

Solving the two equations which contain numerical values for the constants (a,b) yields

$$a = 644.5\ ,\ b = -0.11973$$

The empirical equation obtained by the method of selected points is

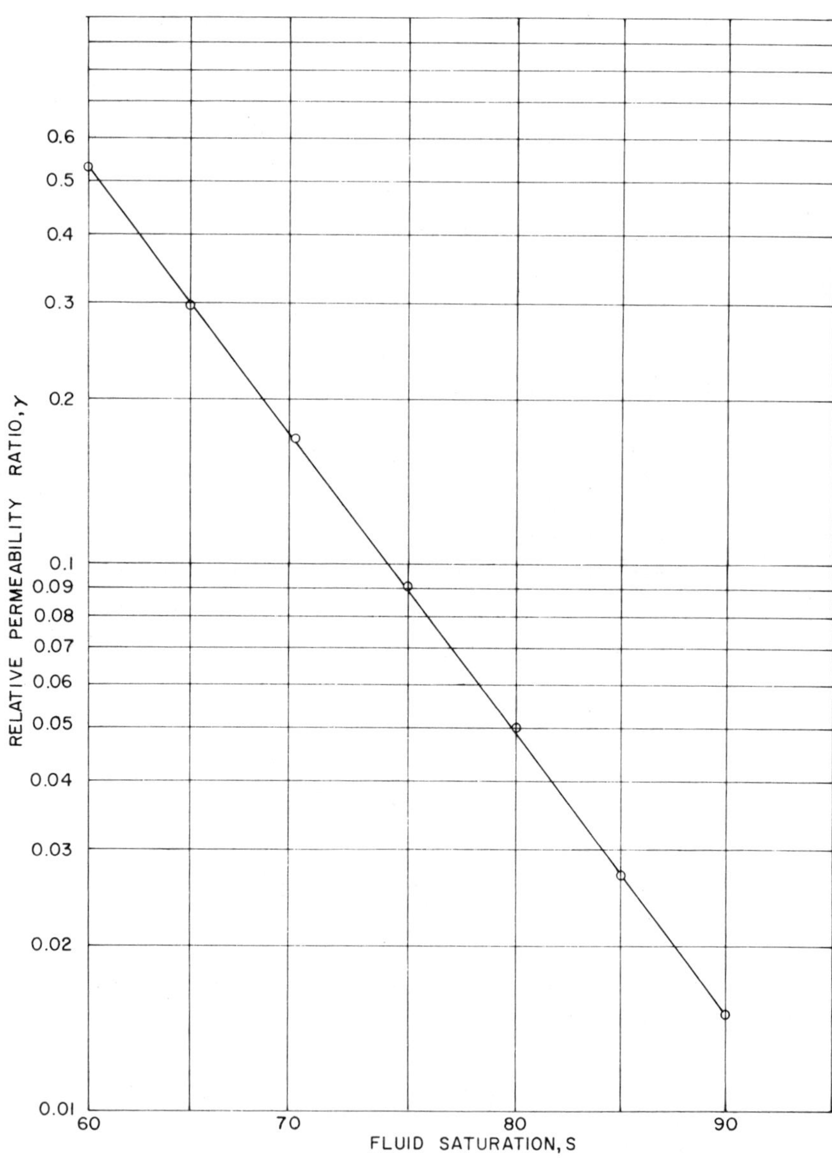

Figure 5-8. A replot of Figure 5-7 in semi-log coordinates.

$$\gamma = 644.5\, e^{-0.11973S}$$

The equation for determining the values for the constants (a,b) by the method of averages is

$$\Sigma \log \gamma = n \log a + (b \log e) \Sigma S \tag{5-27}$$

Dividing the data of Table 5-4 into two groups and substituting the sums in Equation **5-27** yields the two equations:

$$28.41365 - 30 = 3 \log a + (b \times 0.4343)\ 195S$$
$$34.27021 - 30 = 4 \log a + (b \times 0.4343)\ 330S$$

The solution to the two equations yields

$$a = 676.7;\ b = -\ 0.119.$$

The empirical equation by the method of averages is

$$\gamma = 676.7e^{-\ 0.119S}$$

The two equations for determining the values for the constants (a,b) by the method of least squares are

$$\frac{\partial}{\partial \log a} \Sigma\ (\log \gamma - \log a - (b \log e)\ S)^2 = 0$$

and

$$\frac{\partial}{\partial b} \Sigma\ (\log \gamma - \log a - (b \log e)\ S)^2 = 0$$

or

$$\Sigma \log \gamma = n \log a + (b \log e)\ \Sigma S \tag{5-28}$$

and

$$\Sigma S \log \gamma = \Sigma S \log a + b \log e\ \Sigma S^2 \tag{5-29}$$

Substituting the seems from Table 5-4 into Equations **5-28** and **5-29** yields

$$62.68384 - 70 = \log a + 0.4343 \times 525 \times bS$$

and

$$4664.9602 - 5{,}250 = 525 \log a + 0.4343 \times 40{,}075 \times bS$$

Solving the two equations for a and b yields

$$a = 677.3\ ,\ b = -0.11931$$

Checking the empirical equations obtained by the three different methods by calculating the value of γ for $S = 80$ yields

$\gamma = 0.04461$ by the method of selected points

$\gamma = 0.04965$ by the method of averages

and

$\gamma = 0.04848$ by the method of least squares.

The corresponding value for γ in the table is 0.050.

The Hyperbolic Curve, $M = \dfrac{\rho}{a + b\rho}$

The relationship between molecular weight and specific gravity of a stock tank crude may be represented by an empirical curve:[2]

$$M = \frac{\rho}{a + b\rho} \qquad \textbf{5-30}$$

where M is the molecular weight and ρ is the specific gravity. Equation **5-30** represents the ordinary hyperbola with asymptotes, $\rho = a/b$ and $M = 1/b$

Curves for values of $a = -0.5$, 0.5 and $b = 0.5$ are given in Figure 5-9.

Equation **5-30** may be transformed into the forms

$$\frac{\rho}{M} = a + b\rho \qquad \textbf{5-31}$$

and

[2]Cargoe, C. S. "Thermodynamic Properties of Petroleum Products," Bureau of Standards, U.S. Department of Commerce (1929), Miscellaneous Publication No. 97, p. 22.

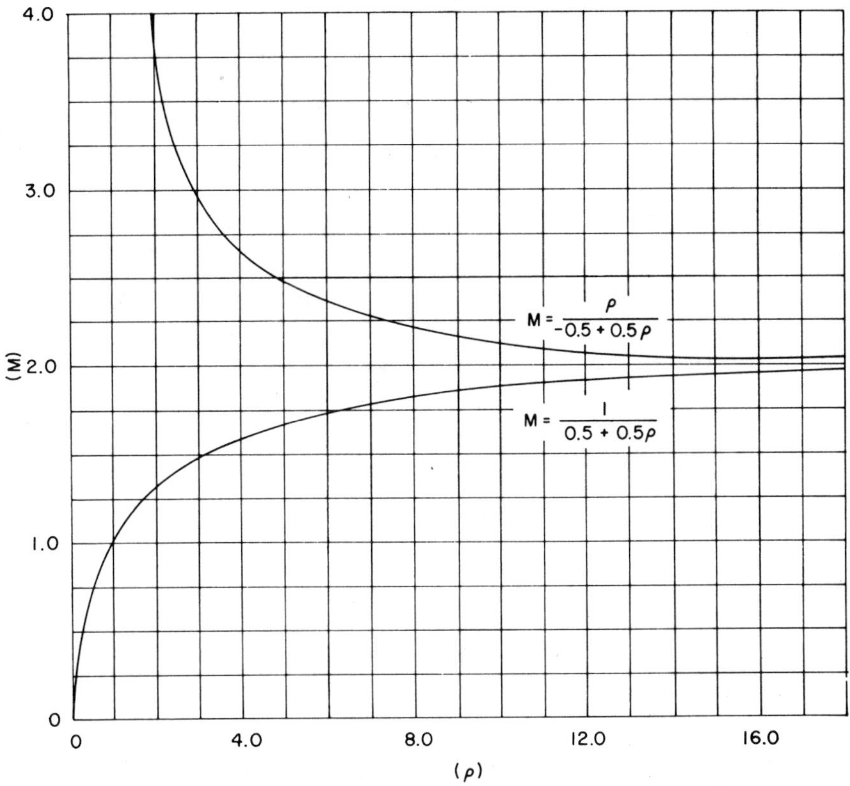

Figure 5-9. Hyperbolic curves of the form $M = \dfrac{\rho}{a + b\rho}$ or $M = \rho/(a + b\rho)$

$$\frac{1}{M} = b + \frac{a}{\rho} \qquad \text{5-32}$$

These two equations indicate that the plots of the data in the form

$$\left(\rho, \frac{\rho}{M}\right) \quad \text{or} \quad \left(\frac{1}{\rho}, \frac{1}{M}\right)$$

are straight lines. Plots of data in the form $(\rho, \frac{\rho}{M})$ for $a = -0.5$, 0.5, $b =$ are straight lines. Plots of data in the form $(\rho,\rho/M)$ for $a = -0.5$, 0.5, $b = 0.5$ are given in Figure 5-10. The data for density and molecular weight of stock crude oils are given in Table 5-5.

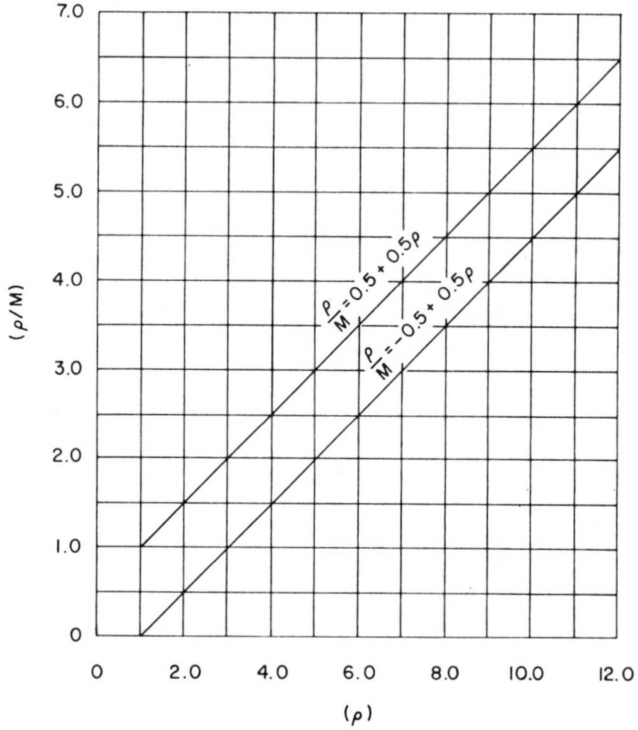

Figure 5-10. **A replot of Figure 5-9 in the form $\rho/M = a + b\rho$.**

Table 5-5
Data for Hyperbolic Curve

Density ρ	Molecular Wt., M	$\frac{\rho}{M}$	$\rho \frac{\rho}{M}$	ρ^2
0.50	41.1	0.012165	0.006083	0.250
0.60	61.2	0.009804	0.005882	0.360
0.70	93.4	0.007496	0.005247	0.490
0.75	121.0	0.006198	0.004649	0.563
0.80	154.0	0.005195	0.004156	0.640
0.85	221.2	0.003843	0.003267	0.723
0.90	300.9	0.002991	0.002692	0.810
0.95	600.0	0.001583	0.001504	0.903
Σ 6.05	Σ 1592.8	Σ 0.049275	Σ 0.033480	Σ 4.739

The curve through the plotted points of the (ρ,M) data on ordinary coordinate paper is given in Figure 5-11. The curve through the plotted points of the $(\rho, \frac{\rho}{M})$ data is given in Figure 5-12. The curve for the (ρ,M) data is not a straight line; however, the curve for the $(\rho, \frac{\rho}{M})$ data approxi-
points of the $(\rho,p/M)$ data is given in Figure 5-12. The curve for the (ρ,M) data is not a straight line; however, the curve for the $(\rho,\rho/M)$ data approxi-
mates a straight line.

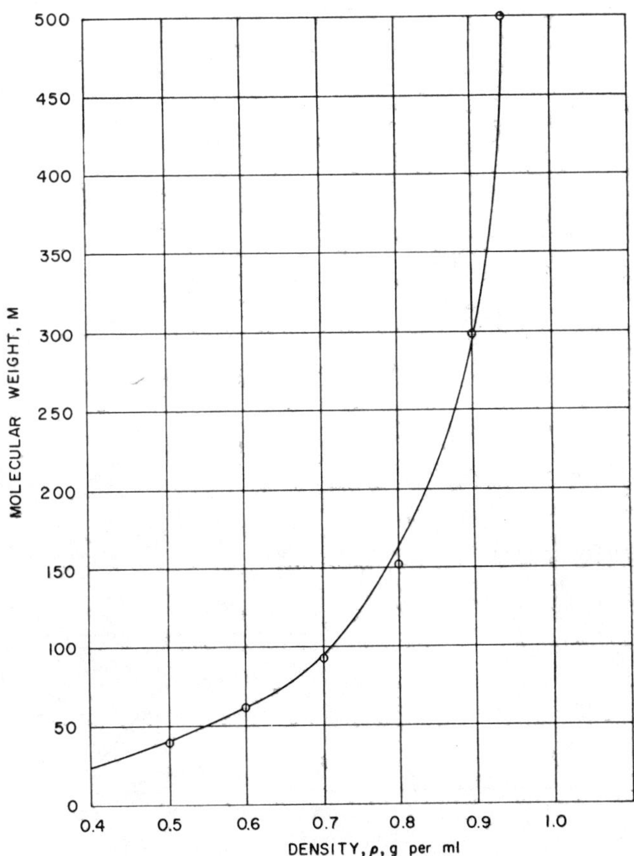

Figure 5-11. A plot of density vs. molecular weight for the data in Table 5-5.

Substituting the numerical values for two coordinate points on the curve of Figure 5-12 into Equation 5-31 gives the two equations:

$$0.01214 = a + 0.5b$$

$$0.00165 = a + 0.9b$$

The solution of the two equations gives

$$a = 0.02527, b = -0.02625.$$

Substituting the values for a and b into Equation 5-30 gives

$$M = \frac{\rho}{0.02527 - 0.02625\,\rho}$$

which is the empirical equation obtained by the method of selected points.

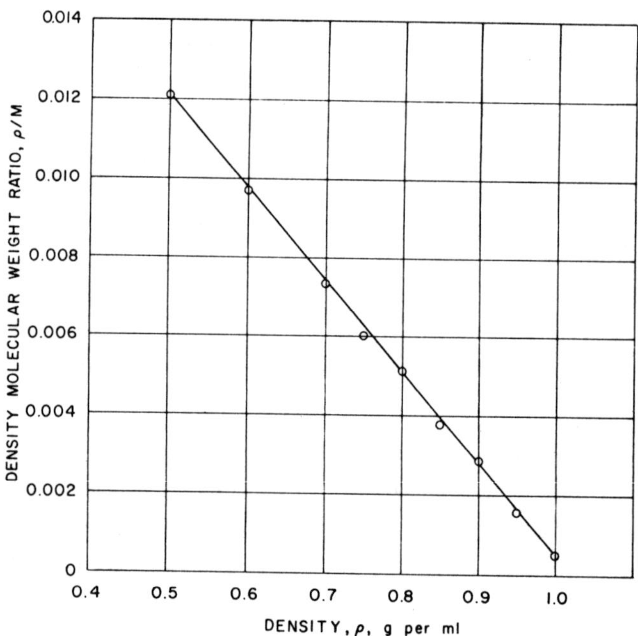

Figure 5-12. A plot of the data from Table 5-5 in the form ($\rho, \rho/M$).

The equation for determining the values of a and b by the method of averages is

$$\Sigma \frac{\rho}{M} = Ma + b\Sigma\rho \qquad \text{5-33}$$

Dividing the data of Table 5-5 into two groups and substituting the sums for the two groups in Equation **5-33** gives the following equations:

$$0.035664 = 4a + 2.55\,b$$
$$0.013615 = 4a + 3.50\,b.$$

The solution of the two equations gives

$$a = 0.002372;\ b = -0.02321$$

Substituting the values of a and b into Equation 5-30 gives

$$M = \frac{\rho}{0.02372 - 0.02321\,\rho}$$

which is the empirical equation obtained by the method of averages.

The equations for determining the values for the constants (a, b) by the method of least squares.

$$\frac{\partial}{\partial a} \Sigma \left(\frac{\rho}{M} - a - b\rho\right)^2 = 0$$

and

$$\frac{\partial}{\partial b} \Sigma \left(\frac{\rho}{M} - a - b\rho\right)^2 = 0$$

or

$$\Sigma \frac{\rho}{M} = na + b\Sigma\rho$$

and

$$\Sigma\rho \frac{\rho}{M} = a\Sigma\rho + b\Sigma\rho^2$$

Substituting the sums from Table 5-5 into Equations **5-34** and **5-35** gives the two equations:

$$0.049275 = 8a + 6.05\,b$$

$$0.033480 = 6.05\,a + 4.739\,b$$

The solution of the two equations gives

$$a = 0.02387\ ;\ b = -0.02196$$

and the empirical equation is

$$M = \frac{\rho}{0.02387 - 0.02196\,\rho}$$

The Hyperbolic or Parabolic Curve, $q = at^b + c$

Often a given set of data may not be fitted to a simple equation with two constants. On the other hand, the given set of data may be fitted by an equation with three constants. The simple (two constant) equation may be modified by the addition of a term containing a third constant. For example, the equation $q = at^b$ may be modified into $q = at^b + c$. For positive values of b, the latter curve represents a parabolic curve with intercept c on the q-axis, and for negative values of b, the latter curve represents a hyperbolic curve with asymptote $q = c$. In Figure 5-13, curves for the equations $q = 2t^{1.5}$; $q = 2t^{1.5}$; $q = 2t^{-1.5}$; and $q = 2t^{-1.5} + 2$ have been sketched in ordinary coordinates. In Figure 5-14, the curves for these equations have been sketched in log-log coordinates paper.

The equation

$$q = at^b + c \qquad \textbf{5-36}$$

may be written as

$$\log\ (q - c) = \log a + b \log t \qquad \textbf{5-37}$$

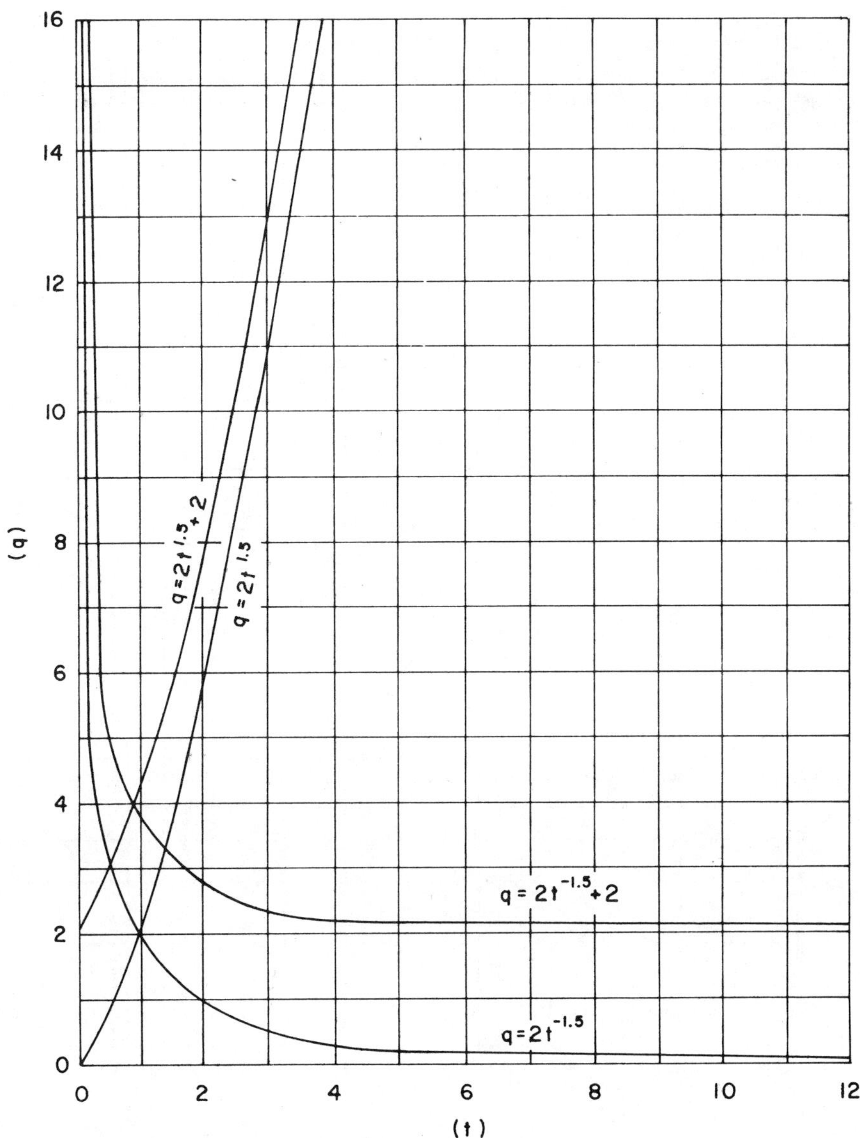

Figure 5-13. Hyperbolic and parabolic curves of the form $q = at^b + c$.

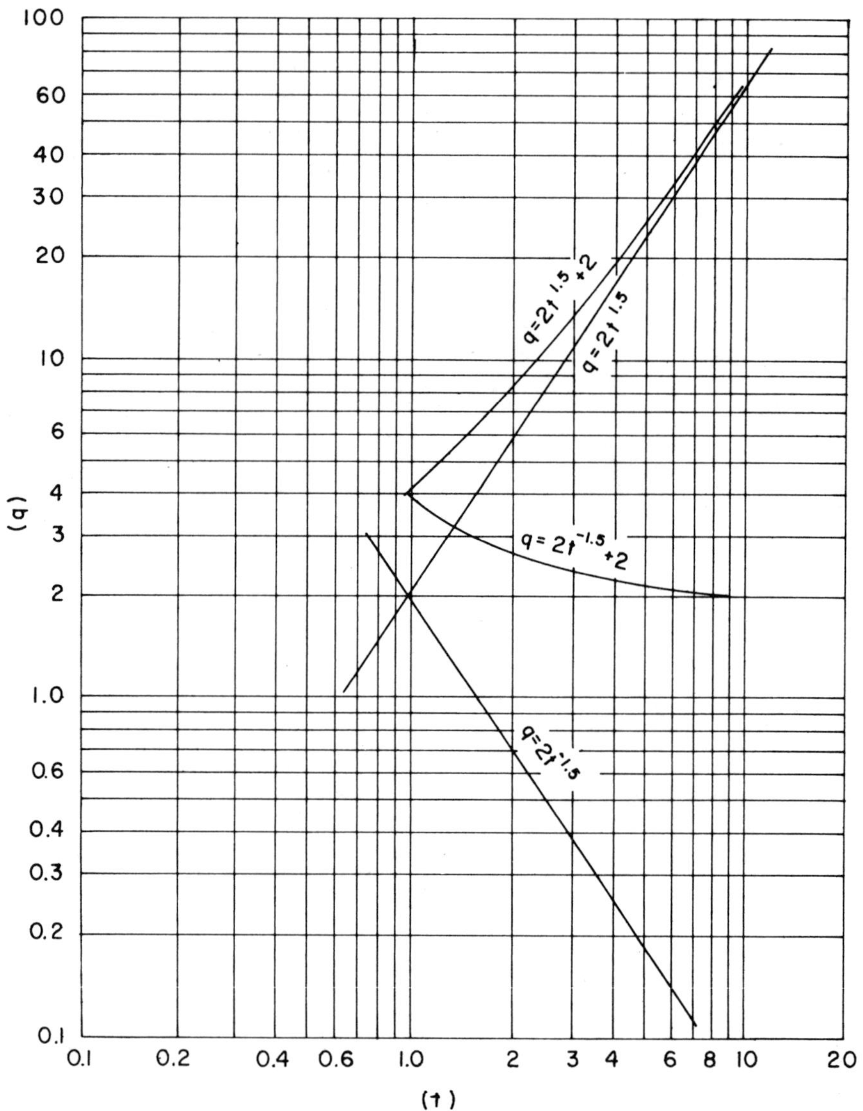

Figure 5-14. A plot of the hyperbolic and parabolic curves in form $\log (q - c) = \log a + b \log t$.

therefore, Equation **5-37** indicates that a plot of $[\log t, \log (q - c)]$ would approximate a straight line.

Petroleum production decline graphs, when plotted from actual well data, on ordinary coordinates are characteristically irregular in appearance for short periods of time. However, it is usually possible to draw smooth curves through the graphs which do not depart widely from the established points. Usually the "decline curves" may be classified as one of three types: exponential, hyperbolic, or loss-ratio.

The production data for a particular oil reservoir is given in Table 5-6. A plot of the (t,q) data of the table is plotted in ordinary coordinates in Figure 5-15. The curve through the plotted points is hyperbolic. A $(\log t, \log q)$ plot does not yield a straight line; therefore, the next step is to determine if the $[\log t, \log (q - c)]$ plot yields a straight line.

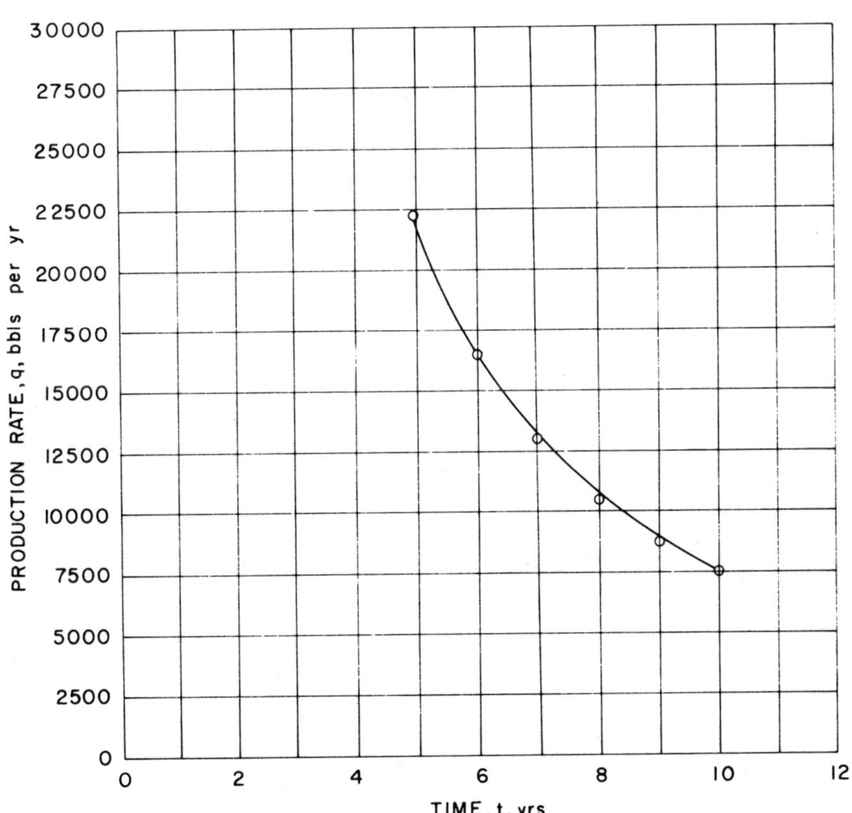

Figure 5-15. A plot of production rate, *bbls/yr*, vs. time, years.

Table 5-6
Data for Hyperbolic Curve

Time years, t	Rate bbls/yr., q	q – c	log (q – c)	log t	log t log (q – c)	(log t)^2
4	32,000	30,260	4.48087	0.60206	2.69775	0.36248
5	22,000	20,260	4.30664	0.69897	3.01021	0.48559
6	16,500	14,760	4.16909	0.77815	3.24418	0.60552
7	13,000	11,260	4.05154	0.84510	3.42396	0.71419
8	10,500	8,760	3.94250	0.90309	3.56043	0.81557
9	8,800	7,060	3.84880	0.95424	3.67279	0.91059
10	7,500	5,760	3.87506	1.00000	3.87506	1.00000
Σ 49	111,300	98,120	28.67450	5.78161	23.48438	4.89392

The procedure for determining values of the constants in the equation follows. Two points, (t_1, q_1) and (t_2, q_2), are selected on the curve of Figure 5-15. Then a third point is determined by setting

$$t_3 = \sqrt{t_1 t_2}$$

For the three points, three equations may be written as

$$q_1 = at_1^b + c\ ;\ q_2 = at_2^b + c\ ;\ q_3 = at_3^b + c.$$

And from these equations it is obvious that

$$t_3 = \sqrt{t_1 t_2}, \qquad t_3^b = \sqrt{t_1^b t_2^b},$$
$$at_3^b = \sqrt{at_1^b at_2^b} \text{ and } q_3 - c = \sqrt{(q_1 - c)(q_2 - c)}$$

Thus, solving for c gives

$$c = \frac{q_1 q_2 - q_3^2}{q_1 + q_2 - 2q_3} \qquad \text{5-38}$$

From the curve in Figure 5-15, the two points $t = 4$, $q = 32{,}000$ and $t = 10$, $q = 75{,}000$ are chosen, and the value is determined as

$$t_3 = \sqrt{4 \times 10} = 6.33$$

Then the value of q_3 is read from the curve; its value is 14,950. The value for c is determined with the use of Equation **5-38** as

$$\frac{32{,}000 \times 7{,}500 - (14{,}950)^2}{32{,}000 + 7{,}500 - 2(14{,}950)} = 1{,}740$$

After determining the value for c, the values for t and $q - c$ are plotted on logarithmic coordinates as shown in Figure 5-16. The curve in Figure 5-16 through the $(t, q - c)$ points approximates a straight line.

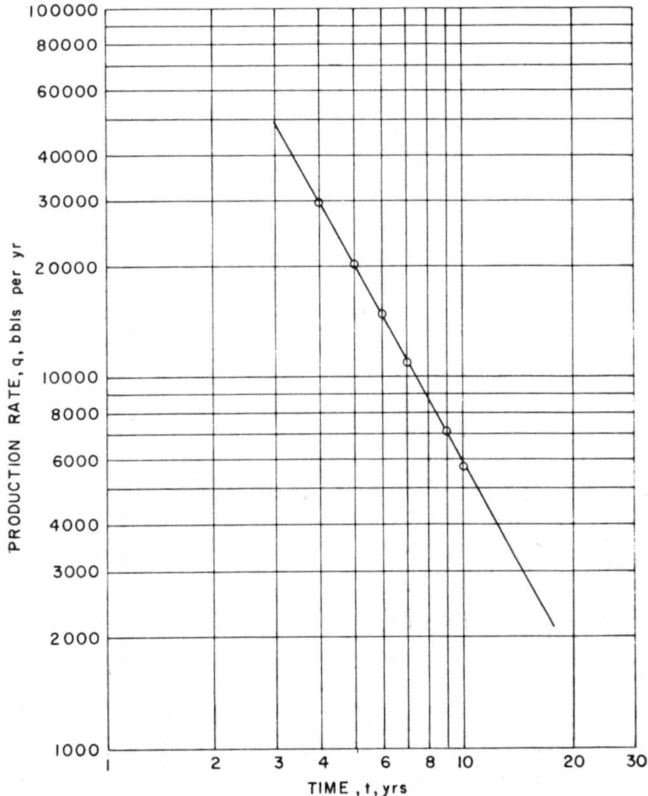

Figure 5-16. A log-log plot of the production rate vs. time.

To determine the values for a and b by the method of selected points, two values for t and $q - c$ are selected from the curve of Figure 5-16, and substituted in Equation **5-37** as follows:

$$\log 30{,}000 = \log a + b \log 4$$

and

$$\log 5{,}800 = \log a + b \log 10$$

The solution of the two equations gives

$$a = 359{,}500 \; ; \; b = -1.7937$$

The empirical equation determined by the method of selected points is

$$q = 359{,}500 t^{-1.7937} + 1{,}740$$

To determine the values for the constants (a, b) by the method of averages the following equation is used:

$$\Sigma \log (q - c) = n \log a = b \Sigma \log t \qquad \textbf{5-39}$$

Dividing the log t and log $(q - c)$ data of Table 5-6 into two portions and substituting the sums into Equation **5-39** gives the two equations:

$$12.98088 = 3 \log a + 2.07918\, b$$
$$15.37023 = 4 \log a + 3.70243\, b.$$

Solving the two equations for a and b gives

$$a = 300{,}600 \; ; \; b = -1.6748$$

and the empirical equation obtained by the method of averages is

$$q = 300{,}600 t^{-1.6748} + 1740$$

To determine the values for the constants (a,b) by the method of least squares, the following two equations are used:

$$\Sigma \log (q - c) = n \log a + b \, \Sigma \log t \qquad \textbf{5-40}$$

$$\Sigma \log t \log (q - c) = \log a \Sigma \log t + b \Sigma (\log t)^2 \qquad \textbf{5-41}$$

Substituting the values for the sums from Table 5-6 into Equations **5-40** and **5-41** gives the two equations:

$$28.67450 = 7 \log a + 5.78161\, b$$
$$23.48438 = 5.78161 \log a + 4.89392\, b$$

The solution to the two equations yields

$$a = 308{,}400; \quad b = -1.6863.$$

and the empirical equation obtained by the method of least squares is

$$q = 308{,}400t^{-1.6863} + 1740$$

The determination of c is partly graphical and its accuracy depends upon the reading of the coordinates of three points on the curve which was sketched to represent the data. Therefore the curve for the data should be drawn as accurately as possible.

The Exponential Curve, $q = ae^{bt} + c$

The exponential curve $q = ae^{bt} + c$ may often be used to represent a given set of data. The exponential curve containing three constants has the asymptote $q = c$. The curves, $q = 2e^{0.2t} + 2$; $q = 2e^{0.2t}$; $q = e^{-0.2t} + 2$, and $q = 2e^{-0.2t}$ are sketched in Fig. 5-17.

The equation

$$q = ae^{bt} + c \qquad \textbf{5-42}$$

may be written as

$$\log (q - c) = \log a + (b \log e)t. \qquad \textbf{5-43}$$

And Equation **5-43** indicates that a plot of $[t, \log (q - c)]$ in rectangular coordinates gives a straight line, or that a plot of $[t, (q - c)]$ in semi-logrithmic coordinates gives a straight line.

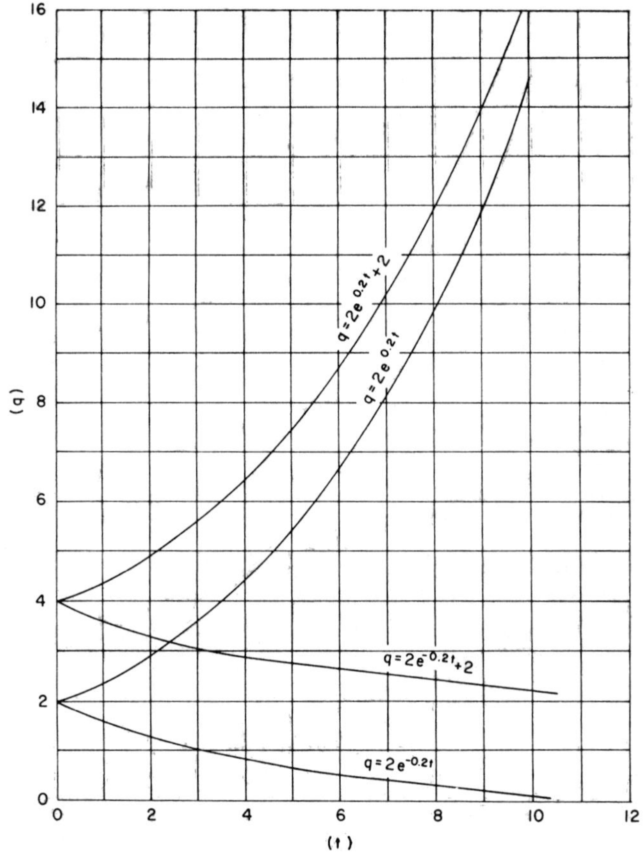

Figure 5-17. Exponential curves of the form $q = ae^{bt} + c$.

The production data for a particular oil reservoir is given in Table 5-7. A curve sketched through the plotted points of the t, q values on rectangular coordinates is given in Figure 5-18.

A curve sketched through the plotted points of the $(t, q - c)$ on semi-lograthmic coordinates is given in Figure 5-19; this curve is a straight line.

The procedure for determining the value for c follows. Two points, (t_1, q_1) and (t_2, q_2), are chosen on the curve which is sketched to represent the data; then a third point is chosen on the curve such that

Table 5-7
Three Constant Exponential Data

Time Years (t)	Production Rate Bbls/Mo. (q)	q – c	log (q – c)
1	595	505.6	2.70381
2	538	448.6	2.65186
3	487	397.6	2.59946
4	444	354.6	2.54974
5	405	315.6	2.49914
6	370	280.6	2.44809
7	341	251.6	2.40071

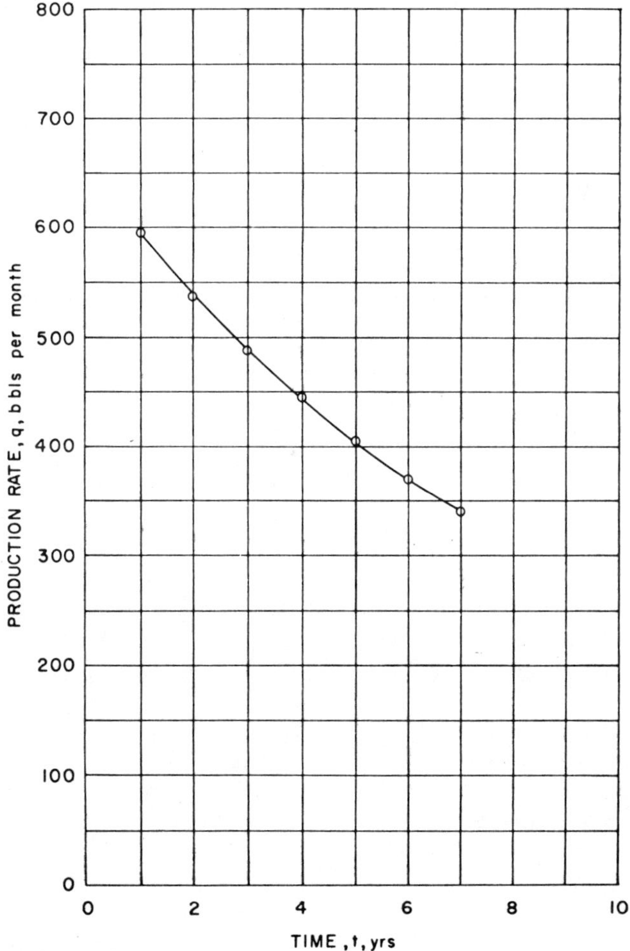

Figure 5-18. A plot of production data from Table 5-7 in rectangular coordinates.

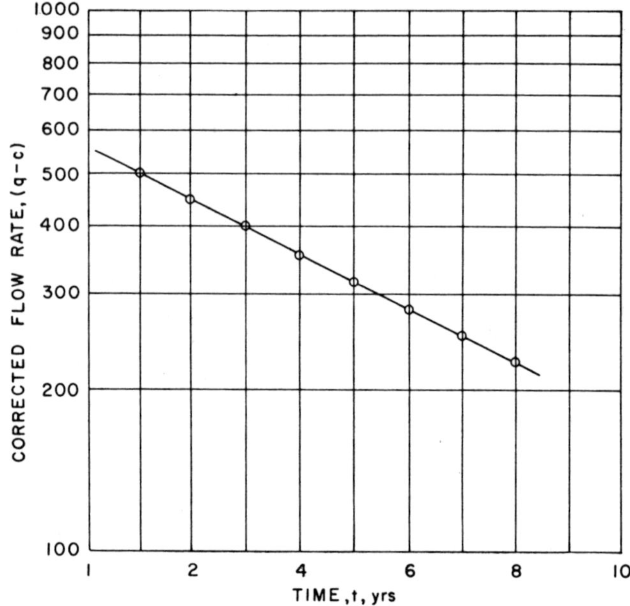

Figure 5-19. A plot of production data from Table 5-7 in semilogarithmic coordinates.

$t_3 = \frac{1}{2}(t_1 + t_2)$, *and* q_3 is measured. Substituting the three sets of values (t_1, q_1), (t_2, q_2), and (t_3, q_3) into Equation 5-42 gives the three equations

$$q_1 = ae^{bt_1} + c \; ; q_2 = ae^{bt_2} + c \; ;$$

$$q_3 = ae^{bt_3} + c.$$

or

$$\log \frac{q_1 - c}{a} = (b \log e)\, t_1 \, , \log \frac{q_2 - c}{a} = (b \log e)\, t_2 \, ,$$

$$\log \frac{q_3 - c}{a} = (b \log e)\, t_3.$$

Since

$$t_3 = \frac{1}{2}(t_1 + t_2)$$

then

$$(b \log e)t_3 = \frac{1}{2}\left[(b \log e)t_1 + (b \log e)t_2\right]$$

and

$$\log \frac{q_3 - c}{a} = \frac{1}{2}\left[\log \frac{q_1 - c}{a} + \log \frac{q_2 - c}{a}\right]$$

$$= \log \sqrt{\frac{q_1 - c}{a} \cdot \frac{q_2 - c}{a}}$$

Hence

$$q_3 - c = \sqrt{(q_1 - c)(q_2 - c)}$$

and

$$c = \frac{q_1 q_2 - q_3^2}{q_1 - q_2 - 2q_3} \quad \text{5-44}$$

From the curve in Figure 5-18 the values $t_1 = 1$, $q_1 = 594$, and $t_2 = 7$, $q_2 = 340$, are taken. The value for t is $\frac{1}{2}(1 + 7) = 4$; then, from the curve, $q = 445$. Substituting these values into Eq. **5-44** gives

$$c = \frac{594 \times 340 - (445)^2}{594 + 340 - 2 \cdot 445} = 89.4$$

To determine the values of the constants $(a, b,)$ by the method of selected points, two sets of $(t, (q\text{-}c))$ values were taken from the semi-logarithmic plot of Figure 5-19 and substituted into Equation **5-43.** The two sets of values are:

$t = 1$, $(q - c) = 502$
$t = 7$, $(q - c) = 253$

The two equations, obtained by the substitution of the values, are:

$2.70070 = \log a + 0.4343b$
$2.40312 = \log a + 0.4343 \times 7b.$

The solution of the two equations gives

$$a = 562.5\ ,\ b - 0.11382$$

And the empirical equation obtained by the method of selected points is

$$q = 562.5e^{-0.11382t} + 89.4$$

The following equation may be used for computing the values for the constants (a,b) by the method of averages:

$$\Sigma \log (q - c) = n \log a + b \log e \Sigma t \qquad \textbf{5-45}$$

To obtain the values for the constants by the method of averages, the data of Table 5-7 was divided into two parts and the sums were substituted into Equation **5-45**; The substitution of the sums gives the equations:

$$10.50486 = 4 \log a + 0.4343 \times 10\, b$$
$$7.34794 = 3 \log a + 0.4343 \times 18\, b.$$

The solution of the two equations is

$$a = 562\ ,\ b = -0.11637.$$

And the empirical equation obtained by the method of averages is

$$q = 562\ e^{-0.11637t} + 89.4$$

The equations for determining the values for the constants (a,b) are

$$\Sigma \log (q\text{-}c) = n \log a + b \log e\Sigma t \qquad \textbf{5-46}$$

and

$$\Sigma\, t \log (q\text{-}c) = \log a\Sigma t + b \log e\Sigma t^2 \qquad \textbf{5-47}$$

If the data are given so that the values of t are equidistant, the values for the constants (a,b) may be determined by a method which involves differences. For the constant difference in the values of $t = h$;

$t_1 = t_0 + h$, $t_2 = t_0 + 2h$, etc., and
$q_0 = ae^{bt_0}$, $q_1 = ae^{b(t_0+h)}$ etc.

The first difference is

$$\Delta q = q_1 - q_0 = ae^{b(t+h)}$$
$$- ae^{bt_0} = ae^{bt_0}(e^{bh} - 1)$$

and the difference written in logarithmic form is

$$\log \Delta q = \log a(e^{bh} - 1) + b(\log e)t \qquad 5\text{-}48$$

The last equation indicates that a plot of $(t, \log \Delta q)$ yields a straight line. After values for the constants (a,b) are determined then the following equation may be applied for determining the value of the c constant:

$$\Sigma q = a\Sigma e^{bt} + nc \qquad 5\text{-}49$$

Of the two methods, the difference method will give the best values for the constants because the determination of the value for c may be found without the use of a curve sketched through plotted points.

The Parabolic Curve, $B = a + bp + cp^2$

The parabolic equation $B = a + bp + cp^2$ may often be used to represent physical relations in both science and engineering. This equation has been used to express the relations between specific heats and temperatures, molal heat capacties and temperatures; gas volume factor and pressure; and between relative permeability ratios and fluid saturation. The curve $B_g = -10 + 0.08p - 0.000004p^2$ is sketched in Figure 5-20. Also, in Figure 5-20 the points $[p, (B_g - 134)/(p - 2000)]$ are plotted from values taken from the plot of the parabolic curve. The plot of the points $[(p, (B_g - 134)/(p - 2000)]$ yields a straight line. The proof that the second plot is a straight line follows.

If any points (p_k, B_{gk}) are chosen on the experimental curve

$B_g = a + bp + cp^2$; then $B_{gk} = a + Bp_k + cp_k$ and $B_g - B_{gk} = b(p - p_k) + c(p^2 - p_k{}^2)$

or

$$\frac{B_g - B_{gk}}{p - p_k} = (b + c_{pk}) + cp \qquad 5\text{-}50$$

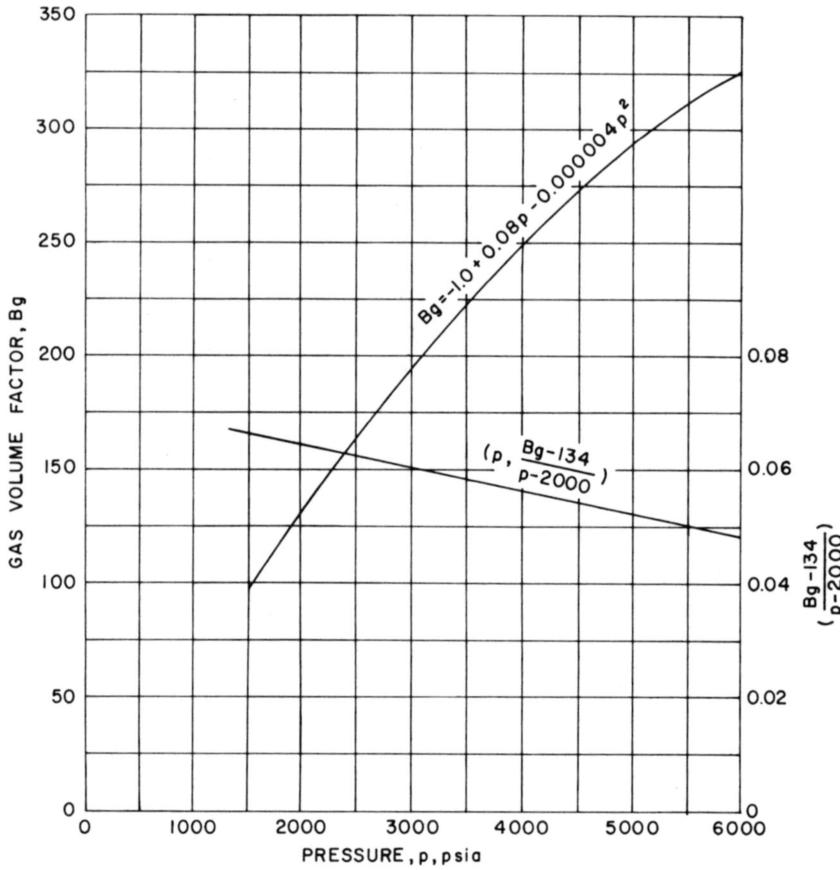

Figure 5-20. Parabolic curves of the form $B_g = a + bp + cp^2$ and plot of the parabolic curve in the form $p = (B_g - 134)/(p - 2000)$.

The last equation is of first degree in p and $(B_g - B_{gk})/(p - p_k)$; therefore, the plot of $[p, (B_g - B_{gk})/(p - p_k)]$ approximates a straight line.

If the values of (p, B_g) are given for equi-increments (h) of p, then p may be replaced by $(p + h)$; therefore, $B_g = a + b\ (p + h) + c(p + h)^2$

and $\Delta B_g = B_g' - B_g = (bh + ch^2) + 2\ chp$ **5-51**

Therefore a plot of $(p, \Delta\ B_g)$ approximates a straight line.

Example 5-1. In Table 5-8, the pressure and gas volume factor data of a reservoir fluid are given. In Figure 5-21 a curve is sketched for the pressure and gas volume values of the table. Using $p_k = 2800$ and $B_{gk} = 170$, a plot of $[(p, B_g - 170)/(p - 2800)]$ is also given in Figure 5-21; the curve for this plot approximates a straight line.

For the method of averages the constants (b,c) may be determined with the use of the equation

$$\Sigma \frac{B_g - B_{gk}}{p - p_k} = n\,(b + cp_k) + \Sigma\, cp \tag{5-52}$$

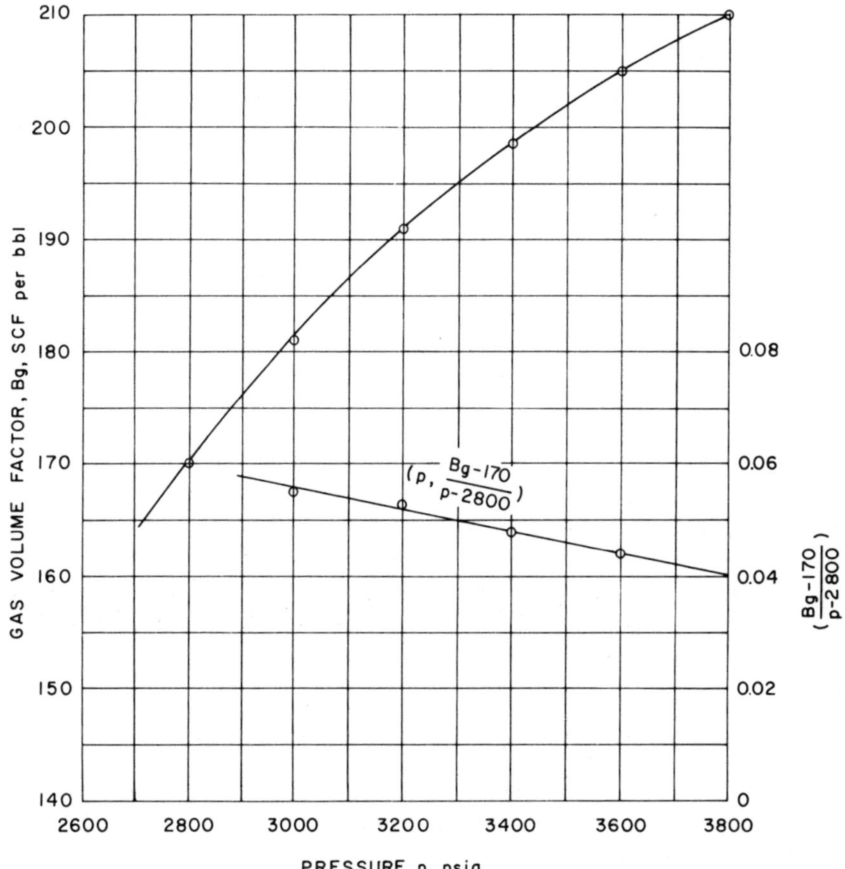

Figure 5-21. A plot of pressure-gas volume factor from data of Table 5-8; and a plot of pressure vs. $(B_g - 170)/(p - 2800)$.

Table 5-8
Gas Volume Factor Data

p	Bg	Bg − 170	p − 2800	$\frac{Bg - 170}{p - 2800}$	p²
2800	170				7,840,000
3000	181	11	200	0.055	9,000,000
3200	191	21	400	0.053	10,240,000
3400	199	29	600	0.048	11,560,000
3600	205	35	800	0.044	12,960,000
3800	210	40	1000	0.044	14,440,000

Dividing the data of Table 5-8 in two groups and substituting the sums into Equation 5-52 gives the two equations:

$$0.108 = 2(b + 2800c) + 6200c$$
$$0.132 = 3(b + 2800c) + 10{,}800c$$

The solution for the two equations gives

$$b = 0.172 \text{ and } c = -0.00002$$

Now using the data of Table 5-8, the calculated values of b and c, and the equation

$$\Sigma B_g = n\,a + b\Sigma p + c\Sigma p^2 \qquad \textbf{5-53}$$

gives

$$1{,}156 = 6a + 0.172 \times 19{,}800 - 0.00002 \times 66{,}040{,}000$$

from which the value of a is computed; the value of a is -154.8.

The empirical equation is

$$B_g = -154.8 + 0.172p - 0.00002p^2.$$

The Hyperbola, B = p/(a + bp) + c. The hyperbola of the type $B = p/(a + bp)$, which has been discussed, may be modified into the equation $B = p/(a + bp) + c$ which contains three constants. The equation with two constants gives $B = 0$ at $p = 0$; but the equation with three constants gives $B = c$ at $p = 0$. A curve for the equation $B = p/(0.5 + 0.5p) + 2$ is sketched in Figure 5-22; also in Figure 5-22 is a curve sketched through the points $(p, (p - 1)/(B - 3))$. The plot through the points $(p, (p - 1)/(B - 3))$ approximates a straight line; the explanation follows:

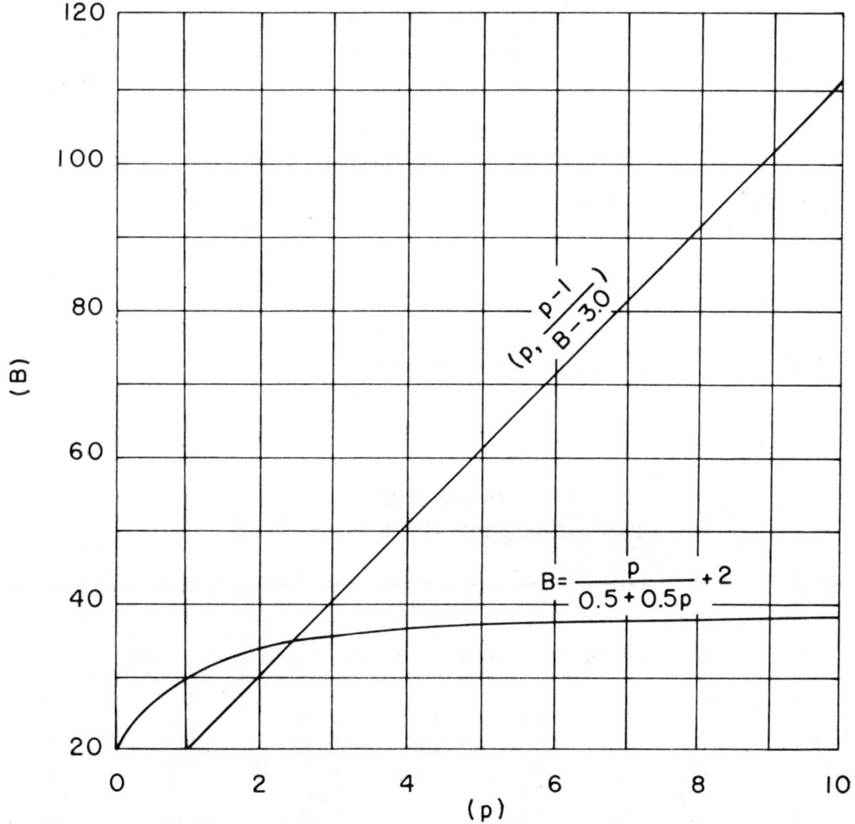

Figure 5-22. **Hyperbolic curve of the form $B = p/(a + bp) + c$; a plot through point (1, 3) in the form of p vs. $(p - 1)/(B - 3.0)$.**

If any point (p_k, B_k) is chosen on the experimental curve, then

$$B_k = \frac{p_k}{a + bp_k} + c,\ B - B_k = \frac{a(p - p_k)}{(a + bp)(a + bp_k)},\ \text{and}$$

$$\frac{p - p_k}{B - B_k} = (a + bp_k) + \frac{b}{a}(a + bp_k)p \qquad 5\text{-}54$$

The last equation is of first degree in p and $(p - pk)/(B - B_k)$; therefore, a plot of $(p, p - p_k/B - B_k)$ is a straight line.

Example 5-2. A plot of the p, B data of Table 5-9 approximates a curve which can be represented by an equation of the type $B = p/(a + bp) + c$. The p, B data of the table is plotted in Figure 5-23. Also in the figure is a plot of $[p, (p - 6.5/B - 1.29)]$. The plot of the p, B data gives a hyperbolic curve; and the plot of $(p, p - 6.5/B - 1.29)$ gives a straight line.

Equation 5-54 may be written as

$$\frac{p - p_k}{B - B_k} = a' + b'p \qquad 5\text{-}55$$

where $a' = (a + bp_k)$ and $b' = b/a\ (a + bp_k)$.

Table 5-9
Three-Constant Hyperbolic Data

p	B	p − 6.5	B − 1.29	$\frac{p - 6.5}{B - 1.29}$	a′ + b′p	$\frac{p}{0.3738 - 0.1584p}$	B_{cal}
6.5	1.29					−9.912	1.28
9.3	2.75	2.8	1.46	1.920	1.929	−8.460	2.73
11.6	3.18	5.1	1.89	2.70	2.569	−7.920	3.27
18.0	4.00	11.5	2.71	4.24	4.347	−7.260	3.93
21.2	4.10	14.7	2.81	5.81	5.238	−7.104	4.08
30.0	4.35	23.5	3.06	7.63	7.684	−6.800	4.29
96.60	19.67					−47.456	

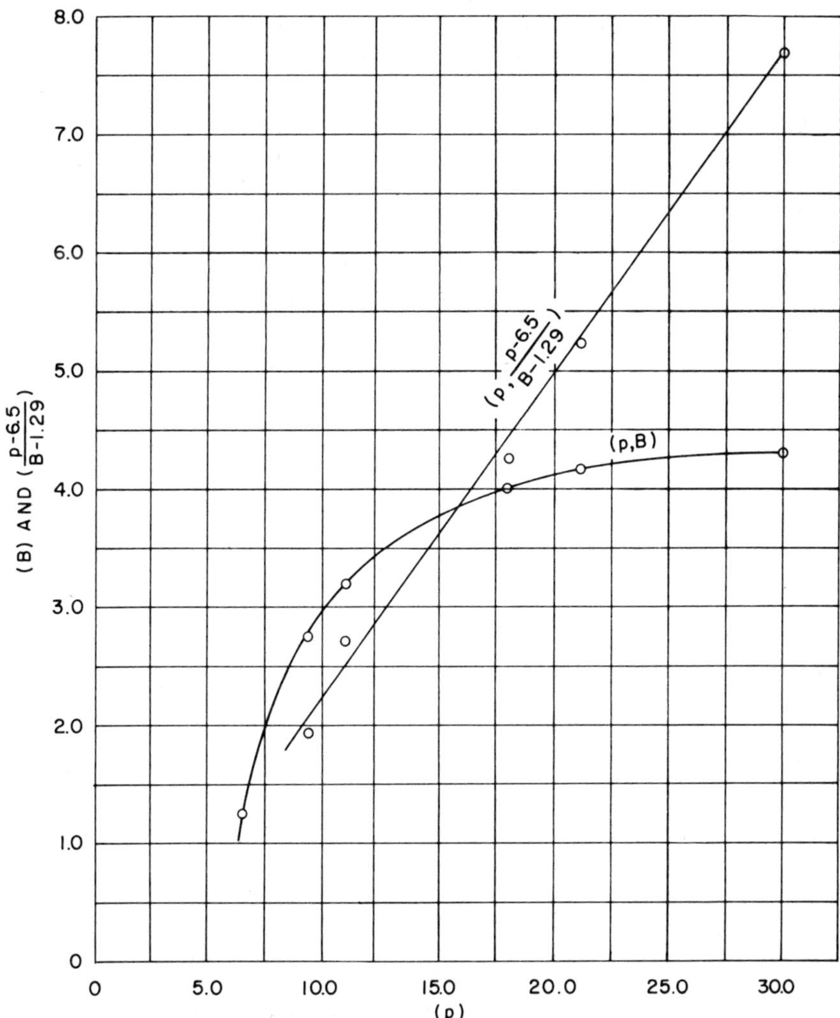

Figure 5-23. A plot of pressure vs. formation volume factor data from Table 5-9; and a plot of *p* vs. $(p - 6.5)/(B - 1.29)$ data from Table 5-9.

The constants a', b' of Equation 5-55 are determined by applying the equation

$$\Sigma \frac{p\ 6.5}{B - 1.29} = \text{n}\ a' + b'\ \Sigma p. \tag{5-56}$$

Dividing the appropriate data of Table 5-9 into two parts and substituting the sums for the data into Equation 5-56 yields the equations

$8.86 = 3a' + 38.9b'$ and $12.91 = 2a' + 51.2b'$. The solution of the two equations gives $a' = 0.656$, and $b' = 0.278$.

Substituting these values for a' and b' into the following equation $p - 6.5/B - 1.29 = a' + b'p$ then computed values for $(p - 6.5)/(B - 1.29)$ are obtained. These computed values are given in the column 6.

A comparison of corresponding values in columns 5 and 6 indicate that acceptable values were obtained for the constants a' and b'.

The values for a and b may be computed with the use of equations

$$a' = (a + bp_k) \text{ and } b' = \frac{b}{a}(a + bp_{k\cdot})$$

Substituting the appropriate values into the two equations give $a + 6.5b = -0.656$ and $\frac{b}{a}(a + 6.5b) = 0.278$.

The solution of the two equations $a = 0.3738$; $b = -0.1584$.

The constant c is determined by the method of averages with the use of the following equation

$$\Sigma B = \Sigma \frac{p}{0.3738 - 0.1584p} + 6c \qquad \textbf{5-57}$$

Substituting the appropriate sums into Equation **5-57** gives $19.67 = -47.456 + 6c$ or $c = 11.1876$.

Therefore the empirical equation for the p,B set of data is

$$B = \frac{p}{0.3738 - 0.1584p} + 11.1876 \qquad 5\text{-}58$$

The computed values for B with the use of the empirical equation are given in column 8 of Table 5-9. A comparison of the corresponding values in columns 2 and 8 indicate that Equation **5-58** adequately represents the p,B data of the table.

The Logarithmic or Exponential Curve

$$\log B = a + bp + cp^2 \text{ or } B = ae^{bp + cp^2}$$

The two equations are modifications of the two equations $\log B = a + bp$ and $B = ae^{bp}$ which contain only two constants. The second equation,

$B = ae^{bp + cp^2}$ may be written as $\log B = \log a + (b \log e)p + (c \log e)p^2$ which is equivalent to $\log B = a' + b'p + cp^2$ because the terms log a, b log e and c log e are constants.

Referring to the section on the discussion of the equation $B = a' + b'p + c'p^2$, it is readily apparent that a plot of $(p, (\log B - \log B_k)/(p - p_k))$ point on the experimental curve will approximate a straight line. Also it is apparent that a plot of $(p, \Delta \log B)$, are the differences in log form for equi-distant incremental vlaues of p, will approximate a straight line.

Example 5-3. Data which when plotted will give a curve that may be represented by the equation $\log B = a + bp + cp^2$ are given in Table 5-10.

The (p, B) and the $(p, \Delta \log B)$ data of the table are plotted in Figure 5-24. The curve sketched through the plot of the (p, B) data is slightly curved whereas a straight line is sketched through the plot of the $(p, \Delta \lg B)$ data.

From the previous discussion of the second order polynomial, it is self-evident that

$$\Delta \log B = b'h + c'h^2 + 2c'hp \qquad \textbf{5-59}$$

or

$$\Delta \log B = a'' + b''p \qquad \textbf{5-60}$$

where $a'' = b'h + c'h^2$ and $b'' = 2c'h$.

Table 5-10
Second-Order Log Polynomial Data

1	2	3	4	5	6	2
p	B	log B	Δ log B	p^2	log B (cals)	B (cals)
0	62.00	1.7924			1.8071	63.2
10	63.10	1.8000	0.0076	100	1.8094	64.48
20	65.00	1.8129	0.0129	400	1.8169	65.60
30	67.78	1.8311	0.0182	900	1.8298	67.58
40	71.55	1.8546	0.0235	1600	1.8481	70.48
50	76.46	1.8834	0.0288	2500	1.8716	74.41
Σ 150		10.9744		5500		

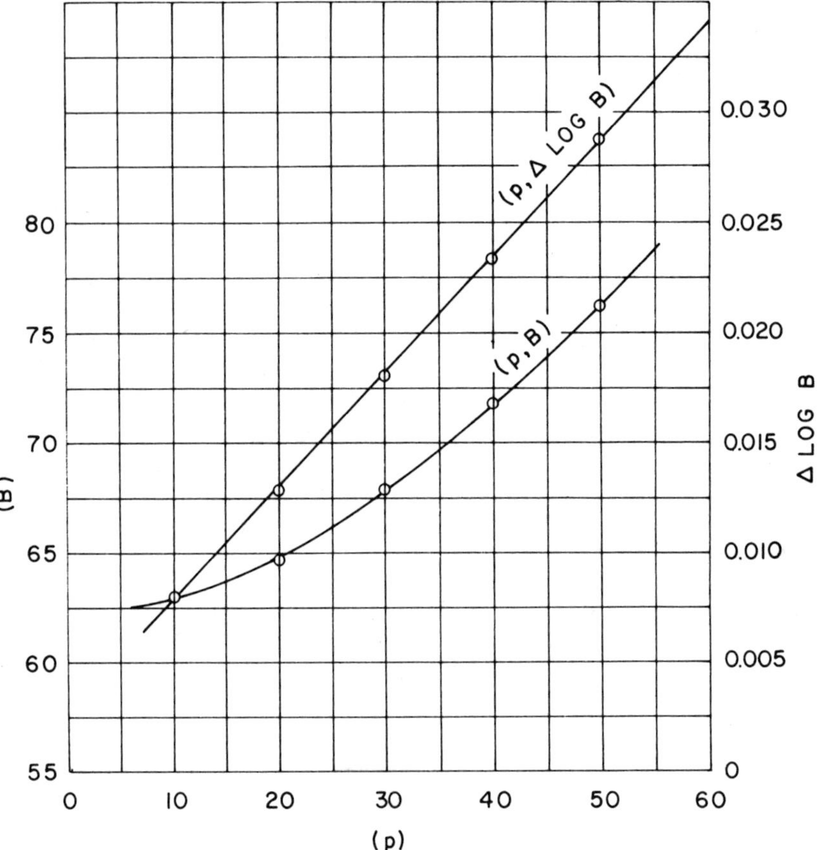

Figure 5-24. A plot of data from Table 5-10 which gives a logarithmic or exponential curve of the form log $B = a + bp + cp^2$; and the plot of the data in the form (p, Δ log B).

To determine the values for the two constants (a', b') by the method of averages the following equation may be employed.

$$\Sigma\ \Delta \log B = n\ a'' + b''\Sigma p \qquad \textbf{5-61}$$

Dividing the appropriate data of Table 5-10 into two groups and substituting the appropriate sums from them yields the two equations. $0.0387 = 3a' + 60b'$ and $0.0523 = 2a' + 90b'$. The solution of the two equations gives

$$a'' = 0.00226\ ,\ b'' = 0.000532$$

Now the values of b' and c' are obtained from the values of a'' and b'' as follows

$$b'' = 2c'h \text{ or } c' = \frac{b''}{2h}$$

or

$$c' = \frac{0.000532}{2 \times 10} = 0.0000266$$

and

$$a'' = b'h + c'h^2$$

or

$$0.00236 = 10b + 100 \times 0.0000266$$

$$b = \frac{0.00226 - 0.00266}{10}$$

$$= -\frac{0.0004}{10} = -0.00004.$$

The constant a may be determined by the method of averages with the use of the following equation:

$$\Sigma \log B = na + b \Sigma p + c \Sigma p^2 \qquad \textbf{5-62}$$

Substituting the appropriate sums from the table into Equation **5-62** gives

$$10.9744 = 6a - 0.00004 \times 150 + 0.0000266 \times 5500$$

or

$$a = \frac{1.843}{6} = 1.8071.$$

The empirical equation for the (p,B) data of the table is:

$$\log B = 1.8071 - 0.00004p + 0.0000266p^2$$

Column 7 Table 5-10 lists the values of B which were computed with the use of the empirical equation.

The Curve, $B = a + bp + ce^{dp}$.

Frequently a portion of a given set of data can be represented approximately by a simple equation after the subtraction of corrections; the remaining portion (residuals) of the data can be represented by an equation of another type. In other words, different portions of the set of data are controlled by different types of equations. The forms ce^{dp} and cp^d are most commonly added to the simple equations. A few suggested forms of the composite equations containing four constants each are:

$$B = a + bp + ce^{dp} \qquad B = a + bp + cp^d$$

$$B = ae^{bp} + ce^{dp} \qquad B = ap^b + cp^d$$

$$B = \frac{p}{a + bp} + ce^{dp} \qquad B = \frac{p}{a + bp} + cp^d$$

In Fig. 5-25 curves of the first composite equation are plotted.

If a portion of a set of data gives a curve which approximates a straight line, it can be fitted by an equation of the form $B = a + bp$. The deviations of the remaining portion of the data can be represented by the residuals, $r = B_o - B_c = B - (a + bp)$. Next, an equation of type $ce^{dp} = r = B - (a + bp)$ can be determined for the remaining portion of the data. Now the entire set of data is represented by an equation of the type

$$B = a + bp + ce^{dp} \qquad \textbf{5-63}$$

A set of data may be fitted by a curve of the type $B = a + bp + ce^{dp}$ although no part of the curve approximates a straight line; this is when the values of the term ce^{dp} are not negligible for values of p. If values of B are taken for equi-incremental value of p the equation for a set of data can be determined as follows if p is replaced by $p + h$.

$$B^1 = a + b(p + h) + ce^{d(p+h)}$$

and for the difference in the values of B

$$\Delta B + B^1 - B = bh + ce^{dp}\,(e^{dh} - 1)$$

If ΔB^1 is the next successive value of ΔB

$$\Delta B^1 = bh + ce^{d(p+h)}\,(e^{dh} - 1)$$

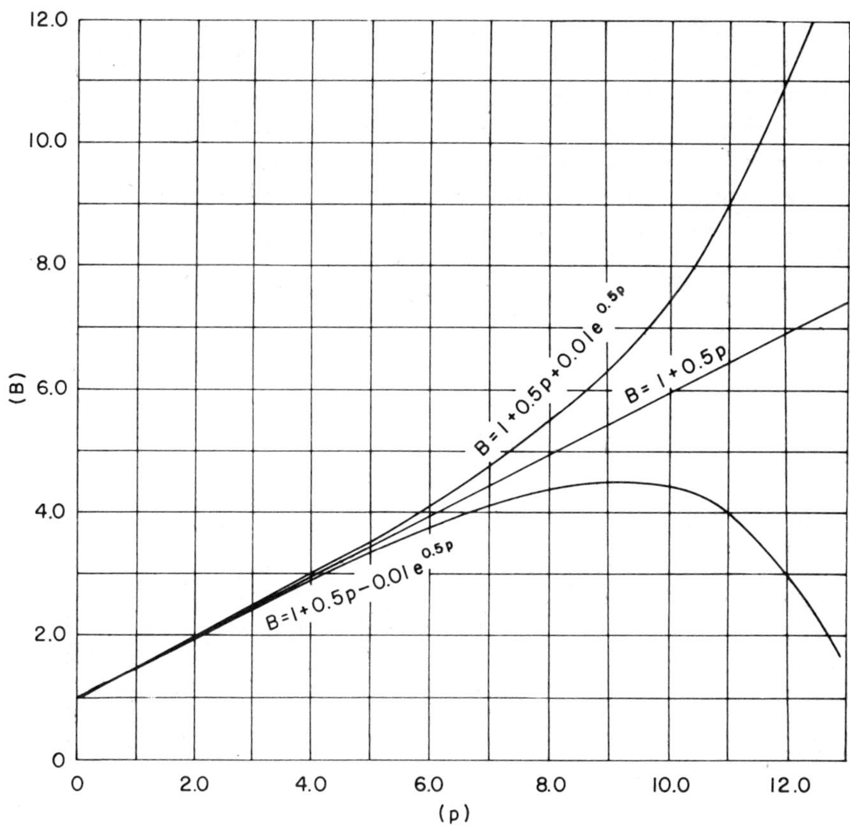

Figure 5-25. Curves of the form $B = a + bp + ce^{dp}$.

and the second difference is

$$\Delta^2 B = \Delta B^1 - \Delta B = ce^{dp}\,(e^{dh} - 1)^2$$

or

$$\log \Delta^2 B = \log c(e^{dh} - 1)^2 + (d \log e)\, p \qquad \textbf{5-64}$$

Equation **5-64** is of first degree in p and $\log \Delta^2 B$ therefore, a plot of $(p, \log \Delta^2 B)$ approximates a straight line. From a plot of $(p, \log \Delta^2 B)$ values for c and d can be determined. After determining the values for c and d, a plot of $(p, B\text{-}ce^{dp})$ can be used for determining the values of a and b for $(B - ce^{dp}) = a + bp$.

Example 5-4. When plotted, the data of Table 5-11 gives a curve which can be represented by the equation $B = a + bp + ce^{dp}$. The data of Table 5-11 is plotted in Figure 5-26. The curve of Figure 5-26 indicates a straight line portion. The data of the table and the curve in Figure 5-26 indicate that the intercept is approximately 1.02; therefore, $a = 1.02$.

The value of b is determined from two sets of values taken from the straight line of Figure 5-26 as

$$b = \frac{2.99 - 1.02}{10 - 0} = 0.197$$

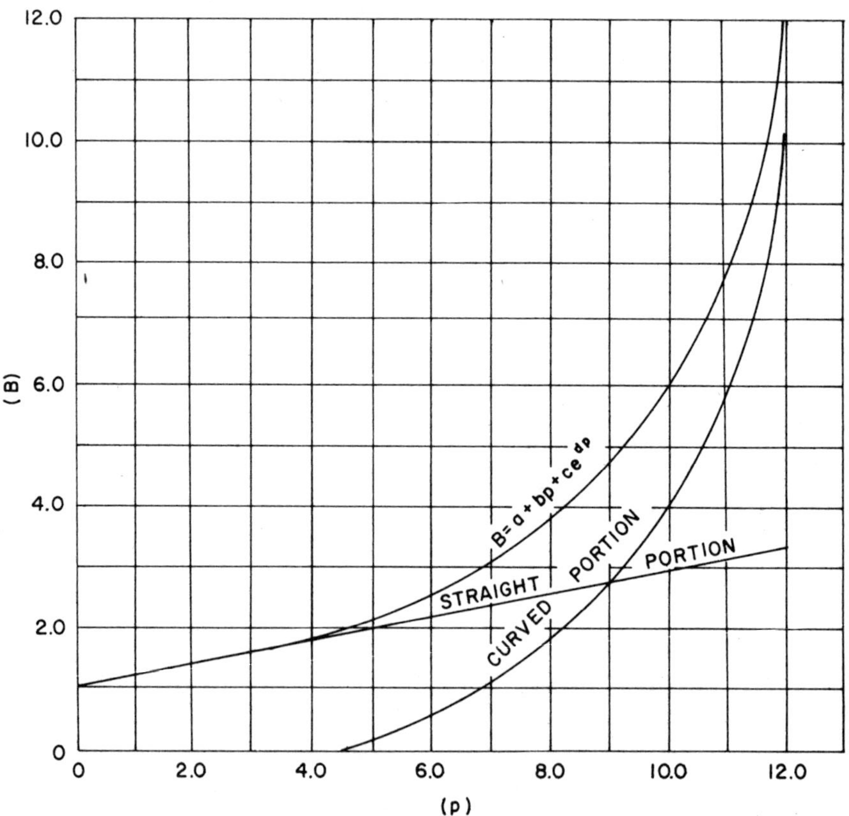

Figure 5-26. A plot of the data from Table 5-11 which gives a curve of the form $B = a + bp + ce^{dp}$.

Table 5-11
Linear-Exponential Curve Data

p	B	B′	B − B′	log (B − B′)
0.0	1.02			
2.0	1.43			
4.0	1.947	1.80	0.147	9.1673 −10
6.0	2.602	2.20	0.402	9.6043 −10
8.0	3.692	2.55	1.142	0.0577
10.0	5.968	2.99	2.978	0.4739
12.0	11.460	3.30	8.130	0.9101

Computed values for the equation $B' = 1.02 + 0.197p$ for values of p greater than 2.0 are given in the third column of Table 5-11. The computed values for $B - B' = r$ (residuals) are given in the fourth column of the table. The residuals when plotted on semi-logarithm paper give a straight line. Using the last four sets of values from the table and the equation

$$\Sigma \log B = n \log c + (d \log e)\Sigma p \qquad \textbf{5-65}$$

the values for c and d are determined as follows:

$$\begin{aligned} 11.3840 - 10 &= 2 \log c + 0.4343 \times 22d \\ 9.6620 - 10 &= 2 \log c + 0.4343 \times 14d \\ \hline 1.7220 \quad &= \quad 0 + 0.4343 \times 8d \end{aligned}$$

$$\text{or } 1.7220/8 \times 0.4343 = 0.495 = d$$

Substituting in the value of d in the first equation and solving for c gives

$$\begin{aligned} 1.3840 &= 2 \log c + 0.4343 \times 22 \times 0.495 \\ \log c &= +1.618 \\ c &= 0.024 \end{aligned}$$

Therefore the empirical equation is

$$B = 1.02 + 0.197p + 0.024e^{0.495p}$$

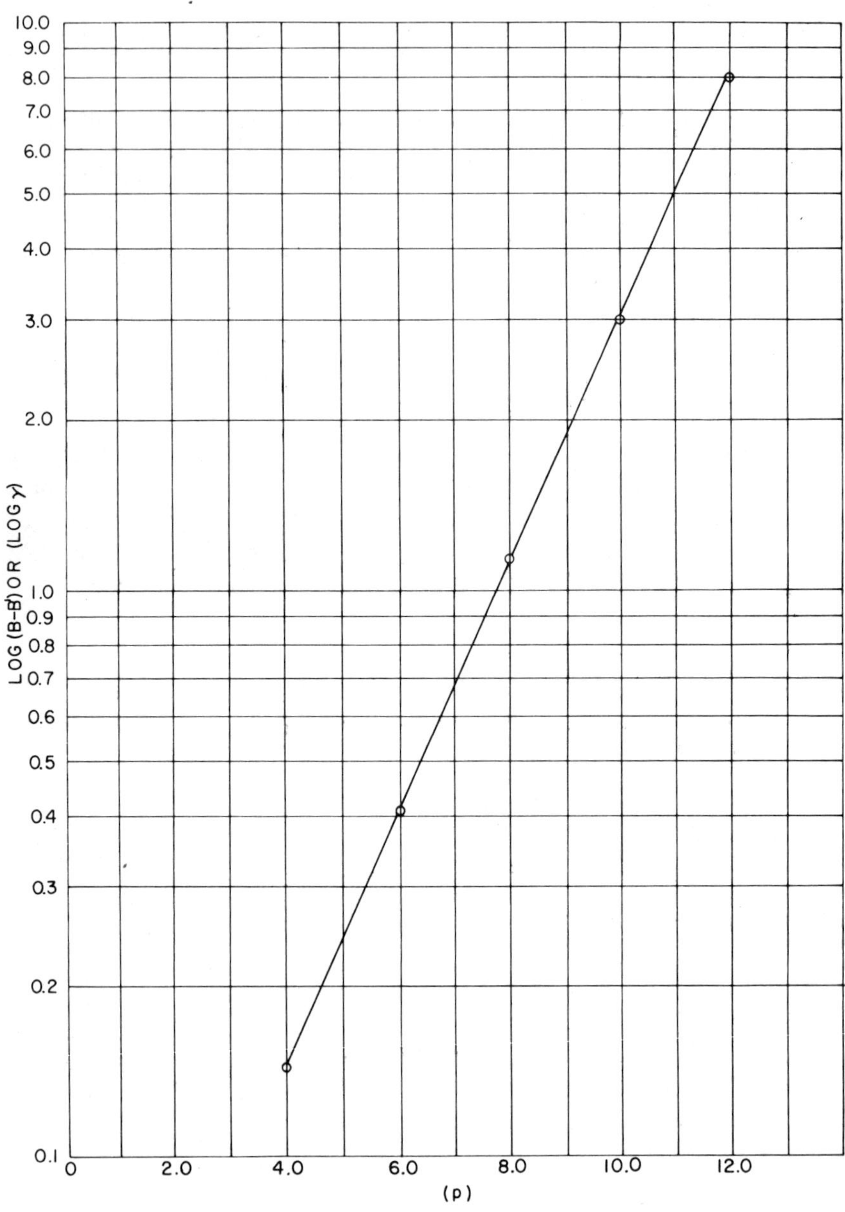

Figure 5-27. A plot of p, log ($B - B'$) values from Table 5-11.

The curve, $B = ae^{bp} + ce^{dp}$

An experimental set of data may be represented by an equation containing two exponential terms. In Figure 5-28 are plotted curves for the equations $B = 0.5e^{p}$, $B = 0.5e^{p} + e^{0.5p}$ and $B = 0.5e^{p} - e^{0.5p}$. The method for determining the values of the constants in the double exponential

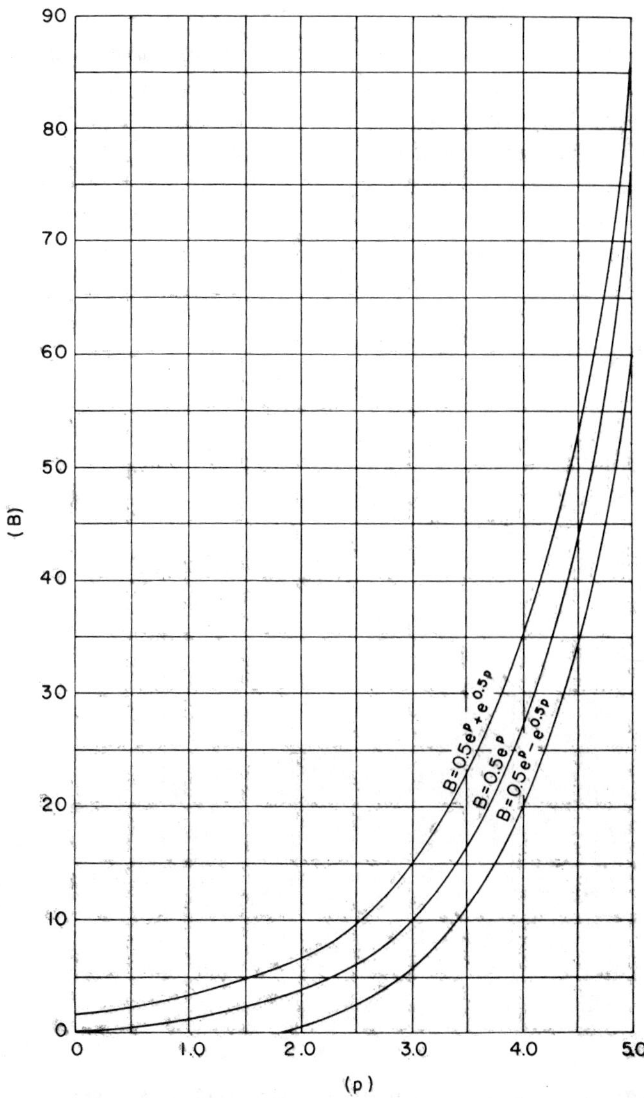

Figure 5-28. Exponential curves of the form $B = ae^{bp} + ce^{dp}$.

equation is similar to the method of determining the values for the constants for the composite equation $B = a + bp + ce^{dp}$. First a portion of the data is used for determining the values of the constants in a single exponential equation, $B = ae^{bp}$. Following the determinations of the values of the constants a and b, the residuals are determined from the equation $r = B_o - B_c = B - ae^{bp}$. Then the values for the constants for the second exponential term are determined from the equation $r = ce^{dp}$.

Example 5-5. The data for the p, B values of Table 5-12 are plotted respectively in rectangular coordinates and semi-logarithm coordinates in Figures 5-29 and 5-30. The semi-logarithm plot indicates that a portion of the curve can be represented by a curve of the type $B = ae^{bp}$. Applying the equation; $\Sigma \log B = n \log a + b(\log c)\ \Sigma p$; dividing the last six sets of data of the tables into two parts; and substituting the appropriate sums into Equation 5-66 gives the equations.

$$19.8819 - 20 = 3 \log a + b\ (0.4343)\ (3.6)$$
$$18.4585 - 20 = 3 \log a + b\ (0.4343)\ (7.5)$$

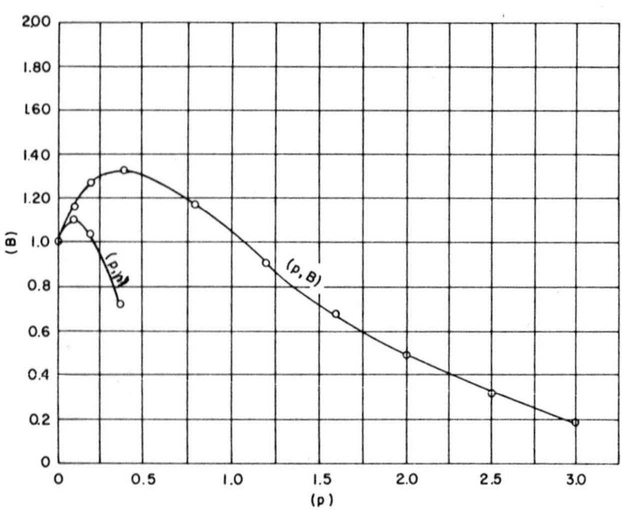

Figure 5-29. A plot of (p, B) values from Table 5-12; and a plot of (p, r) values from Table 5-12.

Table 5-12
Two Exponential Term Curve Data

p	B	log B	r	log r	B calculated
0.0	1.000				0.46
0.1	1.163		1.138	0.05610	1.161
0.2	1.264		0.858	9.9304 −10	1.268
0.4	1.333		0.457	9.6599 −10	1.322
0.8	1.192	0.0763			1.136
1.2	0.933	9.9699 −10			0.870
1.6	0.685	9.8357 −10			0.640
2.0	0.486	9.6867 −10			0.462
2.5	0.308	9.4886 −10			0.305
3.0	0.192	9.2833 −10			0.194

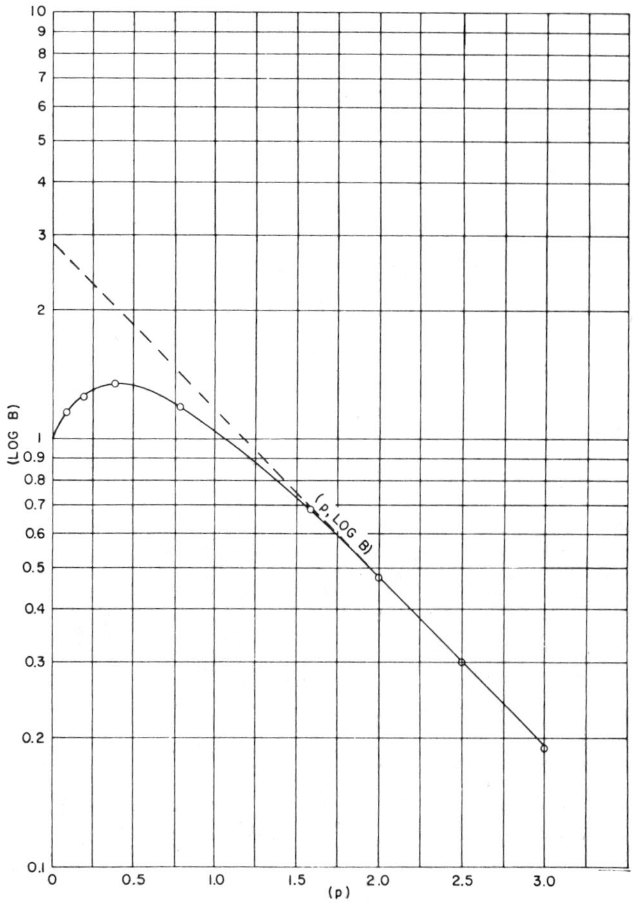

Figure 5-30. A plot of (*p*, log *B*) values from Table 5-12.

Subtracting the second equation from the first gives

$$1.4236 = -(0.4343)\ (3.9)b$$

or

$$b = -\frac{1.4234}{1.6938} = -\ 0.8404$$

Substituting the value of b in the first equation and solving for a gives

$$19.8819 - 20 = 3 \log a + r\ (-\ 0.8404)\ (0.4343)\ (3.6)$$
$$\log\ a = 1.1958/3 = 0.3986$$

or

$$a = 2.504$$

The values for the residuals r were obtained from the equation $r = 2.504e^{-0.8404p} - B$ and listed in the table.

Replacing B in Equation **5-66** with r gives

$$\Sigma \log\ r = n\ \log\ c + d\ (\log\ c)\ \Sigma p \qquad \textbf{5-67}$$

and substituting into Equation **5-67** the appropriate sums from the values of log r and p yields the two equations

$$0.1123 = 2 \log c + d(0.4343)\ (0.2)$$
$$19.5904 - 20 = 2 \log c + d(0.4343)\ (0.6)$$

Solving the two equations for c and d gives: $c = 1.536$ and $d = 3.004$. The empirical equation for the (p, B) data of Table 5-12 is $B = 2.50e^{-0.8404p} - 1.536e^{-3.004p}$. Calculated values of B with the use of the empirical equation are given in the last column of the table. There is a very close agreement between the observed values for B and the computed values for B.

The Curve, $B = a + bp + cp^2 \ldots + kp^n$

The relationships of two variables for sets of scientific and engineering data may often be expressed in the form of nth degree polynomials. For instance the relationships between formation volume factors and pressures or the relationships between molal heat capacities and temperatures may easily be expressed in the form of an nth degree polynomial equation. The degree of the polynomial for equi-incremental sets of data is determined by successive differences in the value of the dependent variable. In the chapter, Selected Topics from Numerical Mathematical Analysis it was shown that the nth difference of a polynomial of degree n is a constant. Therefore, the degree of the polynomial to be used for expressing the relationships between two variables may be determined by setting up a difference table and noting the difference in which all values approach a constant. The plot of the independent variable and the n^{-1} th difference will approximate a straight line.

Several methods for determining the constants of a polynomial for a given set of data have been proposed. In this chapter three methods are discussed and one example is given for each of the methods. With the method of selected points, following the determination of the degree of the polynomial for the given set of data, a curve is plotted. Then $n + 1$ points are selected and the values at these points are substituted into $n + 1$ equation, and finally the $n + 1$ equations are solved for the $n + 1$ constants. The set of $n + 1$ equations are:

$$
\begin{aligned}
B_1 &= a + bp_1 + cp_1^2 + \ldots + kp_1^n \\
B_2 &= a + bp_2 + cp_2^2 + \ldots + kp_2^n \\
B_3 &= a + bp_3 + cp_3^2 + \ldots + kp_3^n \\
&\vdots \\
B_{n+1} &= a + bp_{n+1} + cp^2_{n+1} \ldots + kp_{n+1}^n
\end{aligned}
\qquad \text{5-68}
$$

The constant for an nth degree polynomial may be determined by the method of least mean squares with the use of the following equations.

$$
\begin{aligned}
\Sigma B &= na + b\Sigma p + c\Sigma p^2 \ldots + k\Sigma p^n \\
\Sigma pB &= a\Sigma p + b\Sigma p^2 + c\Sigma p^3 + \ldots + k\Sigma p^{n+1} \\
\Sigma p^2B &= a\Sigma p^2 + b\Sigma p^3 + c\Sigma p^4 + \ldots + k\Sigma p^{n+2} \\
&\vdots \\
\Sigma p^{n-1}B &= a\Sigma p^{n-1} + b\Sigma p^n + c\Sigma p^{n+1} + \ldots + k\Sigma p^{2n-1}
\end{aligned}
\qquad \text{5-69}
$$

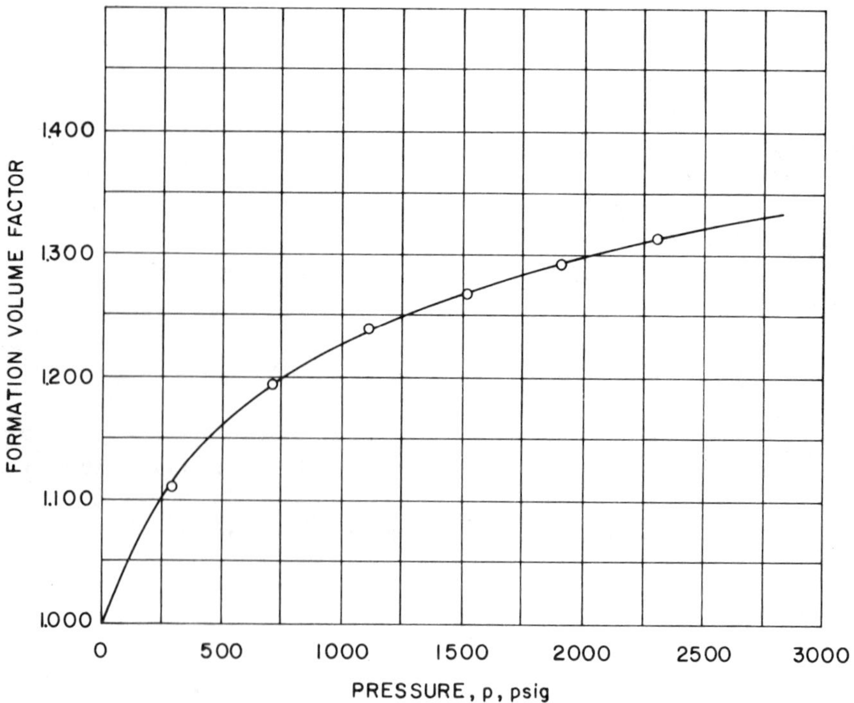

Figure 5-31. A plot of the (*p, B*) values from Table 5-14.

The third method of determining the constants of an *n*th degree polynomial, for $n = 1$ to 5, involves the use of a table of values which was developed by Rachford and Schultz.[3] These values are given in Table 5-13. An example is given which demonstrates the method of using the table for calculating the constants in a 4th degree polynomial.

[3]Rachford, H. H. Jr and Schultz, W. P. "Some Useful Tables for Approximating Smooth Curves by Fifth and Lower Degree Polynomials," Trans, A.I.M.E., (1955), *204* 289-290.

Example 5-6. Given the (p,B) data of Table 5-14, determine a polynomial of degree "n" which gives the relationships between these data. As indicated in Table 5-14 the values for the fourth difference are constant; therefore, a polynomial of 4th degree should adequately represent the relationships for the (p, B) data. A plot of the (p, B) data is given in Figure 5-31. By making the substitution $p_x = (p - 0)/(2500 - 0) = p/2500$ the curve may be plotted p_x versus B for values of p from 0 psig to 2500 psig.

Table 5-13
Numerical Values of Elements for $n = 1, 2, 3, 4,$ and 5

X =	0	.1	.2	.3	.4
			I		
x^0	0.31818182	0.27272727	0.22727273	0.18181818	0.13636364
x^1	−0.45454545	−0.36363636	−0.27272727	−0.18181818	−0.09090909
			II		
x^0	0.58041958	0.37762238	0.20979021	0.07692308	−0.02097902
x^1	−2.20279720	−1.06293706	−0.15617715	0.51748252	0.95804196
x^2	1.74825175	0.69930070	−0.11655012	−0.69930070	−1.04895105
			III		
x^0	0.790210	0.335664	0.055944	−0.083916	−0.118881
x^1	−5.536131	−0.396270	2.288267	3.073038	2.513598
x^2	10.489510	−1.048951	−6.526806	−7.400932	−5.128205
x^3	−5.827506	1.165501	4.273504	4.467754	2.719503
			IV		
x^0	0.916088	0.209791	−0.069930	−0.104896	−0.034965
x^1	−9.906823	3.974339	6.658913	3.801495	−0.400146
x^2	32.342712	−22.902078	−28.379969	−11.043142	9.440552
x^3	−40.792541	36.130538	39.238541	10.295260	−20.590523
x^4	17.482518	−17.482518	−17.482518	−2.913752	11.655012
			V		
x^0	0.973771	0.094415	−0.089159	−0.027978	0.041952
x^1	−14.428829	13.018550	8.166224	−2.228024	−6.429619
x^2	69.600090	−97.416930	−40.799092	38.633413	59.117073
x^3	−145.75444	246.05414	74.225850	−129.65370	−160.53948
x^4	137.66779	−257.85294	−57.544256	157.33315	171.90190
x^5	−48.074109	96.148218	16.024703	−64.098812	−64.098812

Table 5-13 continued

Table 5-13 continued

.5	.6	.7	.8	.9	1.0
			I		
0.09090909	0.04545455	0	−0.04545455	−0.09090909	−0.13636364
0	0.09090909	0.18181818	0.27272727	0.36363636	0.45454545
			II		
−0.08391608	−0.11188811	−0.10489510	−0.06293706	0.01398602	0.12587413
1.16550117	1.13986014	0.88111888	0.38927739	−0.33566434	−1.29370630
−1.16550117	−1.04895105	−0.69930070	−0.11655012	0.69930070	1.74825175
			III		
−0.083916	−0.013986	0.055944	0.090909	0.055944	−0.083916
1.165501	−0.415695	−1.674437	−2.055167	−1.002331	2.039627
−1.165501	3.030303	6.002331	6.293706	2.447552	−6.993007
0	−2.719503	−4.467754	−4.273504	−1.165501	5.827506
			IV		
0.041956	0.069930	0.034965	−0.034965	−0.069930	0.041956
−3.205116	−3.329441	−0.945999	2.315465	3.368307	−2.330994
20.687627	17.599063	2.360142	−15.559437	−19.405598	14.860128
−34.965031	−26.029529	1.359748	30.691527	33.799532	−29.137522
17.482518	11.655012	−2.913752	−17.482518	−17.482518	17.482518
			V		
0.041952	−0.006987	−0.041952	−0.015730	0.045466	−0.015730
−3.205087	2.700070	5.083487	0.808051	−5.675974	2.191151
20.687646	−32.077492	−47.316429	−3.140214	55.109307	−22.397372
−34.965035	113.91956	141.30869	−4.296017	−176.12417	75.824619
17.482476	−148.59201	−163.16058	22.579454	222.88802	−102.70301
0	64.098812	64.098812	−16.024703	−96.148218	48.074109

Table 5-14
Difference Table

p	B	ΔB	Δ^2 B	Δ^3 B	Δ^4 B
300.0	1.120				
		0.070			
700	1.192		−0.023		
		0.047		0.008	
1100	1.239		−0.015		−0.002
		0.032		0.006	
1500	1.271		−0.009		−0.002
		0.023		0.004	
1900	1.294		−0.005		−0.002
		0.018		0.002	
2300	1.312		−0.003		
		0.015			
2700	1.327				

Next the values of B are read from the graph in intervals of 0.10 from 0 to 1.0. These values for p_x and B are recorded respectively in columns 1 and 2 of Table 5-15. Then the values for x^0 from the 4th horizontal set of values of Table 5-13 are listed in column 4 of Table 45-15. The corresponding values for x^1, x^2, x^3, x^4 of Table 5-13 are also listed respectively in columns 6, 8, and 10 and 12 of Table 5-15. Following this listing of x^n values, corresponding values in columns 2 and 4 are multiplied together and the products are listed in the corresponding order in column 5. In a similar manner values in columns 6, 8, 10 and 12 are multiplied by the values in column 2 and the products are listed respectively in columns 7, 9, and 11 and 13. Finally the sums for the values in columns 5, 7, 9, 11 and 13

Table 5-15
Calculation of Constants

Px 1	B 2	Psig 3	x° 4	Bx⁰ 5	x¹ 6	Bx¹ 7
0.0	1.000	000	0.916088	0.916088	−9.906823	−9.906832
0.1	1.105	250	0.209791	0.231819	3.974339	4.391644
0.2	1.166	500	−0.069930	−0.081538	6.658913	7.764292
0.3	1.200	750	−0.104896	−0.125875	3.801495	4.561794
0.4	1.228	1000	−0.034965	−0.042937	−0.400146	−0.491379
0.5	1.250	1250	−0.041956	0.052445	−3.205116	−4.006395
0.6	1.271	1500	0.069930	0.088881	−3.329441	−4.231719
0.7	1.282	1750	0.034965	0.044825	−0.945999	−1.212770
0.8	1.296	2000	0.034965	0.045315	−2.315465	3.000842
0.9	1.312	2250	−0.069930	0.091748	3.368307	4.419218
1.0	1.321	2500	+0.041956	0.055423	−2.330994	−3.079243
				1.002008		1.209461

x² 8	Bx² 9	x³ 10	Bx³ 11	x⁴ 12	Bx⁴ 13	Bcal 14
32.342712	32.342712	−40.792541	−40.792541	17.482518	17.482518	1.002
−22.902078	−25.306796	36.130538	39.924244	−17.482518	−19.318182	1.102
−28.379969	−33.091043	39.238541	45.752138	−17.482518	−20.384615	1.166
−11.043142	−13.251770	10.295260	12.354312	− 2.913752	− 3.496502	1.201
9.440552	11.592997	−20.590523	−25.285162	11.655012	14.312354	1.288
20.687627	25.859533	−34.965031	−43.706288	17.482518	21.853147	1.250
17.599063	22.368409	−26.029529	−33.083531	−11.655012	14.813520	1.267
2.360142	− 3.025702	1.359748	1.743196	− 2.913752	− 3.735430	1.282
−15.559437	−20.165030	30.691527	39.776218	−17.482518	−22.657343	1.297
−19.405598	−25.460144	33.799532	44.344985	−17.482518	−22.937063	1.315
14.860128	19.630229	−29.137523	−38.490667	17.482518	23.094406	1.320
	− 2.455201		2.536904		− 0.9737190	

are found. These sums are respectively the values of the constants in the equation

$$B = a^1 + b^1 p_x + c^1 p_x^2 + d^1 p_x^3 + e^1 p_x^4. \qquad 5\text{-}70$$

Substituting into Equation **5-70** the sums of the products from columns 5, 7, 9, 11 and 13 gives the equation

$$B = 1.002008 + 1.209461 p_x - 2.455201 p_x^2 + 2.536904 p_x^3 - 0.973190 p_x^4 \qquad 5\text{-}71$$

and since $p_x = p/2500$ Equation **5-71** becomes

$$B = 1.002008 + 1.209461 \left(\frac{p}{2500}\right) - 2.455201 \left(\frac{p}{2500}\right)^2 + 2.53694 \left(\frac{p}{2500}\right)^3 - 0.973190 \left(\frac{p}{2500}\right)^4$$

or

$$B = 1.002008 + 4.837844 \times 10^{-4} p - 3.928321 \times 10^{-7} p^2 + 1.623618 \times 10^{-10} p^3 - 2.491366 \times 10^{-4} p^4. \qquad 5\text{-}72$$

The empirical equation is Equation **5-72.** A comparison of the values in columns 2 and 14 of Table 5-15 adequately represents the given (p, B) data.

Periodic Curves

Solutions to scientific and engineering problems are somtimes expressed in series containing trigonometric terms. These solutions may be represented graphically by curves composed of congruent parts at certain intervals. The trigonometric series are called Fourier series and have the form

$$B = f(B) = a_o + a_1 \cos\theta + a_2 \cos 2\theta + \ldots a_n \cos n\theta + b_1 \sin\theta + b_2 \sin 2\theta + \ldots b_n \sin n\theta \qquad 5\text{-}73$$

where θ is expressed in radians and a_k and b_k are coefficients which may be determined if the function is known. The period of the series is 2π.

For known functions, $f(\theta)$, the coefficients may be determined by the equations

$$a_o = \frac{1}{2\pi} \int_0^{2\pi} f(\theta)\, d(\theta)$$

$$a_n = 1/\pi \int_{-\pi}^{\pi} f(\theta) \cos n\, \theta d\theta \quad (n = 0, 1, 2, \ldots\ldots)$$

$$b_n = 1/\pi \int_{-\pi}^{+\pi} f(\theta) \sin n\, \theta\, d\, \theta\ (n = 1, 2 \ldots) \qquad \text{5-74}$$

Often the engineer will be able to collect the data in the form of numerical values for the function $f(\theta)$. The numerical values may be obtained by an oscillograph or by other instruments. After the collection of a given set of values for the function, the problem then becomes that of determining the coefficients of a Fourier Series which will approximately represent the data. Numerical methods of integration (Simpson's rule), utilizing Equation **5-74**, are sometimes used for computing the coefficients a_k and b_k from a set of experimental data. The following sections give two schemes for computing the coefficients of a Fourier Series for sets of experimental data.

Six-Ordinate Scheme. The equation is of the form

$$B = a_o + a_1 \text{cost}\ \theta + a_2 \cos 2\,\theta + a_3 \cos 3\,\theta$$
$$+\, b_1 \sin \theta + b_2 \sin 2\,\theta \qquad \text{5-75}$$

and it has six coefficients. Scheme for easily computing the coefficients have been devised by Runge. The essential steps in developing the six-ordinate scheme follows.

Based upon two theorems in trigonometry the following set of equations have been developed.

$$a_o = 1/n\, \Sigma B_r = 1/n\ (B_o + B_1 + B_2 + \ldots B_{(n-1)})$$
$$a_n = 1/n\, \Sigma B_r \cos n/2\ \theta = 1/n \Sigma B_r$$
$$+ \cos r\pi = 1/n\, \Sigma(B_0 - B_1 + B_2 - B_3 + \ldots \mathrm{B}_{(n-1)})$$
$$a_k = 2/n\, \Sigma B_r \cos k\theta = 2/n\ (B_0 \cos + k\theta_0 + B_1 \cos k\theta_1 + \ldots$$
$$+\, B_{n-1} \cos k\, \theta_{n-1})$$
$$b_k = 2/n\, \Sigma Br \sin k\, \theta = 2/n\ (B_0 \sin k\, \theta_0 + B_1 \sin k\, \theta + \ldots$$
$$+\, B_{n-1} \sin k\, \theta_{n-1}). \qquad \text{5-76}$$

where n is the number of equal intervals into which the interval $\theta = 0$ to $\theta = 2\pi$ was divided and $k = n/2$. The division of the interval is represented by the Table 5-16.

Table 5-16
Six-Ordinate Scheme

		$\frac{2\pi}{n}$	$\frac{4\pi n}{n}$	$\frac{6\pi}{n}$		$\frac{r\,2\pi}{n}$	$\frac{(n-1)\,2\pi}{n}$
θ	θ_0	θ_1	θ_2	θ_3	. . .	θ_r	θ_{n-1}
B	B_0	B_1	B_2	B_3	. . .	B_r	B_{n-1}

Table 5-17
Six-Ordinate Scheme Values

θ	0°	60°	120°	180°	240°	300°
B	B_0	B_1	B_2	B_3	B_4	B_5

Dividing the interval $\theta = 0$ to $\theta = 2\pi$ into six equal intervals and applying equations **5-76** gives the following six equations, with the intervals as represented in the Table 5-17.

$$6\,a_0 = B_0 \quad + B_1 \quad + B_2 \quad + B_3 \quad + B_4 \quad + B_5$$
$$6\,a_3 = B_0 \quad - B_1 \quad + B_2 \quad - B_3 \quad + B_4 \quad = B_5$$

$$3a_1 = B_o \cos 0° + B_1 \cos 60° + B_2 \cos 120° + B_3 \cos 180° + B_4 \cos 240° + B_5 \cos 300°$$
$$3a_2 = B_0 \cos 0° + B_1 \cos 120° + B_2 \cos 240° + B_3 \cos 360° + B_4 \cos 480° + B_5 \cos 600°$$
$$3b_1 = B_0 \sin 0° + B_1 \sin 60° + B_2 \sin 120° + B_3 \sin 180° + B_4 \sin 240° + B_5 \sin 300°$$
$$3b_2 = B_0 \sin 0° + B_1 \sin 120° + B_2 \sin 240° + B_3 \sin 360° + B_4 \sin 480° + B_5 \sin 600°$$

5-76

A scheme for determining the values of the coefficients a_k and b_k of the six equations is now given. The *B*'s are arranged in two rows,

$$\begin{array}{llll} & B_0 & B_1 & B_2 & B_3 \\ & & B_5 & B_4 & \\ \hline \text{sum} & W_0 & W_1 & W_2 & W_3 \\ \text{Diff} & & Z_1 & Z_2 & \end{array} \qquad \text{5-77}$$

where the W's are the sums and the Z's are the differences of the quantities in the same vertical columns. For instance $W_0 = B_0$, $W_1 = B_1 + B_5$, $Z_1 = B_1$ B_5 etc.

Next the W's and Z's are arranged in two rows

$$\begin{array}{llll} & W_0 & W_1 & Z_1 \\ & W_3 & W_2 & Z_2 \\ \hline \text{Sum} & X_0 & X_1 & U_1 \\ \text{Diff} & Y_0 & Y_1 & V_1 \end{array} \qquad \text{5-78}$$

and the sums and differences of the quantities in the same vertical columns are obtained.

Finally the set of equations for calculating the six coefficients are:

$$\begin{aligned} 6\,a_0 &= X_0 + X_1 & \qquad 6\,a_3 &= Y_0 - Y_1 \\ 3\,a_1 &= Y_0 + \tfrac{1}{2}Y_1 & \qquad 3\,a_2 &= X_0 - \tfrac{1}{2}X_1 \\ b_1 &= 0.2888\,U_1 & \qquad b_2 &= 0.28888V_1 \end{aligned} \qquad \text{5-79}$$

Example 5-7. Determine a Fourier series equation which will approximately represent the (p, B) data of Table 5-14 in the interval $p = 0$ to $p = 2500$. The (p, B) data is plotted in Figure 5-32. By setting $\theta = (300p)/(2500)$, Equation **5-75** may be written in terms of B and p as

$$B = a_o + a_1 \cos\frac{(300p)}{(2500)} + a_2 \cos 2\frac{(300p)}{(2500)} + a_3 \cos 3\frac{(300p)}{(2500)} \qquad \text{5-80}$$

$$+ b_1 \sin\frac{(300p)}{(2500)} + b_2 \sin 3\frac{(300p)}{(2400)}$$

The substitution of the values for p in Equation **5-80** converts the trigonometric terms in terms of degrees. The solid line of Figure 5-32 is also a plot of formation volume factor versus degrees.

The coefficients a_k and b_k are computed by substituting the appropriate values from Table 5-18 into Equations **5-77, 5-78** and **5-79** as follows:

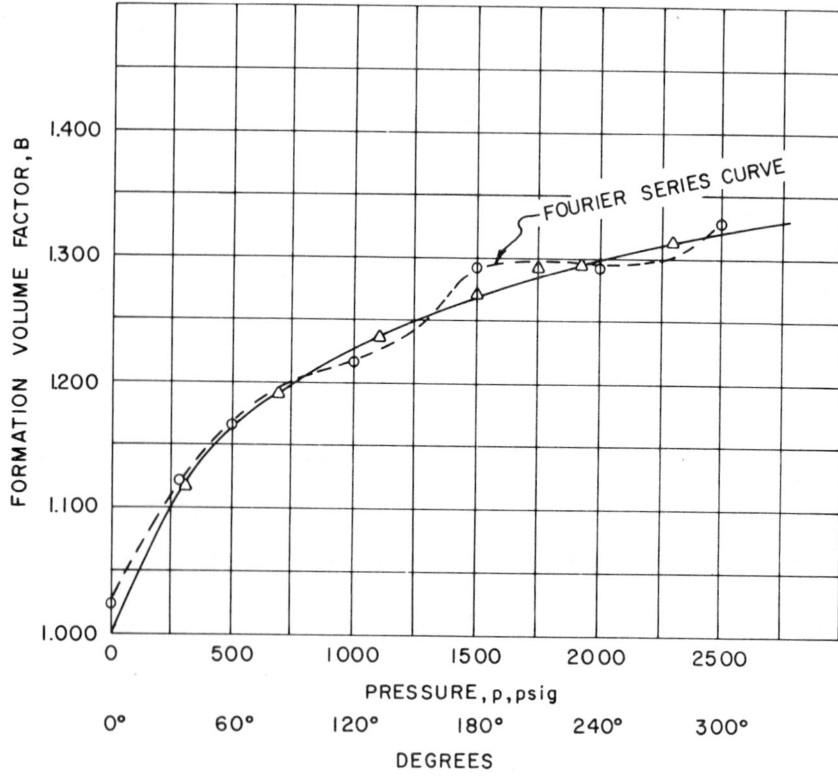

Figure 5-32. A Fourier series curve for formation volume factor.

Table 5-18
Six-Ordinate Scheme Numerical Values

θ	0°	60°	120°	180°	240°	300°
B	1.000	1.157	1.225	1.270	1.300	1.320
p_{psig}	0.00	500	1000	1500	2000	2500
B_{cal}	1.016	1.165	1.216	1.294	1.293	1.329

	1.000	2.477	−0.163
	1.270	2.525	−0.075
Sum	2.270	5.002	−0.238
Diff.	−0.270	−0.048	−0.088

	1.000	1.157	1.225	1.270
		1.320	1.300	
Sum	1.000	2.477	2.525	1.270
Diff.		−0.163	−0.075	

$X_0 = 2.270,$
$X_1 = 5.002,$
$Y_0 = -0.270,$
$Y_1 = -0.048,$
$U_1 = -0.238,$
$V_1 = -0.088$
$6\ a_0 = 0.270 + 0.024$ or $a_0 = 1.212$
$3\ a_1 = -0.270 + 0.024$ or $a_1 = -0.082$
$b_1 = -0.2888 \times (-0.238) = -0.06873$
$b_2 = 0.2888 \times (0.088) = -0.02541$
$6\ a_3 = -0.270 + 0.048$ or $a_3 = -0.037$
$3\ a_2 = 2.270 = 2.501$ or $a_2 = -0.077$

The periodic empirical equation is

$$B = 1.212 - 0.082 \cos \frac{(300p)}{(2500)} - 0.077 \cos 2 \frac{(300p)}{(2500)}$$

$$- 0.037 \cos 3 \frac{(300p)}{(2500)} - 0.06873 \sin \frac{(300p)}{(2500)}$$

$$- 0.02541 \sin 2 \frac{(300p)}{(2500)}$$

Computed values of B with the use of the empirical equation are given in the fourth row of Table 5-18. A plot of the calculated values is represented by the broken line in Figure 5-32. An average curve drawn through the points representing the computed values would approximately coincide with the plot of the (p, B) data.

The reader must bear in mind that the periodic empirical curve will only approximate the experimental values of (p, B) in the intervals from $p = 0$ to $p = 2500$.

Twelve-Ordinate Scheme

The (p, B) data of Table 5-14 could be represented better by a trigonometric curve containing more terms. Schemes containing more terms than six have been developed. A twelve-ordinate scheme follows. The scheme was developed in a similar manner as the six ordinate scheme. The development of the scheme is not given.

The 12 coefficient equation is

$$B = a_0 + a_1 \cos\theta + a_2 \cos 2\theta + a_3 \cos 3\theta + a_4 \cos 4\theta$$
$$+ a_5 \cos 5\theta + a_6 \cos 6\theta + b_1 \sin\theta + b_2 \sin 2\theta$$
$$+ b_3 \sin 3\theta + b_4 \sin 4\theta + b_5 \sin 5\theta$$

where the interval from $\theta = 0$ to $\theta = 360°$ is divide into 12 equal intervals as represented by Table 5-19.

Table 5-19

θ	0°	30°	60°	90°	120°	150°	180°	210°	240°	270°	300°	360°
B	B_0	B_1	B_2	B_3	B_4	B_5	B_6	B_7	B_8	B_9	B_{10}	B_{11}

The scheme is

coordinates	B_0	B_1	B_2	B_3	B_4	B_5	B_6
		B_{11}	B_{10}	B_9	B_8	B_7	
Sum	W_0	W_1	W_2	W_3	W_4	W_5	W_6
Diff.		Z_1	Z_2	Z_3	Z_4	Z_5	
	W_0	W_1	W_2	W_3	Z_1	Z_2	Z_3
	W_6	W_5	W_4		Z_5	Z_4	
Sum	X_0	X_1	X_2	X_3	U_1	U_2	U_3
Diff.	Y_0	Y_1	Y_2		V_1	V_2	
	X_0	X_1			U_1	Y_0	
	X_2	X_3			U_3	Y_2	
Sum	M_0	M_1		Diff.	N_1	N_2	

$$12\,a_0 = M_0 + M_1$$
$$12\,a_6 = M_0 - M_1$$
$$6\,a_1 = Y_0 + 0.866Y_1 + 0.500Y_2$$
$$6\,a_5 = Y_0 + 0.866Y_1 + 0.500Y_2$$
$$6\,a_2 = (X_0 - X_3) + 0.500\,(X_1 - X_2)$$
$$6\,a_4 = (X_0 + X_3) - 0.500\,(X_1 + X_2)$$

$6\,a_3 = N_2$
$6\,b_3 = M_3$
$6\,b_1 = 0.500\,Y_1 + 0.866Y_2 + Y_3$
$6\,b_5 = 0.500\,U_1 - 0.866\,U_2 + U_3$
$6\,b_2 = 0.866\,(V_1 + V_2)$
$6\,b_4 = 0.866\,(V_1 - V_2)$

For other schemes the reader should consult the references given for this chapter.

Exercises

1. Determine by the method of selected points, the method of averages, and the method of least mean squares, the empirical curve, $q = k'p$, for the data:

p	100	200	300	400	500	600	700
q	5.0	9.8	15.0	20.3	24.7	30.2	35.1

2. Determine by the method of selected points, the method of averages, and the method of least mean squares the empirical equation, $q = a + bp$ for the data:

p	5	10	15	20	25	30	35	40
q	3.49	4.98	6.57	7.97	9.61	11.02	12.53	13.95

3. Determine by the three methods stated in Problem 1, the empirical equation $p = ax^b$ for the data:

x	0	1	2	3	4	5
p	0.00	3.00	3.52	3.99	4.21	4.35

4. Determine by the three methods stated in Problem 1, the empirical equation $r = ae^{bs}$ for the data:

s	0.2	0.4	0.6	0.8	1.0	1.2
r	2.26	3.34	4.99	7.43	11.14	16.47

5. By the three methods stated in Exercise 1, determine the empirical equation $M = r/(a + br)$ for the data:

r	1.00	2.00	3.00	4.00	5.00	6.00	7.00	8.00
M	2.11	2.86	3.33	3.65	3.78	3.98	4.14	4.22

6. Determine the best impirical equation $q = ae^{bt} + c$ for the data:

t	0.2	0.8	2.0	6.0	10.0	20.0	40.0
q	1.637	3.337	5.694	11.816	16.664	26.650	42.904

7. Determine the best empirical equation $q = ae^{bt} + c$ for the data:

t	1.0	2.0	4.0	6.0	8.0	10.0	12.5
q	0.988	0.8819	0.5452	0.4437	0.4132	0.4040	0.4009

8. By two methods determine the empirical equation, $B = a + bp + cp^2$, for the data:

p	1.0	1.2	1.4	1.6	1.8	2.0	2.2
B	1.0050	1.0050	1.0042	1.0032	1.0003	1.0000	0.9978

9. Determine the best empirical equation $B = \dfrac{p}{a + bp} + c$ for the data:

p	1.00	2.00	3.00	4.00	5.00	6.00	7.00	8.00
B	2.67	3.00	3.21	3.34	3.43	3.49	3.57	3.61

10. Determine the best empirical equation $B = a + bp + ce^{dp}$, for the data:

p	2.0	4.0	6.0	8.0	10.0	12.0	14.0
q	3.99	6.73	9.82	13.45	17.89	23.52	30.95

11. Determine the best empirical equation $B = ae^{bp} + ce^{dp}$ for the data:

p	2.0	4.0	6.0	8.0	10.0	12.0	14.0
B	2.161	1.669	1.151	0.851	0.638	0.483	0.369

12. Determine an empirical equation in the form of a 5th degree polynomial for the (P, B) data of Table 5-4.

13. For the data of Table 5-8 determine a periodic empirical equation by the twelve-ordinate scheme.

14. For the data of Table 5-4 determine a periodic empirical equation by the six-ordinate scheme.

15. Determine an empirical equation for the data:

p	5000	10,000	15,000	20,000	25,000	30,000
Rs	260	400	490	585	630	690

16-22. Using the best methods, determine the empirical equations for the following sets of (p, B) data:

p	4.22	10.78	15.36	20.4	25.2	37.4	46.3	62.0
B	3.11	5.60	7.34	9.22	11.16	15.66	19.12	25.00

17.

p	29.32	19.08	14.72	12.42	9.79	8.15	7.41	6.15
B	15.2	24.6	32.6	40.9	51.7	63.2	70.4	86.2

18.

p	0.291	0.499	0.572	0.612	0.683	0.706	0.724	0.736
B	5	12	20	25	40	48	56	64

19.

p	1.3	2.0	2.5	4.0	5.6	7.0	9.0	11.9	16.8	23.3
B	0.5	1	1.5	3	5	7	10	15	25	40

20.

p	68.4	64.8	61.3	56.6	54.3	52.3	48.5	46.3	45.0
B	3	4.5	6	8	9	10	12	13.5	14.5

21.

p	0	1	2	3	4	5	6
B	10	4.97	2.47	1.22	0.61	0.30	0.14

22.

p	0°	60°	120°	180°	240°	300°
B	−.85	0.95	0.72	2.75	−1.37	−2.20

Suggested Reading

1. Churchill, R. V., *Fourier Series and Boundary Value Problems,* New York: McGraw-Hill Book Co., Inc., 1941, pp. 53-75.
2. Craft, B. C. and Hawkins, M. F., *Applied Reservoir Engineering*, Englewood Cliffs: Prentice-Hall, Inc., 1959, pp. 66-67, 167-172, 381.
3. Davis, D. S., *Empirical Equations and Nomography,* New York: McGraw-Hill Co., Inc., 1943, pp. 3-90.
4. Glasstone, S., *Text-Book of Physical Chemistry,* New York: D Van Nostrand Co. Inc., 1940, pp. 206-207.
5. Johnson, L. H., *Nomography and Empirical Equation,* New York: John Wiley and Sons, Inc., 1952, pp. 95-145.
6. Lewis, W. K. and Radasch, A. H., *Industrial Stoichiometry,* New York: McGraw-Hill Book Co., Inc., 1926, p. 45.
7. Lipka, J., *Graphical and Mechanical Computations,* New York: McGraw-Hill Book Co., Inc., 1926, p. 45.
8. Rachford, H. H., Jr. and Schultz, W. P., "Some Useful Tables for Approximating Smooth Curves by Fifth-and-Lower Degree Polynomials," Trans. A.I.M.E., pp. 289-290, 204, 1955.
9. Runge, Z. F., *Math u Phys.,* XLVIII (1903), Lii 17 (1905); Erlauterung des Rechnungsformulars, U.S.W., Braunschweig, 1913.
10. Thompson, S. P., "Proc. Phys. Soc., XIX 443, 1905.
11. Thompson, S. P., "The Electrician," May 5, 1905.
12. Uren, L. C., *Petroleum Production Engineering - Oil Field Exploitation,* New York: McGraw-Hill Book Co., Inc., 1953, pp. 104-134.

6 Basic Equations for Fluid Flow Through Porous Media

Although connate water occurs within the pores of an oil-and-gas-producing formation and oil always contains some dissolved gas, the flow of fluids within porous media often reduces to single-phase-flow systems. Single-phase-flow systems, sometimes referred to as homogeneous-fluid systems, are those in which only a single mobile fluid phase is present. A formation producing only free gas will constitute a homogeneous-flow system, even though it may contain 50% connate water—provided the connate water remains immobile. Similarly, a formation producing an undersaturated oil can be treated as a homogeneous-fluid system regardless of the connate water content—provided the water remains immobile.

This chapter deals with the development of the basic equations for single-phase-flow systems through uniform porous producing formations. In the development of the basic single-phase-flow equations, three basic types of physical laws of hydrodynamics are considered. The three basic types of physical laws are: (1) the law of the conservation of matter, (2) the thermodynamic equation of state of the fluid in question and of its condition of flow, and (3) the law of force to which individual fluid elements are subjected. The third law is currently called a momentum balance or an energy balance. The resultant of the three laws defines the structure of the fluid-flow system.

The Equation of Continuity

In terms of fluid-flow systems, the law of conservation of matter may be stated as; the net mass flux, per unit of time, through any infinitesimal volume element is equal to the free volume of the element multiplied by the rate of change of the fluid density within the infinitesimal volume. The *equation of continuity* for porous media may be expressed as

$$\nabla \cdot (\rho v) = \partial/\partial x\ (\rho v_x) + \partial\ \partial y/\ (\rho v_y) + \partial/\partial z\ (\rho v_z) = -\ \phi\ \partial\rho/\partial t \qquad \text{6-1}$$

where v_x, v_y, v_z are fluid velocity components in the cartesian coordinate system (x, y, z), ρ is the fluid density, ϕ is the porosity of the medium at (x, y, z), this time and v is the resultant vector fluid velocity. The symbol div (or $\nabla \cdot$) is the differential operator or div $= \nabla \cdot = \partial/\partial x + \partial/\partial\ y + \partial/\partial z$

The equation of continuity or Equation **6-1** is developed as follows. A rectangular parallelepiped of edges, dx, dy, dz with center at (x, y, z), as given in Figure 6-1 is chosen.

The mass flow into the side dy, dz perpendicular to the x axis at a distance $x - dx/z$ from the yz plane is

(a) $[\rho v_x - \partial/\partial x\ (\rho v_x)\ dz/2]\ dy\ dz$

where $\rho\ v_x$ refers to the point (x, y, z).

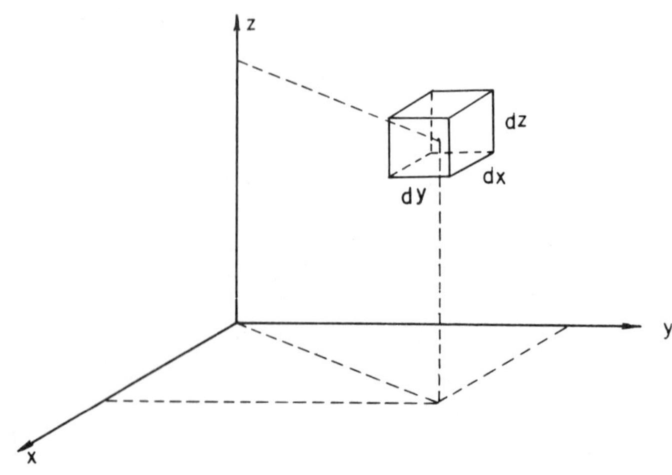

Figure 6-1. Rectangular coordinate system.

The mass flux leaving the parallel side is

(b) $[\rho v_x + \partial/\partial x\ (\rho\, v_x)\ dx/2]\ dydz$

and the net flux leaving the parallelepiped through these two sides is the difference obtained by subtracting (a) from (b) or

(c) $\partial/\partial x\ (\rho\, v_x)\ dxdydz$

In a similar manner, the net flux leaving the parallelepiped through the two sides which are perpendicular to y and z axes are found to be respectively

(d) $\partial/\partial y\ (\rho\, v_y)\ dxdydz$

and

(e) $\partial/\partial z\ (\rho\, v_z)\ dx\ dy\ dz$.

Adding (c), (d) and (e) gives

(f) $[\partial/\partial x\ (\rho\, v_x) + \partial/\partial y\ (\rho\, v_y) + \partial/\partial z\ (\rho\, v_z)\ ]\ dxdydz$

which is the resultant mass flow per unit of time out of the volume element *dxdydz*.

The mass of fluid in the volume element *dxdydz* is

(g) $\phi\ \rho\ dxdydz$

and the loss of mass in the volume element *dxdydz* per unit of time is

(h) $-\ \phi\ \partial\rho/\partial t\ dxdydz$.

By the law of conservation of energy the resultant mass flow per unit of time out of the volume element *dx dy dz* is equal to loss of mass in the volume element *dxdydz* or the expression (f) = (h) or

$$[\partial/\partial x\ (\rho\, v_x) + \partial/\partial y\ (\rho\, v_y) + \partial/\partial z\ (\rho\, v_z)\ ]\ dxdydz = -\ \phi\ \partial\rho/\partial t\ dxdydz \qquad \textbf{6-2}$$

Dividing both sides of the equal sign of Equation **6-2** by *dx dy dz* gives

$$\partial/\partial x\ (\rho\, v_x) + \partial/\partial y\ (\rho\, v_y) + \partial/\partial z\ (\rho\, v_z) = -\ \phi\ \partial\rho/\partial t \qquad \textbf{6-3}$$

which is the *equation of continuity* in rectangular coordinates.

Incompressible Fluid Flow Equation (Laplace Equation)

Thus far in our development of the *equation of continuity* only a material fluid has been considered. To proceed further in the development of fluid flow equations the nature of the fluid involved and the thermodynamic character of its flow must be specified; and, in addition the nature of the flow and the forces acting on the fluid must be specified before completing the fluid-flow system. For an incompressible liquid the *equation of state* is

$$\rho = \text{constant} \tag{6-4}$$

The forces acting on the fluid are of three types: (1) the pressure gradients of the components $\partial p/\partial x$, $\partial p/\partial y$, $\partial p/\partial z$; (2) the external "body forces" such as gravity of the components Fx, Fy, Fz which act on each volume element of the fluid; and (3) the forces which oppose the motion of the fluid are due to the internal resistance or friction experienced by the fluid. The equations which originally were due to Navier and Stokes[1] are the hydrodynamic equations of motion; these equations are:

$$\begin{aligned}
&\rho\,(\partial v_x/\partial t + v_x\,\partial v_x/\partial x + v_y\,\partial v_x/\partial y + v_z\,\partial v_x/\partial z) = \mu\,\nabla^2 v_x\\
&+ 1/3\,\mu\,\partial\theta/\partial x - \partial p/\partial x + F_x\\
&\rho\,(\partial v_y/\partial t + v_x\,\partial v_y/\partial x + v_y\,\partial v_y/\partial y + v_z\,\partial v_y/\partial z) = \mu\,\nabla^2 v_y\\
&+ 1/3\,\mu\,\partial\theta/\partial y - \partial p/\partial y + Fy\\
&\rho\,(\partial v_z/\partial t + v_x\,\partial v_z/\partial x + v_y\,\partial v_z/\partial y + v_z\,\partial v_z/\partial z) = \mu\,\nabla^2 v_z\\
&+ 1/3\,\mu\,\partial\theta/\partial z - \partial p/\partial z + Fz
\end{aligned} \tag{6-5}$$

where

$$\theta = \frac{\partial v_x}{\partial x} + \frac{\partial v_y}{\partial y} + \frac{\partial v_z}{\partial z}$$

The five equations, [Equations **6-3, 6-4** and **6-5**] are sufficient in principle for predicting all the details of motion of a viscous fluid flowing in any container regardless of the shape. The flow of a viscous fluid through a porous media is a special case of the general problem of flow of viscous fluids between impermeable boundaries. Since the left-hand side

[1]Navier, C.L.M.H., Ann Chim Phys; 19 234 (1821); G. G. Stokes, Trans. Cambridge Phil. Soc. 8, 287 (1845).

of Equation **6-5** is non-linear it is almost impossible to solve the three equations except for fluid flowing where the macroscopic geometry of the system is simple and possesses some symmetry. The macroscopic geometry of sand stone containing irregular and tortuous channels is rather complex; even the three equations cannot be applied to the flow of fluids in porous media.

For fluid-flow-systems involving porous media the Stokes-Navier equations are not employed. Instead Darcy's law is applied. And in effect Darcy's law is of the nature of a statistical result which gives the empirical equivalent of the Stokes-Navier equations. Darcy's law states that the velocity of a fluid flowing through a porous medium is directly proportional to the pressure gradient acting on the fluid. In equation form Darcy's law may be written as

$$\begin{aligned} v_x &= -\,k_x/\mu\ \partial p/\partial x \\ v_y &= -\,k_y/\mu\ \partial p/\partial y \\ v_z &= -\,k_z/\mu\ \partial p/\partial z \end{aligned} \qquad \textbf{6-6}$$

where μ is the viscosity of the fluid, k_x, k_y, k_z respectively are the permeabilities in the three directions of the three velocity components v_x, v_y, v_z.

Assuming there are no body forces the fluid-flow equation for a viscous incompressible fluid is found by substituting Equations **6-6** into Equation **6-3** and applying Equation **6-4.** Since the fluid is incompressible there is no change in density with time; therefore the right-hand side of Equation **6-3** becomes zero. Setting the righthand side of Equation **6-3** equal to zero and substituting appropriate values from Equation **6-6** gives

$$\rho\,[\partial/x\,(-\,k_x/\mu\ \partial p/\partial x)\ +\partial/\partial y\,(-\,k_y/\mu\ \partial p/\partial y) + \partial/\partial z\,(-\,k_z/\mu\ \partial p/\partial z)\,] = 0 \qquad \textbf{6-7}$$

Now setting $k_x = k_y = k_z$ and dividing through by $-\,\rho k_x/\mu$ gives

$$\partial/\partial x\,(\partial p/\partial x) + \partial/\partial y\,(\partial p/\partial y) + \partial/\partial z\,(\partial p/\partial z) = 0 \qquad \textbf{6-8}$$

or

$$\frac{\partial^2 p}{\partial x^2} + \frac{\partial^2 p}{\partial y^2} + \frac{\partial^2 p}{\partial z^2} = 0 \qquad \textbf{6-9}$$

Equation **6-9** is often referred to as the Laplace equation; and it is the equation for fluid flow of an incompressible fluid through a porous medium.

Compressible Fluid Flow Equation (Diffusivity Equation)

The equation which characterizes the flow of viscous compressible fluids through porous media is developed by substituting Darcy's equation expressed in terms of density, into the *equation of continuity.* For homogenous fluids of practical interest and all types of viscous flow the relation between density and pressure may be expressed as

$$\rho = \rho_o e^{c(p-p_o)} \qquad \textbf{6-10}$$

where ρ is the density at pressure p, ρ_o is the density at pressure p_o and c is the compressibility of the fluid.

In terms of logarithms Equation **6-10** may be expanded as

$$ln\ \rho = ln \rho_o + c\ p - c\ p_o \qquad \textbf{6-11}$$

And Equation **6-11** may be differentiated with respect to x, y, and z to give the equations.

$$\begin{aligned} 1/\rho\ \partial\rho/\partial x &= c\ \partial p/\partial x \text{ or } \partial p/\partial x = 1/\rho c\ \partial\rho/\partial x \\ 1/\rho\ \partial\rho/\partial y &= c\ \partial p/\partial y \text{ or } \partial p/\partial y = 1/\rho c\ \partial\rho/\partial y \\ 1/\rho\ \partial\rho/\partial z &= c\ \partial p/\partial z \text{ or } \partial p/\partial z = 1/\rho c\ \partial\rho/\partial z \end{aligned} \qquad \textbf{6-12}$$

Replacing pressure gradients of Darcy's equations with their identities as given in Equation **6-12** give the Darcy's equations in terms of density gradients as

$$\begin{aligned} v_x &= -\ k/\mu\ \partial p/\partial x = -\ k/\mu c\rho\ \partial\rho/\partial x \\ v_y &= -\ k/\mu\ \partial p/\partial y = -\ k/\mu c\rho\ \partial\rho/\partial y \\ v_z &= -\ k/\mu\ \partial p/\partial z = -\ k/\mu c\rho\ \partial\rho/\partial z \end{aligned} \qquad \textbf{6-13}$$

Substituting the relations of Equation **6-13** into the *equation of continuity* and assuming that the permeability is identical in all directions gives the equation

$$\begin{aligned} &-\ k/\mu c\ [\partial/\partial x\ (\rho/\rho\ \partial\rho/\partial x) + \partial/\partial y\ (\rho/\rho\ \partial\rho/\partial y) \\ &+ \partial/\partial z\ (\rho/\rho\ \partial\rho/\partial z)\ ] = -\ \phi\ \partial\rho/\partial t \end{aligned} \qquad \textbf{6-14}$$

and further simplification yields the following compressible fluid flow equations.

$$\partial^2\rho/\partial x^2 + \partial^2\rho/\partial y^2 + \partial^2\rho/\partial z^2 = \frac{\phi\mu c}{k}\,\partial\rho/\partial t \qquad \text{6-15}$$

This equation is often referred to as the diffusivity equation written in Cartesian coordinates.

The equation which characterizes the flow of compressible gases through porous media is developed by combining the gas law, Darcy's Law and the *equation of continuity*. In terms of the gas law, density may be expressed as

$$\rho = Mp/zRT \qquad \text{6-16}$$

where M is the molecular weight of the gas, z is the compressibility factor, R is the gas constant, T is the temperature and p is the pressure. The differentiation with respect to time of Equation **6-17** is

$$\partial\rho/\partial t = M/zRT\;\partial p/\partial t \qquad \text{6-17}$$

And Darcy's Law expressed in terms of pressure is given in Equation **6-6.** Combining Equations **6-6, 6-16** and **6-17** and the *equation of continuity* Equation **6-3** gives: (of course k is assumed the same in all directions)

$$k/\mu\; M/zRT\,[\partial/\partial x\,(p\,dp/\partial x) + \partial/\partial y\,(p\,\partial p/\partial y) + \partial/\partial z\,(p\,\partial p/\partial z)\,] = \phi\; M/zRT\;\partial p/\partial t \qquad \text{6-18}$$

Since

$$2p\;\partial p/\partial x = \partial p^2/\partial x;\; 2p\;\partial p/\partial v = \partial p^2/\partial y;\; 2p\;\partial p/\partial z = \partial p^2/\partial z \qquad \text{6-19}$$

Equation **6-20** may be written as

$$\partial/\partial x\,(\partial p^2/\partial x) + \partial/\partial y\,(\partial p^2/\partial y) + \partial/\partial z\,(\partial p^2/\partial z) = 2\phi\mu/k\;\partial p/\partial t \qquad \text{6-20}$$

or

$$\partial^2 p^2/\partial x^2 + \partial^2 p^2/\partial y^2 + \partial^2 p^2/\partial z^2 = 2\;\phi\mu/k\;\partial p/\partial t \qquad \text{6-21}$$

For graphical solutions of the unsteady state gas flow equation, it is often more convenient to express Equation **6-21** as

$$\partial^2 p^2/\partial x^2 + \partial^2 p^2/\partial y^2 + \partial^2 p^2/\partial z^2 = \phi\mu/kp\;\partial p^2/\partial t \qquad \text{6-22}$$

The identities of Equations **6-20** and **6-21** can partially be explained by noting the following identities:

$$\Delta p^2/\Delta x = \frac{p_2^2 - p_1^2}{\Delta x} = \frac{(p_2 + p_1)(p_2 - p_1)}{\Delta x} = 2p\,(\Delta p)/\Delta x \qquad \textbf{6-23}$$

where

$$p = \frac{p_2 + p_1}{2}$$

Fluid Flow Equations in Other Coordinate Systems

More often the petroleum engineer is interested in radial fluid flow problems than in rectangular fluid flow problems. By making simple transformations the fluid flow equations which have been derived for the rectangular coordinate systems may be converted to fluid flow equations in the cylindrical coordinate system. The relations for transforming from the rectangular coordinate system into the cylindrical coordinate system (r,θ,z) is shown in Figure 6-2.

$$r = \sqrt{x^2 + y^2};\ \theta = \tan^{-1} y/x;\ z = z \qquad \textbf{6-24}$$

$$x = r\cos\theta;\ y = r\sin\theta;\ z = z$$

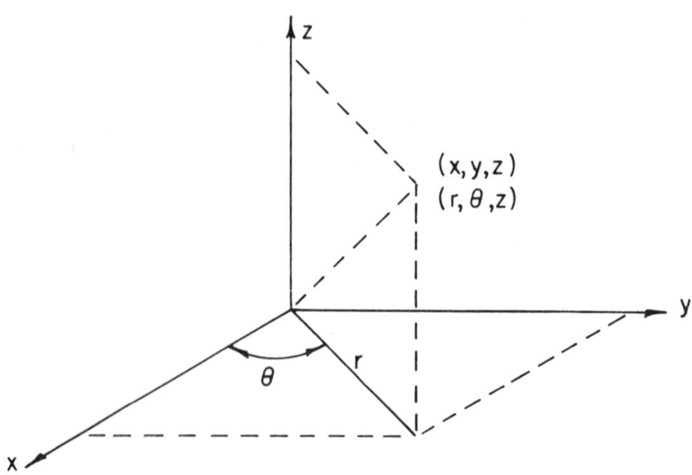

Figure 6-2. Cylindrical coordinate system.

The flow equation for viscous incompressible fluid in the cylindrical coordinate system is:

$$\partial^2 p/\partial r^2 + 1/r\ \partial p/\partial r + 1/r^2\ \partial^2 p/\partial \theta^2 + \partial^2 p/\partial z^2 = 0 \quad \text{6-25}$$

For radial incompressible flow in a plane $z = 0$, $\partial^2 p/\partial\theta^2 = 0$, and Equation **6-25** becomes

$$\partial^2 p/dr^2 + 1/r\ \partial p/\partial r = 0 \quad \text{6-26}$$

Similarly the flow equation for viscous compressible fluids in terms of the cylindrical coordinate system is

$$\partial^2 \rho/dr^2 + 1/r\ \partial\rho/dr + 1/r^2\ \partial^2\rho/\partial\theta^2 + \partial^2\rho/\partial z^2 = \phi\mu c/k\ \partial\rho/\partial t \quad \text{6-27}$$

And Equation **6-27** for flow in a radial plane becomes

$$\partial^2\rho/\partial r^2 + 1/r\ \partial\rho/\partial r = \phi\mu c/k\ \partial\rho/\partial t. \quad \text{6-28}$$

And similarly, the flow equation for gases in terms of the cylindrical coordinate system is

$$\partial^2 p^2/\partial r^2 + 1/r\ \partial p^2/\partial r + 1/r^2\ \partial^2 p^2/\partial\theta^2 + \partial^2 p^2/dz^2 = \phi\mu/\bar{p}k\ \frac{\partial p^2}{\partial t} \quad \text{6-29}$$

And Equation **6-29** for flow in a plane becomes

$$\partial^2 p^2/\partial r^2 + 1/r\ \partial p^2/\partial r = \phi\mu/\bar{p}k\ \partial p^2/\partial t \quad \text{6-30}$$

or

$$\partial^2 p^2/\partial r^2 + 1/r\ \partial p^2/dr = 2\ \phi\ \mu/k\ \partial p/\partial t \quad \text{6-31}$$

The flow equation for incompressible fluids and the flow equation for compressible fluids in the spherical coordinate system (**r**, θ, α) may be developed with the aid of the relations.

$$r = \sqrt{x^2 + y^2 + z^2};\ \theta = \tan^{-1} \sqrt{x^2 + y^2}/z;\ \alpha = \tan^{-1} y/x$$

$$x = r \sin\theta \cos\alpha;\ y = r\sin\theta\sin\alpha;\ z = r\cos\theta \quad \text{6-32}$$

The equations for incompressible and compressible fluids are respectively:

$$\partial/\partial r\ (r^2 \partial p/\partial r) + 1/\sin\theta\ \partial/\partial\theta\ (\sin\theta\ \partial p/\partial\theta) + 1/\sin^2\theta\ \partial^2 p/\partial\alpha^2 = 0 \quad \text{6-33}$$

and

$$\partial/\partial r\,(r^2\,\partial\rho/\partial r) + 1/\sin\theta\,\partial/\partial\theta\,(\sin\theta\,\partial\rho/\partial\theta) + 1/\sin^2\theta\,\partial^2\rho/\partial\alpha^2 = \phi\mu c/k\,\partial\rho/\partial t \qquad \text{6-34}$$

Analogies Between Problems in Fluid Flow Through Porous Media and Other Physical Problems

The flow of fluid through porous media is directly analogous to the flow of heat and the flow of electricity. There follows flow equations which show the analogies between fluid flow through porous media, heat flow and electric flow.

The Darcy equation for the quantitative flow of an incompressible fluid through a porous medium is

$$q = (-\,k/\mu)\,A\,dp/dL \qquad \text{6-35}$$

where q is the quantity of fluid in mls. per second, k/μ is fluid mobility, A is cross-sectional area in cms^2, and dp/dL is the pressure gradient in atmosphere per given length.

The equation for the quantitative flow of heat is

$$q = -k\,A\,dT/dL \qquad \text{6-36}$$

where q is the quantity of heat in B.t.u/hr., k is the thermal conductivity in (B.t.u.) $(\text{hr.})^{-1}$ $(\text{ft.})^{-1}$ $(\text{deg. F.})^{-1}$,A is area in cms.^{2}, and dT/dL is the temperature gradient or the drop in temperature (degree F) per given length.

The equation for the quantitative flow of electricity is:

$$I = A/\rho\,dE/dL \qquad \text{6-37}$$

where I is the quantity of electricity in amperes/second, A is area in cms^2, $1/\rho$ is the specific conductance and dE/dL is the voltage gradient in drop in voltage per given length of the conductor.

The Laplace equation for the flow of a incompressible fluid in a porous medium is:

$$\partial^2 p/\partial x^2 + \partial^2 p/\partial y^2 + \partial^2 p/\partial z^2 = 0 \qquad \text{6-38}$$

The Laplace equation for the steady-state flow of heat is

$$\partial^2 T/\partial x^2 + \partial^2 T/\partial y^2 + \partial^2 T/\partial z^2 = 0 \qquad \text{6-39}$$

The Laplace equation for the flow of electricity is

$$\partial^2 E/\partial y^2 + \partial^2 E/\partial z^2 = 0 \qquad \text{6-40}$$

The diffusivity equation for the flow of a compressible fluid through a porous medium is

$$\partial^2 \rho/\partial x^2 + \partial^2 \rho/\partial y^2 + \partial^2 \rho/\partial z^2 = \phi\ \mu c/k\ \partial \rho/\partial t \qquad \text{6-41}$$

where ρ is the density of the fluid, ϕ is the porosity of the medium, μ is the viscosity of the fluid, c is the compressibility of the fluid, k is the permeability of the medium and t is the time. The term $k/\phi\mu c = \eta$ are often referred to as the hydraulic diffusivity.

And the diffusivity equation for the flow of heat is.

$$\partial^2 T/\partial x^2 + \partial^2 T/\partial y^2 + \partial^2 T/\partial z^2 = \rho c_p/k\ \partial T/\partial t \qquad \text{6-42}$$

where T is the temperature, ρ is the density of the conductor, k is the thermal conductivity of the conductor, c_p is the specific heat, and t is the time. Similarly the terms $k/\rho c_p = \alpha$ are often referred to as thermal diffusivity.

For a given electrical network, there exists an analogy between the resistance to flow of incompressible fluid through a porous medium and the resistance to the flow of electricity; and there exists an analogy between the flow of a compressible fluid through a porous medium and the flow of electrical current. The equation for the resistance to the flow of an incompressible fluid and the resistance to the flow of electricity are respectively:

$$R_{fluid} = \Delta p/q = \mu L/kA \qquad \text{6-43}$$

where R_{fluid} is the resistance to the flow of fluid and the other symbols have already been described for the Darcy equation;
and

$$R_{elec.} = \frac{E_1 - E_2}{I} \qquad \text{6-44}$$

where E is in volts, $R_{elec.}$ is in ohms and I is in amperes.

For the given electrical network the analogy between the flow of a compressible fluid and the flow of electricity is illustrated by the following two equations. The equation for the flow of a compressible fluid is

$$q_2 - q_1 = -\phi V_o c\, dp/dt \qquad \textbf{6-45}$$

where q is the quantity of fluid flowing ϕ is the porosity, V_o is the unit volume and c is the compressibility of the fluid.

And the analogous equation for the flow of electricity is

$$I_2 - I_1 = -C_E dE/dt \qquad \textbf{6-46}$$

where I is current in amperes, C_E is capacitance in microfarads, E is volts, and t is time.

From the analogies it is obvious that much can be learned about the flow of fluids through porous media by studying the flow of heat and the flow of electricity. Flow models which actually involve the flow of fluids are frequently bulky and cumbersome. For these conditions a model which involves the flow of either heat or electricity can be fabricated, which is considerably simpler. Much can be gained by building either an electric or heat transfer model and studying the analogies between fluid flow and either electric or heat flow.

This chapter was devoted primarily to the development of partial differential equations for the flow of fluids through porous media. The partial differential equations which were developed did not take into consideration body forces, such as gravity. These forces will affect the velocity as well as pressure gradients. Solutions for the partial differential equations were not given. However, the next three chapters will be devoted to solutions for the partial differential equations with various boundary conditions.

Exercises

1. For non compressible fluid flow ρ = constant, show that the *equation of continuity* becomes

$$\partial v_x/\partial x + \partial v_y/\partial y + \partial v_z/\partial z = 0$$

2. Determine the dimensions for the hydraulic diffusivity $\eta = k/\phi uc$; also determine the units for the thermal diffusivity $\eta = k/\rho c_p$.
3. With the aid of Equation **6-24** show that Equation **6-9** (incompressible fluid flow) in cylindrical coordinates becomes identical to Equation **6-25.**
4. With the aid of Equation **6-32** show that Equation **6-9** in spherical coordinates becomes identical to Equation **6-33.**
5. With the aid of Equation **6-24,** show that Equation **6-15** (compressible fluid flow) in cylindrical coordinates becomes identical to Equation **6-27.**

Suggested Reading

1. Bruce, W. A., "An Electrical Device for Analyzing Oil-Reservoir Behavior," Trans. A.I.M.E., 1943, 112-114, 151.
2. Carslaw, H. S. and Jaeger, J. C., *Conduction of Heat in Solids,* Oxford: University Press, 1948, pp. 1-32.
3. Craft, B. C. and Hawkins, M. F., *Applied Reservoir Engineering,* Englewood Cliffs: Prentice-Hall, Inc., 1959, p. 296.
4. Jakob, M. and Hawkins, G. A., *Elements of Heat Transfer and Insulation, 2nd ed.,* New York: John Wiley and Sons, Inc., 1954, pp. 6 and 56-65.
5. Katz, D. L., et al., *Handbook of Natural Gas Engineering,* New York: McGraw-Hill Co., Inc., 1959, pp. 403-434.
6. Lamb, H., *Hydrodynamics, 6th ed.,* New York: Dover Publications, 1932, pp. 576-578.
7. Miller, F. H., *Partial Differential Equations,* New York: John Wiley and Sons, Inc., 1941, pp. 71-84.
8. Muskat, M., *The Flow of Homogeneous Fluids Through Porous Media,* Ann Arbor: J. W. Edwards, Inc., 1946, pp. 76-79, 121-146.
9. Sokolnikoff, E. S. and Sokolnikoff, I. S., *Higher Mathematics for Engineers and Physicists, 2nd ed.,* New York: McGraw-Hill Book Co., Inc., 1941, pp. 367-370, 377-380, and 386.

7 Steady-State Fluid Flow Through Porous Media

Few methods for solving the partial differential equations for fluid flow through porous media have been developed. In fact, all of the methods which are discussed in this text were previously developed for the flow of heat by mathematicians and engineers. Engineers have predominantly used five methods for solving the partial differential equations. They are: (1) the method of separation of variables in which Bessel Functions are used to a good advantage, (2) the method of the exponential integral, (3) the method of operations which employs transform calculus, (4) the method of difference equations, and (5) the method of graphs. This chapter and the next chapter emphasize the method of difference equations. Problems which are solvable with the use of difference equations can be programmed for microcomputers and solved on these computers. A few solutions employing difference equations were developed in Chapter 4. In this chapter additional approaches for the numerical solutions are developed.

One-Dimensional Steady-State Fluid Flow

Steady-state fluid flow is another name for incompressible fluid flow. The Laplace equation for one-dimensional steady-state fluid flow through porous media is

$$\partial^2 p/\partial x^2 = 0 \qquad \textbf{7-1}$$

From Equation **4-96** it is obvious that in one dimension the difference equation which is the solution of Equation **7-1** is

$$\partial^2 p/\partial x^2 \approx \frac{p\,(x+h) + p\,(x-h) - 2p\,(x)}{h^2} = 0 \qquad 7\text{-}2$$

or

$$p\,(x) = \frac{p\,(x+h) + p\,(x-h)}{2} \qquad 7\text{-}3$$

In other words Equation **7-3** states that the value of p at any point is the average of the values of the two adjacent points taken at equi-distances from the point, $p\ (x)$; one value of p is taken to the left of $p\ (x)$, that is $p\ (x\text{-}h)$, and the other value of p is taken to the right of $p\ (x)$, that is $p\ (x+h)$. The following example illustrates the use of the difference equation, Equation **7-3.**

Example 7-1. Under steady-state conditions the inlet and outlet pressures for a sand packed pipe are 800 psia and 200 psia at the two ends. Using figure 7-1 determine the pressures in the planes indicated by grid points 2 and 3 by the method of iteration. Obviously, these pressures are 600 psia and 400 psia. The purpose of the example shown in Table 7-1, is to show how the same result would be obtained by means of the iteration method. For the first step p_2 is assumed to be 700 psia and p_3 is assumed to be 300 psia.

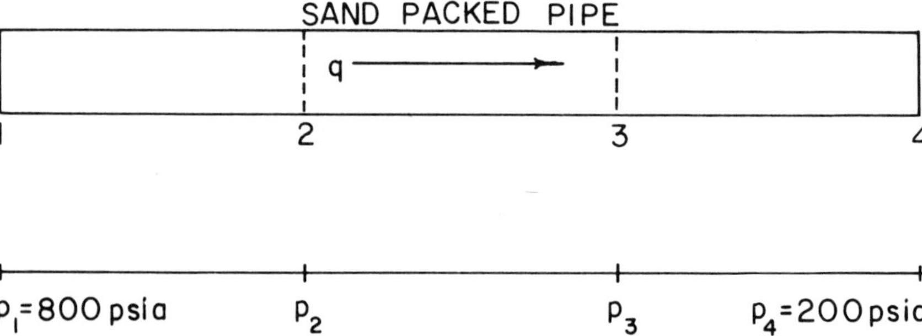

Figure 7-1. Sand packed pipe with four pressure points, $\Delta x = L/3$, illustrating one dimensional fluid flow.

Table 7-1
One-Dimensional Iteration Solution

STEP	p_1	$p_2 = \frac{p_1 + p_3}{2}$	$p_3 = \frac{p_2 + p_4}{2}$	p_4
1	800	700	300	200
2	800	1/2 (800+ 300.0) = 550	1/2 (550 + 200) = 375	200
3	800	1/2 (800+ 375.0) = 587.5	1/2 (587.5 + 200) = 393.8	200
4	800	1/2 (800+ 393.8) = 596.9	1/2 (596.9 + 200) = 399.0	200
5	800	1/2 (800+ 399.0) = 599.5	1/2 (599.5 + 200) = 399.8	200
6	800	1/2 (800+399.8) = 599.9	1/2 (599.9 + 200) = 399.9	200
7	800	1/2 (800+399.9) = 600.0	1/2 (600 + 200) = 400.0	200

The Method of Relaxation

The difference equation which is a solution for the Laplace equation in one dimension can be derived in several ways. Another method for deriving the difference equation follows. This method of derivation makes use of Darcy's equation and is developed independently without the use of the Laplace equation.

Fluid is assumed to flow parallel to the x axis in the homogeneous body into a square grid network (Figure 7-2). A fluid sink is imagined to be at each node (corner) of the lattice. The term "sink" is used here to mean a point of discontinuity where fluid is absorbed. Hence point 2 represents a fluid sink. It is further assumed the operation is such that the lines ab, were at the pressure p, and the line a_2b_2 is at the pressure p_2 and the fluid flow between a_1b_1 and a_2b_2 is q_{12} and the fluid flow between a_2b_2 and a_3b_3 is q_{23} .Then

$$q_{12} = \frac{kz\Delta y}{\mu\Delta x}(p_1 - p_2) \qquad \text{7-4}$$

where z represents the depth of the section. The fluid flow from the sink in the direction from 2 to 3 is

$$q_{23} = \frac{kz\Delta y}{\mu\Delta x}(p_2 - p_3) \qquad \text{7-5}$$

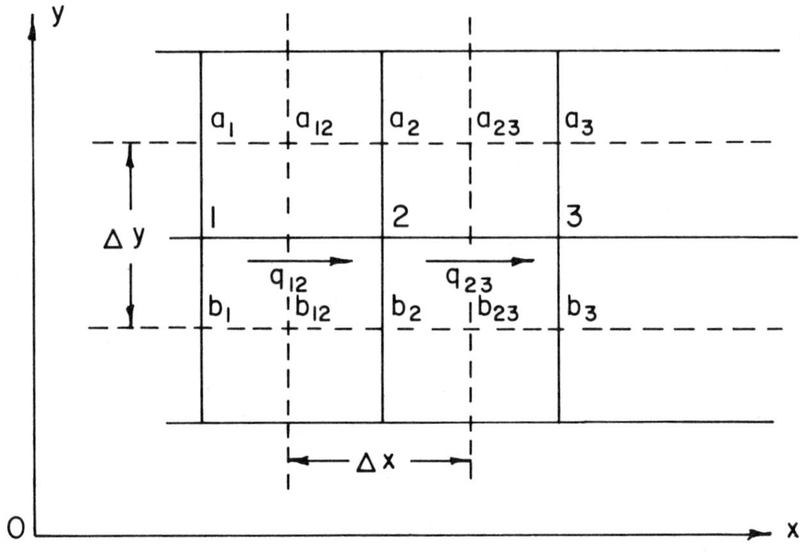

Figure 7-2. Two coordinate system illustrating uni-directional fluid flow.

Subtracting the outgoing from the incoming fluid yields the fluid stored in the sink 2; the difference is

$$q_2 = q_{12} - q_{23} = \frac{kz\Delta y}{\mu\Delta x}(p_1 - p_2) - \frac{kz\Delta y}{\mu\Delta x}(p_2 - p_3) \tag{7-6}$$

Because a square grid was assumed, $\triangle x = \triangle y$. Hence,

$$q_2 = kz/\mu\,(p_1 + p_3 - 2p_2) \tag{7-7}$$

Replacing $q_2\mu/kz$ by the symbol δ_2, which is a pressure difference proportional to the strength of the fluid sink at point 2 in the grid, gives

$$\delta_2 = p_1 + p_3 - 2p_2 \tag{7-8}$$

But under steady-state conditions no fluid can be stored in sink 2; therefore it follows that $q_2 = 0$ and $\delta_2 = 0$. Setting $\delta_2 = 0$ and solving for p_2 gives

$$p_2 = \frac{p_1 + p_3}{2} \qquad 7\text{-}9$$

which is a difference solution similar to Equation **7-3.**

For a place indicated by an integer n (instead of 2) there follows the generalized equation

$$\delta_n = p_{n-1} + p_{n+1} - 2\,p_n \qquad 7\text{-}10$$

and

$$p_n = \frac{p_{n-1} + p_{n+1}}{2} \qquad 7\text{-}11$$

These equations may be used to determine the pressure distribution within a body if steady-state conditions prevail.

The following example illustrates the relaxation method of using Equation **7-10.**

Example 7-2. Solve the sand-packed pipe flow problem of Example **7-1** by the relaxation method.

Solution. The pressures at the sinks 2 and 3, are different from the actual values; these pressures are estimated to be 700 psia and 300 psia as shown in step 1 of Table 7-2. In step 2 the values of δ_2 and δ_3 are calculated with the use of Equation **7-10** (with $n = 2$ and $n = 3$). In step 3, δ_2 is relaxed to zero and p_2 is computed by means of Equation **7-11** (with $n = 2$), and p_3 is kept as found in step 2; in this step a new value is found for δ_3. In step 4, δ_3 is relaxed to zero, a new value of p_3 is calculated and another approximation of δ_2 is obtained. This procedure is continued until both δ_2 and δ_3 equal approximately zero, which in this case is step 11.

Table 7-2
One-Dimensional Relaxation Solution

STEP	p_1	δ_2	p_2	δ_3	p_3	p_4
1	800	$= p_1 + p_3 - 2p_2$	700	$p_2 + p_4 - 2p_3$	300	200
2	800	$= 800 + 300$ $- 2(700) = -300$	700	$= 700 + 200$ $- 2(300) = 100$	300	200
3	800	0	$\frac{800 + 300}{2}$ = ~~1500~~ = 550	$= 550 + 200$ $- 2(300)$ $= 150$	300	200
4	800	$= 800 + 375$ $(-550)\,2 = 75$	550	0	$\frac{550 + 200}{2}$ $= 375$	
5	800	0	587.5	37.5	375	200
6	800	18.8	587.5	0	393.8	200
7	800	0	596.9	9.3	393.8	200
8	800	4.7	596.9	0	398.5	200
9	800	0	599.3	2.3	398.5	200
10	800	1.0	599.3	0	399.6	200
11	800	0	599.8	0.6	399.6	200

Two Zones—Each with a Different Permeability

A sand packed pipe having two zones of unequal permeabilities is shown in Figure 7-3.

The steady-state fluid flow from zone 1 through the interface joining the two zones is equal to the steady-state flow from the interface into zone 2. The steady-state fluid flow in zone 1 is equal to the steady-state fluid flow in zone 2. Applying Darcy's law, the steady-state fluid flow in zone 1 is represented as

$$q_{1\text{-}2} = (k/\mu)_1 \ (dp/dL)_1 \tag{7-12}$$

and the steady-state fluid flow in zone 2 is represented as

$$q_{2\text{-}3} = (k/\mu)_2 \ (dp/dL)_2 \tag{7-13}$$

Equating the two equations gives

$$q_{1\text{-}2} = q_{2\text{-}3} = (k/\mu)_1 \ (dp/dL)_1 = (k/\mu)_2 \ (dp/dL)_2 \tag{7-14}$$

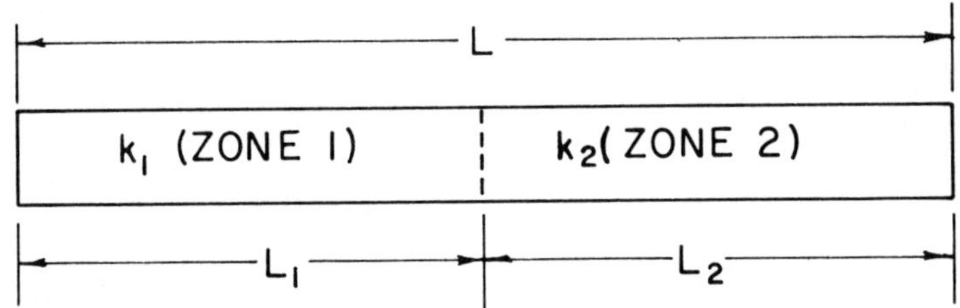

Figure 7-3. Sand-packed pipe containing two zones, each with a different permeability.

or

$$(dp/dL)_1 = (k/\mu)_2 (\mu/k)_1 \ (dp/dL)_2 \qquad \textbf{7-15}$$

and if $\mu_1 = \mu_2$

$$(dp/dL)_1 = k_2/k_1 \ (dp/dL)_2$$

Equation **7-15** gives the value of the pressure gradient in zone 1 in terms of the permeabilities of the two zones and the pressure gradient of zone 2. Similar equations can be derived for any given number of zones, each of which has a different permeability.

Example 7-3. If the pressure gradient in zone 1 is 2 p.s.i. per foot, $k_1 = 200$ mds. and k_2 is 50 mds. find the pressure gradient in zone 2.

Solution. Applying Equation **7-15** gives

$$(dp/dL)_2 = \frac{200 \times 2}{50} = 8 \text{ p.s.i./ft.}$$

Two-Dimensional Steady-State Fluid Flow

The partial differential equation for two-dimensional steady-state flow of fluids in porous media is

$$\partial^2 p/\partial x^2 + \partial^2 p/\partial y^2 = 0 \qquad \textbf{7-16}$$

A solution of this equation by means of a difference equation is developed in Chapter 4. Example 4-9 covers the method of iteration using the difference equation when all exterior points are not given.

Next, a difference equation for using the method of relaxation as a solution to Equation **7-16** is derived. Similar to the one dimensional equation the difference equation is derived without considering the two dimensional Laplace equations. This method is given because with slight modifications of the method for equations for determing the values of points at the boundaries between zones of varying permeabilites, a difference equation can be derived. The method has been used by Dykstra and Parsons[1] for developing relaxation methods which are applicable to oil field research.

The Method of Relaxation

In the development of the equation for the relaxation method, a porous medium is assumed to be replaced by a net of tubes of equal length and a uniform cross-sectional area, as shown in Figure 7-4. The net is also assumed to be fine enough so that it exactly reproduces the porous medium and further it is assumed that the flow from point to point obeys Darcy's law.

Applying Darcy's law, the flow from point 1 to point 0 is

$$q_{1-0} = kA/\mu a\,(p_1 - p_0) \qquad \textbf{7-17}$$

where k is the permeability, A is the cross-sectional area of a tube, μ is the viscosity of the liquid in the porous medium, $(p_1 - p_0)$ is the pressure difference between point 1 and 0, and a is the distance the fluid flowed.

Similarly the flow from point 0 to point 3 is

$$q_{0-3} = kA/\mu a\,(p_0 - p_3) \qquad \textbf{7-18}$$

The fluid remaining at point 0 due to the x-component of velocity is the difference of the quantity of fluid flowing into point 0 and the quantity of fluid flowing out of point 0 or

[1]Dykstra, H. and Parsons, R. L., "Relaxation Methods Applied to Oil Field Research," Trans. A.I.M.E., (1951) *192* pp. 227-232.

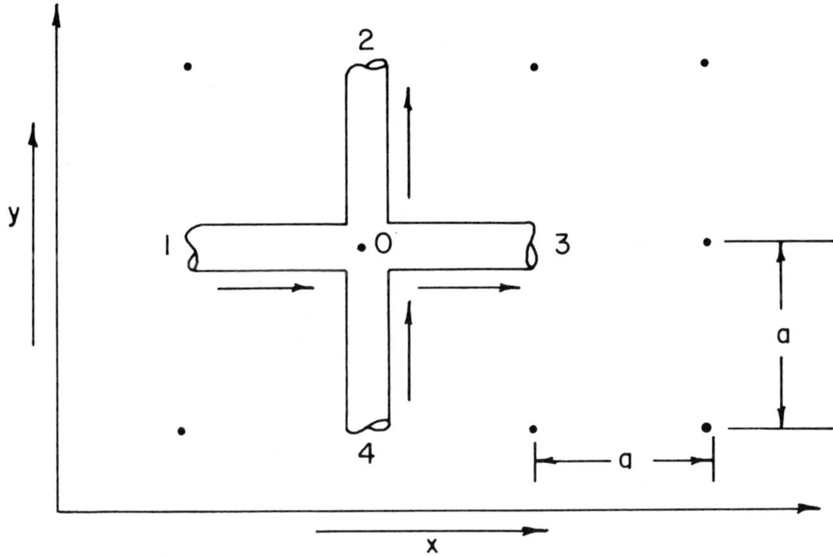

Figure 7-4. Representation of a porous medium by a network of conducting tubes.

$$q_{1-0} - q_{3-0} = kA/\mu a\ (p_1 - p_o) - kA/\mu a\ (p_o - p_3)$$

or

$$\Delta q_x = kA/\mu a\ (p_1 + p_3 - 2p_o) \qquad \textbf{7-19a}$$

In a similar manner the fluid remaining at point 0 due to the y-component of velocity is

$$\Delta q_y = kA/\mu a\ (p_2 + p_4 - 2p_o) \qquad \textbf{7-19b}$$

The net fluid remaining at point 0 is the sum of the remaining fluid due to the two components of velocity.

$$q_o = kA/\mu a\ (p_1 + p_2 + p_3 + p_4 - 4p_o) \qquad \textbf{7-20}$$

For an incompressible fluid the net fluid remaining at point 0 is zero or $q_o = 0$. Since $q_o = 0$ then

$$0 = p_1 + p_2 + p_3 + p_4 - 4p_o \quad \text{7-21}$$

or

$$p_o = \frac{p_1 + p_2 + p_3 + p_4}{4} \quad \text{7-22}$$

The last equation, Equation **7-22,** is identical to the difference equation which was derived as a solution for the two-dimensional Laplace equation.

For cases when Equation **7-21** is not satisfied then it may be written as

$$\delta = p_1 + p_2 + p_3 + p_4 - 4p_o \quad \text{7-23}$$

where δ is a measure of departure from steady-state and is called the residual at point 0. Equation **7-23** shows how much the pressure guess is in error at point 0. A positive residual δ indicates that the pressure at point 0 is low and a negative δ indicates that pressure at point 0 is too high.

In applying the method of relaxation δ is brought to zero by making changes in the pressure guesses. An examination of Equation **7-23** indicates that *a* +1 change at p_o will change the residual at point 0 by −4; and a +1 change in the pressures at any of the four surrounding points will change the residual at point 0 by +1. The object then is to make changes in the pressure guesses so that δ becomes equal to zero. The method of relaxation is illustrated in the example which follows.

Example 7-4. Solve the two-dimensional fluid flow problem given in Example 4-11 by the method of relaxation.

Solution. The values at the boundary and the initial values at the interior points were taken from Example 4-11. Figure 7-5 shows that the values of the residuals at all the interior points were calculated. After calculating the values of the residuals, it was noted that points C-3, D-2, and D-4 had the largest residuals. Since there were very little differences in these three residuals, it was decided to start the relaxation at point C-3 because it was in the center; this point was relaxed by the amount − 1.24/4 = −0.31. Then according to Eq. **7-23,** (−4) (−0.31) was added to the residue at C-3, and −0.31 was added to the residuals at the four adjacent lattice points (C-2, B-3, C-4, and D-3) as indicated in Fig. 7-5.

Next, the other larger residuals D-2 and D-4 were relaxed respectively in a similar manner by the amount −0.31. During each relaxation, the new values of the function were recorded.

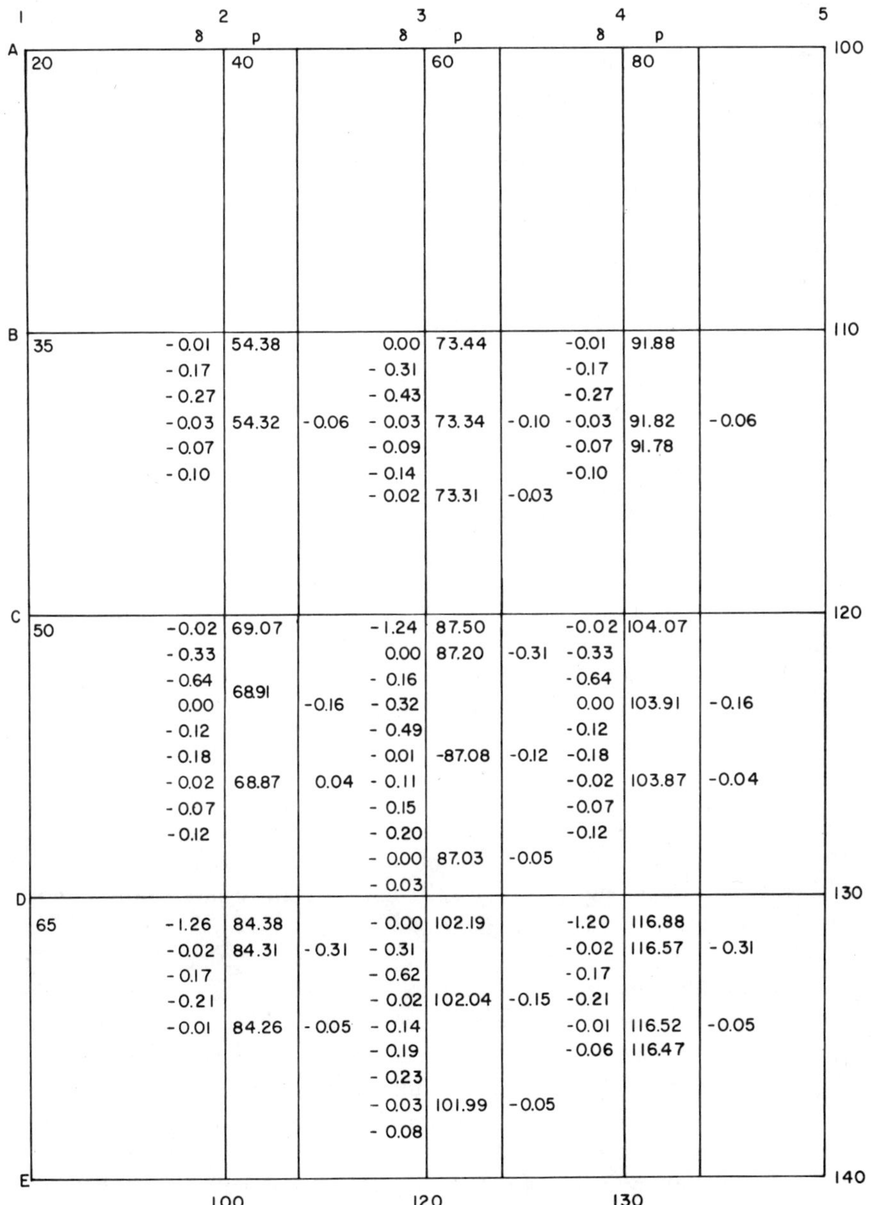

Figure 7-5. Steady state solution of a two-dimensional fluid flow problem by means of the method of relaxation.

After relaxing the three points with the largest residues, the largest residual was at point D-3. This point was then relaxed by the amount −0.23. The procedure of relaxation was continued respectively for points C-2 by −0.16, C-4 by −0.16, C-3 by −0.14, B-3 by −0.11, D-2 by −0.10, D-4 by −0.10, D-3 by −0.09, B-2 by −0.07, B-4 by −0.07, C-2 by −0.08, C-4 by 0.08, C-3 by −0.09, and B-3 by −0.06. At this stage the relaxation procedure was discontinued. For greater accuracy the relaxation procedure should be continued until the residuals at each interior point become smaller. Completion will be left as an exercise of the relaxation procedure.

Use of Images for Calculating Boundary Points

Frequently problems for the flow of noncompressible fluids in porous media occur which cannot be solved analytically because of the nature of the boundary conditions. For these problems, an exact answer is neither necessary nor obtainable because the boundry conditions are not defined or the parameters describing the porous medium are not accurately known. Approximate answers for many of these problems can be obtained by using the method of iteration and assigning images along the boundaries where no values are given. Here, the word image is applied to corresponding points in two different zones or regions separated by an impermeable boundary. Usually one of the zones or regions is considered as imaginary.

The method of solving two dimensional flow problems with limited boundary conditions is illustrated in the example which follows.

Example 7-5. Given the two-dimensional problem as indicated in Figure 7-6, which has given boundary values only on the upper half of the left face and only on the right half of the base; determine by the method of iteration with the aid of images, the approximate values of the interior and boundary points which are not given.

Solution. First, by guess, values are assigned to the points on the boundary which were not originally given; then, with the given boundary values and the assigned boundary values, approximate values for each of the interior points are determined as indicated in Example 4-11. Next, a new set of values for each of the interior points is determined by the method of iteration (see Example 4-11).

To determine new values for the boundary points with assigned values, images along the impermeable boundaries were assigned. The image points were taken perpendicularly to the boundaries and opposite to corresponding perpendicular interior points; for instance, the points B^1—2, B^1—3, B^1—4,

	2'	1	2	3	4	5	4'
B'			34.50 34.50 35.25 ⋮ 36.50 36.52	43.13 43.12 44.07 ⋮ 45.21 45.18	50.00 50.00 50.49 ⋮ 50.34 50.28	55.00 53.73 52.44 ⋮ 51.97 51.93	
A		25.00	30.00 33.00 34.14 ⋮ 35.41 35.42	38.00 41.06 42.51 ⋮ 43.65 43.62	45.00 47.76 48.65 ⋮ 48.70 48.64	50.00 51.13 50.96 ⋮ 50.37 50.30	45.00 47.76 48.65 ⋮ 48.70 48.64
B		25.00	34.50 34.50 34.25 ⋮ 36.50 36.52	43.13 43.12 44.07 ⋮ 45.21 55.18	50.00 50.00 50.49 ⋮ 50.34 50.28	55.00 53.28 52.73 ⋮ 51.97 51.93	50.00 50.00 50.49 ⋮ 50.34 50.28
C		25.00	39.87 39.87 40.12 ⋮ 40.46 40.45	50.00 49.99 50.31 ⋮ 50.20 50.17	56.87 56.87 56.36 ⋮ 55.29 55.25	62.00 59.00 57.84 ⋮ 56.85 56.81	62.00 59.00 57.84 ⋮ 56.85 56.81
D	50.00 50.25 50.01 ⋮ 50.09 50.08	38.00 43.17 43.59 ⋮ 43.81 43.80	50.00 50.25 50.01 ⋮ 50.09 50.08	60.12 60.18 61.45 ⋮ 59.76 59.76	65.50 65.26 64.71 ⋮ 63.72 63.71	69.00 66.13 65.56 ⋮ 64.82 64.81	65.50 65.26 64.15 ⋮ 63.72 63.71
E	63.00 56.37 55.55 ⋮ 56.31 56.30	50.00 47.18 49.36 ⋮ 50.06 50.05	63.00 56.37 55.55 ⋮ 56.31 56.30	75.00	75.00	75.00	75.00
D'		38.00 43.17 43.59 ⋮ 43.81 43.80	50.00 50.25 50.01 ⋮ 50.09 50.08				

Figure 7-6. Numerical solution of a two-dimensional flow problem with all boundaries impermeable except the upper half of the left face and the right half of the base.

and B¹−5 are respectively images of the interior points B−2, B−3, B−4, and B−5. The perpendicular distance of each image point from the impermeable boundary is the same as the perpendicular distance of its corresponding interior point from the boundary. The other image points are A − 4¹, B − 4¹, C−4¹, D−4¹, E−4¹, D¹ − 2, D¹ −1, E−2¹ and D−2¹. Values are assigned to each image point equal to its corresponding image point. With the assigning of the image points, it is readily seen that each point on the impermeable boundary is surrounded by four adjacent points with values; then, by applying the difference equation new values of the boundary points are computed.

The procedure of computing new values for interior points and then computing new values for the boundary points is continued for the required number of cycles; the problem required 8 cycles. The computed values are listed in Figure 7-6.

Equi-Pressure Contours and Streamlines for Porous Media

The two-dimensional difference equation may be used for computing the equi-pressure contours for well networks. Either the method of iteration or the method of relaxation may be used. The streamlines are then obtained by drawing lines perpendicular to the equi-pressure contours.

By utilizing measured boundary values which are determined with the aid of a network of electrical resistances, the values for the interior pressure points are computed by the two dimensional difference equation, utilizing the method of iteration. The results of the calculations are given in Figure 7-7.

From the computed values for the pressure points given in Figure 7-7 the equi-pressure contours were drawn as illustrated in Figure 7-8. Then by drawing lines perpendicular to the equi-pressure contours, the streamlines were obtained; the streamlines are also illustrated in Figure 7-8.

With the aid of an Electronic Computer, the values for the pressure points (both boundary and interior) were determined.[2] The method of iteration was used; and for determining the values of the boundary, images along the boundaries were assigned. The computed values for the pressure points are given in Figure 7-9. The method for computing the values for the pressure points was illustrated in Example 7-5.

Once the equi-pressure contours and streamlines, for a given well network are drawn in the form of a flow-net, the displacement of the fluid along the streamlines as a function of time is easily computed. The development of the equation for computing the displacement of the fluid along the streamlines follows. It is assumed that velocity of a given particle of fluid along a streamline is proportional to the pressure gradient along the streamlines. If the initial position of a particle is known, then after an interval of time the new position of the particle is computed.

[2]Stacy, T. D. "Sweep Efficiency Studies by Use of High Speed Digital Computers" M. S. Thesis, Louisiana Polytechnic Institute (1962).

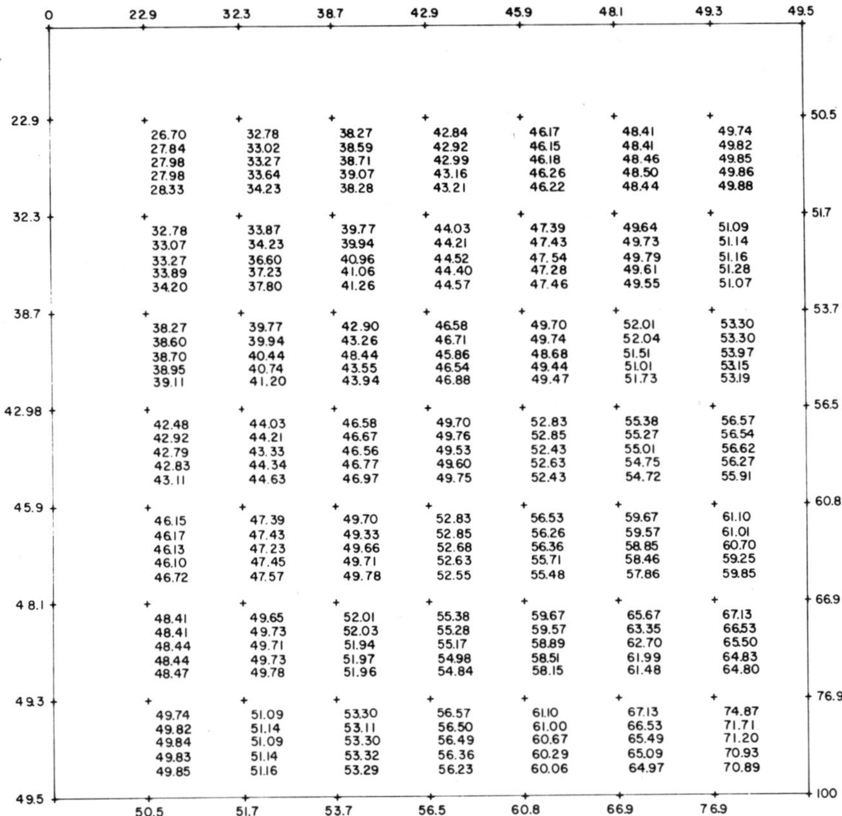

Figure 7-7. Computed potential values from given boundary values.

According to Darcy's Law the velocity of flow is

$$v = -\frac{k}{\mu}\frac{dp}{dL} \tag{7-24}$$

the velocity v is not the true fluid velocity, but is the velocity based upon a cross-sectional area which includes not only the cross-sectional area of the pores but the cross-sectional area of the solid material as well. An effective value for the velocity which includes only the cross sectional area of the pores is

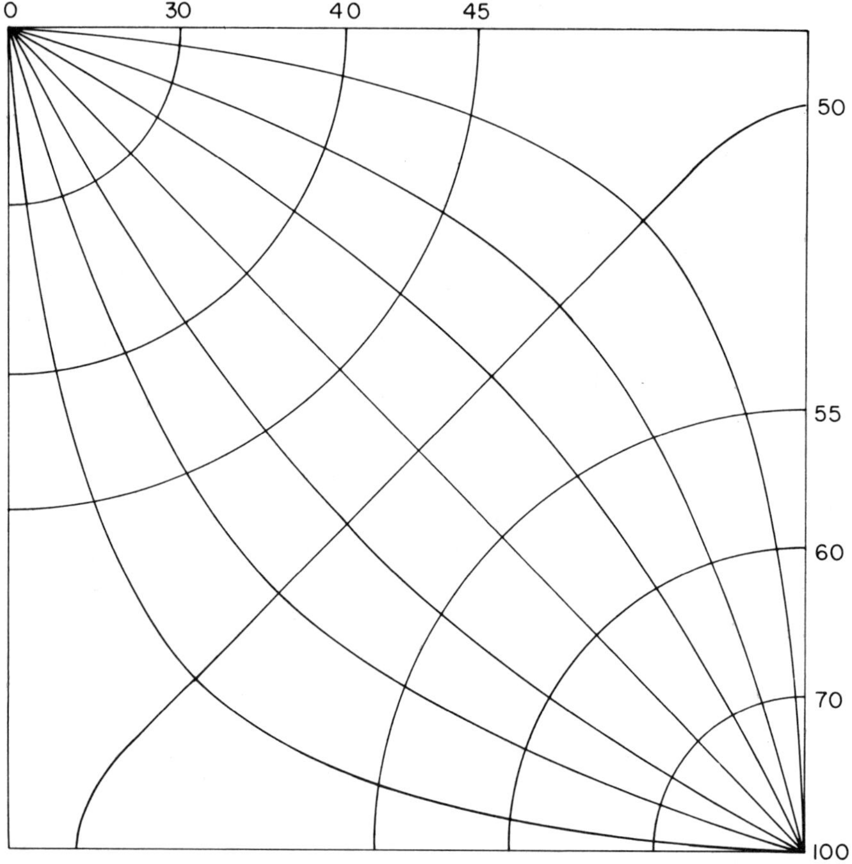

Figure 7-8. Five-spot flow net drawn from computed values of potentials.

$$V_p = \frac{V}{\phi} = -\frac{k}{\mu\phi}\frac{dp}{dL} \tag{7-25}$$

From physics, it was learned that the formula for the true velocity is

$$V_t = \frac{dL}{dt} \tag{7-26}$$

Now setting Equation **7-25** equal to Equation **7-26** gives

$$-\frac{dL}{dt} = \frac{k}{\mu\phi}\frac{dp}{dL} = V_t \tag{7-27}$$

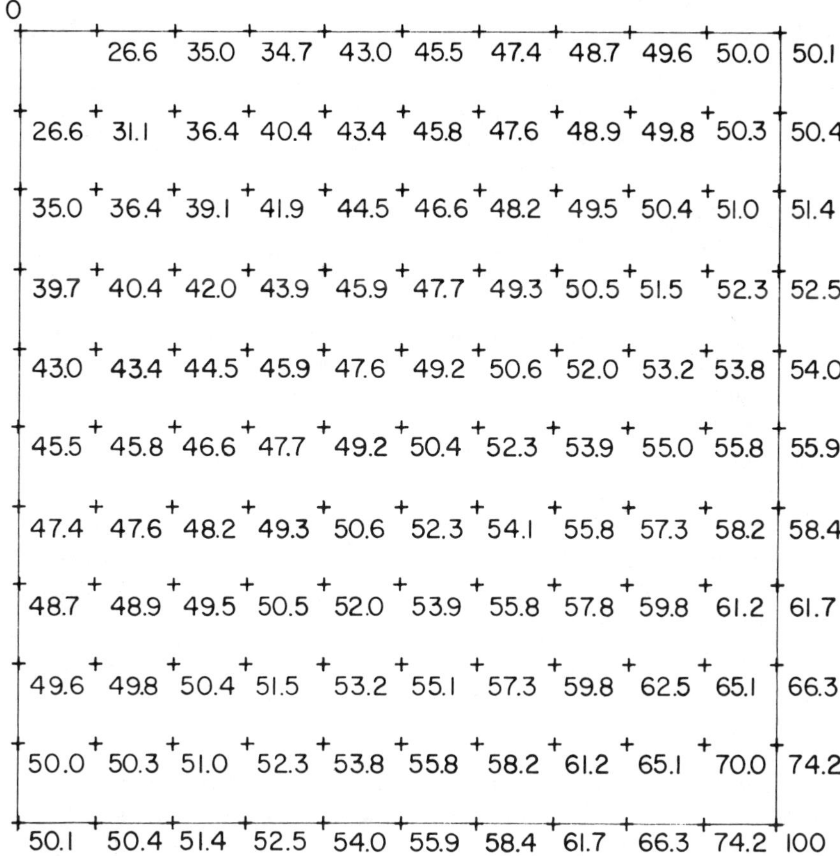

Figure 7-9. Computed potential values for a five-spot flow net by using the method of images for computing boundary values.

or

$$\frac{1}{dp/dL}\,dL = \frac{k}{\phi\mu}\,dt \tag{7-28}$$

and upon integration between limits the following equations are obtained.

$$\int_{L_1}^{L_2} \frac{1}{dp/dL}\,dL = \int_{t_1}^{t_2} \frac{k}{\phi\mu}\,dt \tag{7-29}$$

or

$$\int_{L_1}^{L_2} \frac{1}{dp/dL}\, dL = \frac{k}{\phi\mu}(t_2 - t_1) \qquad \text{7-30}$$

By approximating the integral, Equation **7-30** may be simplified as

$$\frac{(\Delta L)^2}{\Delta p} = \frac{k}{\phi\mu}\,\Delta t \qquad \text{7-31}$$

where k/μ is the mobility of the fluid with respect to the given formation, ΔL is the distance between two pressure points along a streamline, Δp is the pressure difference between two equi-pressure lines and Δt is the time consumed by the particle in moving along the streamline from a higher equi-pressure line to a lower equi-pressure line. The use of Equation **7-31** for computing specific time increments, requires specific values for the terms pressure increment, distance (ΔL), permeability and porosity. Only relative time increments are required for tracing flood fronts in flow nets, since pressures are listed as percent of the largest pressure. For relative time increments Equation **7-31** reduces to the form

$$\Delta t \;\alpha\; \frac{(\Delta L)^2}{\Delta p} \qquad \text{7-32}$$

The use of Equation **7-32** for determining the flood front positions for a five-spot flood pattern is illustrated with the use of Figure 7-10. The values for the relative times on the streamlines in Figure 7-10 are given numbers. The numbers on the streamlines are the summations of the time intervals from the common starting point ($p = 100$). The flood fronts are found by drawing in the contour lines for the constant values of time. These contour lines show the progression of the flood front.

The procedure for the computation of fluid displacement, with the aid of equi-potential and streamline patterns as illustrated in Figure 7-10, only applies for a given mobility ratio of one. In other words, it applies for a fluid of a given mobility displacing a fluid having the same mobility. Mobility ratio is explained by the terms

$$M = \lambda_1/\lambda_2 = \frac{k_1}{\mu_1}\bigg/\frac{k_2}{\mu_2} = \frac{k_1\mu_2}{k_2\mu_1} \qquad \text{7-33}$$

where M represents the mobility ratio, λ_1 is the mobility (k_1/μ_1) of the displacing phase and λ_2 is the mobility (k_2/μ_2) of the displaced phase. For fluid displacement systems with mobility ratios other than one, the pattern

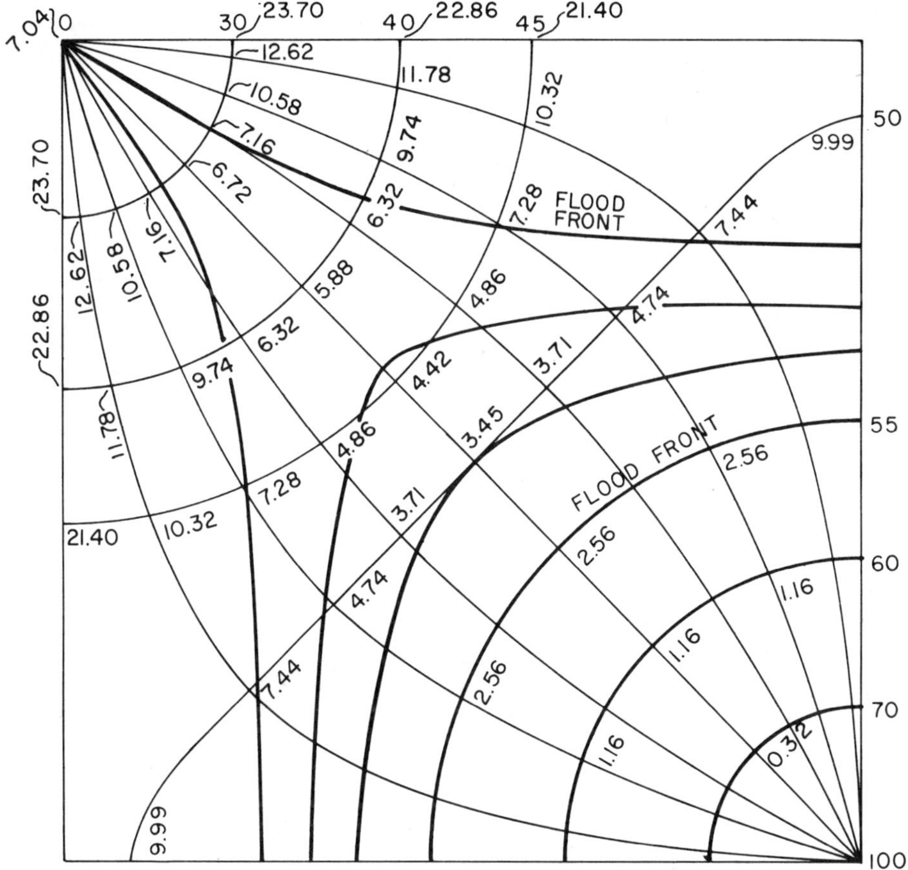

Figure 7-10. Computation of displacement fronts in a five-spot flow net; numbers at intersections of equi-pressure lines and streamlines are only relative values; the heavy lines represent the advances of the flood fronts.

of the equi-potentials and streamlines changes as the flood front progresses toward the producing well.

An approximate method for tracing flood fronts for mobility ratios other than one consists of alternately computing new equi-pressure contours and streamlines for a given flood pattern and advancing the flood front following each calculation of new equi-pressure contours and streamlines for a small increment and then repeating the whole process until there is a break through of the flood front. New equi-pressure contours and new streamlines are computed by dividing the total number of mesh points into two groups or zones. One zone is behind the flood front and the other zone in front of it. The mobility behind the flood front differs from the mobility ahead of it. Either the method of iteration or the method of relaxation are applied for computing the potentials at the mesh points. The boundary points are computed by assigning images. The values of the mesh points on the interface between the two zones are computed by means of equations which are derived in the discussion which follows.

The two equations for computing the values at the interface are derived by representing the interface by a zigzag line which goes through the mesh points as illustrated in Figure 7-11. The mesh points are of two types, that is, either one-fourth of the area for a given mesh point is in one zone and the remaining three-fourths of the area is in the other zone or one-half of the area for a given mesh point is in one zone and the remaining half of the area is in the other zone. The two types of mesh points are represented by mesh points a and b.

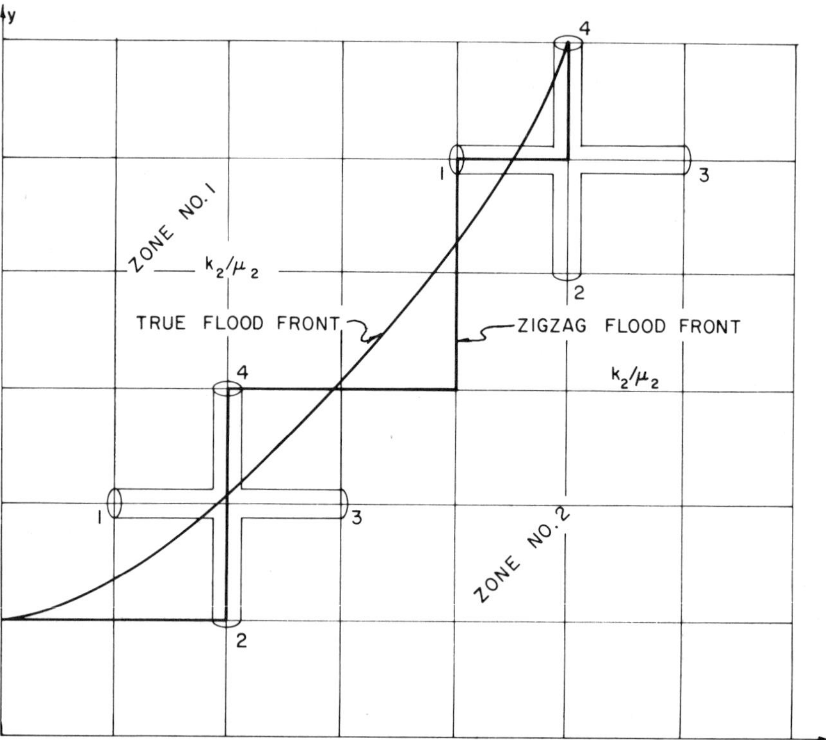

Figure 7-11. Zigzag representation of flood front illustrating two types of mesh points at the interfaces between two zones, each with a different mobility.

The development of the equation for computing the pressure at point a follows. The flow from 1 to a in the x-direction is

$$q_{1\to a} = \frac{1}{2}\left(\frac{k}{\mu}\right)_1 \frac{A(p_1 - p_a)}{\Delta x} + \frac{1}{2}\left(\frac{k}{\mu}\right)_2 \frac{A(p_1 - p_a)}{\Delta x} \qquad \textbf{7-34}$$

and the flow from a to 3 is

$$q_{a\to 3} = \left(\frac{k}{\mu}\right)_2 A(p_a - p_3) \qquad \textbf{7-35}$$

The fluid remaining at point a due to the flow in the x-direction is

$$q_{1\to a} - q_{a\to 3} = \left[\frac{A}{2}\left(\frac{k}{\mu}\right)_1 + \frac{A}{2}\left(\frac{k}{\mu}\right)_2\right]\frac{p_1 - p_a}{\Delta x} - \left(\frac{k}{\mu}\right)_2 \frac{A(p_a - p_3)}{\Delta x} \qquad \textbf{7-36}$$

In the y-direction the flow from 2 to a is

$$q_{2\to a} = \left(\frac{k}{\mu}\right)_2 \frac{A(p_2 - p_a)}{\Delta y} \qquad \textbf{7-37}$$

and the flow from a to 4 is

$$q_{a\to 4} = \frac{A}{2}\left[\left(\frac{k}{\mu}\right)_1 + \left(\frac{k}{\mu}\right)_2\right]\frac{(p_a - p_4)}{\Delta y} \qquad \textbf{7-38}$$

The fluid remaining at point a due to the flow in the y-direction is

$$q_{2\to a} - q_{a\to 4} = \left(\frac{k}{\mu}\right)_2 \frac{A(p_2 - p_a)}{\Delta y} - \frac{A}{2}\left[\left(\frac{k}{\mu}\right)_1 + \left(\frac{k}{\mu}\right)_2\right]\frac{p_a - p_4}{\Delta y} \qquad \textbf{7-39}$$

The total quantity of fluid remaining at point a is the sum of the directional quantities of fluid remaining there. The equation for this quantity of fluid is

$$q_{total} = q_{1\rightarrow a} - q_{a\rightarrow 3} + q_{2\rightarrow a} - q_{a\rightarrow 4} = \frac{A}{2}\left[\left(\frac{k}{\mu}\right)_1 + \left(\frac{k}{\mu}\right)_2\right]\frac{p_1 - p_a}{\Delta x}$$
$$- \left(\frac{k}{\mu}\right)_2 \frac{A(p_a - p_3)}{\Delta x} + \left(\frac{k}{\mu}\right)_2 \frac{A(p_2 - p_a)}{\Delta y}$$
$$- \frac{A}{2}\left[\left(\frac{k}{\mu}\right)_1 + \left(\frac{k}{\mu}\right)_2\right]\frac{p_a - p_4}{\Delta y} \qquad \textbf{7-40}$$

For steady-state fluid-flow the total fluid remaining at point a is zero and Equation **7-40** may be written as

$$0 = \frac{A}{2}\left[\left(\frac{k}{\mu}\right)_1 + \left(\frac{k}{\mu}\right)_2\right]\frac{p_1 - p_a}{\Delta x} - \left(\frac{k}{\mu}\right)_2 \frac{A(p_a - p_3)}{\Delta x}$$
$$+ \left(\frac{k}{\mu}\right)_2 \frac{A(p_2 - p_a)}{\Delta y} - \frac{A}{2}\left[\left(\frac{k}{\mu}\right)_1 + \left(\frac{k}{\mu}\right)_2\right]\frac{p_a - p_4}{\Delta y} \qquad \textbf{7-41}$$

Since the mesh was drawn such that $\Delta x = \Delta y$, Equation **7-41** becomes

$$0 = \frac{1}{2}\left[\left(\frac{k}{\mu}\right)_1 + \left(\frac{k}{\mu}\right)_2\right](p_1 - p_a) - \left(\frac{k}{\mu}\right)_2 (p_a - p_3)$$
$$+ \left(\frac{k}{\mu}\right)_2 (p_2 - p_a) - \frac{1}{2}\left[\left(\frac{k}{\mu}\right)_1 + \left(\frac{k}{\mu}\right)_2\right](p_a - p_4) \qquad \textbf{7-42}$$

and by factoring out p_a from Equation **7-42,** the following equation is obtained.

$$p_a\left[\frac{1}{2}\left(\frac{k}{\mu}\right)_1 + \frac{1}{2}\left(\frac{k}{\mu}\right)_2 + \left(\frac{k}{\mu}\right)_2 + \left(\frac{k}{\mu}\right)_2 + \frac{1}{2}\left(\frac{k}{\mu}\right)_1 + \frac{1}{2}\left(\frac{k}{\mu}\right)_2\right]$$
$$= \frac{1}{2}\left(\frac{k}{\mu}\right)_1 p_1 + \frac{1}{2}\left(\frac{k}{\mu}\right)_2 p_1 + \left(\frac{k}{\mu}\right)_2 p_3 + \left(\frac{k}{\mu}\right)_2 p_2 \qquad \textbf{7-43}$$
$$+ \frac{1}{2}\left(\frac{k}{\mu}\right)_1 p_4 + \frac{1}{2}\left(\frac{k}{\mu}\right)_2 p_4$$

Solving for p_a gives

$$p_a = \frac{\frac{1}{2}\left[\left(\frac{k}{\mu}\right)_1 + \left(\frac{k}{\mu}\right)_2\right]p_1 + \left(\frac{k}{\mu}\right)_2 p_3 + \left(\frac{k}{\mu}\right)_2 p_2 + \frac{1}{2}\left[\left(\frac{k}{\mu}\right)_1 + \left(\frac{k}{\mu}\right)_2\right]p_4}{\left(\frac{k}{\mu}\right)_1 + 3\left(\frac{k}{\mu}\right)_2} \qquad \textbf{7-44}$$

And an inspection of Equation 7-44 indicates that for a mobility of 1, that is $(k/\mu)_1 = (k/\mu)_2$, Equation 7-44 becomes

$$p_a = \frac{p_1 + p_2 + p_3 + p_4}{4} \qquad \text{7-44a}$$

which is the equation for determining the value of a mesh point in a homogeneous medium. If the mobility of one zone is expressed as a multiple of the mobility of the other zone, Equation 7-44 can be simplified further.

Mesh point b is of the second type, that is, one-half of the flow area is in one zone and the other half of the flow area is in the second zone. For the mesh point b the fluid flow in the x-direction from 1 to b

$$q_{1\rightarrow b} = \left(\frac{k}{\mu}\right)_1 \frac{A(p_1 - p_b)}{\Delta x} \qquad \text{7-45}$$

The flow fluid from b to 3 in the x-direction is

$$q_{b\rightarrow 3} = \left(\frac{k}{\mu}\right)_2 \frac{A(p_b - p_3)}{\Delta x} \qquad \text{7-46}$$

And the fluid remaining at b due to flow in the x-direction is

$$q_{1\rightarrow b} - q_{b\rightarrow 3} = \left(\frac{k}{\mu}\right)_1 \frac{A(p_1 - p_b)}{\Delta x} - \left(\frac{k}{\mu}\right)_2 \frac{A(p_b - p_3)}{\Delta x} \qquad \text{7-47}$$

For the mesh point b the fluid flow in the y-direction from 2 to b is

$$q_{2\rightarrow b} = \frac{1}{2}\left[\left(\frac{k}{\mu}\right)_1 + \left(\frac{k}{\mu}\right)_2\right] \frac{A(p_2 - p_b)}{\Delta y} \qquad \text{7-48}$$

The flow in the y-direction from b to 4 is

$$q_{b\rightarrow 4} = \frac{1}{2}\left[\left(\frac{k}{\mu}\right)_1 + \left(\frac{k}{\mu}\right)_2\right] \frac{A(p_b - p_4)}{\Delta y} \qquad \text{7-49}$$

And the fluid remaining at point b due to the flow in the y-direction is

$$q_{2\rightarrow b} - q_{b\rightarrow 4} = \frac{1}{2}\left[\left(\frac{k}{\mu}\right)_1 + \left(\frac{k}{\mu}\right)_2\right] \frac{A(p_2 - p_b)}{\Delta y} - \frac{1}{2}\left[\left(\frac{k}{\mu}\right)_1 + \left(\frac{k}{\mu}\right)_2\right] \frac{A(p_b - p_4)}{\Delta y} \qquad \text{7-50}$$

The total fluid remaining at point b due to the flow in both directions is

$$q_{1\to b} + q_{2\to b} - q_{b\to 4} - q_{b\to 3} = \left(\frac{k}{\mu}\right)_1 \frac{A(p_1 - p_a)}{\Delta x} - \left(\frac{k}{\mu}\right)_2 \frac{A(p_b - p_3)}{\Delta x} + \frac{1}{2}\left[\left(\frac{k}{\mu}\right)_1 + \left(\frac{k}{\mu}\right)_2\right] \frac{A(p_2 - p_b)}{\Delta y} - \frac{1}{2}\left[\left(\frac{k}{\mu}\right)_1 + \left(\frac{k}{\mu}\right)_2\right] \frac{A(p_b - p_4)}{\Delta y} \qquad \textbf{7-51}$$

For steady-state flow the left side of Equation **7-51** is zero and since the mesh was drawn such that $\Delta x = \Delta y$ then the equation becomes

$$0 = \left(\frac{k}{\mu}\right)_1 (p_1 - p_b) - \left(\frac{k}{\mu}\right)_2 (p_b - p_3) + \frac{1}{2}\left[\left(\frac{k}{\mu}\right)_1 + \left(\frac{k}{\mu}\right)_2\right] (p_2 - p_4) - \frac{1}{2}\left[\left(\frac{k}{\mu}\right)_1 + \left(\frac{k}{\mu}\right)_2\right] (p_b - p_4) \qquad \textbf{7-52}$$

Upon solving for p_b, Equation **7-52** becomes

$$p_b = \frac{\left(\frac{k}{\mu}\right)_1 p_1 + \left(\frac{k}{\mu}\right)_2 p_3 + \frac{1}{2}\left[\left(\frac{k}{\mu}\right)_1 + \left(\frac{k}{\mu}\right)_2\right] (p_2 + p_4)}{2\left[\left(\frac{k}{\mu}\right)_1 + \left(\frac{k}{\mu}\right)_2\right]} \qquad \textbf{7-53}$$

Harold B. Janzen[3], with the use of a resistance network traced the flood fronts and determined the flooding sweep efficiencies of a five-spot flood pattern for a number of different mobility ratios. He represented the flood front by a zigzag line which passed though mesh points. The resistances in the areas ahead and behind the flood front were adjusted to be in proportion to the respective mobilities of the two areas. The procedure of alternately determining a new set of pressure contours with respective streamlines and advancing the flood front for a small increment was followed.

T. D. Stacy, with the aid of an Electronic Digital Computer traced the flood fronts and determined the flooding sweep efficiencies of a five-spot flood pattern for a number of different mobility ratios. Numerical methods were used for calculating equi-pressure lines and streamlines. Following the calculation of a new set of equi-pressure contours and streamlines the flood

[3]Janzen, Harold B., "Electric Analogue Studies of Mobility Ratios." M. S. Thesis, Oklahoma State University (1955).

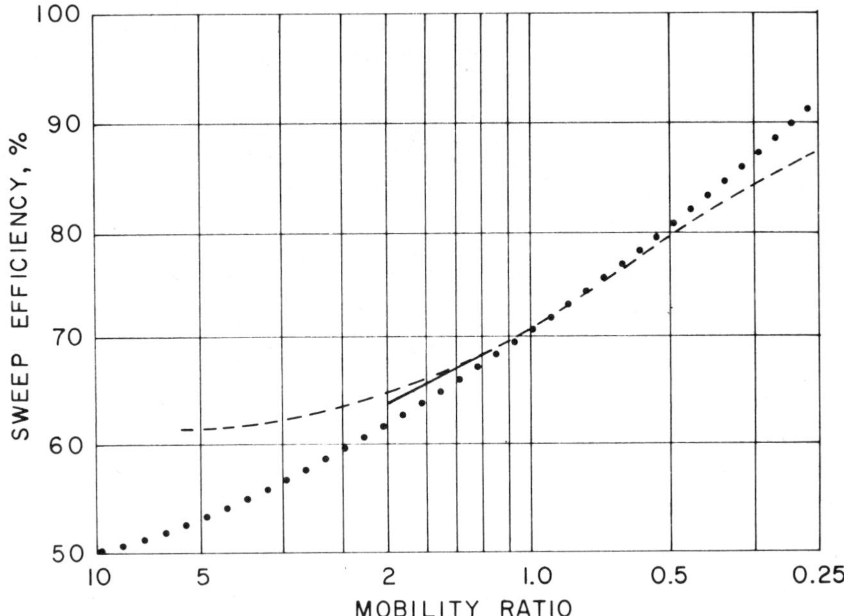

Figure 7-12. Areal sweep efficiency as a function of mobility ratio.

front was advanced for a small increment. This procedure was continued until there was a break through at the producing well. Plots of areal sweep efficiency as a function of mobility ratio are given in Figure 7-12.

The solid line in Figure 7-12 is a plot of the results obtained by Stacy and for computed values of equi-potential contours. The broken line is a plot of results obtained by Janzen and it is for measured boundary values obtained with an electrical resistance network. The dotted line is a plot of the results obtained from X-ray shadowgraph studies by F. F. Craig *et al*[4]. Only five or six steps were used in advancing the flood fronts by Janzen and Stacy, but if more steps had been taken their results would have more nearly approached the results by Craig *et al.*

Fluid Flow through Sections of Complicated Shape

A graphical method based on the construction of curvilinear squares is very useful for computing the flow through porous media for sections of

[4]Craig, F. F. et al, "Oil Performance of Pattern Gas or Water Injection from Model Tests," Trans. A.I.M.E. (1955), *204* p. 7ff.

complicated shapes. To illustrate this method the fluid flowing in a simulated sand packed reservoir will be considered. An assumed reservoir is shown in Figure 7-13.

The depth of the reservoir in the direction perpendicular to the plane of Figure 7-13 is represented by L, and fluid should not flow in this direction. It is assumed that the inner and outer edges of the reservoir are at constant pressures of p_1 and p_2 respectively, where p_2 is greater than p_1. Because of the symmetrical arrangement, no fluid flows across the boundaries AE or DH and only one-eighth of the total section needs to be considered.

As indicated in Figure 7-13, the fluid flowing from the outer to the inner surface will be considered to pass the channels between AC and EG in such a manner that no fluid crosses any of the channel boundaries. For

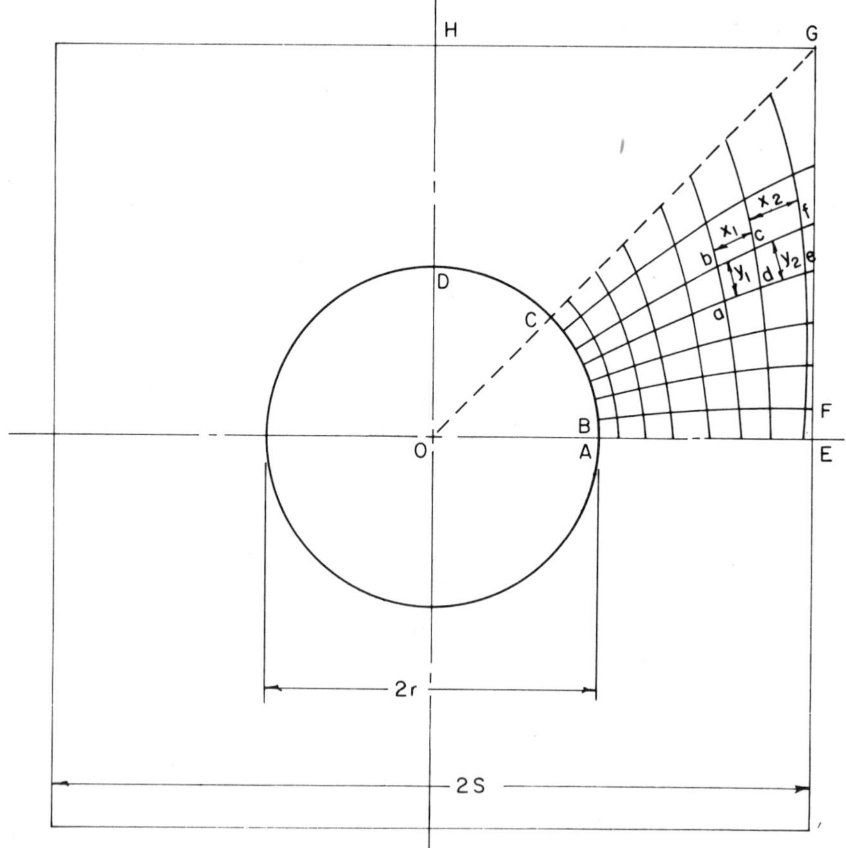

Figure 7-13. Streamlines and isopotential lines in a complicated shape sand packed reservoir.

instance all fluid flows through AB and when flowing through the channel ABFE it will pass the boundary EF. At any position along the channel the rate of fluid flow is the same as that of the entering fluid.

For the section *abcd* in one of the channels, the fluid flows at right angles to the average width of the section. The pressure difference between the isopotential planes *ab* and *cd* flowing through the section *abcd* equals

$$q_1 = \frac{k}{\mu}\frac{Ly_1}{x_1}\Delta p_1 \qquad \text{7-54}$$

where x_1 is the average path length of the fluid flowing through this section. Similarly, the fluid transmitted through section *cdfe* equals

$$q_2 = \frac{k}{\mu}\frac{Ly_2}{x_2}\Delta p_2 \qquad \text{7-55}$$

If the average path length of each section is selected so that each section has an equal pressure drop, then $\Delta p_1 = \Delta p_2$ and considering steady-state conditions, $q_1 = q_2$. Now equating the two equations for q_1 and q_2 yields

$$\frac{y_1}{x_1} = \frac{y_2}{x_2} \qquad \text{7-56}$$

and more generally

$$\frac{y_n}{x_n} = \text{Constant} \qquad \text{7-57}$$

where the subscript refers to any such section *abcd*. If these sections are constructed such that y_n and x_n are equal then the fluid flowing through the channel is

$$q = \frac{kL\,(p_2 - p_1)}{\mu N} \qquad \text{7-58}$$

In the expression, N represents the number of sections between the inner and outer surfaces; that is, the sum of the full sections and any fractional part of a section which may occur. Designating the number of channels (such as *ABFE*) by M, then the total fluid flowing becomes

$$q_{\text{total}} = \frac{MkL\,(p_2 - p_1)}{\mu N} \qquad \text{7-59}$$

In the solution of a given problem the value for the number of channels, M, may be fixed. Isopotentials and streamlines are then constructed by trial and error so as to produce a pattern of lines intersecting at right angles and forming individual sections having equal ratio values of x_n and y_n at a given section. The quadrilaterals formed are called curvilinear squares. In some cases the solution becomes easier if an integer number M of flow channels is obtained from the graph. The important items to remember are that the channel flow lines and isopotential lines must intersect at right angles in such a way that curvilinear squares are formed.

Steady-State Radial Flow Equation

In terms of cylindrical coordinates, the two-dimensional Laplace equation shown in Equation 7-25 is

$$r^2 \frac{\partial^2 p}{\partial r^2} + r \frac{\partial p}{\partial r} + \frac{1}{r^2} \frac{\partial^2 p}{\partial \theta^2} = 0$$

or

$$\frac{\partial^2 p}{\partial r^2} + \frac{1}{r} \frac{\partial p}{\partial r} + \frac{\partial^2 p}{\partial \theta^2} = 0 \qquad \textbf{7-60}$$

By letting $u = \ln r$ or $r = e^u$, Equation **7-60** may be transformed into a form which can be solved by a rectangular mesh. The procedure for transforming Equation **7-60** into a rectangular form follows. From calculus the identities are obtained.

$$\frac{\partial p}{\partial r} = \frac{\partial p}{\partial u} \frac{\partial u}{\partial r} = \frac{\partial p}{\partial u} \cdot \frac{1}{r} \qquad \textbf{7-61}$$

and

$$\frac{\partial^2 p}{\partial r^2} = \frac{\partial p}{\partial u}\left(-\frac{1}{r^2}\right) + \frac{1}{r} \frac{\partial^2 p}{\partial u^2}\left(\frac{1}{r}\right) = \frac{1}{r^2}\left[\left(\frac{\partial^2 p}{\partial u^2}\right) - \left(\frac{\partial p}{\partial u}\right)\right] \qquad \textbf{7-62}$$

The substitution of Equation **7-61** and Equation **7-62** into Equation **7-60** gives:

$$r^2 \left(\frac{1}{r^2}\right)\left(\frac{\partial^2 p}{\partial u^2} - \frac{\partial p}{\partial u}\right) + r\left(\frac{1}{r}\right)\left(\frac{\partial p}{\partial u}\right) + \frac{\partial^2 p}{\partial \theta^2} = 0$$

$$(\Delta r)_1 = 3(e^{0.1733} - 1) = 0.569$$
$$(\Delta r)_2 = 3(e^{0.3466} - 1) = 1.230$$
$$(\Delta r)_3 = 3(e^{0.5199} - 1) = 2.042$$

The Δr's are measured from the point where $r = 3.0$ on the line $\theta = 0$.

Steady-State Gas Flow

The steady-state fluid-flow equations for incompressible liquids do not hold true for steady-state gas flow because as the pressure on a gas is reduced there is an appreciable expansion of the gas. In considering the steady-state flow of gases the basic flow equations for incompressible fluids must be altered. For the steady-state flow of gases considered in this chapter, the equations derived for the flow of gases contain terms involving the square of pressure.

The basic Darcy equation is altered for the flow of gases by applying the ideal gas law for isothermal conditions. In equation form the law is

$$p_1 v_1 = p_2 v_2 = pv$$

or in petroleum engineering notation

$$p_1 q_1 = p_2 q_2 = pq \qquad \textbf{7-65}$$

where q refers to the volume of gas for a given time interval at pressure p, and q_1 and q_2 refer to the volumes of a gas for a given time interval at pressures p_1 and p_2. For the development of the equations for fluid flow of compressible gases, the basic Darcy equation is

$$q = -A\,\frac{k\,dp}{\mu\,dL} \qquad \textbf{7-66}$$

By the substitution of Equation **7-65** into the basic Darcy equation, Equation **7-66** becomes

$$q = \frac{q_2 p_2}{p} = -A\,\frac{k\,dp}{\mu\,dL} \qquad \textbf{7-67}$$

Linear Gas Flow Equation—Darcy's Law

For pressure limits of p_1 and p_2 where p_1 is the greater pressure, Equation **7-67** may be written in integral form as

$$q_2 p_2 \int_0^L dL = - A \frac{k}{\mu} \int_{p_1}^{p_2} p dp \qquad \textbf{7-68}$$

Upon integration Equation **7-68** becomes

$$q_2 = A \frac{k}{\mu} \frac{1}{p_2} \frac{p_1^2 - p_2^2}{2L} \qquad \textbf{7-69}$$

and by rewriting Equation 7-69 as

$$q_2 = A \frac{k}{\mu} \frac{1}{p_2} \frac{(p_1 + p_2)}{2} \cdot \frac{(p_1 - p_2)}{L} \qquad \textbf{7-70}$$

it can be further reduced to a simpler form as

$$q_2 = A \frac{k}{\mu} \frac{p_m}{p_2} (p_1 - p_2) \qquad \textbf{7-71}$$

where $p_m = \dfrac{p_1 + p_2}{2}$

Radial Gas Flow Equation—Darcy Equation

For radial flow the cross sectional area A becomes $(2\pi rh)$, where h is the thickness, and the limits of the radii and pressures are r_e and r_w, and p_e and p_w; the flow rate into the well bore is q_w for a given increment of time. The subscript w refers to the limits at the well bore. The substitution of these limits in Equation **7-68** gives

$$q_w p_w \int_{r_w}^{r_e} \frac{dr}{r} = 2\pi h \frac{k}{\mu} \int_{p_w}^{p_e} p dp \qquad \textbf{7-72}$$

and the integration of Equation **7-72** gives.

$$q_w = 2\pi h \frac{k}{\mu} \frac{1}{p_w} \frac{p_e^2 - p_w^2}{2 ln\, r_e/r_w} \qquad \textbf{7-73}$$

Then the substitution of

$$\frac{p_e^2 - p_w^2}{2} = \frac{(p_e + p_w)}{2} (p_e - p_w) = p_m (p_e - p_w)$$

into Equation 7-73 gives

$$q_w = 2\pi h \frac{k}{\mu} \frac{p_m}{p_w} \frac{p_e - p_w}{ln\, r_e/r_w} \qquad \textbf{7-74}$$

Gas Flow Equation—Rectangular Coordinates

In Chapter 6 the continuity equation, the Laplace equation and the diffusivity equation were derived for the flow of fluids. The unsteady state flow equation **6-22** was derived in rectangular coordinates for the flow of gases.

The unsteady-state gas flow equation in rectangular coordinates becomes the steady-state gas flow equation when the right-hand side of the equal sign is set equal to zero. With the suggested changes and for two-dimensions Equation **6-22** becomes

$$\frac{\partial^2 p^2}{\partial x^2} + \frac{\partial^2 p^2}{\partial y^2} = 0 \qquad \textbf{7-75}$$

This steady-state equation for the flow of gases may be solved numerically by the methods previously presented for solving the two-dimensional Laplace equation. In the numerical solutions of Equation **7-75** the values of the pressure squared are placed at the mesh points.

Gas Flow Equation—Cylindrical Coordinates

In a similar manner as discussed in the section *Steady-State Radial Flow Equation* Equation **6-29** may be transformed by setting $u = \ln r$ into the rectangular form.

$$\frac{\partial^2 p^2}{\partial u^2} + \frac{\partial^2 p^2}{\partial \theta^2} = 0 \qquad \textbf{7-76}$$

Equation **7-76** may be solved numerically in a manner similarly to the methods suggested for Equation **7-63**. In the solutions, the values of pressure squared are used at the mesh points instead of pressure as was suggested for Equation **7-63**.

Exercises

1. Oil is flowing through a long homogeneous sand-packed pipe. The inlet pressure is 500 psig and the outlet pressure is atmospheric pressure. If the pipe is divided into four equal segments, determine the pressures at the grid points by means of the method of iteration.
2. Repeat Exercise 1 by using the method of relaxation for determining the values of the pressures at the grid points.
3. Using the method of relaxation determine the pressure values at the interior points for the following mesh.
4. Using the method of iteration with images, determine pressure values at the boundary and interior points for the following mesh.
5. Using the method of iteration with images determine the pressure values at the boundary points and at the interior points of a five-spot flood pattern. Hint: to avoid an excessive number of iterations, use only one-eighth five-spot mesh as indicated below. Use assumed boundary values of pressures which drop somewhat logarithmically near the injection well and near the production well.

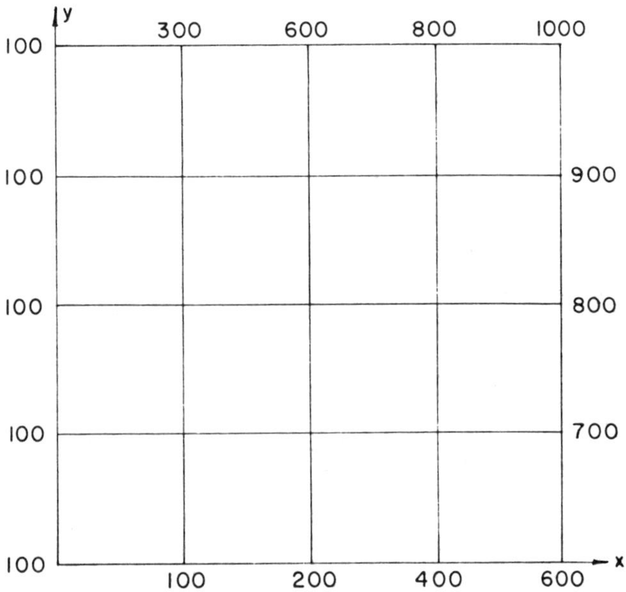

Figure 7-16. Pressure values at the boundary of a square reservoir.

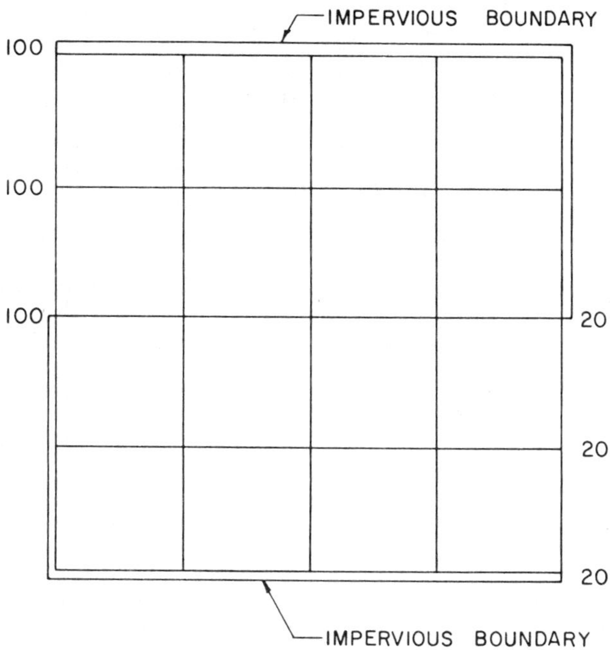

Figure 7-17. Selected pressure values at the boundary of a square reservoir containing impervious boundaries.

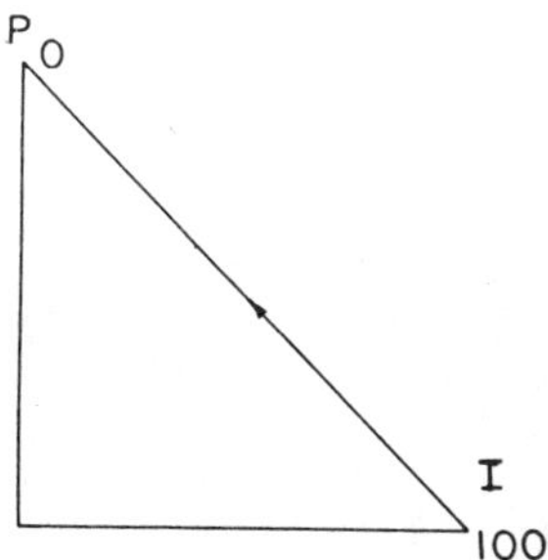

Figure 7-18. One-eighth section (triangular section) of a five-spot flood pattern.

6. For the computed five-spot pattern of Exercise 5, trace a flood front for a mobility ratio of one.
7. Use graphical methods for determining the quantity of water flowing underneath a foot section (length) of a dam as indicated in the following sketch. Take k = 1.00 Darcy and viscosity as 1.00 centipoise.
8. Using graphical methods compare the relative flow rates of water into a circular oil reservoir when the oil reservoir is at the center of a homogeneous water drive reservoir which has a perfect square perimeter. Assume uniform and identical thicknesses for both oil reservoir and water drive reservoir. Also assume that the

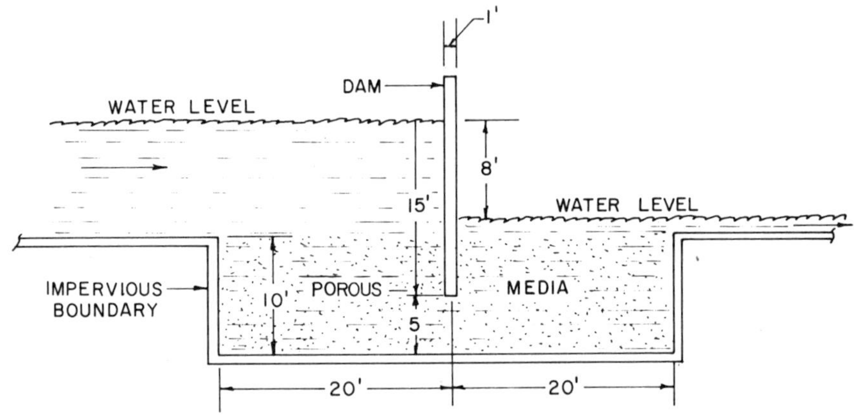

Figure 7-19. Water reservoir which contains a porous medium under a dam.

radius of the oil reservoir is one-half that of the outside radius of the water drive reservoir; and also assume that the diameter of the circular oil reservoir is one-half that of one side of the square reservoir. In addition assume that the pressures are uniform on the perimeter of the oil reservoir and on the perimeter of the water drive reservoir.

9. Transform the results from Exercise 3 (rectangular coordinates) to cylindrical coordinates and take $\theta = 30°$.

Suggested Reading

1. Craig, F. F., Jr., Giffen, T. M., and Morse, R. A., "Oil Recovery Performance of Pattern Gas or Water Injection Operations from Model Tests," Trans. A.I.M.E., 1955, p. 7ff, 244.
2. Dystra, H., and Parsons, R. L., "Relaxation Methods Applied to Oil Field Research," Trans. A.I.M.E., pp. 227-232, 92.
3. Jakob, M., and Hawkins, G. A., *Elements of Heat Transfer and Insulation*, New York: John Wiley and Sons, 1959, pp. 38-46.
4. Katz, D. L., et al., *Handbook of Natural Gas Engineering*, New York: McGraw-Hill Co., Inc., 1959, pp. 403-434.
5. Muskat, M., *The Flow of Homogeneous Fluids Through Porous Media*, Ann Arbor: J. W. Edwards Inc., 1946, pp. 76-79, 121-146.
6. Nobles, M. A. and Janzen, H. J., "Application of a Resistance Network for Studying Mobility Ratio Effects," Trans. A.I.M.E., 1958, pp. 356-358, 213.
7. Pirson, S. J., *Oil Reservoir Engineering, 2nd ed.*, New York: McGraw-Hill Co., Inc., 1958, pp. 392-397.
8. Scarborough, James B., *Numerical Mathematical Analysis*, Baltimore: The Johns Hopkins University Press, 1950, pp. 325-333.
9. Stacy, T. D., "Sweep Efficiency Studies by Use of High-Speed Digital Computers," Master's Thesis, Louisiana Polytechnic Institute, 1963, pp. 2-23.

8 Unsteady-State Fluid Flow Through Porous Media

Since reservoir fluids are slightly compressible, often the flow of the fluids in the reservoir is unsteady-state. The compressibility of water ranges from 2.5×10^{-6} to 3.6×10^{-6} volume/volume psi. This range represents the compressibilities of water for various temperatures and various pressures. The compressibilities of reservoir oils are from 10 to 100 times more than the compressibilities of water. The greater the compressibility of the fluid, the more pronounced the unsteady-state effect of the reservoir fluid. Scientists and engineers use the diffusivity equation for studying the unsteady-state flow of fluids in porous media.

Discussion in this chapter is centered around explicit solutions, by means of difference equations, for the linear and radial parabolic diffusivity equations as applied to unsteady-state flow of viscous compressible fluids and gases through porous media. Several example problems are given; a complete solution of any one of the problems requires thousands of arithmetic computations which with a desk calculator would require an enormous number of hours. A microcomputer is suitable for solving most of the problems in this chapter.

In the remaining examples the calculations for the solutions to the problems are given for only a few steps; the calculations have been carried out to an extent that it is evident that the arithmetic computations are converging toward an answer. The calculations are carried far enough to give complete information for solving the problems. All but one example are concerned with build-up pressures. The principles illustrated in the example for numerically solving the diffusivity equations can also be applied to other problems such as draw down pressures, water encroachment, and allowance for variation in thicknesses of oil reservoirs.

Emphasis is placed upon solutions to the diffusivity equations for various boundary conditions. The example problems require the use of the basic units, gram, seconds, centimeter, atmosphere, centipoise, and Darcy. The use of the practical units, feet, milli-Darcy, pounds, pounds per square inch, barrels, days, hours, etc. are discussed in later chapters.

One Dimensional Unsteady-State Flow of Viscous Compressible Fluids

Linear Diffusivity Equation

The three-dimensional diffusivity equation as developed in Equation **6-15** in terms of density is $\frac{\partial^2 \rho}{\partial x^2} + \frac{\partial^2 \rho}{\partial y^2} + \frac{\partial^2 \rho}{\partial z} = \frac{\phi \mu c}{k} \frac{\partial \rho}{\partial t}$.

Where k is the permeability, Darcy; ρ is the density, grams per cubic millimeter; μ is the viscosity, centipoise; ϕ is the porosity, fraction of pore space per unit volume; c is compressibility, volume/volume atmosphere; and t is time, seconds.

Often it is preferred to use the diffusivity equation in terms of pressure instead of in terms of density. The devlopment of the diffusivity equation in terms of pressure follows. The development of the relations between first and second derivatives of pressure and the first and second derivatives of density are determined from the differentiation of the equation for the relation between density, pressure and compressibility. The equation for this relation is $\rho = \rho_0 e^{-c(p_0 - p)}$. Utilizing the differential calculus identity

$$\frac{d}{dx} e^u = e^u \frac{du}{dx}$$

the first differentiation of Equation **6-10** with respect to x is

$$\frac{\partial \rho}{dx} = \rho_0 e^{-c(p_0 - p)} \left(c \frac{\partial p}{\partial x} \right) = \rho c \frac{\partial p}{\partial x} \qquad \textbf{8-1}$$

and the second differentiation of Equation **6-10** with respect to x is

$$\frac{\partial^2 \rho}{\partial x^2} = \rho_o e^{-c(p_o - p)} \left(c \frac{\partial p}{\partial x}\right)\left(c \frac{\partial p}{\partial x}\right) + \rho_o e^{-c(p_o - p)} \left(c \frac{\partial^2 p}{\partial x^2}\right) = \rho c^2 \left(\frac{\partial p}{\partial x}\right)^2 + \rho c \left(\frac{\partial^2 p}{\partial x^2}\right) \quad \textbf{8-2}$$

In a similar manner the second derivatives of density with respect to y and z and the first derivative of density with time are determined to be

$$\frac{\partial^2 \rho}{\partial y^2} = \rho c^2 \left(\frac{\partial p}{\partial y}\right)^2 + \rho c \left(\frac{\partial^2 p}{\partial y^2}\right)$$

$$\frac{\partial^2 \rho}{\partial z} = \rho c^2 \left(\frac{\partial p}{\partial z}\right)^2 + \rho c \left(\frac{\partial^2 p}{\partial z^2}\right)$$

$$\frac{\partial \rho}{\partial t} = \rho c \left(\frac{\partial p}{\partial t}\right) \quad \textbf{8-3}$$

The substitution of Equations **8-2** and **8-3** into Equation **6-15** gives

$$\rho c^2 \left[\left(\frac{\partial p}{\partial x}\right)^2 + \left(\frac{\partial p}{\partial y}\right)^2 + \left(\frac{\partial p}{\partial z}\right)^2\right] + \rho c \left(\frac{\partial^2 p}{\partial x^2} + \frac{\partial^2 p}{\partial y^2} + \frac{\partial^2 p}{\partial x^2}\right) = \rho c \frac{\phi \mu c}{k} \frac{\partial p}{\partial t} \quad \textbf{8-4}$$

or

$$c \left[\left(\frac{\partial p}{\partial x}\right)^2 + \left(\frac{\partial p}{\partial y}\right)^2 + \left(\frac{\partial p}{\partial z}\right)^2\right] + \frac{\partial^2 p}{\partial x^2} + \frac{\partial^2 p}{\partial y^2} + \frac{\partial^2 p}{\partial z^2} = \frac{\phi \mu c}{k} \frac{\partial p}{\partial t}$$

Since the value of c is in the range of 10^{-3} to 10^{-4} volume/volume atmosphere, the ommission of the terms in the bracket of Equation **8-4** may be deleted from the equation without it becoming affected appreciably. With these terms deleted Equation **8-4** becomes

$$\frac{\partial^2 p}{\partial x^2} + \frac{\partial^2 p}{\partial y^2} + \frac{\partial^2 p}{\partial z^2} = \frac{\phi \mu c}{k} \frac{\partial b}{\partial t} \quad \textbf{8-5}$$

For unsteady-state fluid-flow in one-dimension (linear flow) Equation **8-5** becomes

$$\frac{\partial^2 p}{\partial x^2} = \frac{\phi \mu c}{k} \frac{\partial p}{\partial t} \quad \textbf{8-6}$$

Often, the hydraulic diffusivity η is used for the terms $k/\phi\mu c$; with the use of the hydraulic diffusivity notation Equation **8-6** becomes

$$\frac{\partial^2 p}{\partial x^2} = \frac{1}{\eta}\frac{\partial p}{\partial t} \qquad \textbf{8-7}$$

where the terms are the same as those given for Equation **6-15**; the pressure is in atmospheres.

Numerical Solution for Linear Build-Up Pressure

A difference equation solution for the linear diffusivity equation, written in terms of pressure is

$$p(x, t + k) = \frac{p(x + h, t) + p(x - h, t)}{2} \qquad \textbf{8-8}$$

when $\eta k/h^2$ was arbitrarily taken as ½.

With use of the difference, Equation **8-8**, the values of pressure at the interior points of a pressure-time mesh are determined as illustrated in Chapter 4, Example 4-10. However, the determination of the values of pressure at the boundary points of the mesh where the values of pressure are constantly changing requires the use of images. For determining the pressure values at the boundary, image values are assigned at the imaginary portion of the pressure-time mesh as indicated by the broken-line portion of Figure 8-1; then Equation **8-8** is applied for calculating the new boundary value.

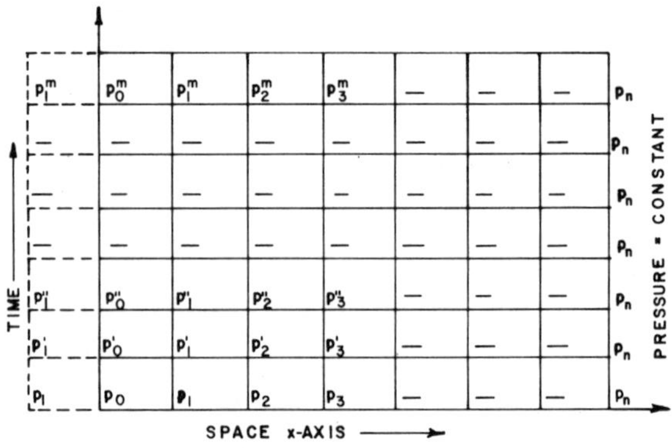

Figure 8-1. Pressure time mesh illustrating images.

By carefully observing, it is obvious that the changing boundary value p_0^m is identical with the adjacent value p_1^m on the same row. The procedure to follow in filling in the pressure values at the mesh points is first, complete a set of pressure values for a row of interior mesh points with the use of the difference equation; then for a given row, set the boundary value equal to the value of its adjacent interior mesh point. Repeat this procedure for each row of the pressure-time mesh.

A numerical problem for calculating build-up pressures in a sand-packed pipe is presented in Example 8-1. The numerical values for the build-up pressures were obtained with a computer program written in machine language for IBM 650 Computer and solved thereon.

Example 8-1.* Compute the build-up pressures for a homogeneous sand-packed pipe which has a total length of 10,000 centimeters, if initially there is steady-state fluid-flow through the sand-packed pipe and suddenly a valve is closed at the low-pressure end of the pipe and the pressure at the other end is kept at the higher pressure. Additional data are: $\phi = 0.20$; $p_o = 0.00$ atmosphere gauge; $p_n = 100.00$ atmospheres gauge; $k = 1.00$ Darcys; $\mu = 1.00$ centipoise; and $c = 5.00 \times 10^{-4}$ volume/volume atmosphere.

Solution. The pressure p_o will build up at the end of the packed pipe where the valve is suddenly closed. The boundary values are:

$$p(0,0) = p_o$$

$$p(x,0) = p_o + (p_n - p_o)\frac{x}{L}$$

$$p(L,t) = p_n = \text{constant}$$

$$p(x,\infty) = p_n$$

$$p_x(0,t) = 0$$

For 30 equal length subdivisions of the pipe

$$h = \frac{10{,}000}{30} = 333.33 \text{ cms}$$

$$n = \frac{1}{0.20 \times 1.00 \times 5.00 \times 10^{-4}} = 1.0 \times 10^4$$

*See Chapters 10 and 11 for computer solutions to linear build-up problems.

and since $k\eta/h^2 = 1/2$

$$k = \frac{1}{2} \frac{(333.33)^2}{1 \times 10^4} = 5.55 \text{ sec.}$$

The computed values for build-up pressure are represented by the curve with the triangles in Figure 8-2. Also in Figure 8-2 is a plot of computed values of pressure obtained from a classical solution of the linear diffusivity equation for the boundary conditions given for the example. The classical solution of the linear diffusivity equation for the given boundary conditions is

$$p(x, t) = p_n - \frac{8(p_n - p_o)}{\pi^2} \sum_{n=1}^{\infty} \frac{1}{(2n-1)^2} \cos \frac{(2n-1)\,\pi x}{2L}$$

$$\cdot e - \frac{-(2n-1)^2\, n\pi^2 t}{4L^2}$$

Graphical Solution for Linear Build-Up Pressure

The difference equation solution **8-8** for the linear diffusivity equation indicates that the pressure value after an increment of time $(t + \Delta t)$ is the average of two pressure values at time t prior to the increment of time. This suggests that an average for the pressure values could be found from a pressure-distance graph as illustrated in Figure 8-3. The average pressure value is determined by taking the value at the midpoint of a straight line drawn between two different pressure values.

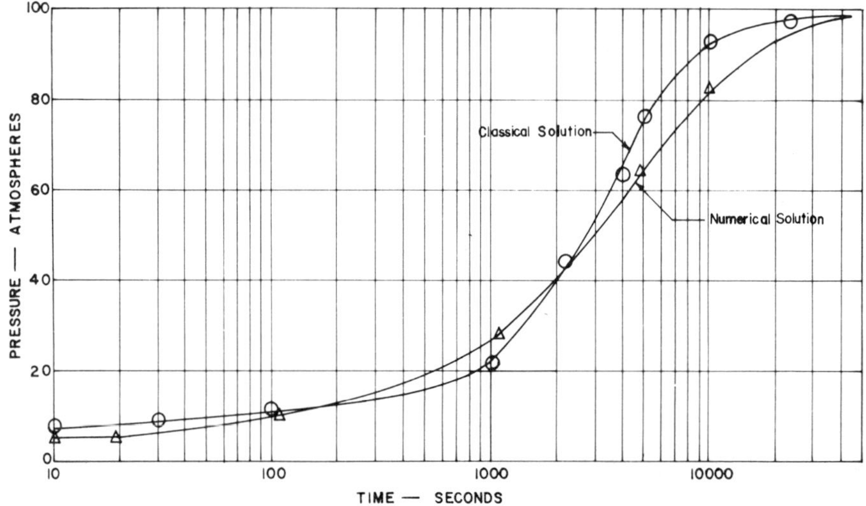

Figure 8-2. Build-up pressure as a function of time in a sand-packed pipe.

In Figure 8-3 the solid curved line represents a plot of the pressure at time t versus distance and the dashed lines were drawn vertically at equal intervals (Δx or h) along the x-axis. The points of intersection between the vertical lines and the pressure versus distance curve are designated by both letters of the alphabet (a, b, c, d) and pressure values $p(x, t)$. If a line is drawn from a to c, it is obvious that the pressure value at b' is the average of the two pressures $p(x-h,\ t)$ and $p(x+h,\ t)$ or $(p(x-h,\ t) + p(x+h, t))/2$. This mathematical notation is equal to the right-hand side of Equation **8-8**. In the figure $p(x, t+k)$ is represented by $p(x, t')$. In a similar manner it is obvious that the pressure value at c' is the average of the two pressures $p(x, t)$ and $p(x+2h, t)$.

The procedure for graphically using the difference equation **8-8** as a solution of the linear diffusivity equation is:

1. Plot pressure values versus distance on rectangular coordinates with x-axis as the abscissa.
2. At equal intervals (Δx or h) along the x-axis construct vertical lines.
3. For a given value of Δx or h compute the value of Δt or k from the equation $k = \frac{1}{2}h^2/\eta = \Delta t = \frac{1}{2}(\Delta x)^2/\eta$.
4. With straight lines connect alternate intersecting points (points of intersection of pressure-distance curve and vertical lines) as indicated in Figure 8-3.
5. Draw a smooth curve through the points formed by the intersection of the vertical lines and the connecting lines (see step 4); this smooth curve is a plot of pressure versus distance at time $(t + \Delta t)$.

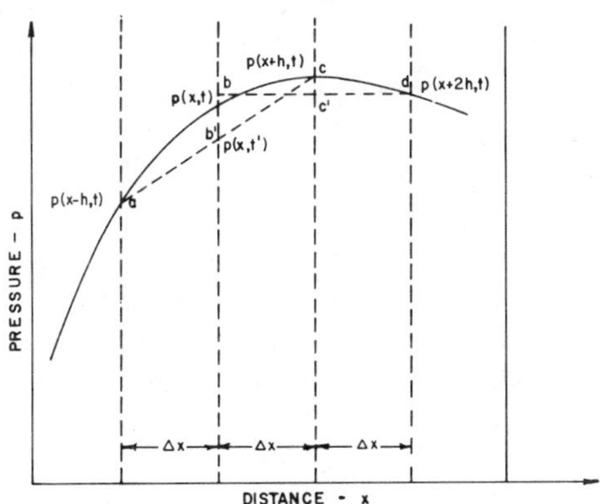

Figure 8-3. Schematic drawing which illustrates graphically how to solve the linear diffusivity equation.

6. With the second curve repeat steps 4 and 5 to obtain the third pressure versus distance curve for time $(t + 2\Delta t)$; then repeat steps 4 and 5 to obtain the fourth pressure versus time curve for time $(t + 3\Delta t)$ and until the $n + 1$ curve is obtained for time $(t + n\Delta t)$.

The procedure for graphically using the difference equation **8-8** as a solution of the linear diffusivity equation is illustrated with an example.

Example 8-2. Graphically by means of a difference solution of the linear diffusivity equation, compute six build-up pressures for a homogeneous sand-packed pipe 600 centimeters in length, if, initially there is steady-state fluid-flow through the sand-packed pipe and suddenly a valve is closed at the low pressure end of the pipe and the pressure at the other end remains constant at the higher pressure. Additional data are: $\phi = 0.20$; $p_0 = 00.0$ atmospheres gauge; $p_n = 600.00$ atmospheres gauge; $k = 1.00$ Darcys; $\mu = 1.00$ centipoise; and $c = 5.00 \times 10^{-4}$ volume/volume atmosphere.

Solution. The pressure p_o will build up at the end of the sand-packed pipe where the valve is closed. The boundary values are identical to those given for Example 8-1.

For six equal length subdivisions of the pipe $h = 600/6 = 100$ cms and $\eta = 1/0.20 \times 1.00 \times 5.00 \times 10^{-4} = 1.00 \times 10^4$; and since $k\eta/h^2 = 1/2$ and $k = 1/2\ (100)^2/1.0 \times 10^4 = 1.0 \times 10^4/2.0 \times 10^4 = 0.5$ seconds.

The graphical solution for the example is given in Figure 8-4. Since there was initially steady-state fluid-flow through the pipe, the plot of pressure versus distance gave a straight line. This plot is represented by the solid diagonal line in the figure. The first value of the build-up pressure p_0' is the average value of the pressure at the origin and the values of the pressure at point b. This average value of the two pressures is the value of the pressure at point a. A solid horizontal line is projected back to the ordinate for $x = 0$. The value of p_0' is indicated by the horizontal line. The solid lines in the figure represent pressure distributions at various time increments and the broken lines represent connecting lines between two alternate different pressures on the pressure-distant curves. When a section of the pressure-distance curve is a straight line any straight connecting lines drawn between any two pressure values on this section of the curve will coincide with the pressure-distance curve.

The uppermost solid line represents the pressure-distribution in the sand-packed pipe after the valve has been closed for 3.0 seconds or six time increments. The values of the build-up pressure are given in Table 8-1.

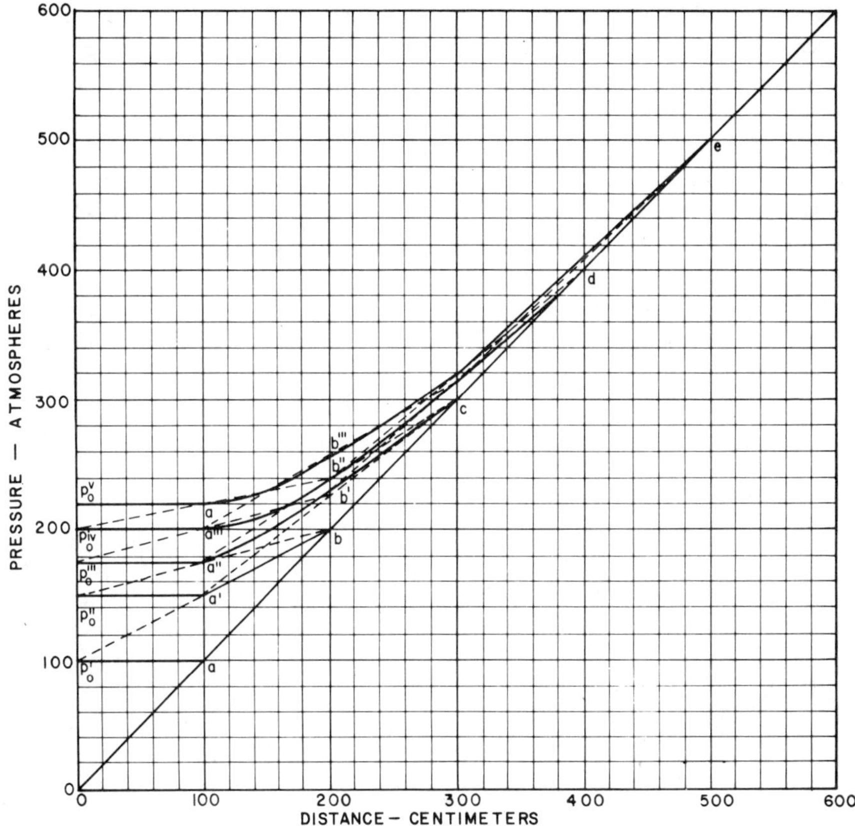

Figure 8-4. Graphical solution of the linear diffusivity equation for build-up pressure.

Table 8-1
Values for Pressure in Sand-Packed Pipe

Time Interval	Pressure Designation	Time (nΔ') Seconds	Pressure-Gauge Atmospheres
1	p_o'	0.5	100
2	p_o''	1.0	155
3	p_o'''	1.5	180
4	p_o^{IV}	2.0	207
5	p_o^{V}	2.5	227
6	p_o^{VI}	3.0	247

Numerical Solution for Linear Build-Up Pressure—Two Zones Each with a Different Permeability

For computing build-up pressure in linear fluid flow for systems which contain two homogeneous zones, each with a different permeability, the difference equation solution of the linear diffusivity equation may be used for computing the build-up pressures for the interior mesh points. However, this equation is not sufficient for computing pressure values at the junction between the different zones. The derivation of an equation for computing the pressure values at the junction between two zones, each with a different permeability, makes use of the principle that at the junction between the two zones the pressure gradient times its respective mobility of one zone is equal to the pressure gradient times its mobility of the other zone or mathematically

$$\left(\frac{k_1}{\mu_1}\right)\left(\frac{\partial p}{\partial x}\right)_1 = \left(\frac{k_2}{\mu_2}\right)\left(\frac{\partial p}{\partial x}\right)_2$$

In Figure 8-5, which is a space-time graph containing two zones (1 and 2), each with a different mobility $(k/\mu)_1$ and $(k/\mu)_2$, the increments of time

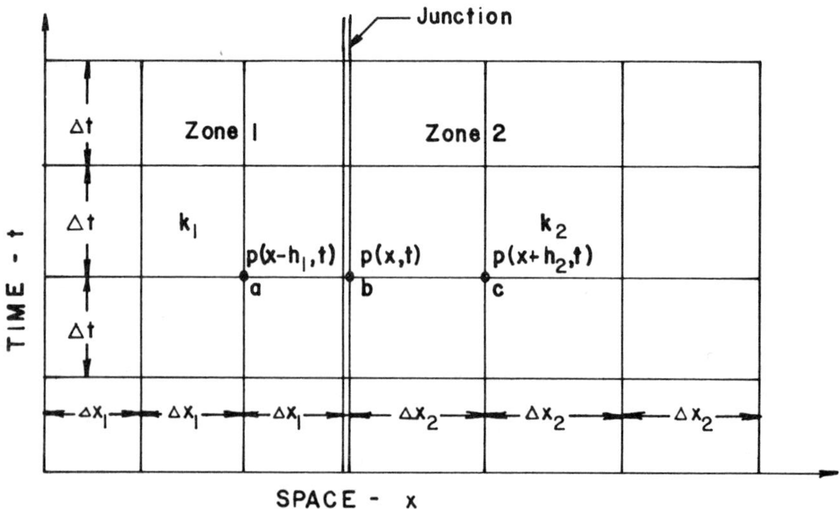

Figure 8-5. Space-time coordinate system for linear fluid flow illustrating two zones each with a different permeability.

are the same for both zones, but the increments of space for the two zones are different, but the increments of space for the two zones were adjusted so that each zone could have the same increment of time. Considering the values of pressure at the three points *a, b, c,* with point *b* being a junction point, Equation 8-9 may be written as

$$\frac{k_1}{\mu_1}\frac{p(x,t)-p(x-h_2,t)}{h_1}=\frac{k_2}{\mu_2}\frac{p(x+h_2,t)-p(x,t)}{h_2} \qquad \textbf{8-10}$$

Solving for $p(x,t)$, the junction pressure, gives

$$p(x,t)=\frac{\dfrac{k_1}{\mu_1 h_1}p(x-h_1,t)+\dfrac{k_2}{\mu_2 h_2}p(x+h_2,t)}{\left(\dfrac{k_1}{\mu_1 h_1}\right)+\left(\dfrac{k_2}{\mu_2 h_2}\right)} \qquad \textbf{8-11}$$

In computing the pressure values for the interior points of the space-time mesh Equation **8-8** is used. Following the computation of a row of interior pressure values Equation **8-11** is applied for computing the pressure value at the junction between the two zones for the given row. Example 8-3 illustrates the use of Equation **8-11** for computing pressure values at the junction between zones with different permeabilities.

Example 8-3. For a sand-packed pipe 1000 centimeters long containing two homogeneous zones of equal lengths but of different permeabilities, compute twelve values of build-up pressure, if, initially there is steady-state flow of fluid through the pipe and suddenly a valve at the low pressure end is closed and the pressure at the other end remains constant at the higher pressure. Additional data are:

$\phi = 0.20$
$p_0 = 0.0$ atmosphere gauge
$p_n = 600$ atmospheres gauge
$c = 5.0 \times 10^{-4}$ volume/volume atmosphere
$k_1 = 1.00$ Darcy
$k_2 = 0.50$ Darcy
$L_1 = 500$ cms
$L_2 = 500$ cms
$\mu_1 = 1.00$ centipoise
$\mu_2 = 1.00$ centipoise

Solution. The boundary conditions are

Zone 1

$$p(0,0) = p_0$$
$$p(x,0) = p_0 + (p_b - p_0)\ (x/L_1)$$
$$p_b = \text{pressure at junction at } t = 0 \text{ and } x = L_1$$
$$p(x,\infty) = P_n$$
$$p_x(0,t) = 0$$

Zone 2

$$p(x_b,0) = p_b$$
$$p(x,0) = p_b + (p_n - p_b)\ (x - x_b)/L_2 \text{ where } x > L_1$$
$$p(x,\infty) = p_n$$
$$(k_1/\mu_1)\ p_{x1}(x_b,t) = (k_2/\mu_2)\ p_{x2}(x_o,t)$$

The pressure will build up at the end where the valve is closed suddenly; the pressures at this end are known as build-up pressures. Using the relation $\eta = k/\phi\mu c$ the two hydraulic diffusivities are computed thus

$$\eta_1 = \frac{1}{0.20 \times 1.0 \times 5.00 \times 10^{-4}} = 1.0 \times 10^4$$

$$\eta_2 = \frac{0.50}{0.20 \times 1.0 \times 5.00 \times 10^{-4}} = 5.0 \times 10^3$$

Assuming the increments of space for the first zone to be 100 cms the increment of time is determined from the relation $\Delta t = \frac{1}{2}\ (\Delta x)^2/\eta_1$; and found to be $\Delta t = \frac{1}{2}\ (100)^2/1 \times 10^4 = 0.50$ second.

Having determined the value for the increment of time the increment of space for the second zone is determined by $(\Delta x)^2 = 2\Delta t\ \eta_2$ or $\Delta x_2 = \sqrt{0.5 \times 2.0 \times 5.0 \times 10^3} = 70.7$ cms.

And with the two values of increments of space Δx_1 and Δx_2, and the increment of time, a space-time mesh was constructed as given in Figure 8-6. Since there was steady-state fluid-flow at the time t_o and since the mobility for fluid flow was 1.0 for the first zone and 0.5 for the second zone, the pressure drop in the second zone was twice that in the first zone; the pressure at the junction was one-third the total pressure, $600/3 = 200$ atmospheres. The number of subdivisions in the first zone is $500/100 = 5$; and the incremental pressure drops in the first zone are $200/5 = 40$ atmospheres.

m (TIME - SECONDS)	Zone 1 k=1.00 Darcy					Junction, Zone 2 k=0.50 Darcy							
	00	100	200	300	400	500	571	641	712	783	854	925	1000
12	131	131	141	158	185	219	266	317	371	427	483	540	600
11	126	126	136	155	182	215	264	316	370	426	483	540	600
10	121	121	131	150	178	213	262	315	370	426	483	540	600
9	115	115	126	146	175	209	262	315	370	426	483	540	600
8	109	109	120	142	171	208	260	315	370	426	483	540	600
7	102	102	115	137	169	205	259	314	370	426	483	540	600
6	96	96	108	133	165	204	258	313	370	426	483	540	600
5	88	88	103	128	163	202	257	313	370	426	483	540	600
4	80	80	95	125	160	200	257	313	370	426	483	540	600
3	70	70	90	120	160	200	257	313	370	426	483	540	600
2	60	60	80	120	160	200	257	313	370	426	483	540	600
1	40	40	80	120	160	200	257	313	370	426	483	540	600
0 (Δt)	00	40	80	120	160	200	257	313	370	426	483	540	600

DISTANCE - CENTIMETERS (x)

Figure 8-6. Space-time mesh for build-up pressure in a sand-packed pipe with two zones each with a different permeability.

The number of subdivisions in the second zone is $500/70.7 = 7.07$ or seven whole subdivisions and 0.07 of one subdivision. The incremental pressure drops in the second zone are $(600 - 200)/7.07 = 56.6$ atmospheres. The second zone was divided into six increments of length, 70.7 cms each, and one increment of 75.8 cms. The larger of the increments was placed at the higher pressure end of the sand-packed pipe since this was the end at which there was the least pressure change of the adjacent interior mesh points, the pressure at mesh point x_{n-1} can be obtained approximately from the equation

$$p(x_{n-1}, t + \Delta t) = \frac{\Delta x_2}{\Delta x_2 + \Delta x_n} \left[p(x_n, t) + p(x_{n-2}, t \right] + p(x_{n-2}, t) \qquad \text{8-12}$$

where Δx_n is the length of the last increment of distance.

An inspection of the space-time mesh reveals that there was not a need for applying Equation **8-11** until the fifth increment of time. The junction pressure for the fifth increment of time is

$$\frac{k_1}{\mu_1 h_1} = \frac{1.00}{1.00 \times 100} = 0.01$$

$$\frac{k_1}{\mu_2 h_2} = \frac{0.50}{1.00 \times 70.7} = 0.0071$$

$$p_b = \frac{0.01 \times 163 + 0.0071 \times 257}{0.01 + 0.0071} = 202 \text{ atms}$$

Also, an inspection of Figure 8-6 indicates that there was not a need to compute a pressure value at the interior mesh point next to the higher pressure end of the mesh until sometime after the twelfth increment of time. The equation for calculating this pressure is

$$p(x_{n-1}, t) = \frac{70.7}{70.7 + 75.8}\left[p(x_n, t) + p(x_{n-2}, t)\right] + p(x_{n-2}, t)$$

The method for computing pressure values at $x = 0$ was discussed in Example 8-1.

Numerical Solution for Linear Pressure Decline

Pressure decline for unsteady-state linear fluid-flow can be determined with the use of the difference equation **8-8** if the pressure gradient is known for the end of the system where the pressure is declining the most, and if this pressure gradient is assumed to remain constant over a length Δx next to the point of most pressure decline. In calculating pressure decline, the interior points on a given row of a distance-time mesh are computed with the use of Equation **8-8**. Then the pressure decline $p_o{}^m$ at the end is taken as the pressure $p_1{}^m$ at the adjacent mesh point minus the pressure gradient times the increment of space Δx or

$$p_o{}^m = p_1{}^m - \left(\frac{dp}{dx}\right)\Delta x \qquad \textbf{8-13}$$

For a constant flow rate of fluid from the end where the pressure is declining the most, the pressure gradient may be computed by means of Darcy's equation

$$q = \frac{kA}{\mu}\frac{dp}{dx}$$

or

$$\frac{dp}{dx} = \frac{q\mu}{kA} \tag{8-14}$$

A numerical solution for calculating pressure decline in a sand-packed pipe is illustrated in Example 8-4.

Example 8-4. If initially the pressure in a homogeneous fully saturated sand packed pipe is at 1000 atmospheres gauge pressure and suddenly from one end of the packed pipe the fluid is permitted to flow out at a rate of 0.10 ml per second and the pressure at the other end is maintained constant at 1000 atmospheres gauge; then compute the time it takes for the pressure to decline to 682 atmospheres gauge at the partially opened end. Additional data are:

$k = 0.20$ Darcy $\mu = 2.0$ centipoises $L = 1000$ centimeters
$p_n{}^m = 1000$ atmospheres $q = 0.10$ ml/second $\phi = 0.20$
$c = 5.00 \times 10^{-4}$ atmospheres^{-1} $A = 1.00$ cm^2

Solution. The boundary conditions are:

$$p(x,0) = 1000 \text{ atmospheres gauge}$$
$$px(0,t) = \text{constant} = \frac{\partial p}{\partial x} = \frac{q\mu}{kA}$$
$$p(L,t) = 1000 \text{ atmospheres gauge}$$
$$p(0,\infty) = 0$$

The pressure gradient is

$$\frac{dp}{dx} = \frac{\mu q}{kA} = \frac{2.0 \times 0.10}{0.20 \times 1.00} = 1.00 \text{ atmosphere/cm}$$

If Δx is assumed to be 100 cms the pressure drop over the first increment is $1.00 \times 100 = 100$ atmospheres.

The hydraulic diffusivity is

$$\eta = \frac{k}{\phi\mu c} = \frac{0.20}{0.20 \times 2.0 \times 5.00 \times 10^{-4}} = 1.00 \times 10^3$$

and

$$\Delta t = \frac{(\Delta x)^2}{2\eta} = \frac{(100)^2}{2 \times 1.00 \times 103} = 5.0 \text{ seconds}$$

The space-time mesh is given in Figure 8-7. Applying Equation **8-13,** the pressure valve at the open end of the sand-packed pipe following the first increment of time is $p_o' = 1000 - 1.0 \times 100 = 900$ atmospheres gauge. With the pressure values for the row in the space-time mesh corresponding to the first increment of time, the pressure values for the interior mesh points of the row for the second increment of time are calculated by means of Equation **8-8;** then by applying Equation **8-13** the pressure value at the open end of the pipe after the second increment of time is computed as $p_o'' = 950 - 1.0 \times 100 = 850$ atmospheres gauge.

The same procedure as described for computing pressure values for the first and second increments of time is used for computing pressure values for the remaining mesh points of the space-time mesh.

The space-time mesh indicates that it took 12 increments of time for the pressure to decline to 682 atmospheres at the partially opened end. Therefore, the total time required for the pressure to be reduced from 1000 to 682 atmospheres gauge is

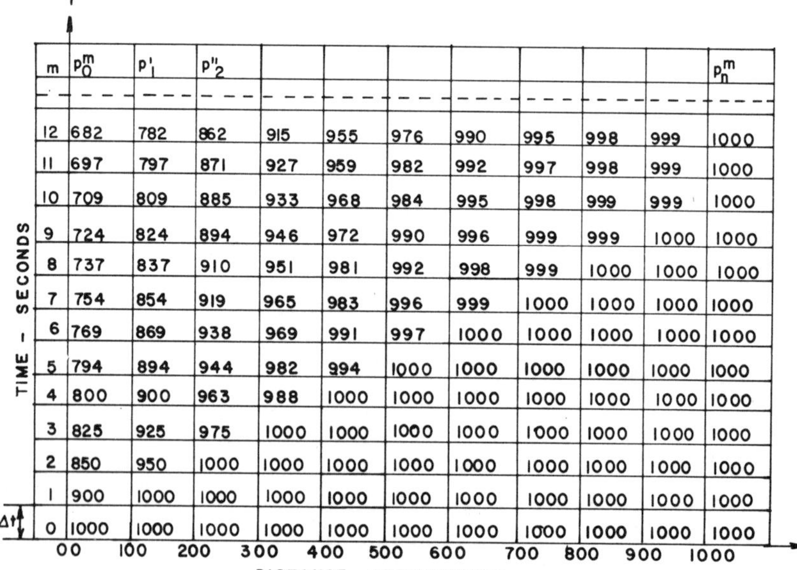

m	p_0^m	p'_1	p''_2								p_n^m
12	682	782	862	915	955	976	990	995	998	999	1000
11	697	797	871	927	959	982	992	997	998	999	1000
10	709	809	885	933	968	984	995	998	999	999	1000
9	724	824	894	946	972	990	996	999	999	1000	1000
8	737	837	910	951	981	992	998	999	1000	1000	1000
7	754	854	919	965	983	996	999	1000	1000	1000	1000
6	769	869	938	969	991	997	1000	1000	1000	1000	1000
5	794	894	944	982	994	1000	1000	1000	1000	1000	1000
4	800	900	963	988	1000	1000	1000	1000	1000	1000	1000
3	825	925	975	1000	1000	1000	1000	1000	1000	1000	1000
2	850	950	1000	1000	1000	1000	1000	1000	1000	1000	1000
1	900	1000	1000	1000	1000	1000	1000	1000	1000	1000	1000
0	1000	1000	1000	1000	1000	1000	1000	1000	1000	1000	1000

Figure 8-7. Space-time mesh for pressure decline in a sand-packed pipe.

$m \times \Delta t = 12 \times 5 = 60$ seconds or one minute

The example could have been worked graphically with the use of a distance-pressure graph as was illustrated in Example 8-2.

Radial Unsteady-State Flow of Fluids

Radial Diffusivity Equation

From Equation 8-5 it can be derived that the equation for fluid flow in an x-y plane is

$$\frac{\partial^2 p}{\partial x^2} + \frac{\partial^2 p}{\partial y^2} = \frac{\phi \mu c}{k} \frac{\partial p}{\partial t} \quad \text{8-15}$$

In terms of cylindrical coordinates and for radial flow, $\partial^2 p / \partial \theta^2 = 0$, Equation 8-15 becomes

$$\frac{\partial^2 p}{\partial r^2} + \frac{1}{r} \frac{\partial p}{\partial r} = \frac{\phi \mu c}{k} \frac{\partial p}{\partial t}$$

or

$$\frac{\partial^2 p}{\partial r^2} + \frac{1}{r} \frac{\partial p}{\partial r} = \frac{1}{\eta} \frac{\partial p}{\partial t} \quad \text{8-16}$$

where $k/\phi\mu c = \eta$, the hydraulic diffusivity.

Numerical Solution For Radial Build-Up Pressure

By considering a space-time mesh the first and second difference quotients for the derivatives in Equation 8-16 are:

$$\frac{\partial p}{\partial r} = \frac{p(r + h, t) - p(r - h, t)}{2\Delta r} \quad \text{8-16a}$$

$$\frac{\partial p}{\partial t} = \frac{p\,(r,t + k) - p\,(r,t)}{\Delta t} \quad \text{8-16b}$$

$$\frac{\partial^2 p}{\partial r^2} = \frac{p\,(r + h,t) + p\,(r - h,t) - 2p\,(r,t)}{(\Delta r)^2} \quad \text{8-16c}$$

Two increments of distance were taken for the first difference quotient, so that the difference solution of Equation **8-16** would not contain a $p(r,t)$ term. The substitution of the three difference quotients into Equation **8-16** gives

$$\frac{p(r+h,t) + p(r-h,t) - 2p(r,t)}{(\Delta r)^2} + \frac{p(r+h,t) - p(r-h,t)}{2r\Delta r} = \left(\frac{1}{\eta}\right)\frac{p(r,t+k) - p(r,t)}{\Delta t}$$

or

$$\left(1+\frac{\Delta r}{2r}\right)p(r+h,t) + \left(1-\frac{\Delta r}{2r}\right)p(r-h,t) = \frac{(\Delta r)^2}{\eta\Delta t}\left[p(r,t+k) - p(r,t)\right] + 2p(r,t) \qquad \textbf{8-17}$$

and if $(\Delta r)^2/\eta\ \Delta t = 2$, then Equation **8-17** becomes

$$p(r,t+k) = \frac{\left(1+\frac{\Delta r}{2r}\right)p(r+h,t) + \left(1-\frac{\Delta r}{2r}\right)p(r-h,t)}{2} \qquad \textbf{8-18}$$

It is known that $dr/r = d\ 1nr$ and $\Delta r/r = \Delta(1nr)$; therefore the terms $\Delta r/2r$ in Equation **8-18** suggest the use of logarithm of space versus time graph for graphically using Equation **8-18.** However, for a numerical solution with the use of Equation **8-18** only a space-time mesh is needed; this is illustrated in Example 8-5.

Example 8-5. For a homogeneous circular reservoir containing a well drilled in its center, calculate the pressure in the well bore after 14 increments of time if initially there is steady-state fluid-flow into the well and suddenly the well bore is plugged and pressure at the perimeter of the oil reservoir is kept constant by a large aquifier. Additional data are:

$r_0 = 10$ cms
$r_n = 10{,}000$ cms

$p_o = 00$ atms gauge initially
$p_n = 500$ atms gauge
$k = 1.0$ Darcy
$\phi = 0.20$
$\mu = 1.00$ centipoise
$c = 5.00 \times 10^{-4}$ atm^{-1}

Solution. The boundary conditions are:

$$p\,(0,0) \;=\; p_0$$

$$p\,(r,0) \;=\; p_0 + \frac{q\mu}{2\pi hk} \ln r/r_0$$

$$p\,(r_n,t) \;=\; p_n \;=\; \text{constant}$$

$$p\,(r,\infty) \;=\; p_n$$

$$p_{\ln r}(0,t) \;=\; 0$$

A space-time mesh as seen in Figure 8-8 is used in the solution of this problem. The space coordinate for the mesh is the ratio of the radius r to the radius r_o of the well bore. For ten equal subdivisions of space, the increment of space is $\Delta r = 10{,}000/10 = 1000$ cms and the increment for the ratio $\Delta r/r_o$ is 100.

The hydraulic radius is

$$\eta \;=\; \frac{1}{0.2 \times 1.00 \times 5.00 \times 10^{-4}} \;=\; 1.00 \times 10^{4}$$

and the increment of time is

$$\Delta t \;=\; \frac{(\Delta r)^2}{2 \times \eta} \;=\; \frac{1.00 \times 10^{6}}{2 \times 1.00 \times 10^{4}} \;=\; \frac{100}{2} \;=\; 50 \text{ sec.}$$

The initial values of pressure for the space-time mesh were taken from a plot of pressure versus log r/r_o as seen in Figure 8-9. These values of pressure can also be calculated as follows:

$$\Delta p \;=\; p_n - p_o \;=\; \frac{2.303\ \mu q}{2\ \pi hk} \log r/r_o$$

m	$\Delta t/2r$	0.5	0.25	0.167	0.125	0.10	0.083	0.071	0.063	0.055	
14	436	436	439	444	451	458	468	476	484	492	500
13	433	433	436	442	449	458	466	476			500
12	430	430	434	440	448	456	466	475			500
11	427	427	431	438	446	456	465	475			500
10	423	423	428	435	444	454	465	475	484		500
9	420	420	424	432	442	453	464	475	483		500
8	416	416	421	428	440	452	464	474			500
7	411	411	417	427	439	451	463	474			500
6	405	405	412	424	437	451					500
5	399	399	406	421	437	450					500
4	399	399	401	419	433						500
3	380	380	397	413	433						500
2	371	371	383	413	433						500
1	333	333	383	413	433						500
0	000	333	383	413	435	450	463	474	483	492	500
r/r_o	1.0	100	200	300	400	500	600	700	800	900	1000

TIME — SECONDS (t)

DISTANCE — r/r_o

Figure 8-8. Space-time mesh for build-up pressure in a one-well oil reservoir.

If $p_o = 0.00$ atm gauge at $r_o = 10$ cms and $p_n = 500$ atm gauge at $r_n = 10{,}000$ cms and letting $K' = 2.303\ \mu q/2\pi hk$; then the equation for Δp becomes $500 = K' \log 10{,}000/10 = 3.00\ K'$ or $K' = 166.7$.

At $r = 1{,}000$ cms

$$p = 166.7 \log 1000/10 = 2.0000 \times 166.7 = 333.7 \text{ atms}$$

at $r = 2{,}000$ cms

$$p = 166.7 \log 2000/10 = 2.301 \times 166.7 = 383.5 \text{ atms}$$

and at $r = 3{,}000$ cms

$$p = 166.7 \log 3000/10 = 2.4471 \times 166.7 = 412.9 \text{ atms}$$

Calculated pressure values are preferred to those obtained from a graph if the problem can be solved by a microcomputer.

In the solution of the problem, it was assumed that the pressure gradient at the well bore was zero and that the pressure in the well bore was equal to the pressure at the first increment of space after the first increment of time and for each increment of time thereafter and that

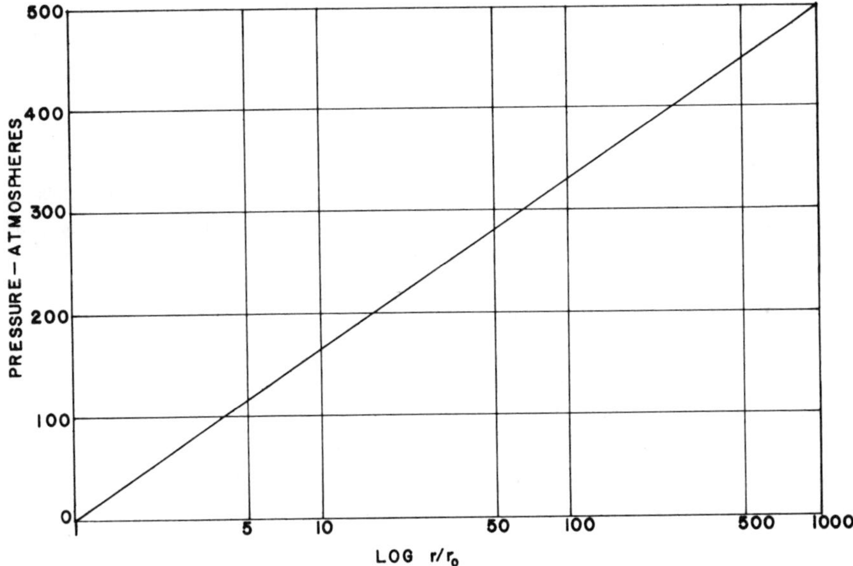

Figure 8-9. Distance-pressure graph for reading pressure values at steady-state conditions.

$$p(x^m{}_0, t+nk) = p(x^m{}_1, t+nk)$$

Sample calculations, by the use of Equation **8-18,** for the first five pressure values for the tenth increment of time are

$$p_1^{10} = \frac{0.50 \times 420 + 1.50 \times 424}{2} = 423 \text{ atms gauge}$$

$$p_0^{10} = p_1^{10} = 423 \text{ atms gauge}$$

$$p_2^{10} = \frac{0.75 \times 420 + 1.25 \times 432}{2} = 428 \text{ atms gauge}$$

$$p_3^{10} = \frac{0.833 \times 424 + 1.167}{2} = 435 \text{ atms gauge}$$

$$p_4^{10} = \frac{0.875 \times 432 + 1.125}{2} = 444 \text{ atms gauge}$$

The distance-time mesh indicates that the pressure reached 436 atms gauge at the well bore after 14 increments of time or $14 \times 50 = 700$ seconds or 11.4 minutes.

Graphical Solution For Radial Build-Up Pressure

The difference equation solution, Equation 8-8 of the radial diffusivity equation contains the terms $\Delta r/2r$ which suggests logarithms; therefore, this difference equation may be solved graphically by means of a logarithm of distance versus time graph. Since at steady-state fluid-flow the pressure at a given radius from the well bore is a function of the logarithm of the ratio of the given radius to the radius of the well bore, the logarithm of this ratio is plotted instead of the given radius. The graphical method for determining radial build-up pressure is illustrated in Example 8-6.

Example 8-6. Solve Example 8-5 by means of a logarithmic ratio log r/r_o versus pressure and compare the results with those of Example 8-5.

Solution. The boundary conditions are identical to those of Example 8-5. A plot of log r/r_o versus pressure in semi-logarithmic coordinates at $t = 0$ is given in Figure 8-10. The plot is a straight diagonal line. The solid horizontal lines in Figure 8-10 represent the setting of $p_1{}^m = p_0{}^m$ and the broken lines represent the joining of alternate pressure values at a given time. The example is worked in the same manner as the example for the graphical solution of linear build-up pressure; the only difference is in the use of semi-logarithmic coordinates instead of rectangular coordinates. The

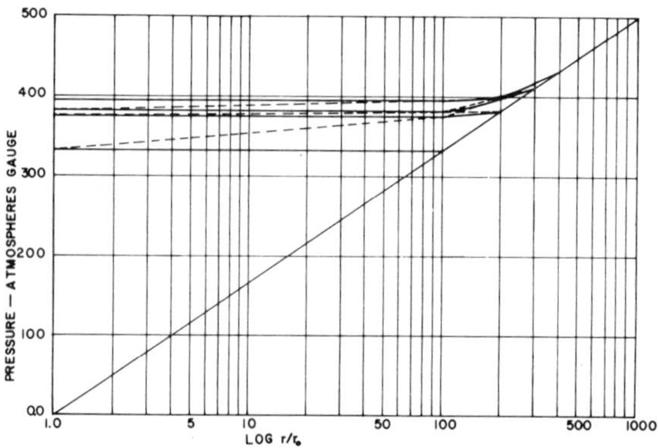

Figure 8-10. Graphical solution for build-up pressure in a circular reservoir.

increment of time for this example is the same as that for Example 8-5. The determinations of build-up pressure were discontinued after five increments of time because thereafter it became very difficult to sketch straight lines on the small graph.

Table 8-2 gives the build-up pressure values obtained by the two different methods.

The discrepancies between the sets of pressure values are due to the difficulties in reading pressure values from the graph of Figure 8-11.

Numerical Solution For Radial Build-Up Pressure—Two Zones Each with a Different Permeability

For radial fluid-flow systems containing two concentric homogeneous zones, each with a different permeability, the difference equation solution for the radial diffusivity equation can be used for computing pressure values for all interior points of a space-time mesh except the pressure values corresponding to mesh points at the junction between the two zones. An approximate equation can be derived for calculating these pressure values at the junction between the two zones, if it is assumed that the quantity of mass accumulating at the junction is negligible for a small increment of time Δt. If $p(r_j,t)$ represents the pressure value at the junction for time t, $p(r_j - h_1,t)$ represents the pressure value at time t at one mesh point adjacent to the junction and $p(r_j + h_2,t)$ represents the pressure value at time t at the other mesh point adjacent to the mesh point at the junction. The fluid-flow into and from the mesh point at the junction can be expressed in terms of Darcy's equation as follows:

Table 8-2
Numerical and Graphical Solution for Build-Up Pressure Values

Time		Pressure Values, Atms	
Increment	Seconds	Computed	Graphically
$t + \Delta t$	50	333	332
$t + 2\Delta t$	100	371	372
$t + 3\Delta t$	150	380	382
$t + 4\Delta t$	200	393	396
$t + 5\Delta t$	250	399	401

$$q_2 = \frac{k_2}{\mu} \frac{2\pi h \left[p\,(r_j + h_2, t) \;-\; p\,(r_j, t)\right]}{\ln\,(r_j + h_2)/r_j}$$

$$q_1 = \frac{k_1}{\mu} \frac{2\pi h \left[p\,(r_j, t) \;-\; p\,(r_j - h_1, t)\right]}{\ln r_j/(r_j - h_1)}$$

For negligible fluid accumulation at the junction between the two zones q_1 equals q_2. By equating the identities for q_1 and q_2 and solving for $p(r_j, t)$ the following equation is obtained

$$p(r_j,t) = \frac{\dfrac{k_1 p(r_j - h_1, t)}{\ln r_j/(r_j - h_1)} + \dfrac{k_2 p(r_j + h_2, t)}{\ln\,(r_j + h_2)/r_j}}{\dfrac{k_1}{\ln r_j/(r_j - h_1)} + \dfrac{k_2}{\ln\,(r_j + h_2)/r_j}} \qquad \textbf{8-19}$$

Equation **8-19** is applicable for computing pressure values at mesh points at the junction between two homogeneous zones in radial fields when each zone has a different permeability. The computation of the pressure value at the junction between the two zones follows after the computation of the pressure values for the other interior mesh points for a given row of a space-time mesh. The application of Equation **8-19** is illustrated in Example 8-7.

Example 8-7. For a circular oil reservoir containing a well drilled in its center and two concentric homogeneous zones, each with a different permeability, calculate the pressure value in the well-bore if initially there is steady-state fluid-flow into the well-bore and suddenly the well-bore is plugged and the pressure at the perimeter of the oil reservoir is maintained constant by a large aquifier. Additional data are:

$r_o = 10$ cms
$r_j = 3{,}000$ cms
$r_n = 10{,}000$ cms
$p_o = 00$ atms gauge initially
$p_n = 500$ atms gauge
$k_1 = 1.00$ Darcy (zone in the center with the well)
$k_2 = 0.50$ Darcy
$\phi = 0.20$
$\mu = 1.00$ centipoise and
$c = 5.00 \times 10^{-4}$ atm^{-1}

Assume the thickness of the reservoir remains constant throughout.

Solution. The boundary conditions are:

Zone 1

$$p_o(0.0) = 0$$

$$p(r,0) = p_o + \frac{q\mu}{2\pi hk_1} \ln r/r_o \text{, from } r = r_o \text{ to } r = r_j$$

$$p(r,\infty) = p_n$$

$$p_{1nr}\,(0,t) = 0$$

Zone 2

$$p(r_j,0) = p_j$$

$$p(r,0) = p_j + \frac{q\mu}{2\pi hk_2} \ln r/r_b \text{, from } r = r_j \text{ to } r = r_n$$

$$p(r,\infty) = p_n$$

$$p(r_n,t) = p_n$$

A space-time mesh is used in the solution of this problem (see Figure 8-11). The space coordinate for the mesh is the ratio of the radius r to the radius r_o of the well bore. For the first zone, an increment of space is assumed to be 1000 cms, and this will cause $3000/1000 = 3$ increments in the first zone. In the first zone the increment of the ratio $\Delta r/r$ is 100.

The hydraulic diffusivity of the first zone is

$$\eta_1 = \frac{1}{0.2 \times 1.00 \times 5.00 \times 10^{-4}} = 1.00 \times 10^4$$

and the increment of time is

$$\Delta t = \frac{(\Delta r_1)^2}{2 \times \eta} = \frac{1.00 \times 10^6}{2 \times 1.00 \times 10^4} = 50 \text{ sec}$$

The hydraulic diffusivity of the second zone is

$$\eta_2 = \frac{0.50}{0.2 \times 1.00 \times 5.00 \times 10^{-4}} = 5.00 \times 10^3$$

Zone 1 $k_1 = 1.00$ | Junction | Zone 2 $k_2 = 0.50$

TIME — SECONDS

m	Δr/2r	0.5	0.25	0.167	.095	.080	.069	.061	.054	.049	.046	.041	.038	
14	383	383	387	394	402	412	422	435	447	459	472	481	491	500
13	379	379	384	391	399	410	422	433	445					500
12	376	376	380	388	397	409	420	432						500
11	372	372	377	385	394	406	420							500
10	368	368	373	382	392	405	418							500
9	363	363	369	379	390	403	417							500
8	358	358	364	376	387	401	417							500
7	352	352	359	371	384	400	416							500
6	346	346	354	366	381	399				459				500
5	339	339	349	363	379	397				458				500
4	335	335	341	358	376									500
3	326	326	338	350										500
2	318	318	326	350										500
1	283	283	326	350										500
0	0	283	326	350	378	397	416	432	445	458	472	481	491	500
DISTANCE — r/r_o	1.0	100	200	300	371	441	513	582	654	724	795	866	936	1000

Figure 8-11. Distance-time mesh for radial build-up pressure in a circular reservoir containing one well and two concentric zones each with a different permeability.

To simplify the solution of the problem, the increment of time for the second zone is assumed to be the same as for the first zone. The increment of space for the second zone is $(\Delta r_2)^2 = 2\,\Delta t\,\eta_2$ or $\Delta r_2 = \sqrt{2 \times 50 \times 10^3} = 707$ cms.

The initial pressure value at the junction of the two zones is easily determined because initially the fluid-flow in the reservoir was steady-state; therefore the radial rate of fluid-flow into the well-bore of the first zone is equal to the radial rate of fluid-flow across the boundary between the two zones. Assuming this initial boundary pressure is p_j, its value is computed as

$$q_1 = q_2$$

or

$$q_1 = \frac{k_1\, 2\pi h\, (p_j - p_o)}{\mu\, 2.303 \log 3000/10}$$

and

$$q_2 = \frac{k_2\, 2\pi h\, (p_n - p_j)}{\mu\, 2.303 \log 10000/3000}$$

Equating the two expressions and solving p_j gives

$$p_j = \frac{250}{0.5224(1/2.4771 + 0.5/\ 0.5224)}$$

$$= 350 \text{ atms gauge}$$

To obtain the initial pressure values for the space-time mesh, a graph of log r_n/r_o versus pressure was constructed with the use of semi-logarithmic coordinate paper (see Figure 8-12). Log r_n/r_o was taken as the abscissa and pressure was taken as the ordinate. In the graph a straight line was drawn from the origin to the coordinate point (300,350) and a straight line was drawn from this point to the coordinate point (1000,500).

The incremental distances on the space-time mesh were 100 cms for the first zone (1.00 to 3000 cms) and 707 cms for the second zone as is

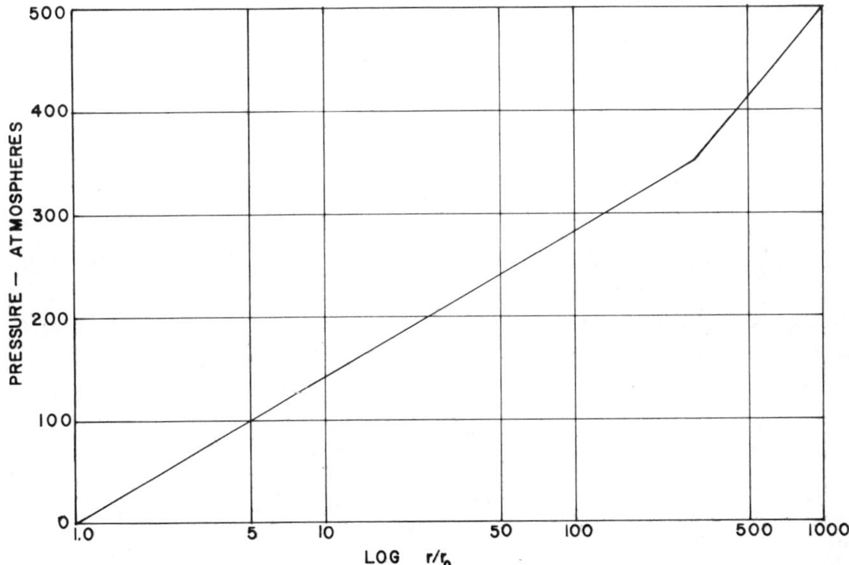

Figure 8-12. Distance-pressure graph for reading pressure values at steady-state conditions in a circular reservoir with two concentric zones, each with a different permeability.

illustrated in Figure 8-11. The number of increments of distance in the second zone is 7000/707 = 6.93.

All pressure values for the interior points of the space-time mesh except the pressure values at the junction were computed with the use of Equation **8-18**; the pressure values at the junction between the two zones were computed with the use of Equation **8-19**; and the pressure values at the well-bore were taken as equal to the pressure values at their adjacent interior mesh point for corresponding increments of time. Examples of calculations for mesh points for five increments of time are:

$$\frac{k_1}{\log r_j/r_j - h_1} = \frac{1.00}{\log 3000/2000} = \frac{1.00}{\log 1.5} = 5.67$$

$$\frac{k_2}{\log r_j + h_2/r_j} = \frac{0.50}{\log 3707/3000} = \frac{0.50}{\log 1.236} = 5.43$$

At $m = 4$ or time $t + 4\Delta t$

$$p_j = \frac{5.67 \times 341 + 5.43 \times 376}{5.67 + 5.43} = 358 \text{ atms gauge}$$

at $m = 5$ or time $t + 5\Delta t$

$$p_j = \frac{5.67 \times 349 + 5.43 \times 379}{5.67 + 5.43} = 363 \text{ atms gauge}$$

at $m = 6$ or time $t + 6\Delta t$

$$p_j = \frac{5.67 \times 354 + 5.43 \times 379}{5.67 + 5.43} = 366 \text{ atms gauge}$$

at $m = 7$ or time $t + 7\Delta t$

$$p_j = \frac{5.67 \times 359 + 5.43 \times 384}{5.67 + 5.43} = 371 \text{ atms gauge}$$

and at $m = 8$ or time $t + 8\Delta t$

$$p_j = \frac{5.67 \times 364 + 5.43 \times 387}{5.67 + 5.43} = 376 \text{ atms gauge}$$

From the space-time mesh (Figure 8-11), it is noted that the pressure reached 383 atmospheres gauge at the well bore after 14 increments of time or $14 \times 50 = 700$ seconds or 11.67 minutes.

A glance at Figure 8-11 indicates that at the end of the 14th increment of time there was no need for computing a new value of pressure for the r_{n-1} mesh point. Since the value of $\Delta r/2r$ is very small for this mesh point and its pressure value would change very little with each increment of time, Equation **8-12** could be used in computing new pressure values for the r_{n-1} mesh point for increments of time which require the computation of such values.

For unsteady-state flow of compressible fluids in systems containing two zones, each with a different permeability, it is possible to adjust the value of the terms $(\Delta x)^2/\eta\Delta t$ to a value other than 2 and derive a difference equation solution, containing a term $p(x, t)$, for the linear diffusivity equation which would have equal increments of time and equal increments of space in both zones. This is also possible for the radial diffusivity equation.

Unsteady-State Flow of Gases

Unsteady-State Linear Flow of Gases

Equation **6-30** developes the gas diffusion equation and for linear flow (one-dimension) it becomes

$$\frac{\partial^2 p^2}{\partial x^2} = \frac{\phi\mu}{k\bar{p}} \frac{\partial p^2}{\partial t}$$

or

$$\frac{\partial^2 p^2}{\partial x^2} = \frac{1}{\alpha} \frac{\partial p^2}{\partial t}$$

where $\alpha = kp/\phi\mu$

The linear gas diffusivity equation is similar to the linear, viscous, compressible fluid diffusivity equation. The differences between the two diffusivity equations are: the viscous compressible fluid equation contains partial derivatives with pressure raised to the first power and the gas equation contains derivatives with the pressure squared. The viscous compressible fluid equation contains a fluid compressibility term c and the gas equation contains a term of average pressure instead of the compressibility term. Since the two diffusivity equations are very similar, it appears that the diffusivity equation can be solved numerically by means of a difference equation containing terms with the pressure squared. The difference quotients for the partial derivatives of Equation **8-20** are

$$\frac{\partial p^2}{\partial t} = \frac{p^2(x,\ t + \Delta t) - p^2(x,\ t)}{\Delta t}$$

and

$$\frac{\partial^2 p^2}{\partial x^2} = \frac{p^2(x + \Delta x,\ t) + p^2(x - \Delta t,\ t) - 2p^2(x,\ t)}{(\Delta x)^2} \qquad \textbf{8-20}$$

Replacing the partial derivatives of Equation **8-20** by their respective difference quotients gives

$$\frac{p^2(x + \Delta x, t) + p^2(x - \Delta x,\ t) - 2p^2(x,\ t)}{(\Delta x)^2}$$

$$= \frac{1}{\alpha \Delta t}\,[p^2(x, t + \Delta t) - p^2(x, t)]$$

or

$$p^2(x,\ t + \Delta t) = \frac{p^2(x + \Delta x,\ t) + p^2(x - \Delta x, t)}{2} \qquad \textbf{8-21}$$

where $(\Delta x)^2/\Delta t\alpha = 2$

The equation **8-21** may be used with a space-time mesh in a similar manner as was done for the difference equation solution for viscous compressible fluids. In the space-time mesh, pressure squared is used instead of just pressure. The computed pressure values obtained for gases are not as accurate as computed pressure values for compressible fluids, because an average value is used for pressure and an average value is used for viscosity. Viscosity of gases change appreciably with pressure. For linear flow of gases in homogeneous media, values of Δt change with changing values of pressure and changing values of viscosity. As the average value of pressure changes appreciably the increment of time changes. The overall time is a summation of the individual increments of time (Δt's). The use of a space-time mesh for determining linear build-up pressure is illustrated in Example 8-8.

Example 8-8. Compute the build-up pressures of a gas in a homogeneous sand-packed pipe which has a total length of 10,000 cms, if initially there is approximately steady-state flow through the sand-packed pipe and suddenly a valve is closed at the low pressure end of the pipe and the pressure at the other end is kept constant at the higher pressure. Additional data are: $\phi = 0.20$, $p_o = 1.00$ atm gauge; $p_n = 500$ atms gauge; $k = 0.50$ Darcy; and $\mu = 0.10$.

Solution. The pressure p_o will build-up at the end of the sand-packed pipe where the value is suddenly closed. The boundary conditions are:

$$p^2(0,0) = p_o^2$$

$$p^2(x,0) = p_o^2 + (p_n^2 - p_o^2)\frac{x}{L}$$

$$p^2\ (L,t) = p_n^2 = \text{constant}$$

$$p^2(x,\infty) = p_n^2$$

$$px^2(0,t) = 0$$

For 10 equal space subdivisions

$$\Delta x = \frac{10{,}000}{10} = 1{,}000 \text{ cms.}$$

The initial values of pressure squared are obtained from the formula

$$p^2 = p_o^2 + (p_n^2 - p_o^2)\frac{x}{L}$$

for $x = 1{,}000$ cms

$$p^2 = 1 + (250{,}000 - 1)\frac{1{,}000}{10{,}000} = 25{,}000 \text{ atm}^2$$

and for $x = 2{,}000$

$$p^2 = 1 + (250{,}000 - 1)\frac{2{,}000}{10{,}000} = 50{,}000 \text{ atm}^2$$

The computed values of pressure and pressure squared are given in the Table 8-3.

Table 8-3
Values of Pressure and Pressure Squared

x	p atms	p^2 atms^2
0	1	
1,000	158	25,000
2,000	224	50,000
3,000	274	75,000
4,000	316	100,000
5,000	354	125,000
6,000	387	150,000
7,000	418	175,000
8,000	447	200,000
9,000	474	225,000
10,000	500	250,000

The first increment of time is computed as

$$p = \frac{1 + 500}{2} = 250.5 \text{ atm}$$

$$\Delta t = \frac{(\Delta x)^2 \mu \phi}{2kp} = \frac{(1{,}000)^2 \times 0.2 \times 0.1}{2 \times 0.50 \times 251} = 79.7 \text{ sec}$$

The space-time mesh for this problem is given in Figure 8-13. At the mesh points are placed values of pressure squared; the interior mesh points were computed by the difference equations. The values in the mesh are pressure squared divided by 1,000.

$$p^2 (x, t + \Delta t) = \frac{p^2 (x + \Delta x, t) + p^2 (x - \Delta x, t)}{2}$$

as (for $x = 2\,\Delta x$ at $6\,\Delta t$)

$$p^2 (2\,\Delta x, t + 6\,\Delta t) + \frac{56 + 80}{2} = 68 \text{ atms}^2$$

For corresponding times the values of $x = \Delta x$ were set equal to the values at $x = 0$ or $(p_o{}^m)^2 = (p_1{}^m)^2$.

Table 8-4 gives the values of Δt for each increment of time. The value for the viscosity of the gas was assumed to remain constant as 0.10 centipoise for all pressures.

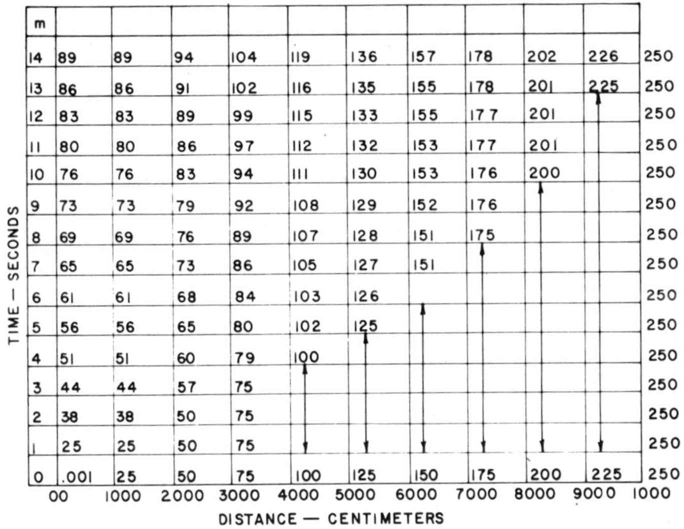

Figure 8-13. Space-time mesh for build-up pressure of gas in a sand-packed pipe.

Table 8-4
Values of Δt for Each Increment of Time

m	$(p_o^m)^2$ atms²/1,000	p_o^m atms	p atms	Δt sec.
0	.001	1.0	251	79.7
1	25	158	329	60.8
2	38	195	348	57.5
3	44	210	355	56.3
4	51	220	364	55.0
5	56	237	369	54.2
6	61	247	374	53.5
7	65	255	378	53.0
8	69	263	382	52.4
9	73	270	385	52.0
10	76	276	388	51.6
11	80	283	392	51.0
12	83	288	394	50.8
13	86	297	398	50.4
14	89	298	399	50.4
			$\Sigma\Delta t$	=634.7

The gas pressure at the partially opened end of the sand-packed pipe built up to 298 atms in 634.7 sec. or 634.7/60 = 10.55 minutes.

Pressure values for mesh points corresponding to the junction between two zones, each with a different permeability, can be computed with the use of Equation **8-23**

$$p_j^2 = \frac{\left(\frac{k}{\mu}\right)_1 \frac{p^2_{j-1}}{\Delta x_1} + \left(\frac{k}{\mu}\right)_2 \frac{p^2_{j+1}}{\Delta x_2}}{\left(\frac{k}{\mu}\right)_1 \frac{1}{\Delta x_1} + \left(\frac{k}{\mu}\right)_2 \frac{1}{\Delta x_2}} \qquad \textbf{8-23}$$

Unsteady-State Radial Flow of Gases

The radial diffusivity equation for the flow of gases given in Equation **6-30** can be written as in Equation **8-24.**

$$\frac{\partial^2 p^2}{\partial^2 r^2} + \frac{1}{r}\frac{\partial p}{\partial r} = \frac{1}{\alpha}\frac{\partial p^2}{\partial t} \qquad \textbf{8-24}$$

where $\bar{p}k/\phi\mu = \alpha$

The difference quotients for the partial derivatives of Equation **8-24** are

$$\frac{\partial p^2}{\partial t} = \frac{p^2 (x, t + \Delta t) - p^2 (r, t)}{\Delta t}$$

$$\frac{\partial p^2}{\partial r} = \frac{p^2 (r + \Delta r, t) - p^2 (r - \Delta r, t)}{2\Delta r}$$

$$\frac{\partial^2 p^2}{\partial r^2} = \frac{p^2 (r + \Delta r, t) + p^2 (r - \Delta r, t)\ 2p^2 (r, t)}{(\Delta r)^2} \qquad \textbf{8-25}$$

The first difference quotient of pressure squared with respect to distance was taken in terms of two increments of space instead of one increment of space so as to eliminate a $p^2(x, t)$ term from the difference solution for the radial diffusivity equation.

The substitution of the difference quotients into the radial diffusivity equation **8-23** gives

$$\frac{p^2 (r + \Delta r, t) + p^2 (r - \Delta r, t) - 2p^2 (r, t)}{(\Delta r)^2} + \frac{p^2 (r + \Delta r, t) - p^2 (r - \Delta r, t)}{2r\Delta r} = \frac{1}{\alpha}\frac{p^2 (r, t + \Delta t) - p^2 (r, t)}{\Delta t} \qquad \textbf{8-26}$$

or

$$p^2(r+\Delta r,t)\left(1+\frac{\Delta r}{2r}\right)$$
$$+\,p^2(r-\Delta r,t)\left(1-\frac{\Delta r}{2r}\right)=\frac{(\Delta r)^2}{\alpha\Delta t}\left[p^2(r,t+\Delta t)-p^2(r,t)\right]-2p^2(r,t)$$

8-27

Letting $(\Delta r)^2/\alpha\Delta t = 2$ and solving for $p^2(r, t + \Delta t)$ gives

$$p^2(r,t+\Delta t)=\frac{\left(1+\frac{\Delta r}{2r}\right)p^2(r+\Delta r,t)+\left(1-\frac{\Delta r}{2r}\right)p^2(r-\Delta r,t)}{2} \quad \text{8-28}$$

which is a difference equation solution for the radial diffusivity equation.

An equation for calculating the pressure at the junction between two concentric radial zones, each with a different permeability, is derived by assuming that the quantity of gas accumulating at the junction between the two zones is negligible, that is, the quantity of gas flowing from the junction is equal to the quantity of gas flowing into the junction. In Chapter 7, Equations **7-65** and **7-73** are discussed for steady-state radial flow of gases. The application of these steady-state equations for a point at the junction between two concentric zones, each with a different permeability, gives the pressure for the radial flow of gas from zone 2 into a point at the junction between the two zones.

$$q_j=\frac{\pi h}{p_j}\frac{k_2}{\mu_2}\frac{(p_{j+1}^2-p_j^2)}{\ln r_{j+1}/r_j} \quad \text{8-29}$$

And the application of the steady-state equation for a point at the junction between two concentric zones, each with a different permeability, gives for the radial flow of gas from the junction between the two zones into zone 1

$$q_j=\frac{\pi h}{p_j}\frac{k_1}{\mu_1}\frac{(p_j^2-p_{j-1}^2)}{\ln r_j/r_{j-1}} \quad \text{8-30}$$

Equating Equations **8-29** and **8-30** and solving for p_j^2 gives

$$p_j^2 = \frac{\dfrac{k_2}{\mu_2} \dfrac{p_{j+1}^2}{\ln r_{j+1}/r_j} + \dfrac{k_1}{\mu_1} \dfrac{p_{j-1}^2}{\ln r_j/r_{j-1}}}{\dfrac{k_2}{\mu_2} \dfrac{1}{\ln r_{j+1}/r_j} + \dfrac{k_1}{\mu_1} \dfrac{1}{\ln r_j/r_{j-1}}} \qquad \textbf{8-31}$$

which is the equation for computing values of gas pressures at the junction between two concentric zones each with a different permeability.

Applications of Equations **8-28** and **8-29** for the radial flow of a gas are illustrated in Example 8-9.

Example 8-9. For a circular gas reservoir containing a well drilled in its center and two concentric homogeneous zones, each with a different permeability, calculate build-up pressure values at the well bore if initially there is steady-state gas flow into the well bore and suddenly the well bore is plugged and the pressure at the perimeter of the gas reservoir is maintained constant by a large aquifer. Additional data are:

$r_o = 10$ cms
$r_j = 3{,}000$ cms
$r_m = 10{,}000$ cms
$p_o = 1$ atm gauge
$p_n = 500$ atm gauge
μ (at 251 atm) $= 0.10$ centipoise
μ (at 375 atm) $= 0.20$ centipoise
$k_1 = 1.00$ Darcy
$k_2 = 0.5$ Darcy

Assume the thickness is constant throughout the gas reservoir.

Solution. The boundary conditions are:

Zone 1

$$p^2(0, 0) = p_o{}^2$$

$$p^2(r, 0) = p_o{}^2 + \frac{q_o p_o \mu_1}{k_1} \ln \frac{r}{r_o} \text{ from } r = r_o \text{ to } r = r_j$$

$$p^2(r, \infty) = p_n$$

$$p^2{}_{1nr}(0, t) = 0$$

Zone 2

$$p^2(r_j, 0) = p_j$$

$$p^2(r, o) = p_j^2 + \frac{q_j p_j \mu_2}{h_2} \ln \frac{r}{r_j} \text{ from } r = r_j \text{ to } r = r_n$$

$$p^2(r, \infty) = p_n$$

$$p^2(r_n, t) = p_n$$

A space-time mesh for this problem is given in Figure 8-14. The space coordinates for the mesh are the ratio of the radius r to the radius r_o of the well bore; the mesh points contain values of pressure squared divided by 1000. The initial value of the pressure at the junction between the two zones was computed as

Zone 1 k = 1.0 Darcy | Junction | Zone 2 k = 0.5 Darcy

TIME — SECONDS

m	Δr/2r	0.5	0.25	.167	.090	.077	.066	.059	.052	.047	.043	.040	.037	
14	154	154	158	163	168	174	184	194	203	212	222	232	238	250
13	151	151	155	161	168	173	183	193	203					250
12	149	149	152	158	164	172	181	191	203					250
11	145	145	149	156	162	171	179	191	202					250
10	142	142	146	153	160	169	179	190	201					250
9	138	138	143	151	158	167	178	189						250
8	134	134	139	148	157	166	177	189						250
7	129	129	135	144	153	164	177	189						250
6	125	125	130	140	150	163	175	188						250
5	119	119	127	138	148	160								250
4	114	114	120	134	147									250
3	109	109	115	126	137									250
2	106	106	110	120	137									250
1	94	94	110	120	137									250
0	0.001	94	110	120	137	160	175	188	201	212	222	232	238	250
	1.0	100	200	300	366	432	498	564	630	696	762	828	894	1000

DISTANCE — RADIUS / RADIUS OF WELL-BORE

Figure 8-14. Distance-time mesh for radial build-up pressure of gas in a circular reservoir containing one well and having two concentric zones each with a different permeability.

$$p_j^2 = \frac{\dfrac{k_1}{\mu_1} \dfrac{p_o^2}{\ln r_j/r_o} + \dfrac{k_2}{\mu_2} \dfrac{p_n^2}{\ln r_n/r_j}}{\dfrac{k_1}{\mu_1 \ln r_j/r_o} + \dfrac{k_2}{\mu_2 \ln r_n/r_j}}$$

$$p_j^2 = \frac{\dfrac{1.0 \times 1.0}{0.1 \ln \dfrac{3,000}{1,000}} + \dfrac{0.5 \times (500)^2}{0.2 \ln \dfrac{10,000}{3,000}}}{\dfrac{1.0}{0.1 \ln \dfrac{3,000}{1,000}} + \dfrac{0.5}{0.2 \ln \dfrac{10,000}{3,000}}} = 120,000 \text{ atms}^2$$

With the value of p_j^2, a graph was sketched on semi-logarithm coordinate graph paper. On semi-logarithm paper, pressure squared was plotted versus the ratio of the radius to the radius of the well bore as r/r_o. The coordinate point $(r_j/r, p_j^2)$ or (300,120000) was located, and from this coordinate point to the coordinate point (1.0, 1.0), a straight line was drawn; then from the coordinate point (300, 120000) to the coordinate point (1,000, 250000) another straight line was drawn. The values of pressure squared for the initial mesh points were taken from the graph (Figure 8-15).

The values of pressure squared for the interior mesh points, not counting those mesh points representing the junction between the two different zones, were computed with the aid of Equation **8-28** as for $m = 5$ and $r/r_o = 200$

$$p_2^5 = \frac{(1 - 0.25)\, 11.4 + (1 + 0.25)\, 134}{2} = 125 \times 10^3 \text{ atms}^2$$

The increment of distance Δr was taken as 1,000 cms in zone 1.

The values of pressure squared for the boundary points $r/r_o = 1.0$ were taken as equal to corresponding values of pressure squared for the interior mesh points $r/r_o = 100$ or

$$(p_o^m)^2 = (p_1^m)^2$$

Since the value of the time increment for the first zone will change considerably for the first few increments of time and since during this time the changes in values of the pressure squared are negligible for the second zone, the increment of space for the second zone was not computed until after the second increment of time.

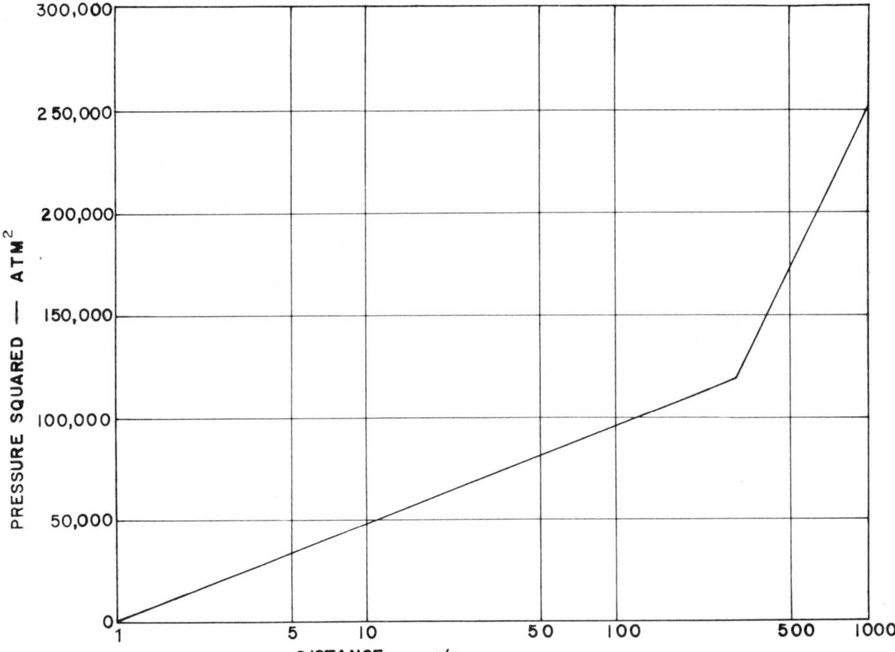

Figure 8-15. Pressure-distance graph for determining initial values of pressure squared in a circular reservoir having two zones each with a different permeability.

The first increment of time for zone 1 is

$$\Delta t = \frac{(\Delta r_1)^2 \, \mu_1 \phi}{2p \, k_1} = \frac{(1{,}000)^5 \times 0.1 \times 0.2}{2 \times 173 \times 1.0} = 58.0 \text{ sec.}$$

The second increment of time for zone 1 is (μ_1 was taken as 0.14 centipoise)

$$\Delta t = \frac{(1{,}000)^2 \times 0.14 \times 0.20}{2 \times 340 \times 1.0} = 41.2 \text{ sec.}$$

With the second value for the increment of time, the increment of space (Δr_2) was computed for the second zone as

$$(\Delta r_2)^2 = \frac{2p\, k_2\, \Delta t}{\mu_2 \phi}$$

or

$$\Delta r_2 = \sqrt{\frac{2 \times 423 \times 0.5 \times 41.2}{0.2 \times 0.2}} = 660 \text{ cms}$$

The number of increments of space in the second zone is 7,000/660 = 10.6, and the initial values for the mesh points of the second zone are shown in Table 8-5.

The values of pressure squared for the mesh points at the junction between the two zones were computed as

$$p_j^2 = \frac{\dfrac{k_1}{\mu_1} \dfrac{p_j^2}{\log \dfrac{3{,}000}{2{,}000}} + \dfrac{k_2}{\mu_2} \dfrac{p_{j+1}^2}{\log \dfrac{3{,}660}{3{,}000}}}{\dfrac{k_1}{\mu_1} \dfrac{1}{\log \dfrac{3{,}000}{2{,}000}} + \dfrac{k_2}{\mu_2} \dfrac{1}{\log \dfrac{3{,}660}{3{,}000}}}$$

For these values of pressure squared, μ_1 was taken as equal to μ_2; the constants in the previous equation are

Table 8-5
Initial Values for the Mesh Points of the Second Zone

r/r_o	p^2 (atms^2)	p (atms)
366	137,000	368
432	163,000	404
498	175,000	418
564	188,000	434
630	201,000	448
696	213,000	462
762	222 000	471
828	232,000	482
894	238,000	488
1,000	250,000	500

$$\frac{1}{\log 1.5} = 5.67$$

$$\frac{0.5}{\log 1.22} = 5.80$$

Therefore, p_j^2 at $m = 4$ is

$$\frac{5.68 \times 120 + 5.80 \times 147}{5.68 + 5.80} = 134{,}000 \text{ atms}^2$$

and p_j^2 at $m = 5$ is

$$\frac{5.68 \times 127 + 5.80 \times 148}{11.48} = 140{,}000 \text{ atms}^2$$

The fourteenth increment of time is

$$\Delta t = \frac{(1{,}000)^2 \times 0.15 \times 0.20}{2 \times 401 \times 1.00} = 37.5 \text{ sec}$$

Therefore, the average value of Δt for the last 13 increments of time is $(37.5 + 41.2)/2 = 39.4$ sec. and the total time for 14 increments of time is $39.4 \times 13.0 + 58.0 = 560$ sec. $560/60 = 9.33$ minutes.

As the time increments in the first zone change appreciably, new space-time meshes are constructed with appropriate values of the space increments Δr_2 in the second zone; this adjustment is necessary because the average pressures in both zones increase with time.

Summarizing, the gas pressure in the well-bore built-up to 394 atms gauge in 9.33 minutes.

Exercises

1. The build-up pressure Examples 8-3, 8-5, 8-7, 8-8 and 8-9 can best be solved with
. the aid of a microcomputer. If one is skilled sufficiently with the methods of using
. a microcomputer, he should write computer programs for the problems in these
. examples and solve the problems. In writing the computer programs the number
5. of increments of space (Δx or Δr) should be increased by at least three-fold and the computations should be carried out until $p_n^m - p_o^m$ is less than two atms pressure.
6. If initially the pressure in homogeneous fully saturated sand-packed pipe is at 100 atms gauge and suddenly from one end of the sand-packed pipe the fluid is permitted to flow out at a rate of 0.20 ml per second and the other end of the

pipe is maintained constant at 100 atms gauge, compute the time it takes for the pressure to decline to one atm gauge at the partially opened end. Additional data are: $k = 0.1$ Darcy; $\mu = 2.00$ centipoises, $L = 1000$ cms, $q = 0.20$ ml/sec; $\phi = 0.20$; $c = 2.00 \times 10^{-4}$ atm^{-1}; $A = 1.00$ cm^2.

7. For a homogeneous circular reservoir containing a well drilled in its center, calculate the time it takes for the pressure in the well bore to decline to 10 atms gauge if initially the reservoir is at a constant pressure of 300 atms gauge. Assume that the pressure at the perimeter of the reservoir is kept constant by a large aquifer. Additional data are: $r_o = 10$ cms; $r_n = 10{,}000$ cms; $k = 0.30$ Darcy; $\phi = 0.20$; $\mu = 3.00$ centipoise; $c = 8.00 \times 10^{-4}$; and $dp/dr = 1.00$ atm/cm at well bore.
8. A circular oil reservoir with a radius of 10,000 cms is bounded with a circular aquifer which has a radius of 100,000 cms. The original field pressure was 500 atms gauge. The oil field pressure was 500 atms. gauge and it was developed very rapidly; almost instantaneously the field pressure reached 100 atm gauge. It is intended to keep the field pressure at the lower pressure by adjusting the fluid withdrawals. Additional data are: $k = 0.20$ Darcy; $c_w = 3 \times 10^{-5}$ atms^{-1}; $\mu_w = 0.60$ centipoise; $\phi = 0.20$; and $h = 100$ cms. For one week determine.
   1. The pressure distribution in the aquifer
   2. The daily rate of water influx into the oil reservoir.
9. A circular oil reservoir with a radius of 10,000 cms is bounded by an aquifer which keeps the pressure constant at the perimeter of the oil reservoir. Assume that the thickness of the reservoir decreases rather markedly at a radius of 5,000 cms to one-half of its thickness in the center zone; for a well in the center of the oil reservoir compare build-up pressures with those if the reservoir thickness was constant throughout.

Suggested Reading

1. Bruce, G. H., Peaceman, D. W., and Rachford, Jr., H. H., "Calculation of Unsteady-State Gas Flow Through Porous Media," Trans. AIME, 1953, pp. 79-92, 1953.
2. Cornell, D., and Katz, D., "Pressure Gradients in Natural Gas Reservoirs," Trans. AIME, 1953, pp. 61-70, 198.
3. Douglas, Jim, Jr., Peaceman, D. W., and Rachford, Jr., H. H., "Calculation of Unsteady-State Gas Flow Within a Square Drainage Area," Trans. AIME, 1955, pp. 190-195, 204.
4. Jakob, M., and Hawkins, G. A., *Elements of Heat Transfer and Insulation,* New York: John Wiley and Sons, 1959, pp. 77-80.
5. Muskat, M., *The Flow of Fluids Through Porous Media,* New York: McGraw-Hill Book Co., Inc., 1937, pp. 131-146.
6. Pirson, S. J., *Oil Recovery Engineering,* New York: McGraw-Hill Co., Inc., 1958, pp. 418-423.
7. Yang, Tsuo-I, "Transient Temperature Lags in Slabs," Master's Thesis, Tennessee Technological University, 1968, pp. 20-22.

9 Classical Methods for Fluid Flow Through Porous Media

The previous chapters of this book deal with the applications of numerical methods for fluid and gas flow through porous media. The numerical methods are those ordinarily introduced in mathematical courses such as "Numerical Methods" or "Finite Difference Methods." In the two previous chapters difference equation solutions for the Laplace equation and the diffusivity equation are derived, and examples applying the difference equation solutions are given. The difference equations require a vast number of arithmetic manipulations before numerical solutions are obtained. To perform the vast number of arithmetic manipulations, microcomputers are often required.

This chapter examines classical solutions to both the Laplace equation and the diffusivity equation. Generally, the classical solutions are straightforward and require fewer arithmetic manipulations for obtaining numerical solutions. Often, it is exceedingly difficult and sometimes impossible to obtain a classical solution for fluid-flow problems involving nonhomogeneous media. For these systems, the numerical solutions are preferred. In the classical solutions considered in this chapter, wells are treated as points (either as sources or sinks), meaning that the radii of the wells are assumed to be negligible.

Steady-State Flow of Fluids Through Porous Media Classical Solutions

Single-Well Solutions

The Laplace equation for steady-state radial flow in a plane where there is no change in pressure in the radial direction is shown in Equation **7-60.**

$$\frac{\partial^2 p}{\partial r^2} + \frac{1}{r}\frac{\partial p}{\partial r} = 0 \qquad \textbf{9-1}$$

However Equation **9-1** can be written as

$$\frac{\partial}{\partial r}\left(r\,\frac{\partial p}{\partial r}\right) = 0 \qquad \textbf{9-2}$$

and the first integration of Equation **9-2** is

$$\left(r\,\frac{\partial p}{\partial r}\right) = C_1 \qquad \textbf{9-3}$$

The second integration of Equation **9-2** is

$$\int dp = C_1 \int dr/r \qquad \textbf{9-4}$$

or

$$p = C_1 \ln\ r + C_2$$

Evaluating the constants in Equation **9-4** for $p = p_o$ at $r = r_o$ and $p = p_n$ at $r = r_n$ gives

$$p_n = C_1 \ln r_n + C_2$$
$$p_o = C_1 \ln r_o + C_2$$

or

$$C_1 = \frac{p_o - p_n}{\ln r_o/r_n}$$

and

$$C_2 = p_n - \left(\frac{p_o - p_n}{\ln r_o/r_n} \ln r_n\right)$$

The substition of the values for the constants into Equation 9-4 gives

$$p = \frac{p_o - p_n}{\ln r_o/r_n} \ln r + p_n - \left(\frac{p_o - p_n}{\ln r_o/r_n} \ln r_n\right)$$

and dividing the above equation by ln r_n gives

$$p = p_n + \frac{p_o - p_n}{\ln r_o/r_n} \ln r/r_n \qquad \textbf{9-5}$$

From the radial flow equation

$$q = \frac{2\pi hk}{\mu} \frac{(p_n - p_o)}{\ln \frac{r_n}{r_o}} = \frac{2\pi hk\,(p_o - p_n)}{\mu \ln r_o/r_n}$$

or

$$\frac{p_o - p_n}{\ln r_o/r_n} = \frac{q\,\mu}{2\pi hk}$$

And by substitution, Equation 9-5 becomes

$$p = p_n + \frac{q\,\mu}{2\pi hk} \ln r/r_n \qquad \textbf{9-6}$$

which is a solution of the Laplace equation for radial horizontal flow of incompressible fluid. The equation may be used for determining the pressures radially from a well in the center of a radial reservoir with a radius r_n. The application of Equation 9-6 is illustrated in Example 9-1.

Example 9-1. For a circular homogeneous reservoir with a radius of one mile, determine the reservoir pressure at radial distances of ¼ and ⅛ mile from the well bore. Assume that the pressure p_n at the perimeter (r_n = constant) of the reservoir is constant. Additional data are:

q = 100 bbls/day
p_n = 200 psia
r_n = 1.00 mile
k = 0.100 Darcy
μ = 4.00 centipoises
B = 1.00
h = 10 ft.

Solution. Since the units in the example are practical units a correction constant is determined so that the practical units may be substituted directly into Equation 9-6. This constant is computed from the radial Darcy equation as

$$q_{ml/sec} = \frac{2\,\pi\,h_{cm}\,(p_n - p)\ \text{atms}}{\mu\ \text{centipoise B}\ \ln r/r_n}$$

or

$$q_{bbls/day} = \frac{2\,\pi \times 24 \times 3600 \times 30.5\,h_{ft}\,k_{Darcy}\,(p_n - p)\ \text{psi}}{5.615 \times (30.5)^3 \times 14.7 \times \mu\ \text{centipoise B}\ \ln r/r_n}$$

or

$$q_{bbls/day} = \frac{7.08\,h_{ft}\,k_{Darcy}\,(p_n - p)}{\mu\ \text{centipoise B}\ \ln r/r_n}$$

With this correction constant Equation 9-6 becomes

$$p = p_n + \frac{q\,\mu\,B}{7.08\,hk}\ \ln r/r_n$$

or

$$p = p_n - \frac{q\,\mu\,B}{7.08\,hk}\ \ln r_n/r \qquad \text{9-7}$$

The pressure at ¼ mile distance from the well bore is

$$p = 200 - \frac{100 \times 4.0 \times 1.00}{7.08 \times 10 \times 0.10}\ \ln 4.0 = 121.5\ \text{psia}$$

and the pressure at ⅛ mile radial distance from the well bore is

$$p = 200 - \frac{100 \times 4.00 \times 1.00}{7.08 \times 10 \times 0.10}\ \ln 8.0 = 83\ \text{psia.}$$

Two-Well Solution

From well interference phenomena, it is known that the pressure at any given point in a reservoir is the resultant of the pressure effects caused by all wells in the reservoir. No single well contributes all the effects required for fixing the pressure at the given point. Also, the pressure at the given point is not determined wholly by a multiple of the effect of a single well. By applying the superposition principle, a solution for the Laplace equation for a two-well system may be obtained. The superposition principle may be stated as: the sum of two or more solutions to the Laplace equation is also a solution of the Laplace equation; and since constants are solutions of the Laplace equation, the sum of one or more solutions to the Laplace equation plus one or more constants is also a solution of the Laplace equation. With the aid of Equation 9-6 and applying the superposition principle a solution of the Laplace equation for two wells may be written as

$$p_{(x,\,y)} = p_n + \frac{q_1 \mu B}{2\,\pi\, hk} \ln r_1/r_n + \frac{q_2 \mu B}{2\,\pi\, hk} \ln r_2/r_n \qquad \text{9-8}$$

In Equation 9-8, $p_{(x,y)}$ represents the pressure at any point in the reservoir, q_1 represents the flow rate of Well No. 1 and q_2 represents the flow rate of Well No. 2, r_1 is the distance from Well No. 1 to coordinate point (x, y) and r_2 is the distance from Well No. 2 to coordinate point (x, y); r_n is the radius of the reservoir and p_n is the pressure at the boundary of the reservoir. For the equation to apply, the distances of the wells from the center of the reservoir must be small in comparison with the radius r_n.

After a solution to a partial differential equation has been found, the solution must be checked for uniqueness. The uniqueness check includes a check to determine whether the solution will satisfy the original partial differential equation and to determine whether the boundary conditions of the physical problem fits the conditions of the mathematical solution. By direct differentiation Equation 9-8 may be shown to be a solution of the Laplace equation 9-2. The boundary conditions of the problem are:

at r_1 and $r_2 = r_n$, $p_{x,y} = p_n$
at $r_1 = a_1$ (radius of bore of Well No. 1) $q = q_1$
and at $r_2 = a_2$ (radius of bore of well No. 2) $q = q_2$

By observing Equation 9-8 it is obvious that as the values of r_1 and r_2 approach the values of r_n the logarithmic terms in the equation approach zero and the value of $p_{(x,y)}$ approaches the value of p_n (the value of the pressure at the boundary of the reservoir).

At the radius of the bore of Well No. 1, the flow into the well is

$$q r_1 \rightarrow a_1 = \int_0^{2\pi} \frac{k}{\mu} \frac{\partial p\,(x, y)}{\partial r_1}\, r_1\, h\, d\,\Theta \qquad \text{9-9}$$

The differentiation of Equation **9-8** with respect to r_1 is

$$\frac{\partial p\,(x, y)}{\partial r_1} = \frac{q_1 \mu}{2\,\pi\, hk} \frac{1}{r_1} \quad \textbf{9-10}$$

The substitution of Equation **9-10** into Equation **9-9** gives

$$q r_1 \rightarrow a_1 = \int_o^{2\pi} \frac{k\, q_1 \mu\, h}{2\,\pi\,\mu\, hk} \frac{1}{a_1}\, a_1\, d\,\Theta$$

or

$$q r_1 \rightarrow a_1 = q_1 \quad \textbf{9-11}$$

In a similar manner, the flow rate into Well No. 2 may be shown to be equal to q_2.

Example 9-2. For a homogeneous uniform thickness semi-circular reservoir containing two wells as indicated in Figure 9-1, determine the perpendicular distance from Well No. 1 to the fault. For the problem, it is assumed that Well No. 2 has been shut in until there is steady-state flow into Well No. 1. At that time the following data are applicable: $q_1 = 100$ bbls/day, $\mu = 4.00$ centipoise, $k = 0.10$ Darcy, $h = 10.00$ ft, $B = 1.10$, $p_{(x,y)}$ (at Well No. 2) = 158 psia, $p_n = 400$ psia, $r_n = 5{,}000$ ft.

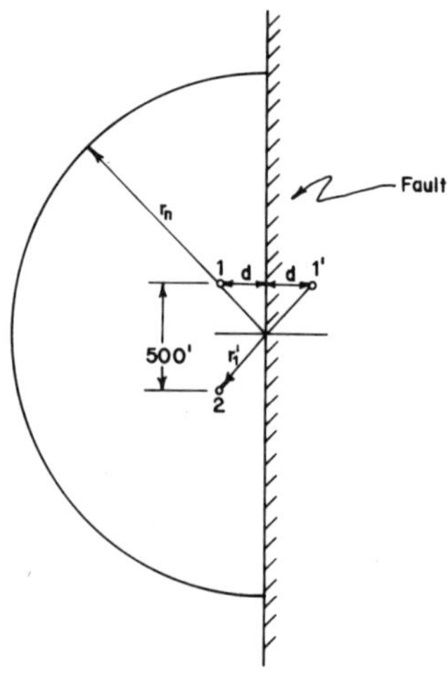

Figure 9-1. Semicircular reservoir adjacent to a fault containing two wells.

Solution. An image Well No. 1′ is taken of Well No. 1; the distance of the image well from the fault is the same as that of Well No. 1 from the fault, *d*. As shown in the figure, the perpendicular distance from Well No. 2 to the fault is also *d;* therefore, a right triangle is formed by the points 1, 1′, and 2. The object of the problem is to first determine the value of r'_1; then from the value or $r'_{1'}$ and r_1, the value of *d* can be determined.

The practical form of Equation **9-8** is shown in Equation **9-12.**

$$p_{(x,\,y)} = p_n + \frac{q_1 \mu B}{7.08\, hk} \ln r_1/r_n + \frac{q'_1 \mu B}{7.08\, hk} \ln r_1'/r_n \qquad \textbf{9-12}$$

For this problem $q_1 = q'_1$; assuming $r'_1 = 1{,}200$ ft, the computed pressure at Well No. 2 is

$$p_{(x,\,y)} = 400 - \left(\frac{100 \times 4.00 \times 1.10}{7.08 \times 10 \times 0.10}\right)\left(\ln \frac{5000}{500} + \ln \frac{5000}{1200}\right)$$

$$= 400 - 232 = 168 \text{ psia}$$

And assuming $r'_1 = 1{,}000$ ft, the computed pressure at Well No. 2 is

$$p_{(x,\,y)} = 400 - \left(\frac{100 \times 4.00 \times 1.10}{7.08 \times 10 \times 0.11}\right)\left(\ln \frac{5000}{500} + \ln \frac{5000}{1000}\right)$$

$$= 400 - 242 = 158 \text{ psia}$$

Therefore, the true value of r'_1 is 1,000 ft, and the value of *d* is

$$2\,d = \sqrt{(1000)^2 - (500)^2}$$

$$= 866 \text{ ft.}$$

or

$$d = 433 \text{ ft.}$$

Multiple-Well Solution

In a similar manner as was used for deriving the two-well solution, a multiple-well solution can be derived by applying the superposition principle to Equation **9-6.** The multiple-well solution is shown in Equation **9-13.**

$$p_{(x,y)} = p_n + \frac{q_1 \mu}{2\pi h k} \ln r_1/r_n + \frac{q_2 \mu}{2\pi h k} \ln r_2/r_n$$

$$+ \frac{q_3 \mu}{2\pi h k} \ln r_3/r_n + \cdots \frac{q_m \mu}{2\pi h k} \ln r_m/r_n$$

or

$$p_{(x,y)} = p_n + \sum_{m=1}^{m} \frac{q_m \mu}{2\pi h k} \ln \frac{r_m}{r_n} \tag{9-13}$$

or the practical equation is

$$p_{(x,y)} = p_n + \sum_{m=1}^{m} \frac{q_m \mu B}{7.08\, h k} \ln \frac{r_m}{r_n} \tag{9-14}$$

where q_m is the flow rate for Well No. m and r_m is the distance from Well No. m to the coordinate point (x, y). The other items in Equation **9-14** have the same meaning as their corresponding terms in Equation **9-12**. By similar methods as applied for the two-well solution, Equation **9-13** can be shown to be unique.

Experiments conducted with an electrical resistance network indicated that differences between experimentally measured values of potentials and computed values of potentials for steady-state multi-well systems, by means of Equation **9-14,** were never greater than 14% when the distances of the wells were never over $\frac{1}{2} r_n$ from the center of the reservoir. The greater the symmetry of the wells about the center of the reservoir, the smaller the differences between experimental and calculated values of potentials. The experiments indicated that computed values of potentials tended to be lower than measured values of potentials for wells near the center of a circular reservoir. However, the reverse was found to be true for wells near the boundary of the reservoir.

The use of Equation **9-14** for computing the flow of oil across a lease line is illustrated in Example **9-3.**

Example 9-3. Compute the flow rate of incompressible oil across the lease line at point $p(x, y)$ in a homogeneous, constant thickness, circular reservoir as is indicated in Figure 9-2. Assume that the wells are located near the center of the reservoir and that the flow rate at point $p(x, y)$

remains approximately constant for 25 ft. on either side of that point along the lease line. Additional data are:

$q_1 = 100$ bbls/day, $q_2 = 75$ bbls/day, $q_3 = 50$ bbls/day, $k = 0.20$ Darcy, $\mu = 5.00$ centipoise, $B = 1.20$, $r_n = 50{,}000$ $p_n = 600$ psia, $h = 10$ ft

Solution. The pressure values at points 25 ft to the right and 25 ft to the left of point (x, y) are computed. The value of r_1 at 25 ft to the right of point (x, y) is 325 ft and its value at 25 ft to the left of point (x, y) is 275 ft. The corresponding values for r_2 are 275 ft and 325 ft respectively. The value of r_3 at 25 ft to the right of point (x, y) is $r_3 = \sqrt{[(325)^2 + (800)^2]} = 864$ ft and the value of r_3 at 25 ft to the left of point (x, y) is $r_3 = \sqrt{[(275)^2 + (800)^2]} = 846$ ft.

With the appropriate values of r_1, r_2, and r_3 the pressure value at 25 ft to the right of point (x, y) is

$$
\begin{aligned}
p_{(x,y)} &= 600 - \frac{100 \times 5.0 \times 1.20}{7.08 \times 10 \times 0.20} \ln 50{,}000/325 \\
&\quad - \frac{75 \times 5.0 \times 1.20}{7.08 \times 10 \times 0.20} \ln 50{,}000/275 \\
&\quad - \frac{50 \times 5.0 \times 1.20}{7.08 \times 10 \times 0.20} \ln 50{,}000/864 \\
&= 600 - (213 + 165 + 86) = 136 \text{ psia}
\end{aligned}
$$

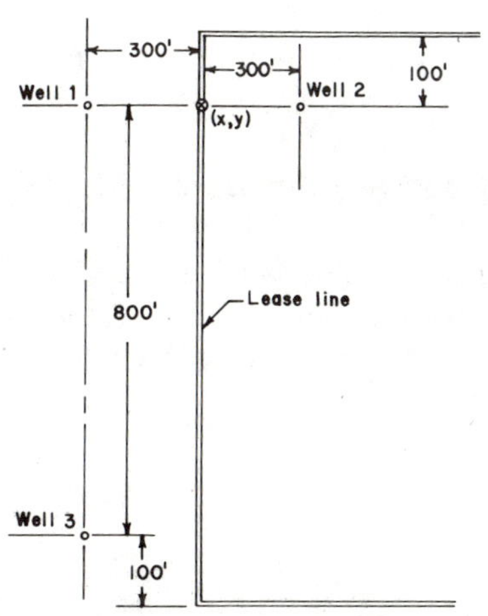

Figure 9-2. Three-well reservoir with a lease line.

And with the appropriate values of r_1, r_2 and r_3 the value of pressure at 25 ft to the left of point (x, y) is

$$p_{(x,y)} = 600 - \frac{100 \times 5.0 \times 1.20}{7.08 \times 10 \times 0.20} \ln 50{,}000/275$$

$$- \frac{75 \times 5.0 \times 1.20}{7.08 \times 10 \times 0.20} \ln 50{,}000/325$$

$$- \frac{50 \times 5.0 \times 1.20}{7.08 \times 10 \times 0.20} \ln 50{,}000/846$$

$$= 600 - (215 + 160 + 92) = 133$$

Since the pressure is the greatest to the right of point (x, y), the incompressible fluid will flow across the lease line toward Wells No. 1 and 3. Applying the steady-state Darcy equation for the flow across the lease line for 50 ft of the lease line gives

$$q = \frac{1.127 \times \mathrm{k} \times A\,(\Delta p)}{\mu \times \mathrm{B} \times (\Delta L)}$$

or

$$q = \frac{1.127 \times 0.20 \times 10 \times 50 \times (136 - 133)}{5.0 \times 1.20 \times 50}$$

$$= 1\,.127 \text{ bbls/day}$$

The procedure is continued for other 50-ft sections of the lease line, and the total flow across all the lease line is the sum of the values obtained for the flow across each 50-ft section.

Incompressible Fluid Build-Up Pressure

An equation for calculating the build-up pressure in a well for incompressible fluid can be derived with the aid of the single well solution in Equation **9-7** or it can be written as in Equation **9-15.**

$$p_o = p_n - \frac{q\,\mu}{7.08\,h\,k} \ln r_n/r_o \qquad \textbf{9-15}$$

where p_o is the pressure in the well bore and r_o is the radius of the well bore.

Solving Equation 9-15 for q gives Equation **9-16**

$$q = \frac{7.08\, h\, k\, (p_n - p_o)}{\mu \ln r_n/r_o} \quad \text{9-16}$$

If the pressure is building up in the well bore due to the rising of fluid in the well, then (see Fig. 9-3)

$$q = dV/dt \quad \text{9-17}$$

Equating Equations **9-16** and **9-17** gives Equation (9-18)

$$\frac{dV}{dt} = \frac{7.08\, h\, k\, (p_n - p_o)}{\mu \ln r_n/r_o} \quad \text{9-18}$$

But

$$dV = \frac{A\, dh_o}{5.616} \text{ bbls} \quad \text{9-19}$$

where A is the cross sectional area of the well in square feet and h_o is the height of the fluid in the well. In terms of pressure

$$dh_o = \frac{dp_o}{0.433\, \rho_o} \quad \text{9-20}$$

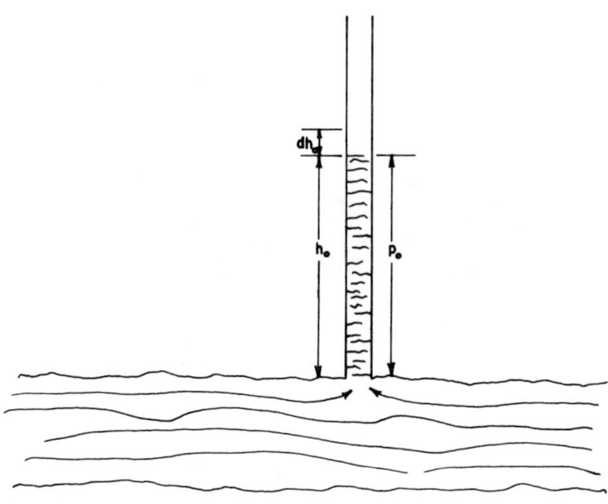

Figure 9-3. A vertical section of a reservoir illustrating the flow of fluid up the well bore.

where p_o is the pressure (psi) in the well bore, ρ_o is the specific gravity of the fluid in the well bore and 0.433 is pressure (psi) of one foot of water in height.

The substitution of Equations **9-19** and **9-20** into Equation **9-18** gives

$$\frac{dp_o}{dt} = \frac{7.08\, h\, k \times 0.433\, \rho_o \times 5.616\,(p_n - p_o)}{A\, \mu\, \ln r_n/r_o}$$

or

$$\frac{dp_o}{dt} = \frac{17.22\, h\, k\, \rho_o\,(p_n - p_o)}{A\, \mu\, \ln r_n/r_o}$$

or

$$\frac{dp_o}{dt} = C(p_n - p_o) \qquad \textbf{9-21}$$

where

$$C = \frac{17.22\, h\, k\, \rho_o}{A\, \mu\, \ln r_n/r_o}$$

The integration of Equation **9-21** gives

$$\int_{p_i}^{p} \frac{dp_o}{p_n - p_o} = C \int_{o}^{t} dt$$

or

$$\ln\left[\frac{p_n - p_i}{p_n - p}\right] = Ct \qquad \textbf{9-22}$$

or

$$\frac{p_n - p_i}{p_n - p} = e^{C_t} \qquad \textbf{9-23}$$

where p_i is any pressure after the cessation of production.

Equation **9-22** may be written as $2.303 \log_{10} [p_n - p/p_n - p_i] = - Ct$. And if $(p_n - p)$ versus t is plotted on semi-logarithmic paper the result is a straight line with a slope of $-\ (C/2.303)$.

With the value of the slope, the value of C can be determined and by applying the relation for C the permeability of the reservoir can be determined from the relation

$$k = \frac{C A \mu \ln r_n/r_o}{17.22\ h\ \rho_o} \qquad 9\text{-}24$$

Also with the value of C and the relation for C, a value for the effective radius of a homogeneous circular reservoir may be determined from build-up pressure by the relations

$$\ln r_n/r_o = \frac{17.22\ hk\ \rho_o}{C \mu A}$$

or

$$r_n = r_o\, e^{\frac{17.22\ hk\ \rho_o}{C \mu A}} \qquad 9\text{-}25$$

The computation of the effective drainage radius is illustrated in Example 9-4.

Example 9-4. Compute the radius of drainage for a single well located near the center of a homogeneous circular reservoir. When the well was closed in for several days the bottom hole pressure reached a maximum of 500 psia. The build-up pressure obtained during the test is given in Table 9-1.

Additional data are

h 20 ft
k 0.02 Darcy

Table 9-1
Build-Up Pressure Data

Time-Days	Pressure Differential ($p_n - p$) psi
0.2	440
0.5	330
1.0	230
2.0	110
4.0	22
6.0	5

$\mu = 4.0$ centipoise
$B = 1.2$
$\rho_0 = 0.85$ gram/ml 60° C
$r_0 = 3.0$ in
$A = 0.1964$ ft^2

Solution. A semi-logarithmic plot of build-up data pressure from Table 9-1 is given in Figure 9-4. The slope of the curve in the figure is $[2.00/(8.00 - 2.10)] = 0.3389$.

And from the value of the slope, the value of the constant C is $C = 2.303 \times 0.3389 = 0.7805$.

Appling Equation 9-25, the value of the drainage radius is

$$r_n = 0.25\, e^{\frac{17.22 \times 20 \times 0.02 \times 0.85}{0.7805 \times 4.0 \times 0.1964}}$$

$$= 0.25\, e^{9.549} = 3{,}506 \text{ ft.}$$

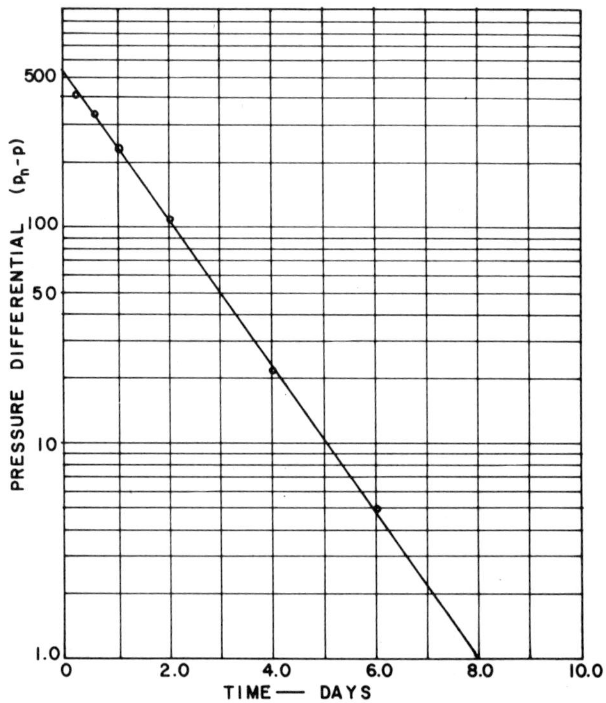

Figure 9-4. Build-up pressure in a well.

Unsteady-State Flow of Fluids Through Porous Media—Classical Solution

Single-Well Solution

In this section, the single-well solution for the unsteady-state flow of fluids through homogeneous porous media deals with an exponential integral solution of the radial diffusivity equation

$$K\left[\frac{\partial^2 p}{\partial r^2} + \frac{1}{r}\frac{\partial p}{\partial r}\right] = \frac{\partial p}{\partial t} \qquad \text{9-26}$$

where K has the dimensions of $k/\mu c\phi$. The boundary conditions for the solution are

$$p(\infty, t) = p_n \qquad \text{9-27a}$$

$$p(r,0) = p_n \qquad \text{9-27b}$$

$$\text{Flow} = q(\text{at the well bore}) = 2\pi r_w h \frac{k}{\mu}\frac{\partial p}{\partial t} \qquad \text{9-27c}$$

where r_w is the radius of the well bore and h is the thickness around the well bore which is being considered.

The boundary conditions are for a well of radius r_w drilled through a constant thickness, homogeneous horizontal oil producing zone in a circular reservoir with an infinite radius and initially at a uniform pressure of $p = p_n$. The rate of production of the well is constant and is designated by q.

The assumed exponential integral solution to Equation **9-26** is

$$p = p_n - \frac{\mu}{4\pi kh}\int_0^t qe^{-r^2/4Kt}\frac{dt}{t} \qquad \text{9-28}$$

The proof that the exponential integral is a solution of Equation **9-28** follows. The first differentiation of Equation **9-28** with respect to r is

$$\frac{\partial p}{\partial r} = -\frac{\mu}{4\pi hk}\int_0^t qe^{-r^2/4Kt}\,[-2r/4Kt]\frac{dt}{t} \qquad \text{9-29}$$

and the second differentiation of Equation **9-28** with respect to r is

$$\frac{\partial^2 p}{\partial r^2} = \frac{\mu}{8Kt^2\pi hk}\int_0^t qe^{-r^2/4Kt}\left[1 - \frac{r^2}{2Kt}\right]dt \qquad \text{9-30}$$

and the differentiation of Equation **9-28** with respect to t is

$$\frac{\partial p}{\partial t} = \frac{\mu}{t^2 4\pi hk} \int_0^t q e^{-r^2/4Kt} \left[1 - \frac{r^2}{4Kt}\right] dt \qquad \textbf{9-31}$$

The substitution of Equations **9-29, 9-30** and **9-31** into Equation **9-26** gives

$$K\left[\frac{\partial^2 p}{\partial r^2} + \frac{1}{r}\frac{\partial p}{\partial r}\right] = \frac{\partial p}{\partial t}$$

$$= \frac{\mu}{t^2 4\pi kh} \int_0^t q e^{-r^2/4Kt} \left[\frac{1}{2} + \frac{1}{2} - \frac{r^2}{4Kt}\right] dt$$

$$= \frac{\mu}{t^2 4\pi hk} \int_0^t q e^{-r^2/4Kt} \left[1 - \frac{r^2}{4Kt}\right] dt \qquad \textbf{9-32}$$

Equation **9-32** indicates that the appropriate differentiations of the exponential integral and the substitutions into the diffusivity equations, satisfies the radial diffusivity equation.

An examination of the exponential integral equation, Equation **9-28** indicates that as r approaches infinity the exponential integral approaches zero. Therefore, the first boundary condition is satisfied; that is $p = p_n$.

An examination of the exponential integral equation when t approaches zero, indicates that the integral vanishes because the upper limit becomes identical with the lower limit. Therefore, the second boundary condition is satisfied.

The substitution of Equation **9-29** into the third boundary condition and letting r approach r_w gives

$$q = -\frac{2\pi r_w hk\mu}{4\mu\pi hk} \int_0^t q e^{-r_w^2/4Kt} \left[-\frac{2r_w}{4Kt}\right] \frac{dt}{t} \qquad \textbf{9-33}$$

For a constant flow rate, q comes out of the integral and Equation **9-33** becomes

$$\text{flow} = q = -q \int_0^t e^{-r_w^2/4Kt} \left[-\frac{r_w^2}{4Kt}\right] \frac{dt}{t} \qquad \textbf{9-33a}$$

or

$$\text{flow} = -q\, e^{-r_w^2/4Kt}\Big]_0^t \qquad \text{9-33b}$$

because the integrand in Equation **9-33a** is a perfect differential.

And for appreciable values of t and since K is always larger than r_w^2; then flow $= q$. Therefore the third boundary condition is satisfied.

For simple use of Equation **9-28** a change of variables is made. The single variable x is substituted for $r^2/4Kt$ and the differential of x is $dx = r^2 dt/4Kt$.

And the substitution of x and dx into Equation **9-28** gives

$$p = p_n - \frac{\mu}{4\pi hk}\int_0^t q e^{-r^2/4Kt}\, dt/t$$

or

$$p = p_n - \frac{\mu}{4\pi hk}\int_x^\infty q e^{-x}\frac{dx}{x}$$

or

$$p = p_n - \frac{\mu}{4\pi hk}\int_{r^2/4Kt}^\infty q e^{-r^2/4Kt}\frac{dt}{t} \qquad \text{9-34}$$

The values for the integral of Equation **9-35** may be found in *Tables of Sine, Cosine, and Exponential Integrals,* Vols. I and II[1], and in *Handbook for Mathematics.*[2]

$$Ei(-x) = \int_x^\infty e^{-x}\frac{dx}{x} \qquad \text{9-35}$$

[1]Tables of Sine, Cosine, and Exponential Integrals, Vols, I and II, Federal Agency, W.P.A. for the City of New York, Sponsored by U.S. National Bureau of Standards. Available from Supt. of Documents, Washington 25, D.C.

[2]Weast, R. C. and Selby, S. M., "Handbook for Mathematics," Chemical Rubber Co. Cleveland, Ohio, pp. 687-697 3rd. (1967)

A graph of exponential values is given in Fig. 9-5. Also a small table of exponential values is given in Table 9-2. Exponential values taken from the graph can be used when accuracy to two significant figures is adequate.

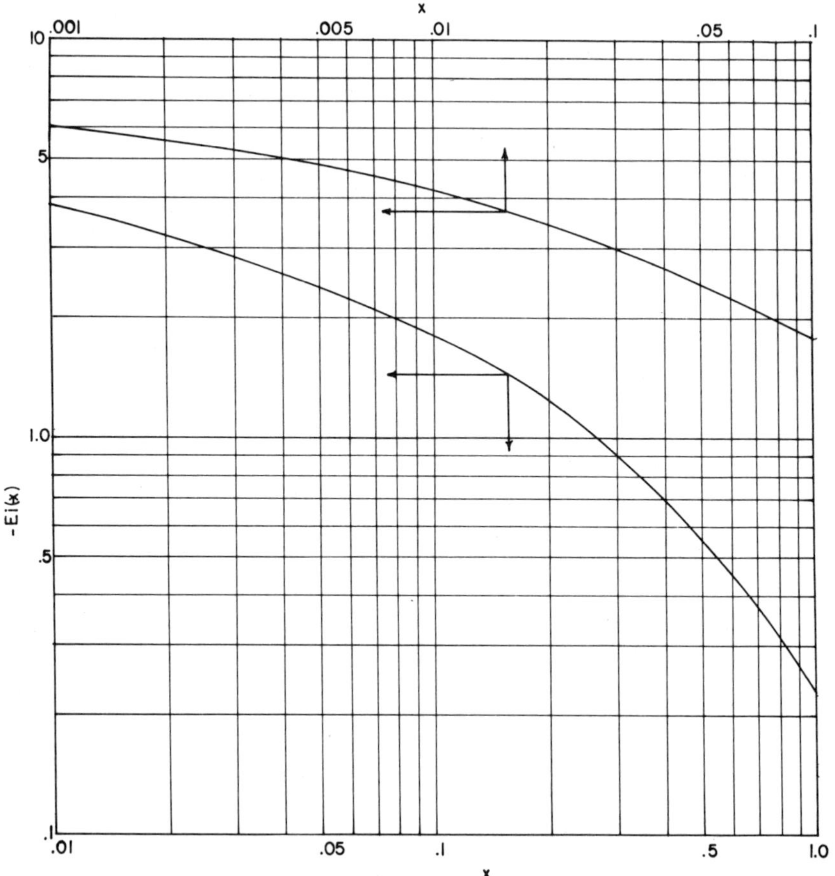

Figure 9-5. Graph of the function $f(x) = -Ei(-x)$.

Table 9-2
Table of Exponential Values

x	0	1	2	3	4	5	6	7	8	9
0.000	∞	6.332	5.639	5.235	4.948	4.726	4.545	4.392	4.259	4.142
0.010	4.038	3.944	3.858	3.779	3.705	3.637	3.574	3.514	3.458	3.405
0.020	3.355	3.307	3.261	3.218	3.176	3.137	3.098	3.062	3.026	2.992
0.030	2.959	2.927	2.897	2.867	2.838	2.810	2.783	2.756	2.731	2.706
0.040	2.681	2.658	2.634	2.612	2.590	2.568	2.547	2.527	2.507	2.487
0.050	2.468	2.449	2.431	2.413	2.395	2.378	2.360	2.344	2.327	2.311
0.060	2.295	2.280	2.265	2.249	2.235	2.220	2.206	2.192	2.178	2.164
0.070	2.151	2.138	2.125	2.112	2.099	2.087	2.074	2.062	2.050	2.039
0.080	2.027	2.016	2.004	1.993	1.982	1.971	1.960	1.950	1.939	1.929
0.090	1.919	1.909	1.899	1.889	1.879	1.870	1.860	1.851	1.841	1.832
0.100	1.823	1.814	1.805	1.796	1.788	1.779	1.770	1.762	1.754	1.745
0.110	1.737	1.729	1.721	1.713	1.705	1.697	1.690	1.682	1.675	1.667
0.120	1.660	1.652	1.645	1.638	1.631	1.623	1.616	1.609	1.603	1.596
0.130	1.589	1.582	1.576	1.569	1.562	1.556	1.549	1.543	1.537	1.530
0.140	1.524	1.518	1.512	1 506	1.500	1.494	1.488	1.482	1.476	1.470
0.150	1.465	1.459	1.453	1.448	1.442	1.436	1.431	1.425	1.420	1.415
0.160	1.409	1.404	1.399	1.393	1.388	1.383	1.378	1.373	1.368	1.363
0.170	1.358	1.353	1.348	1.343	1.338	1.333	1.329	1.324	1.319	1.315
0.180	1.310	1.305	1.301	1.296	1.292	1.287	1.283	1.278	1.274	1.269
0.190	1.265	1.261	1.256	1.252	1.248	1.244	1.239	1.235	1.231	1.227
0.200	1.223	1.219	1.215	1.211	1.207	1.203	1.199	1.195	1.191	1.187

x	0	1	2	3	4	5	6	7	8	9
0.00	∞	4.0380	3.3548	2.9592	2.6813	2.4680	2.2954	2.1509	2.0270	1.9188
0.10	1.8230	1.7372	1.6596	1.5890	1.5242	1.4645	1.4092	1.3578	1.3099	1.2649
0.20	1.2227	1.1830	1.1454	1.1099	1.0763	1.0443	1.0139	0.9850	0.9574	0.9310
0.30	0.9057	0.8816	0.8584	0.8362	0.8148	0.7943	0.7745	0.7555	0.7372	0.7195
0.40	0.7024	0.6860	0.6701	0.6547	0.6398	0.6354	0.6114	0.5979	0.5848	0.5721
0.50	0.5598	0.5479	0.5363	0.5350	0.5141	0.5034	0.4931	0.4830	0.4732	0.5721
0.60	0.4544	0.4454	0.4366	0.4281	0.4197	0.4116	0.4036	0.3959	0.3884	0.3810
0.70	0.3738	0.3668	0.3600	0.3533	0.3468	0.3404	0.3342	0.3281	0.3221	0.3163
0.80	0.3107	0.3051	0.2997	0.2944	0.2892	0.2841	0.2791	0.2742	0.2695	0.2648
0.90	0.2602	0.2558	0.2514	0.2471	0.2429	0.2388	0.2348	0.2308	0.2270	0.2232
1.00	0.2194	0.2158	0.2122	0.2087	0.2053	0.2019	0.1986	0.1954	0.1922	0.1891
1.10	0.1861	0.1831	0.1801	0.1772	0.1744	0.1716	0.1689	0.1662	0.1636	0.1610
1.20	0.1585	0.1560	0.1536	0.1512	0.1488	0.1465	0.1442	0.1420	0.1398	0.1377
1.30	0.1355	0.1335	0.1314	0.1294	0.1274	0.1255	0.1236	0.1217	0.1199	0.1181
1.40	0.1163	0.1146	0.1129	0.1112	0.1095	0.1079	0.1063	0.1047	0.1032	0.1016
1.50	0.1002	0.0987	0.0972	0.0958	0.0944	0.0930	0.0917	0.0904	0.0890	0.0878
1.60	0.0865	0.0852	0.0840	0.0828	0.0816	0.0805	0.0793	0.0782	0.0771	0.0760
1.70	0.0749	0.0738	0.0728	0.0718	0.0708	0.0698	0.0679	0.0669	0.0669	0.0660
1.80	0.0651	0.0642	0.0633	0.0624	0.0616	0.0607	0.0599	0.0591	0.0583	0.0575
1.90	0.0567	0.0559	0.0552	0.0545	0.0537	0.0530	0.0523	0.0516	0.0509	0.0503
2.00	0.0496	0.0490	0.0483	0.0477	0.0471	0.0465	0.0459	0.0453	0.0448	0.0442

The exponential integral can be solved by first expanding the exponential term into an infinite series as shown in Equation **9-36.**

$$Ei\,(-x) = -\int_x^\infty e^{-x}\frac{dx}{x} = -\int_x^\infty \frac{1}{x}\left(1 - x + \frac{x^2}{2!} - \frac{x^3}{3!} + \cdots\right) dx$$

$$= -\int_x^\infty \left(\frac{1}{x} - 1 + \frac{x}{2!} - \frac{x^2}{3!} + \cdots\right) dx$$

$$= \ln x - x + \frac{x^2}{2 \times 2!} - \frac{x^3}{3 \times 3!} + \cdots \Bigg|_x^\infty \qquad \textbf{9-36}$$

and since

$$\lim_{x \to \infty} \left[\ln x - x + \frac{x^2}{2 \times 2!} - \frac{x^3}{3 \times 3!} + \cdots\right] = 0.57722$$

Equation **9-36** can be written as

$$Ei\,(-x) = \ln x - x + \cdots + 0.57722 \qquad \textbf{9-37}$$

For cases when r approaches r_w and time t is appreciable, the expression $e^{-r^2/4Kt}$ approaches 1.00 and Equation **9-37** reduces to

$$Ei\,(-x) = \ln x + 0.57722 \qquad \textbf{9-38}$$

For computing pressures at the bore hole such as draw down pressures or build-up pressures, Equation **9-38** may be used. However, if practical units are used and logarithms to the base 10 are used, constants for Equation **9-38** must be evaluated. Darcy units apply to Equation **9-38.** Table 9-3 lists Darcy units and practical units.

Table 9-3
Darcy and Practical Units for Parameters in the Exponential Solution of the Diffusivity Equation

Parameter or Variable	Darcy Units	Practical Units
c	vol/vol/atm	vol/vol/psi
ϕ	porosity	porosity
h	cm	ft
k	Darcy	milli Darcies
μ	centipoise	centipoise

In the appropriate symbols Equation **9-38,** the exponential integral solution of the diffusivity equation can be written as

$$p = p_n - \frac{q\mu}{4\pi kh}\left(\ln \frac{kt}{\phi c\mu r^2} + \ln \frac{4}{\gamma}\right) \tag{9-39}$$

where γ is determined as

$$0.57722 = \ln e^{0.57722} = \ln 1.78$$

or

$$\gamma = 1.78$$

And the substitution of the value for γ into Equation **9-39** gives

$$p = p_n - \frac{q\mu}{4\pi kh}\left(\ln \frac{kt}{\phi\mu c r^2} + 0.8907\right) \tag{9-40}$$

The dimensionless time $\bar{t} = kt/\phi\mu c r^2$ is converted to practical units as

$$\bar{t} = (kt/\phi\mu c r^2)_{Darcy} = \frac{3600}{1000 \times (30.5)^2 \times 14.7}(kt/\phi\mu c r^2)_{Practical}$$

or

$$\bar{t} = 0.000264\,(kt/\phi\mu c r^2)_{Practical} \tag{9-41}$$

The quantity $q\mu/4\pi kh$ is converted to practical units as

$$(q\mu/4\pi kh)_{Darcy} = \frac{(30.5)^2 \times 5.62 \times 14.7 \times 10^3}{4\pi \times 3600 \times 24 \times 30.5}(q\mu B/kh)_{Practical}$$

$$= 70.6\,(q\mu B/kh)_{Practical} \tag{9-42}$$

The substitution of Equations **9-41** and **9-42** into Equation **9-39** and simplifying gives

$$p = p_n - 70.6\,\frac{q\mu B}{kh}\left(\ln \frac{0.000264\,kt}{\phi\mu c r^2} + \ln \frac{4}{1.78}\right) \tag{9-43}$$

or in terms of logarithms to the base ten Equation **9-43** becomes

$$p = p_n - \frac{162.6\, q\mu B}{kh}\left(\log \frac{kt}{\phi\mu c r^2} - 3.23\right) \qquad \textbf{9-44}$$

The use of Equation **9-44** for calculating the pressure at the well bore is illustrated in the following example.

Example 9-5.

Compute values for a draw down pressure curve for a single well in the center of a circular reservoir of infinite drainage radius of uniform thickness. Assume the following data:

$q = 50$ bbls/day
$t = 0.01$, 0.1, 1, 5, 12, 36, and 100 hours
$c = 5.0 \times 10^{-5}$
$\mu = 1.50$ centipoise
$r_w = 3$ inches
$\phi = 0.20$
$k = 300$ milli Darcies
$h = 10$ feet
$B = 1.31$
$p_n = 300$ psia

Plot the computed drawn down pressures.

Solution.

$$\frac{162.6\, q\mu B}{kh} = \frac{162.6 \times 50 \times 1.50 \times 1.31}{10 \times 300} = 5.32$$

For $t = 0.01$ hour
$p = 300 - 5.32\ (\log 3.2 \times 10^6 - 3.23) = 282$ psia
For $t = 0.1$ hour
$p = 300 - 5.32\ (\log 3.2 \times 10^7 - 3.23) = 277$ psia
For $t = 1.0$ hour
$p = 300 - 5.32\ (\log 3.2 \times 10^8 - 3.23)$
$= 300 - 5.32\ (8.5051 - 3.23) = 272$ psia
For $t = 5.0$ hours
$p = 300 - 5.32\ (\log 5.0 \times 3.2 \times 10^8) = 268$ psia
For $t = 12$ hours
$p = 300 - 5.32\ (\log 12 \times 3.2 \times 10^8 - 3.23) = 266$ psia
For $t = 36$ hours
$p = 300 - 5.32\ (\log 36 \times 3.2 \times 10^8 - 3.23) = 264$ psia

For $t = 100$ hours

$$p = 300 - 5.32\ (\log 3.2 \times 10^{10} - 3.23)\ 261 \text{ psia}$$

As expected, the pressure at the well bore versus logarithm of time is a straight line; this is indicated in Fig. 9-6.

Applying Equation 9-35 for constant flow rate, Equation 9-28 can be written in practical units in the form of Equation 9-45.

$$p = p_n + \frac{70.6\ \mu Bq}{kh}\left[-\ Ei\left(-\ \frac{948.4\ c\mu\phi r^2}{kt}\right)\right] \qquad 9\text{-}45$$

because

$$\left(\frac{c\mu\phi r^2}{4\ kt}\right)_{Darcy} = \frac{1000 \times (30.48)^2 \times 14.7}{3600 \times 4}\left(\frac{c\mu\phi r^2}{kt}\right)_{Practical}$$

$$= 948.4\left(\frac{c\mu\phi r^2}{kt}\right)_{Practical}$$

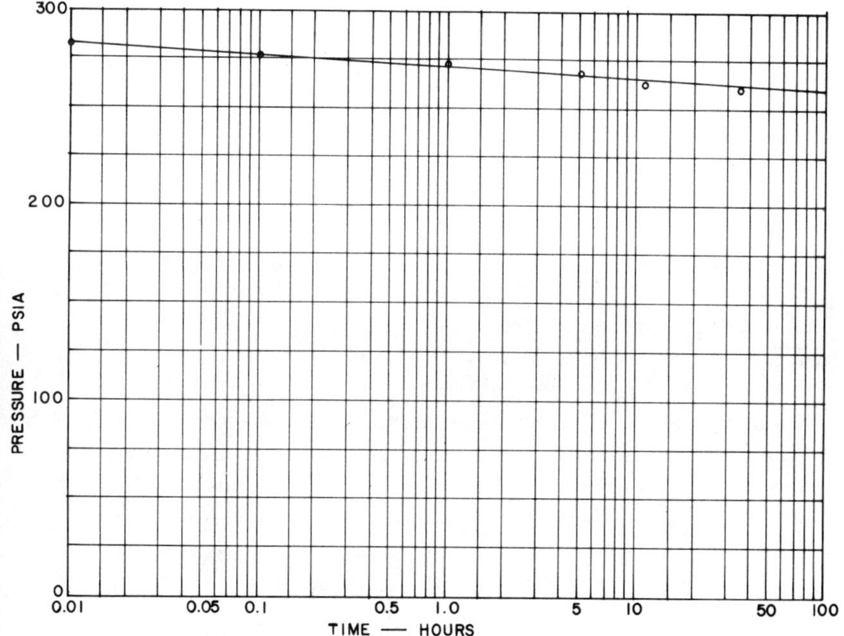

Figure 9-6. Drawn-down pressure curve for a circular reservoir.

For cases where r is large, and t is small relative to r, Equation **9-45** can be used to compute the pressure at large radial distances from the well bore. The use of Equation **9-45** for computing pressures at large distances from the well bore is illustrated in the example which follows.

Example 9-6. Compute the pressure at 500 feet, 100 hours after production began for a single well in the center of a circular reservoir of infinite drainage radius and uniform thickness. Assume the data:

$q = 100$ bbls/day
$\mu = 3.0$ centipoise
$\phi = 0.22$
$k = 100$ milli darcies
$h = 10$ feet
$B = 1.31$
$p_n = 300$ psia
$c = 5.00 \times 10^{-5}$ psi^{-1}

Solution.

$$p = 300 + \frac{70.6 \times 3.0 \times 1.31 \times 100}{10.0 \times 100}$$

$$\left[- Ei \left(- \frac{948.4 \times 5.0 \times 10^{-5} \times 3.0 \times 0.22 \times (500)^2}{100 \times 100} \right) \right]$$

$$= 300 - 2.74 \; [Ei \; (-0.784) \;]$$

And from Table 9-2, $Ei(-0.78) = 0.322$ and $Ei(-0.79) = 0.316$; therefore $Ei(-784) = 0.322 \; -0.4(0.322 - 0.316) = 0.3196$. Now $p = 300 - 2.74 \times 0.3196 = 299$ psia when rounded off.

Build-up pressures for a single well in a homogeneous reservoir with an infinite radius and with constant thickness can be computed with the aid of Equation **9-44** and the principle of superposition. In the development of the equation, the single well is treated both as a producing well and as an injection well with both injection and producing rates the same. It is treated as an injection well for only the time of the shut-in. If t is the production time and $\triangle$ is the shut-in time for a well, Equation **9-44** is altered to Equation **9-46** for computing build-up pressures.

$$p = p_n - \frac{162.6 \, q\mu B}{kh} \left[\left(\log \frac{k \, (t + \Delta)}{\phi \mu c r_w{}^2} - 3.23 \right) - \left(\log \frac{k\Delta}{\phi u c r_w^2} - 3.23 \right) \right]$$

or

$$p = p_n - \frac{162.6\, q\mu B}{kh} \log \frac{t + \Delta}{\Delta} \qquad \textbf{9-46}$$

The permeability of a formation can be determined with the use of Equation **9-46** by plotting build-up pressure versus log $(t + \Delta)/\Delta$ and determining the slope of the curve. The slope of the straight line is equal to $(162.6 q\mu B)/kh$.

Example 9-7. Compute the pressure at the well bore for the well of Example 9-5 by assuming that the well produced for 100 hours and was shut-in for 25 hours.

Solution. From computed data of Example 9-5

$$\begin{aligned} p &= 300 - 5.32(\log 125/25) \\ &= 300 - 5.32 \log 5 \\ &= 300 - 5.32 \times 0.699 = 300 - 3.72 \end{aligned}$$

or

$$p = 296 \text{ psia approximately.}$$

The rate of production of a well varies. However, a variable production rate can be approximated by a series of constant production rates as is indicated by the dotted stairsteps in Fig. 9-7. Applying the principle of

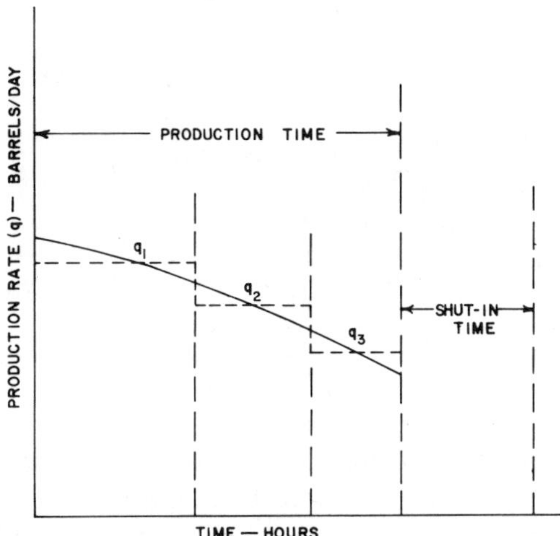

Figure 9-7. Variable flow rate represented by a series of constant flow rates.

superposition a build-up pressure equation for a variable flow rate can be developed with the use of Equation **9-46** In the development of the equation, a series of terms are written for the constant flow rates. Each term is for a given flow rate and it is treated as a production well and as an injection well. For instance, the term for production rate (q_1) has a time for production of $(t_3 + \triangle)$; and a time for injection of $(t_3 - t_1 + \triangle)$; and the time of production for rate q_2 is $(t_3 - t_i + \triangle)$ and its time of injection is $(t_3 - t_2 + \triangle)$. In a similar manner, a term for q_3 can be written. The equation for the build-up pressure curve for three constant flow rates is

$$p = p_n - \frac{162.6\,\mu B}{kh}\left[q_1 \log \frac{(t_3 + \Delta)}{(t_3 - t_1 + \Delta)} + q_2 \log \frac{(t_3 - t_1 + \Delta)}{(t_3 - t_2 + \Delta)} + q_3 \log \frac{(t_3 - t_2 + \Delta)}{\Delta}\right] \quad \textbf{9-47}$$

And the equation for pressure at the well bore at (t_3) for the variable rate of production is

$$p = p_n - \frac{162.6\,\mu B}{kh}\left[q_1 \log \frac{t_3}{t_3 - t_1} + q_2 \log \frac{t_3 - t_1}{t_3 - t_2} + q_3 (\log (t_3 - t_2) - 3.23)\right] \quad \textbf{9-48}$$

The exponential integral form of Equation **9-48** is

$$p = p_n - \frac{70.6\,\mu B}{kh}\left\{q_1\, Ei\left(-\frac{948.4\, c\mu\phi r^2}{kt_3}\right) - q_1\left[Ei\left(-\frac{948.4\, c\mu\phi r^2}{k\,(t_3 - t_1)}\right)\right] + q_2\left[Ei\left(-\frac{948.4\, c\mu\phi r^2}{k\,(t_3 - t_1)}\right)\right] - q_2\left[Ei\left(-\frac{948.4\, c\mu\phi r^2}{k\,(t_3 - t_2)}\right)\right] + q_3\left[Ei\left(-\frac{948.4\, c\mu\phi r^2}{k\,(t_3 - t_2)}\right)\right]\right\} \quad \textbf{9-49}$$

For large radial distances from the well bore Equation **9-49** is applicable for computing pressures; but Equation **9-48** is preferred for computing pressures at the well bore. Equations containing a series of terms with various constant rates of production are rather cumbersome to use. Experience has indicated that acceptable build-up pressures can be obtained with Equation **9-46** when q is taken as the ratio of total production divided by total time of production.

Exact solutions to the radial diffusivity equation are available for the case of a single well in the center of a finite circular reservoir. These solutions are complicated to have other than theoretical interest. An approximate solution has been derived for a finite reservoir by modifying Equation **9-28.** In the modification, it is assumed that the pressure drop of a well in a finite reservoir is greater than the pressure drop of a well in an infinite reservoir. This is caused by the fluid which has flowed across a circle of radius (r_n) equal to that of the finite reservoir.

For a finite reservoir, Equation **9-28** becomes

$$p = p_n + \frac{q\mu B}{4\pi kh}\left[Ei\left(-\frac{\phi\mu c r^2}{4kt}\right) - y\left(\frac{\phi\mu c r_n^2}{4kt}\right)\right] \quad \textbf{9-50}$$

Where the y-function is defined as $y(x) = Ei\ (-x) + \frac{1}{x} e^{-x}$

In practical units Equation **9-50** becomes

$$p = p_n + \frac{70.6 q\mu B}{kh}\left[Ei\left(-\frac{948.4\phi\mu c r^2}{kt}\right) - y\left(\frac{948.4\phi\mu c r_n^2}{kt}\right)\right] \quad \textbf{9-51}$$

and the build-up pressure equation **9-46** becomes as follows for a finite reservoir.

$$= p_n - \frac{162.6 q\mu B}{kh}\left[\log\frac{t+\Delta}{\Delta} + 0.434\, y\left(\frac{948.4\phi\mu c r_n^2}{k(t+\Delta)}\right) - 0.434\, y\left(\frac{948.4\phi\mu c r_n^2}{k\Delta}\right)\right] \quad \textbf{9-52}$$

A graph of the y-functions is given in Fig. 9-8. The following example illustrates the use of Equation **9-52** for computing build-up pressures.

Example 9-8. Repeat Example 9-7 for a finite reservoir with a radius of 10,000 feet and a radius of 1000 feet.

Solution.

$$\frac{162.6 q\mu B}{kh} = \frac{162.6 \times 50 \times 1.50 \times 1.31}{300 \times 10} = 5.32$$

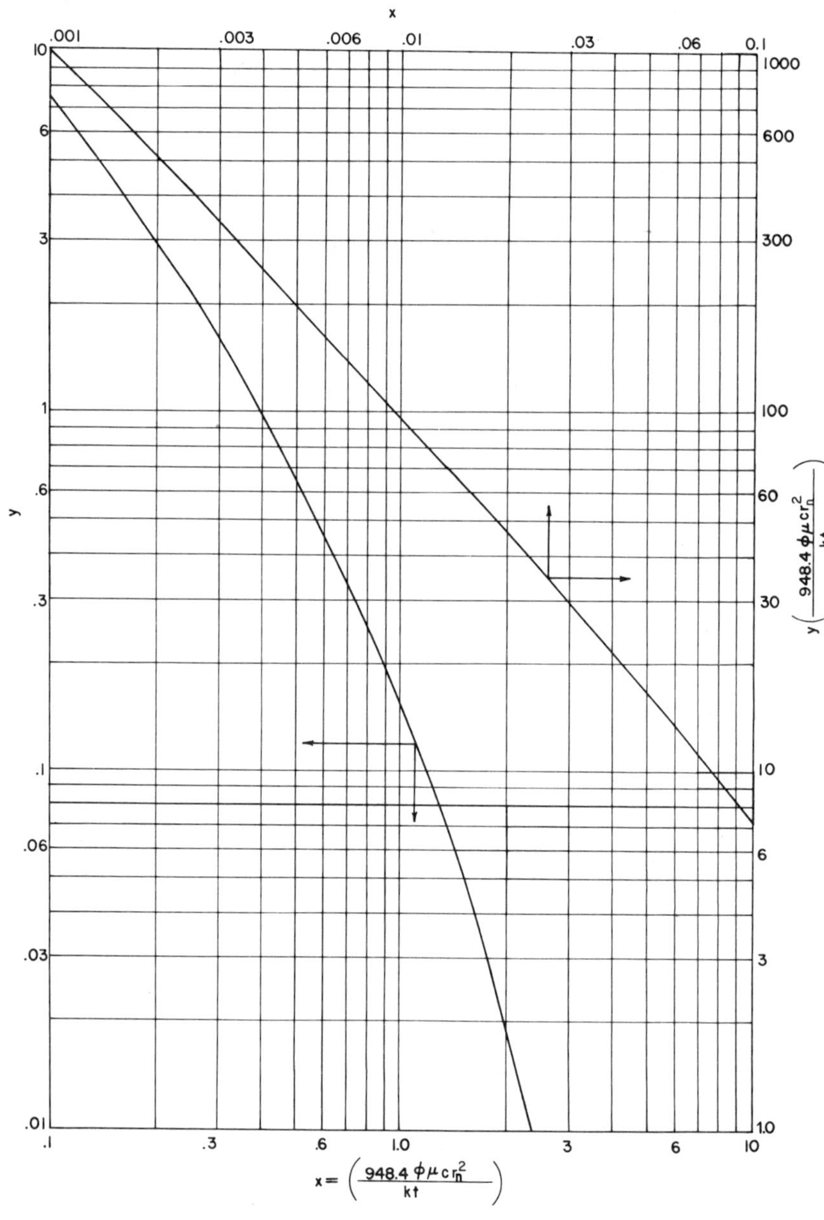

Figure 9-8. A graph of y-factors.

For $r_n = 10{,}000$ feet,

$$\frac{948.4\phi\mu c r_n^2}{k(t+\Delta)} = \frac{948.4 \times 0.2 \times 1.50 \times 5 \times 10^{-5} \times 10^8}{300 \times 125} = 38$$

$$\frac{948.4\phi\mu c r_n^2}{k(\Delta)} = \frac{948.4 \times 0.2 \times 1.50 \times 5 \times 10^{-5} \times 10^8}{300 \times 25} = 190$$

From Figure 9-8, it is noted that the values of the y-function are negligible. Therefore, the infinite radius equation holds for the large radius of 10,000 feet. This value was given in Example 9-7. For $r_n = 1000$ feet

$$\frac{948.4\phi\mu c r_n^2}{k(t+\Delta)} = \frac{948.4 \times 0.2 \times 1.50 \times 5 \times 10^{-5} \times 10^6}{300 \times 125} = 0.38$$

$$\frac{948.4\phi\mu c r_n^2}{k(\Delta)} = \frac{948.4 \times 0.2 \times 1.50 \times 5 \times 10^{-5} \times 10^6}{300 \times (25)} = 1.90$$

From Figure 9-8,

$$y(0.38) = 1.0 \text{ and } y(1.90) = 0.190$$

therefore

$$p = 300 - 5.32\ [0.699 + 0.434(1.00 - 0.190)]$$

or

$$p = 294 \text{ psia when rounded off to three significant figures}$$

Discussion has indicated that a straight line will be obtained when build-up pressure is plotted versus log $(t + \Delta)/\Delta$. In actual practice this is not always true because the permeability of the zone surrounding the well bore may be different from the permeability of the remaining portion of the reservoir. For small periods of pressure build-up time, the variation of the permeability of the zone surrounding the well bore will alter the shape of the build-up pressure curve, that is, it will cause a deviation from the straight line, as shown in Figure 9-9. After an appreciable build-up time, the pressure build-up curve becomes a straight line. Some of the factors which cause reductions of permeability in the zone surrounding the well bore are invasion by drilling fluids, presence of a mud cake and cement, dispersion of clays, limited perforations, and high gas saturation.

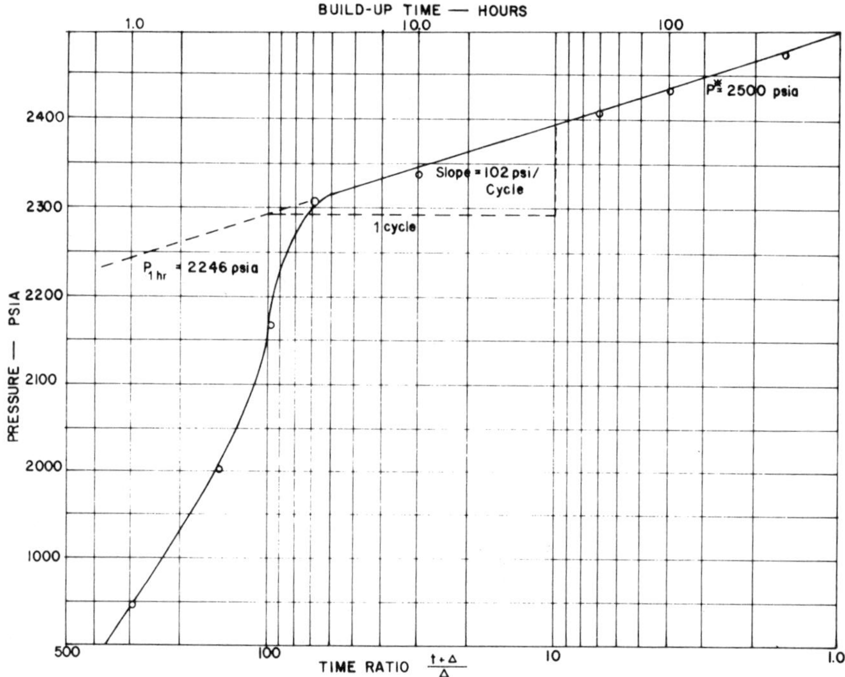

Figure 9-9. Unsteady-state build-up pressure curve.

For quantitatively evaluating and reporting the effects of the permeability changes near the well bore Van Everdingen,[3] has proposed the term "Skin Factor". The skin factor S relates the increase or decrease in the pressure in the skin surrounding the well bore to the dimensionless rate of flow as shown in Equation 9-53a.

$$\Delta p_{skin} = S(q\mu/2\pi\, kh) = p_w - p_s \qquad \textbf{9-53a}$$

A relationship between the radius r_s of the "skin" zone around the well bore, the permeability k_s of the zone and the skin factor has been derived by Hawkins[4]. In the derivation of the skin factor, steady-state fluid-flow was assumed in the skin zone around the well bore. For the derivation, the increased or decreased pressure drop across the skin is.

[3]Van Everdingen, A. F.: "The Skin Effect and Its Influence on the Productive Capacity of a Well", Trans. AIME (1953) 198, pp. 171-176.

[4]Hawkins, M. F.: "A Note on the Skin Effect", Trans. AIME (1956) 207, pp. 356-357.

$$\Delta p_s = \frac{q\mu B}{2\pi k_s h} \ln \frac{r_s}{r_w} - \frac{q\mu B}{2\pi k_e h} \ln \frac{r_s}{r_w}$$

$$= \frac{q\mu B}{2\pi k_e h} \left[\frac{k_e - k_s}{k_s} \ln \frac{r_s}{r_w} \right] \qquad \text{9-53b}$$

where k_s is the permeability of the skin zone and k_e is the permeability of the reservoir other than that of the skin zone. The terms in brackets of Equation **9-53** are a relation for skin factor—note Equation **9-54.**

$$S = \frac{k_e - k_s}{k_s} \ln \frac{r_s}{r_w} \qquad \text{9-54}$$

Applying Equation **9-40**, the total pressure drop in a flowing well is

$$\Delta p_t = (p_e - p_w) + \Delta p_s = \frac{q\mu B}{4\pi k_e h} \left[\ln \frac{4kt}{\phi \mu c r_w^2} + 2\left(\frac{k_e - k_s}{k_s}\right) \ln \frac{r_s}{r_w} + 0.8907 \right]$$

or

$$\Delta p_t = \frac{q\mu B}{4\pi k_e h} \left(\ln \frac{4kt}{\phi \mu c r_w^2} + 2S + 0.8907 \right) \qquad \text{9-55}$$

For short periods of shut-in time the term $(t + \Delta)/\Delta$ becomes t/Δ in the ideal build-up pressure equation as in Equation **9-56.**

$$p_{wf} = p_e - \frac{q\mu B}{4\pi k_e h} \left(\ln \frac{t}{\Delta} \right) \qquad \text{9-56}$$

Equation **9-57** combines Equation **9-55** and **9-56** and applies the principle of superposition to form an equation for the relationship between the build-up pressure for a shut-in time t/Δ and the pressure in the well bore at the time of shut-in.

$$p_\Delta - p_{wf} = (p_e - p_{wf}) - (p_e - p_\Delta)$$

$$= \frac{q\mu B}{4\,\pi\, k_e h} \left[\ln \left(\frac{4k_o t}{\phi \mu c r_w^2} \right) + 2S + 0.8907 \right]$$

$$- \left[\frac{q\mu B}{4\,\pi\, k_e h} \ln t - \ln \Delta \right]$$

or

$$p_\Delta - p_{wf} = \frac{q\mu B}{4\pi k_e h}\left[\ln\left(\frac{4 k_e \Delta}{\phi\mu c r_w^2}\right) + 2S + 0.8907\right] \qquad \textbf{9-57}$$

Following the same procedure for converting from Darcy units to practical units that was used for Equations **9-43** and **9-44,** Equation **9-57** becomes

$$p_\Delta - p_{wf} = \frac{162.6\, q\mu B}{k_e h}\left(\log\frac{k_e\Delta}{\phi\mu c r_w^2} - 3.23 + 0.868S\right) \qquad \textbf{9-58}$$

And writing $m = \dfrac{162.6\, q\mu B}{k_e h}$

$$p - p_{wf} = m\left(\log\frac{k_e\Delta}{\phi\mu c r_w^2} - 3.23 + 0.868S\right)$$

or

$$S = 1.151\left[\frac{p_\Delta - p_{wf}}{m} - \log\frac{k_e\Delta}{\phi\mu c r_w^2} + 3.23\right] \qquad \textbf{9-59}$$

For determining skin effects, Δt has been chosen as 1.0 hour[5]; therefore, Equation **9-59** may be written as

$$S = 1.151\left[\frac{p_{1\,hr} - p_{wf}}{m} - \log\frac{k_e}{\phi\mu c r_w^2} + 3.23\right] \qquad \textbf{9-60}$$

As an index for reporting the efficiency of a well, the term "flow efficiency" has been given the following expression[6]:

$$\text{flow efficiency} = \frac{p^* - p_{wf} - \Delta p_s}{p^* - p_{wf}} \qquad \textbf{9-61}$$

[5]Mid-Continent District Study Committee on Completion Practices, "Selection and Evaluation of Well Completion Methods," Drilling and Production Practice, API (1955), p. 421.

[6]Matthews, C. S., and Russell, D. G.: "Pressure Build-Up and Flow Tests in Wells", Monograph Vol. I AIME—Henery L. Daugherty Series (1967) p. 21, Society of Petroleum Engineers of AIME, Dallas, Texas.

where p^* is the extrapolated pressure at $(t + \Delta)/\Delta = 1$ and p_{wf} is the pressure just prior to shut-in. Equation **9-52** can be used for calculating p_s when it is corrected for practical units, as in Equation **9-62.**

$$p_s = S\,(q\mu/2\,\pi\,k_e h)_{Darcy} = S\,(141.2\,q\mu B/k_e h)_{Practical}$$

$$= S\,(m)\frac{2}{2.303} = 0.87m \qquad \textbf{9-62}$$

A positive skin factor indicates a damaged zone around the well bore, whereas a negative skin factor indicates an increase in permeability around the well bore. Example 9-9 illustrates the use of skin factor equations.

Example 9-9. For the following build-up data

Pressure, psia	Build-Up Time Hours (Δ)
1845	1
2000	2
2170	3
2310	5
2341	10
2410	50
2437	100
2487	500

$P_{wf} = 1730$ psia
$h = 20$ ft
$\mu = 10$ c.p.
$\phi = 0.2$
$r_w = 0.25$ ft
$B = 1.2$
$k_e = 122$ mds
$t = 300$ hrs.

1. Plot build-up pressure versus log $(t + \Delta)/\Delta$
2. Determine the slope of the straight line portion of the curve
3. From the slope, compute the effective flow rate of the well
4. Calculate the skin factor
5. Find p^*
6. Calculate the flow efficiency of the well.

Solution.

1. The complete data for plotting the graph is given in the Table 9-3.

Table 9-3
Build-Up Pressure Data

Pressure psia	Build-Up Time hours	t + Δ hours	$\frac{t+\Delta}{\Delta}$
1845	1	301	301
2000	2	302	151
2170	3	303	101
2310	5	305	61
2341	10	310	31
2410	50	350	7
2437	100	400	4
2480	500	800	1.6

See Figure 9-9 for the plot of the data from the table.

2. Values taken from the graph indicate the slope is $m = (2394 - 2292) = 102$ psi/cycle

3. The flow rate is computed from the slope formula as

$$m = \frac{162.6\, q\mu B}{kh}$$

or

$$q = \frac{mkh}{162.6\,\mu B} = \frac{102 \times 122 \times 20}{162.6 \times 10 \times 1.2} = 128 \text{ bbls/day}$$

4. The skin factor is computed with the aid of Equation **9-60.** From the graph $p_{1\,hr}$ was found to be 2246 psia. Therefore the skin factor is

$$S = 1.151\left[\frac{2246 - 1730}{102} - \log\frac{122}{0.2 \times 10 \times 1.5 \times 10^{-5} \times (0.25)^2} + 3.23\right]$$

$$= 1.151\,(8.29 - 7.83) = 0.553$$

5. The effective pressure of the reservoir at infinite build-up time is taken from the graph by extrapolating the curve until $(t + \triangle)/\triangle = 1$. At this point the pressure is noted to be approximately 2500 psia or $p^* = 2500$ psia.

6. The flow efficiency is computed from Equation **9-61** with the aid of Equation **9-62.**

$$\triangle p_s = 0.87\, m = 0.87 \times 102 \text{ psi} = 89 \text{ psi}$$

Substituting this incremental value into Equation **9-61** gives the flow efficiency as

$$\frac{p^* - p_{wf} - \Delta p_s}{p^* - p_{wf}} = \frac{2500 - (1730 + 89)}{2500 - 1730} = \frac{681}{770} = 0.886$$

Two-Well Solution—Unsteady-State

The principle of superposition can be applied to solutions of the diffusivity equation because of the linearity of the diffusivity equation. This means that for a number of wells the effects of each separate well or sink in a system can be added algebraically. The first example of applying the principle of superposition was in the derivation of the build-up pressure Equation **9-46**, where the single well was treated as a production well and as an injection well. In this section unsteady-state fluid-flow equations are developed for two producing wells in a homogeneous reservoir with an infinite drainage radius. In Figure 9-10, the boundary conditions are:

1. $p_e - p = 0$ at $r_1 = \infty, r_2 = \infty$
2. $p_e - p = 0$ at $t = 0$, for all values of r_1 and r_2
3. Flow into well No. 1 $= q_1$
4. Flow into wall No. 2 $= q_2$

Applying the principle of superposition to Equation **9-28**, an equation relating the pressure at point p and time t may be written as in Equation **9-63.**

$$p = p_e - \frac{\mu}{4\,\pi\,kh}\left[\int_0^t q_1 e^{-r_1/4Kt}\frac{dt}{t} + \int_0^t q_2 e^{-r_2/4Kt}\frac{dt}{t}\right] \qquad \text{9-63}$$

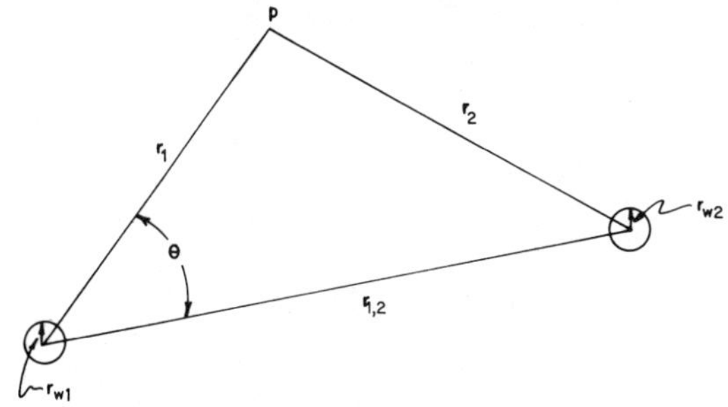

Figure 9-10. Infinite reservoir containing two wells.

The proof that Equation **9-63** satisfies Equation **9-26**, the original diffusivity equation, can be obtained by the appropriate differentiations and the substitution into Equation **9-26** as carried out for the single well solution. See Equations **9-28** through **9-32**. Boundary conditions 1 and 2 can be proved in a similar manner as was carried out for the corresponding boundary conditions of the single-well solution. See Equations **8-27a** and **8-27b.**

To prove the third boundary condition for Equation **9-63** it must be shown that the flow into well No. 1 is

$$\text{Flow into well No. 1} = k/\mu \int_0^{2\pi} r_1 h\,(\partial p/\partial r_1)d\theta = q_1 \qquad \textbf{9-64}$$

By a process of reasoning similar to that used for the third boundary condition, Equation **9-27c**, of the single-well solution for the diffusivity equation, it can be shown that when the first integral term of Equation **9-63** is differentiated with respect to r_1 and substituted into Equation **9-64** this equation reduces as r_1 approaches r_{w1} to flow in well No. 1 $= q_1$. Now it remains to show that the second integral term of Equation **9-63**, when differentiated with respect to r_1 vanishes. First in the proof, it must be shown that r_2 is a function of r_1. This is evident because by the law of cosines $r_2^2 = r_1^2 + r_{1,2}^2 + 2r_1r_{1,2}\cos\theta$. The substitution of the cosine relation into the second integral term of Equation **9-63** gives

$$\frac{\mu}{4\pi kh}\int_0^t q_2 e^{-[r_1^2 + r_{1,2}^2 - 2r_1r_{1,2}\cos\theta]/4Kt}\,\frac{dt}{t} \qquad \textbf{9-65}$$

The differentiation with respect to r_1 of Equation **9-65** is

$$\frac{\mu q_2}{4\pi kh}\lim_{r_1\to r_{w1}}\int_0^t e^{-[r_{w1}^2 + r_{1,2}^2 - 2r_{w1}r_{1,2}\cos\theta]/4Kt}\left(\frac{2r_{w1} - 2r_{1,2}\cos\theta}{4Kt^2}\right)dt \qquad \textbf{9-66}$$

And the substitution of **9-66** into Equation **9-64** gives

$$\frac{1}{4\pi}\int_0^{2\pi}\int_0^t q_2 e^{-[r_{1,2}(r_{w1}^2/r_{1,2} + r_{1,2} - 2r_{w1}\cos\theta)]/4Kt}$$

$$2r_{w1}\left(\frac{r_{w1} - r_{1,2}\cos\theta}{4Kt^2}\right)dt\,d\theta \qquad \textbf{9-67}$$

For most well spacings, $r_{1,2}$ is very large as compared to r_w and $r_{1,2} - r_{w1}$ approaches $r_{1,2}$; therefore Equation **9-67** vanishes and the third boundary condition is satisfied. In a similar manner, boundary condition No. 4 can be proved to be satisfied.

Applying the principle of superposition an equation for determining the build-up pressure in a two-well homogeneous reservoir with an infinite drainage radius can be written in practical units for the times and rates. Well No. 1 at rate q_1 for $t + \triangle$ hours, but well No. 2 produced at a rate of q_2 barrels for t hours and was shut in for $\triangle$ hours. Using Equation **9-46** for the build-up pressure in well No. 2 and Equation **9-45** for the draw down pressure in well No. 1, the two-well build-up equation is

$$p_{w2} = p_n + \frac{70.6\,\mu B q_1}{kh}\,Ei\left(-\frac{948.4\,c\mu\phi r_{1,2}^{\,2}}{k\,(t+\Delta)}\right) - \left(\frac{162.6\,q_2\mu B}{kh}\right)\log\frac{t+\Delta}{\Delta} \qquad \textbf{9-68}$$

Example 9-10 illustrates the use of an image well for locating the distance from a well to a fault.

Example 9-10. For a two-well system in a homogeneous reservoir with infinite radius of drainage away from the fault as is indicated in Fig. 9-11, determine the distance of Well No. 1 from the fault. Consider the conditions that both wells are closed in until the pressure in the reservoir is stabilized at a uniform pressure throughout; then well No. 1 is permitted to produce for 100 hours, at which time the pressure in well No. 2 is measured and found to be 987 psia.

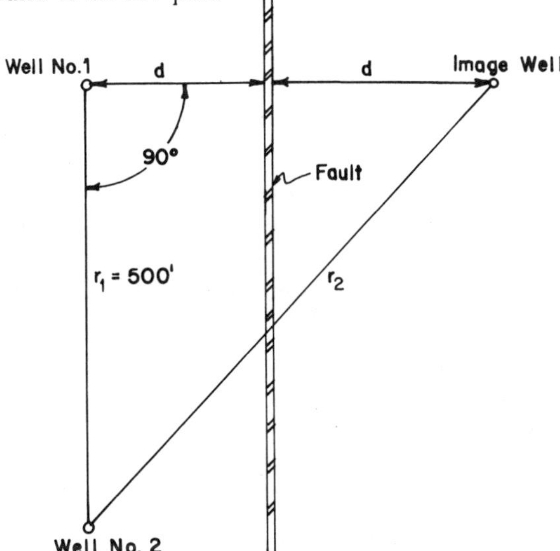

Figure 9-11. Semi-infinite reservoir containing two wells and bounded by a fault.

Additional data are:

$q_1 = 200$ bbls/day
$\mu = 3.00$ c.p.
$k = 200$ milli-darcies
$h = 30$ ft.
$c = 1.5 \times 10^{-5}$ psi^{-1}
$\phi = 0.15$
$p_n = 1017$ psia
$B = 1.30$

Solution. Utilizing Equation **9-45,** the two-well equation (Well No. 1 and image well) becomes

$$p = p_n + \frac{70.6\ \mu B q_1}{kh}\left[Ei\left(-\frac{948.4\ c\mu\phi r_1^2}{kt}\right) + Ei\left(-\frac{948.4\ c\mu\phi r_2^2}{kt}\right)\right] \quad \textbf{9-69}$$

The solution is obtained by assuming values of r_2 and computing the pressure for p_2 until a value of 987 psia is obtained. As a first trial, r_2 is assumed to be 1000 ft.

At $r_2 = 1000$ ft.

$$\frac{70.6\ \mu B q_1}{kh} = \frac{70.6 \times 3.00 \times 1.3 \times 200}{200 \times 30} = 9.18$$

$$\frac{948.4\ c\mu r\phi_1^2}{kt} = \frac{948.4 \times 1.5 \times 10^{-5} \times 3.00 \times 0.15 \times (500)^2}{200 \times 100} = 0.08$$

$$\frac{948.4\ c\mu\phi r_2^2}{kt} = \frac{948.4 \times 1.5 \times 10^{-5} \times 3.00 \times 0.15 \times (1000)^2}{200 \times 100} = 0.32$$

From Figures 10-13 and 10-14, $Ei(-0.08) = 2.02$ and $Ei(-0.32) = 0.86$ or

$$p_2 = 1017 - 9.18(2.02 + 0.86) = 1017 - 26.4 = 991 \text{ psia}$$

Since p_2 is larger than the measured value, an assumption of a smaller value for r_2 is in order.
At $r_2 = 800$ ft

$$\frac{948.4\ c\mu\phi r_2^2}{kt} = \frac{948.4 \times 1.5 \times 10^{-5} \times 3.00 \times 0.15 \times (800)^2}{200 \times 100} = .0.21$$

From Figure 10-13, $Ei(-0.21) = 1.22$

$$p_2 = 1017 - 9.18\,(2.0 + 1.22) = 987 \text{ psia}$$

Therefore,

$$2d = \sqrt{800^2 - 500^2}$$

or

$$d = \frac{\sqrt{39 \times 10^4}}{2} = \frac{624}{2} = 317 \text{ ft.}$$

The fault is located 317 feet from Well No. 1. This does not necessarily mean that the fault lies exactly perpendicular to the line joining Well No. 1 and its image.

Multi-Well Solution—Unsteady State

By a process of reasoning similar to the one used for the two-well system, a multi-well integral solution for the diffusivity equation can be written for the boundary conditions:

1. $p = p_n$ at $r_1, r_2, r_3 \cdots r_n$ ------ ∞
2. $p = p_n$ at $t = 0$ for all values of $r_1, r_2, r_3 \cdots r_n$
3. Flow at wells 1, 2, 3 $\cdots n$ are respectively $q_1, q_2, q_3 \cdots q_n$

The multi-well system for $n = 5$ is represented as indicated in Figure 9-12, where r_1, r_2, r_3, r_4, r_5 are the distances from the respective wells to point P. Point P represents the place where it is desired to compute the pressure.

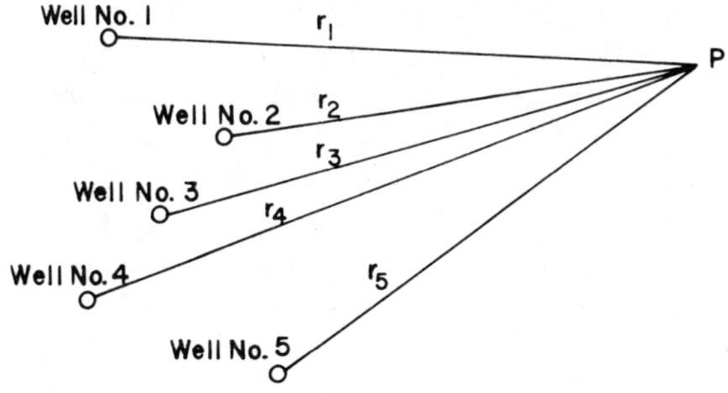

Figure 9-12. Infinite reservoir containing five wells.

Equation **9-70** gives the multi-well solution of the diffusivity equation for the given boundary conditions.

$$p = p_n + \frac{\mu}{4\,\pi\,kh}\left[\int_0^t q_1 e^{-r_1^2/4Kt}\frac{dt}{t} + \int_0^t q_2 e^{-r_2^2/4Kt}\frac{dt}{t}\right.$$

$$\left. + \int_0^t q_3 e^{-r_3^2/4Kt}\frac{dt}{t} + \cdots \int_0^t q_n e^{-r_n/4Kt}\frac{dt}{t}\right]$$

or

$$p = p_n + \frac{\mu}{4\,\pi\,kh}\sum_{i=1}^{n} q_i \int_0^t e^{-r_i^2/4Kt}\frac{dt}{t} \qquad \textbf{9-70}$$

The proof that this solution is unique follows the same procedure as was given for the two-well solution of the diffusivity equation.

Applying the principle of superposition with the aid of the build-up pressure equation **9-46** and the draw down pressure equation **9-45** for single wells, a build-up pressure equation for a multi-well system can be written as

$$p = p_n - \frac{162.6\,q_b\mu B}{kh}\log\frac{t+\Delta}{\Delta} + \frac{70.6\,\mu B}{kh}\sum_{i=1}^{n} q_i\,Ei\left(-\frac{948.4\,c\mu\phi r_i^2}{k\,(t+\Delta)}\right) \qquad \textbf{9-71}$$

where q_b is the flow rate of the well for which its build-up pressures are desired and q_i and r_i refer to the other wells which are not shut in. The previous equation refers to a well system of $n + 1$ wells in a homogeneous reservoir with infiinite drainage radius; intially all wells are shut in until the reservoir pressure reached equilibrium p_n throughout. Then n wells were allowed to produce at rates q_i for time $(t + \Delta)$; but well b was only allowed to produce at a rate of q_b for time t before it was shut in for time Δ. If the times of production varied for the n wells, the term $t_i + \Delta$ under the sigma sign and $t_b + \Delta$ is substituted for the well which was shut in.

Should the n+1 wells have been located somewhat near the center of a finite reservoir, an approximate build-up pressure equation could be obtained by adding the y-factor term. See Equation **8-52** to Equation **9-71.**

If the flow rates of the wells in a multi-well system vary, adjustments for these flow rates can be made by representing the variable flow rates as a series of constant flow rates. See Equations **8-48** and **8-49.** For n large and an appreciable variation in flow rates the multi-well equation becomes rather lengthy and a solution involving the use of the equation would require the aid of a high speed electronic digital computer.

The flow across a lease line can be computed by using Equation **9-70** for computing the pressures at incremental distances on both sides of the lease line, and then applying the linear Darcy equation for various segments along the lease line. The method for computing the flow across a lease line for a three-well system is illustrated in Example 9-10.

Example 9-10. Compute the flow gradient for the point on the lease line which is on a line joining Wells Nos. 2 and 3 as given in Figure 9-13, for the following data:

$q_1 = 100$ bbls/day
$q_2 = 250$ bbls/day
$q_3 = 200$ bbls/day
$t_1 = 300$ hrs
$t_2 = 200$ hrs
$t_3 = 250$ hrs
$p_n = 1525$ psia

$k = 200$ milli-Darcies
$\mu = 3.0$ centipoises
$h = 300$ ft
$B = 1.30$

$$\frac{70.6\,\mu B}{kh} = 4.8 \times 10^{-1} \quad \text{and} \quad \frac{948.4\,c\mu\phi}{k} = 3.3 \times 10^{-1}$$

Solution. The exponential solution of the diffusivity equation for the three-well system is

$$p = p_n + \frac{70.6\,\mu B}{kh}\left[q_1\,Ei\left(-\frac{948.4\,c\mu\phi r_1^2}{kt_1}\right) + q_2\,Ei\left(-\frac{948.4\,c\mu r_2^2}{kt_2}\right) + q_3\,Ei\left(-\frac{948.4\,c\mu r_3^2}{kt_3}\right)\right] \qquad \textbf{9-72}$$

Referring to Fig. 9-13, point No. 1 is 10 feet to the left of the lease line on a line joining Well No. 2 and Well No. 3; and point No. 2 is 10 feet to the right of the lease line on the line joining the two wells. The radial distances from point No. 1 to the respective wells are:

$$r_1 = \sqrt{(390)^2 + (800)^2} = 890 \text{ ft.}$$

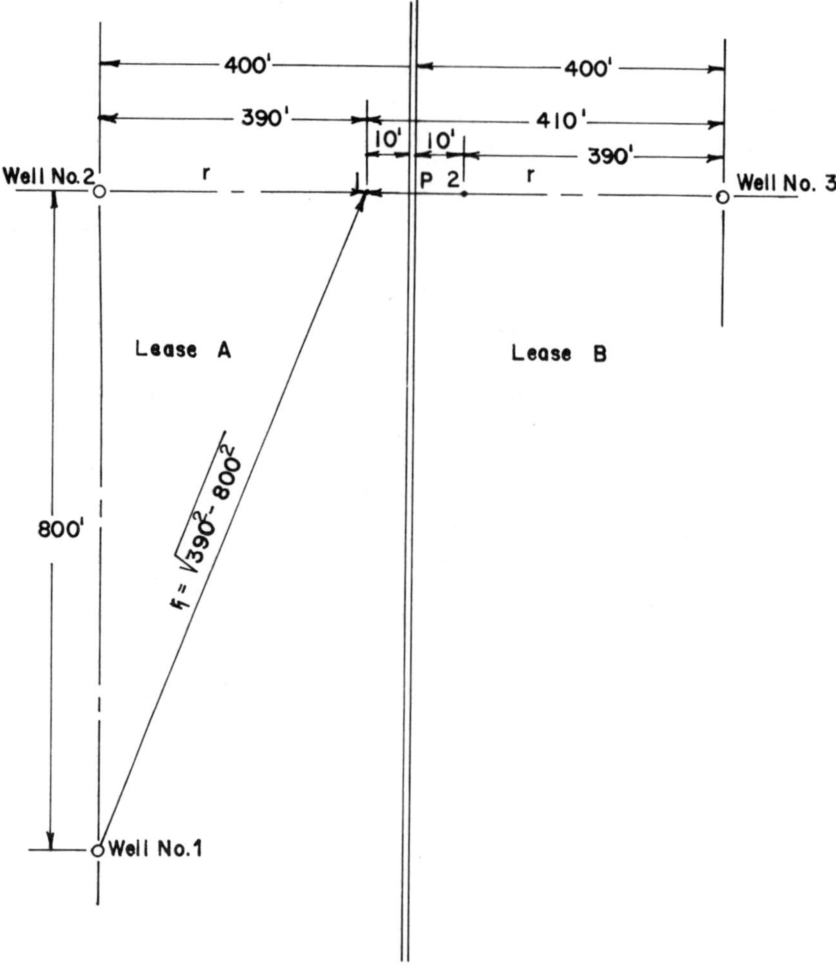

Figure 9-13. Infinite reservoir containing three wells and a lease line.

$r_2 = 390$ ft.
$r_3 = 410$ ft.

And the radial distances from point No. 2 to the respective wells are:

$r_1 = \sqrt{(410)^2 + (800)^2} = 899$ ft.
$r_2 = 410$ ft.
$r_3 = 390$ ft.

Substituting the appropriate values into Equation **9-72,** the pressure at point No. 1 is:

$$p = 1525 + 4.8 \times 10^{-1} \left[100\ Ei \left(-\frac{3.3 \times 10^{-5} \times 890^2}{300} \right) \right.$$
$$\left. + 250\ Ei \left(-\frac{3.3 \times 10^{-5} \times 390^2}{200} \right) + 200\ Ei \left(-\frac{3.3 \times 10^{-5} \times 410^2}{250} \right) \right]$$

$$p = 1525 + 4.8 \times 10^{-1} [100\ Ei\ (-\ 0.0871) + 250\ Ei\ (-\ 0.0251)$$
$$+ 200\ Ei\ (-\ 0.0222)]$$

And substituting in the values for the exponential integrals which were taken from Figure 10-13.

$$p = p_n - 4.8 \times 10^{-1} [100 \times 1.90 + 250 \times 3.20 + 200 \times 3.3]$$
$$= 1525 - 795 = 730 \text{ psia}$$

The pressure at point 2 is

$$p = 1525 + 4.8 \times 10^{-1} \left[100\ Ei \left(-\frac{3.3 \times 10^{-5} \times 899^2}{300} \right) \right.$$
$$\left. + 250\ Ei \left(-\frac{3.3 \times 10^{-5} \times 410^2}{200} \right) + 200\ Ei \left(-\frac{3.3 \times 10^{-5} \times 390^2}{250} \right) \right]$$

or

$$p = 1525 - 4.8 \times 10^{-1} [100 \times 1.80 + 250 \times 3.25 + 200 \times 3.3]$$
$$= 1525 - 793 = 732 \text{ psia}$$

Using the linear Darcy equation, the flow gradient at point P is

$$q/\text{ft.}^2 =$$
$$-\frac{1.127\,k}{\mu B}\frac{dp}{dL} = \frac{1.127 \times 0.2 \times 2(732 - 730)}{3 \times 1.30 \times 20} = 0.00587 \text{ bbls/day/ft.}^2$$

Since $h = 300$ ft., the flow rate per foot of lease line is $300 \times 0.00587 = 1.761$ barrels per day.

To obtain the total flow across the lease line, in a similar manner the flow rates per unit section of lease line would have to be determined for various points on the lease line. Then the area under a plot of these flow rates per unit section of lease line versus distance along the lease line would be equal to the total flow rate across the lease line. It is obvious that the computation of flow across a lease line for a multi-well system is rather time consuming. Problems of this kind should be programmed and solved with the aid of a high speed digital computer.

Reservoir Size

The reservoir engineer is often interested in determining the size of an oil reservoir. Many benefits would be gained by management if the size of an oil reservoir could be determined from the early performance of the discovery well. This would be true even if the order of magnitude of the size of the oil reservoir could be determined.

The basic compressibility equation can be used for determining the approximate size of an oil reservoir when written in the form

$$N = \frac{N_p}{c\Delta p} \qquad \textbf{9-73}$$

where

N = number of barrels of reservoir initially in place
N_p = cumulative number of barrels of reservoir oil produced
c = compressibility of the reservoir oil
Δp = drop in reservoir pressure due to the production of N_p barrels of reservoir oil.

To use Equation **9-73** the pressure drop must be measured after the reservoir reaches equilibrium conditions following the shutting in of all wells. After a number of barrels of initial reservoir oil has been produced, one must have available the porosity of the reservoir, the formation volume factor of the reservoir oil, and the average thickness of the reservoir before an estimate of the size of the reservoir can be computed.

A method for estimating the size of the effective radius of a reservoir utilizes Equation **9-50.** This method is applicable for the discovery well. In the process for estimating the size of the reservoir various performance curves are computed with the use of the equation for a number of r_n values. The computed performance curves are then compared with the experimentally measured performance curve, The r_n value for the reservoir is taken as the one used in computing the performance curve which most nearly coincides with the experimentally measured curve[7].

[7]Sweeney, Robert J., Jr., "A Critical Analysis of Methods for Determining Reservoir Size and Limits from Pressure Behavior of the Discovery Well", Thesis, Louisiana Polytechnic Institute, Ruston, Louisiana (1962).

Exercises

1. For a single-well system in a homogeneous oil reservoir, calculate the pressure at the well bore if the data applies:

$r_n = 2000$ ft.	$r_w = 0.25$ ft.
$\mu = 5.00$ centipoises	$h = 30$ ft.
$q = 150$ bbls/day	$B = 1.45$
$p_n = 1200$ psia	$k = 50$ milli-darcies

2. For a two-well system in a homogeneous circular oil reservoir, calculate the pressure at point $P(1000,1000)$ if well No. 1 is located at $P(300,300)$ and well No. 2 is located at $P(-200,10)$ and if the respective flow rates are 100 and 125 bbls/day. Additional data are $r_n = 20{,}000$ ft., $\mu = 3.00$ centipoise, $h = 20$ ft., $B = 1.32$, $k = 50$ milli-darcies and $p_n = 2000$ psia.
3. Calculate the pressure drop $(p_n - p)^{*}$ at point $P(1000,1500)$ in a homogeneous circular oil reservoir for a four-well system if the well locations are $P(-200,200)$, $P(-200,-200)$, $P(200,-200)$, $P(200,200)$ and the respective flow rates are 100, 200, 150, 175 bbls/day. Additional data are $r_n = 20{,}000$ ft., $\mu = 5.00$ centipoises, $h = 30$ ft., $k = 200$, $B = 1.35$.
4. Re-work Example 9-5 by using Equation **9-45** and Figure 9-5. Also show that for small values of Ei $(-x)$, Equation **9-37** suffices.
5. From the following data and a plot of build-up pressures, determine the permeability of the formation (infinite reservoir).

Build-up Time-hours	Pressure, psia
1.0	1559
3.0	1560
5.0	1510
10	1576
30	1680
50	1720
80	1766
100	1786
300	1860
500	1890

$t = 500$ hrs.	$P_n = 1960$ psi
$h = 30$ ft.	$q = 175$ bbls/day
$\mu = 6.0$ centipoises	$B = 1.40$
$\phi = 0.25$	$\rho = 0.6$
$r_w = 0.5$ ft.	$c = 5.0 \times 10^{-5}$ psi^{-1}

6. Compute the pressure drop $(p_n - p_w)$ at the well bore after the well had produced for 450 hours. There follows a listing of instantaneous flow rates and other data for the well which is in a homogeneous circular reservoir with an infinite radius.

Time Hours	Instantaneous Flow Rate bbls/day
50	238
100	225
150	213
200	195
250	170
300	140
350	130
400	80
450	40

$k = 128$ millidarcies
$h = 30$ ft.
$\mu = 5.0$ centipoise
$c = 5 \times 10^{-5}$ psi^{-1}

$p_n = 2160$ psia
$B = 1.38$
$\phi = 0.25$
$r_w = 0.5$ ft.

7. If the well in problem 9-6 was shut in after 450 hours of production, what would be the pressure at the well bore after a build-up time of 100 hours?

8. For a homogeneous circular reservoir with a finite radius, derive an equation for computing build-up pressures at the well bore if the flow rate varies.

9. Referring to the Darcy gas equation, derive an equation for computing build-up pressures in gas wells.

10. Using Cartesian coordinates, rewrite Equation 9-45 with r^2 expressed in terms of x,y. Assume the well is located at $P(x_w,y_w)$ and $P(x_p,y_p)$ are the coordinates at which the pressure is desired.

11. If a single-well in a homogeneous circular reservoir with a finite radius of 10,000 ft produced oil for 100 hours and was then shut in for 25 hours, what is the pressure at the well bore? Additional data are:

$q = 100$ STB/day
$\mu = 2.0$ centipoise
$r_n = 15{,}000$ ft.
$r_w = 0.25$ ft.
$\phi = 0.23$
$k = 350$ millidarcies
$h = 20$ ft.
$p_n = 1000$ psia
$c = 2.0 \times 10^{-5}$
$B = 1.42$

12. For the data which follows:
    a. Plot build-up pressure versus log $(t+\Delta)/\Delta$
    b. Determine the permeability of the formation
    c. Find p* from the build-up pressure curve
    d. Calculate the skin factor
    e. Calculate the flow efficiency of the well.

Pressure, psia	Build-Up Time, hours
1350	1
1410	2
1440	3
1680	5
1780	10
1855	50
1886	100
1920	500

$h = 20$ ft.
$r_w = 0.25$ ft.
$\mu = 9.0$
$\phi = 0.21$
$P_{wf} = 1052$ psia
$q = 126$ STB/day
$t = 500$ hrs.
$B = 1.22$

13. In terms of Cartesian coordinates, there is reason to believe that a fault runs parallel to the y-axis, and two wells are located at $P(200,1000)$ and $P(20,100)$, calculate the distance the well which is located at $P(200,1000)$ is from the fault, if the two-well system is shut in until equilibrium pressure. Then the well at $P(200,1000)$ is permitted to produce for 500 hours while the other well remains shut in; at this time the pressure is measured and found to be 1195 psia. Additional reservoir data are:

$q_1 = 150$ STB/day
$\mu = 3.00$ cp
$k = 228$ millidarcies
$c = 2.1 \times 10^{-5}$ psi^{-1}
$\phi = 0.15$
$P_n = 1223$ psia
$B = 1.31$
$h = 10$ ft.

14. Complete the solution of Example 9-10 by determining the flow gradients for various locations along the lease-line, plotting these data and taking the area under the curve.

15. Determine the pressure drop $(p_n - p)$ for the four-well system at point $P(1500,1000)$ if the wells are located at points $P(-200,-400)$, $P(-400,250)$, $P(700,800)$ and $P(500,-300)$ and their flow rates are respectively 300, 200, 370, and 400 STB/day; the respective times are 100, 320, 75, and 62 hours. Additional reservoir data are:

$\mu = 10.0$ centipoises
$B = 1.45$
$k = 345$ milli-Darcies
$\phi = 0.22$
$c = 1.0 \times 10^{-5}$ psi^{-1}
$h = 30$ ft.

Suggested Reading

1. Craft, B. C. and Hawkins, M. F., *Applied Reservoir Engineering,* Englewood Cliffs: Prentice-Hall, Inc., 1959, pp. 259-339.
2. Hawkins, M. F., *A Note on the Skin Effect,* Trans. A.I.M.E. 1956, pp. 356-357, 207.
3. Matthews, C. S. and Russell, D. C., "Pressure Build-Up and Flow Tests in Wells," *Monograph Vol. I A.I.M.E.,* Henry L. Daugherty Series, Dallas: Society of Petroleum Engineers of A.I.M.E., 1967, p. 21.
4. Mid-Continent District Study Committee on Completion Practices. "Selection and Evaluation of Well Completion Methods," *Drilling and Production Pratcices,* p. 421. A.P.I., 1955.
5. Muskat, M. and Wyckoff, R. D., *The Flow of Homogeneous Fluids Through Porous Media,* 1st ed., Ann Arbor: J. W. Edwards, Inc., 1946, pp. 150-257.
6. Nisle, R. G., "How to Use the Exponential Integral," *The Petroleum Engineer,* August, 156, p. B-171.
7. Sweeney, R. J., Jr., "A Critical Analysis of Methods for Determining Reservoir Size and Limits from Pressure Behavior of the Discovery Well," Master's Thesis, Louisiana Polytechnic Institute, 1962.
8. *Tables of Sine, Cosine and Exponential Integral,* Vols. 1 and 2, Federal Agency, W.P.A. for the City of New York, sponsored by U.S. National Bureau of Standards, available from Supt. of Documents, Washington 25, D.C.
9. Van Everdingen, A. F., "The Skin Effect and Its Influence on Productivity Capacity of a Well, Trans. A.I.M.E. 1953, pp. 171-176, 198.
10. Weast, R. C. and Selby, S. M., *Handbook for Mathematics,* 3rd ed., Cleveland: Chemical Rubber Co., 1967, pp. 687-697.

10 FORTRAN Programs and Examples

The reader may be wondering just how all the material of the previous chapters fits into the solving of petroleum engineering problems with the aid of electronic digital computers. This chapter presents and discusses in detail examples of computer programs. These examples were selected to show how the material presented in the previous chapters is used in solving problems in petroleum engineering with the aid of electronic digital computers and to illustrate computer programming in FORTRAN.

The eleven examples of FORTRAN computer programs become progressively more difficult. If one understands these examples thoroughly, he has the necessary foundation for programming and solving a large number of petroleum engineering problems with the aid of an electronic digital computer.

The examples can be used as guides in writing other computer programs. Flow charts are given with most of the examples. Each of the computer programs were written in FORTRAN IV and tested on an IBM 360, Model 40. Some of the FORTRAN programs were further tested on a Xerox Sigma VI. The FORTRAN programs worked equally as well without alterations on both of the computers.

Although the computer programs in this chapter were written in FORTRAN IV (ANSI 66), most of the computer programs will run on computers with FORTRAN 77 (ANSI 77) compilers. Differences between FORTRAN IV and FORTRAN 77 were discussed in Chapter 2.

In the commentaries about the FORTRAN programs, duplications of comments on individual FORTRAN statements were minimized. For instance, if the reader wishes comments about a particular FORTRAN statement in Example 10-11, he may have to refer to another example for the particular comment.

In connection with Example 10-7 there is a brief discussion on the use of dimensionless variables.

Programmed FORTRAN Examples

Eleven examples of FORTRAN programs follow and, as the examples are studied, the section on FORTRAN programming in Chapter 2 should be referred to freely. Also, in the study of the examples the details in which the FORTRAN statements are written should be noted with care; that is, commas, apostrophes, parentheses, periods, etc. should be noted. Each of the characters in a FORTRAN statement has a purpose, and these characters must be placed in an exact manner. Each character takes up one space or column in a card.

Example 10-1. Sums of Number Sets I

Write a FORTRAN program for computing the sums for 15 sets of two numbers as

$$Z_i = X_i + Y_i \qquad i = 1 \text{ to } 15$$

by means of counter looping and by using a READ statement and an IF statement.

Commentary

The flow diagram is given in Figure 10-1. A FORTRAN program is written on a FORTRAN coding form as illustrated in Figure 10-2. There are 80 columns on an IBM card; therefore, the numbered 80 columns on the FORTRAN coding form correspond to the 80 columns on an IBM card. After completion of the FORTRAN program, an IBM card is punched for each of the FORTRAN statements given on the FORTRAN coding form. As the FORTRAN statement is written in Figure 10-2, an IBM card is punched for each line on the FORTRAN coding form. On the IBM cards, the column in which a character is punched on a card must correspond to the column in which the character is listed in the FORTRAN coding form.

FORTRAN comment statements may be written, but they are prevented from being executed on the computer by punching a *C* in the first column

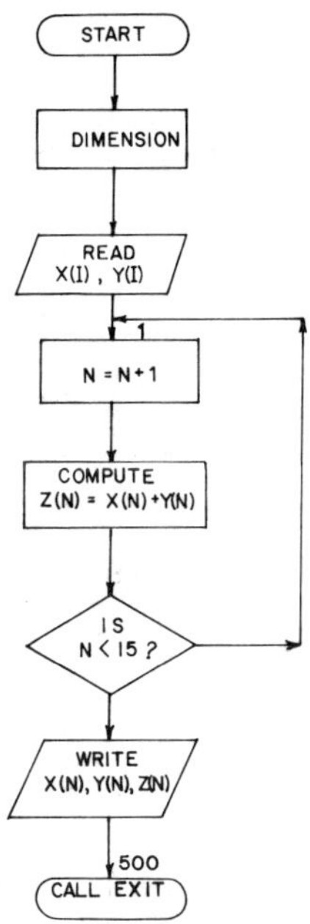

Figure 10-1. Flow diagram of the FORTRAN program for computing the sums of number sets; counter looping.

of a card. The FORTRAN comment statements are, however, printed on the print-out sheet from the computer (see Figure 10-3). A card with only a *C* in the first column causes a blank line on the printout sheet from the computer. Usually only FORTRAN statements which are to be referred to in a program are given numbers. However, if it is necessary to number a FORTRAN statement, the number of the statement is written in columns 2 through 5. The FORTRAN statements are written in columns 7 through 72. But should a FORTRAN statement require more than 66 columns, it is continued on the next line. In this case the next line must have a number in column 6. As a rule of thumb, this number should be the number corresponding to the number of the continue card. Columns 73 to 80 are reserved for identification sequences. Characters written in these columns have no effect on a FORTRAN statement.

Following the preparation of the FORTRAN coding form, IBM cards are prepared. Three cards (JOB CARD, EXEC FORTRAN, SYSIN) are placed in order as written on top of the FORTRAN program card deck and a DELIMITER CARD is placed at the end of the FORTRAN program card deck. Then all the previous cards are placed on top of the data cards.

FORTRAN CODING FORM

```
C     STEPHEN OWENS          SAMPLE PROBLEM ONE        COUNTER LOOPING
C
C     RESERVE STORAGE REGISTERS FOR X AND Y VALUES
      DIMENSION X(25),Y(25),Z(25)
C     READ DATA CARDS AND STORE X AND Y VALUES IN RESERVED SPACE
      READ(5,101) (X(I),Y(I),I=1,15)
  101 FORMAT(2F10.3)
      N=0
    1 N=N+1
      Z(N)=X(N)+Y(N)
      IF(N.LT.15)GO TO 1
      WRITE(6,102)
      WRITE(6,103)(I,X(I),Y(I),Z(I),I=1,15)
      GO TO 500
  102 FORMAT(2X,'I',5X,'X(I)',8X,'Y(I)',8X,'Z(I)'//)
  103 FORMAT(1X,I3,3F12.2)
  500 CALL EXIT                                     NOTE -- MOST COMMENT STATEMENTS
      END                                                  HAVE BEEN OMITTED

521.41     -666.66                                         FIRST DATA CARD
```

Figure 10-2. FORTRAN coding form with FORTRAN statements for computing the sums of number sets; counter looping.

```
C     STEPHEN OWENS   SAMPLE PROBLEM 1  COUNTER LOOPING
C
C     RESERVE STORAGE REGISTERS FOR X AND Y VALUES.
      DIMENSION X(25),Y(25),Z(25)
C     READ DATA CARDS AND STORE X AND Y VALUES IN RESERVED SPACE
      READ (5,101) (X(I),Y(I),I=1,15)
C     FLOATING POINT FORMAT FOR X AND Y
  101 FORMAT(2F10.3)
C     INITIALIZE VALUE OF N
      N=0
C     COUNTER FOR X AND Y SUBSCRIPTS
    1 N=N+1
C     COMPUTES VALUE OF NUMBER SET (N)
      Z(N)=X(N)+Y(N)
C     COUNTER CHECK--HAS THE LAST VALUE OF Z BEEN CALCULATED
      IF (N.LT.15) GO TO 1
C     PRINTS OUT HEADINGS FOR DATA I,X(I),Y(I), AND Z(I)
      WRITE (6,102)
C     PRINTS OUT DATA I,X(I),Y(I), AND Z(I)
      WRITE (6,103) (I,X(I),Y(I),Z(I),I=1,15)
C     INSTRUCT COMPUTER TO EXECUTE STATEMENT NO. 500
      GO TO 500
C     FORMAT FOR DATA HEADINGS
  102 FORMAT (2X,'I',5X,'X(I)',8X,'Y(I)',8X,'Z(I)'//)
C     FORMAT FOR DATA
  103 FORMAT (1X,I3,3F12.2)
C     RETURN COMPUTER CONTROL TO MONITOR
  500 CALL EXIT
C     INSTRUCT FORTRAN COMPILER THAT THE END OF THE SOURCE
C     PROGRAM HAS BEEN ENCOUNTERED.
      END
```

I	X(I)	Y(I)	Z(I)
1	521.41	-666.66	-145.25
2	425.67	222.22	647.89
3	333.32	333.33	666.65
4	222.22	424.67	646.89
5	333.33	523.46	856.79
6	444.44	-643.21	-198.77
7	555.55	543.21	1098.76
8	411.11	748.72	1159.83
9	123.45	343.45	466.90
10	678.90	678.88	1357.78
11	167.54	424.21	591.75
12	432.31	667.78	1100.09
13	677.42	-24.64	652.78
14	777.71	334.21	1111.92
15	111.11	897.47	1008.58

Figure 10-3. Printout of a FORTRAN program for computing the sum of number sets, counter looping.

A DELIMITER CARD is placed at the bottom of the combined card deck. Since the computer processes the cards in the order in which it receives them, it is important that care be taken in arranging the cards in the order as indicated by the FORTRAN coding form; otherwise, the program will not be processed correctly. Finally, the composite deck of cards is given to the computer center. The special cards will vary for different computers. However, the FORTRAN statements are the same for the different computers which process FORTRAN IV programs.

These special cards are required by the computer before it will process the FORTRAN program. The first part of the JOB CARD is used for bookkeeping, and the last part of the card informs the computer that a job is coming. The EXEC FORTRAN CARD identifies the type of job the computer is to perform and instructs the computer to compile and execute the FORTRAN program. The SYSIN CARD indicates to the computer that the FORTRAN program is to follow. The DELIMITER card instructs the computer that the end of the input program is being executed.

Information as printed by the computer is given in Figure 10-3. First in the figure is a print of the FORTRAN program as the FORTRAN statement was punched on IBM cards. As previously stated, the first three statements are comment statements. The first comment statement identifies the writer of the FORTRAN program and the problem. The second and third comment statements cause two lines to be skipped before printing out the FORTRAN statements which are required for the program to solve the problem. Fourteen FORTRAN statements were required for the FORTRAN program to solve the problem. The 14 FORTRAN statements were given consecutive numbers by the computer as they were printed, but these numbers are not a part of the FORTRAN program and are different from the numbers assigned by the programmer.

The first FORTRAN statement, DIMENSION *X*(25),*Y*(25),*Z*(25), causes the FORTRAN compiler to reserve 75 storage registers for the values of the variables, whereas 45 storage registers would have been sufficient. The surplus storage registers are not used during the execution of the object program.

The second FORTRAN statement, READ (5,101)*X*(*I*),*Y*(*I*),*I* = 1,15), causes the computer to read one new card at a time for 15 consecutive times. The first number, 5, enclosed in the first parentheses designates the type of input device of the computer for receiving the program. The number, 5, designates a card reader. The second number, 101, enclosed in the parentheses refers to the number of the format statement which specifies how the data are punched in the cards. Within the next set of parentheses of the FORMAT statement are the two subscripted variables, *X* (*I*) and *Y*(*I*), and an indication of the values of the subscripts, *I* = 1,15.

The third FORTRAN statement, 101 FORMAT(2F10.3), was assigned a number, 101, it was referred to by the READ statement, The term, 2F10.3, within the parentheses of the format statement specifies that there are two numerical values on each data card (a value for *X*(*I*) and a value for *Y*(*I*). The numerical values are punched in decimal notation. The ,*F*, denotes floating-point arithmetic and is to be used in computing the sums. The term, 2F10.3, indicates that two numerical values (an *X* and a *Y*) are to be punched in each data card; the width of the field is 10 columns. Three digits are to the right of the decimal point; an X value is punched in the first ten columns of the data card, and a Y value is punched in the second ten columns of the same data card.

The fourth FORTRAN statement, $N = 0$, initializes the counter, causing N to have an initial value of zero.

The fifth FORTRAN statement ,1 $N = N + 1$, causes N to be increased by one at the beginning of each cycle of the looping. The value of N also sets the subscript values for $X(N)$, $Y(N)$, $Z(N)$ prior to the arithmetic operation. A number was assigned to this FORTRAN statement because the IF statement (seventh statement) instructs the computer to return to this FORTRAN statement when N is less than 15; this statement is the beginning statement for a cycle of looping.

The sixth FORTRAN statement, $Z(N) = X(N) + Y(N)$, causes a value of $Z(N)$ to be computed and stored in the storage register reserved for $Z(N)$ by the DIMENSION statement.

The seventh FORTRAN statement, IF($N.LT.$15) GO TO 1, is a *logic statement.* If after checking N is found to be less than 15, the next statement to be executed is the FORTRAN statement which was assigned 1 as its number (its listed number is 5). If N is equal to 15, the eighth FORTRAN statement is executed.

The eighth FORTRAN statement, WRITE (6,102), causes the computer to print out the headings for the data as given in Figure 10-3 according to the Format statement with an assigned number of 102. The number 6 in the WRITE statement refers to the type of device to be used for the output of data from the computer. The output device to be used for this computer set up is a printer.

The ninth FORTRAN statement, WRITE (6,103) (I,$X(I)$,$Y(I)$,$Z(I)$, $I = 1,15$), causes the input data and the computed data to be printed out according to the Format statement with the assigned number of 103. Following the printing out of the input and computed data, the tenth FORTRAN statement, GO TO 500, causes the thirteenth FORTRAN statement, 500 CALL EXIT, to be executed. This FORTRAN statement causes a termination in the execution of the Object Program. The last FORTRAN statement (14 th), END, indicates to the FORTRAN Compiler that the end of the source program has been encountered.

Referring to the eleventh FORTRAN statement, 102 FORMAT (2*X*, '*I*',5*X*,'*X*(*I*)',8*X*,'*Y*(*I*)',8*X*'*Z*(*I*)'//), an explanation of the terms enclosed within the parentheses follows. The term, 2X, causes the first two columns of the heading for the data to remain blank. Apostrophes before and after a variable automatically cause sufficient columns to be reserved for the variable. In this case one column is sufficient to be reserved for the variable I. The I is printed in the third column. Following the I, five columns are left blank because of the term $5X$; then X(I) is printed in columns 9-12. The next eight columns are left blank because of the term $8X$. $Y(I)$ is printed in columns 20-23 and again eight columns are left blank because of the second $8X$ term. Finally $Z(I)$ is printed in columns 31-34. The two slashes ,//, cause two lines to be skipped before the data is printed.

The twelfth Format statement, 103 FORMAT ($I3$,3F12.2), causes three columns to be reserved for printing out the values of I. Then twelve columns are reserved for the printing out the values of $X(I)$ and twelve columns are reserved for printing out the values of $Y(I)$; then, twelve columns are reserved for printing out the values of $Z(I)$. The number 3 before F12.2 means that there are three values to be printed on a line. In this case the three values are for a $X(I)$, a $Y(I)$, and a $Z(I)$. The I values are printed in columns 1-3; the $X(I)$ values are printed in columns 4-15; the $Y(I)$ values are printed in columns 16-27; and the $Z(I)$ values are printed in columns 28-39.

The print out of the data with headings are given in the lower portion of Figure 10-3. The print out of the data with headings indicate that with appropriate FORMAT statements data from a computer can be printed out in table form.

Example 10-2. Sums of Number Sets II

Write a FORTRAN program for computing the sums of 15 sets of two numbers as

$$Z_i = X_i + Y_i \qquad i = 1 \text{ to } 15$$

by means of Do Looping and a CONTINUE statement.

Commentary

The flow diagram of the computer solution is given in Fig. 10-4. The flow diagram indicates that at the beginning of the processing of the Object Program, the computer prints out the headings I, $X(I)$, $Y(I)$, and $Z(I)$. Next, the computer reads a value of an X and a Y and computes their sum; then the computer prints out the values for N, $X(I)$, $Y(I)$, and $Z(I)$ under the appropriate headings. Afterwards it tests to determine whether 15 cards have been read for X and Y values. If 15 cards have not been read, the cycle of reading a card and computing a sum, etc. is performed again. The cycles are repeated until fifteen cards have been read before the processing of the Object Program is ceased.

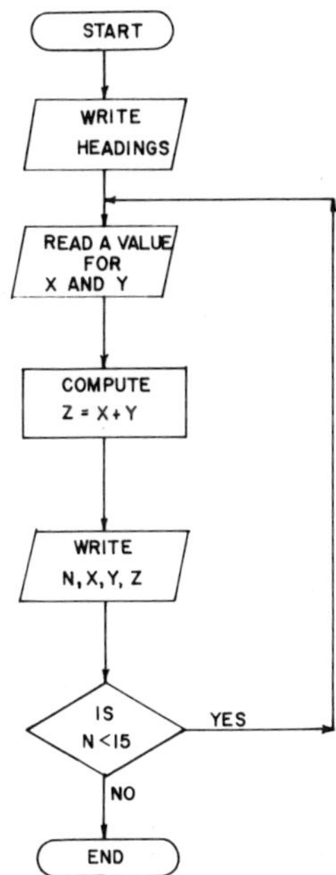

Figure 10-4. Flow diagram of the FORTRAN program for computing the sums of number sets; DO looping.

The print-out of the FORTRAN program with data is given in Figure 10-5. The FORTRAN program for this example contains two less FORTRAN statements than the FORTRAN program for the previous example, although the computed results of the two programs are identical.

The first FORTRAN statement, WRITE (6,102), causes the headings for the data to be printed out according to the FORMAT statement with assigned No. 102 prior to computing the sums for the *X*'s and *Y*'s, because only one working storage register is reserved for each of the three variables for all the computations. These three working storages are used over and over again, once during each cycle. The values of the three variables are printed out immediately following the respective computation; otherwise, the data would be lost. The FORTRAN program in the previous example caused all the values of the *X*'s, *Y*'s, and *Z*'s to be stored in the computer and when the computations of the sums were completed, then the *X*'s, *Y*'s, and *Z*'s were printed out. The FORTRAN program containing the DO looping requires fewer storage registers than the FORTRAN program containing the COUNTER looping.

```
C       STEPHEN OWENS            SAMPLE PROBLEM  2  DO-LOOPING
C
C       PRINT OUT HEADINGS FOR DATA
        WRITE (6,102)
C       DO-LOOP EXECUTE STATEMENTS THROUGH NO. 3 FOR 15 TIMES
        DO 3 N=1,15
C       READ ONE CARD AT A TIME WITH X AND Y VALUES
        READ (5,101) X,Y
        Z=X+Y
        WRITE (6,103) N,X,Y,Z
C       CONTINUE THE COMPUTATION OF Z VALUES
      3 CONTINUE
C       AFTER THE COMPUTATION OF 15 SUMS EXECUTE STATEMENT NO. 500
        GO TO 500
    101 FORMAT(2F10.3)
    102 FORMAT (2X,'I',5X,'X(I)',8X,'Y(I)',8X,'Z(I)'//)
    103 FORMAT (1X,I3,3F12.2)
    500 CALL EXIT
        END
```

I	X(I)	Y(I)	Z(I)
1	521.41	-666.66	-145.25
2	425.67	222.22	647.89
3	333.32	333.33	666.65
4	222.22	424.67	646.89
5	333.33	523.46	856.79
6	444.44	-643.21	-198.77
7	555.55	543.21	1098.76
8	411.11	748.72	1159.83
9	123.45	343.45	466.90
10	678.90	678.88	1357.78
11	167.54	424.21	591.75
12	432.31	667.78	1100.09
13	677.42	-24.64	652.78
14	777.71	334.21	1111.92
15	111.11	897.47	1008.58

Figure 10-5. Printout of a FORTRAN program for computing the sums of number sets; DO looping.

The second FORTRAN statement, DO 3 $N = 1{,}15$, causes all the FORTRAN statements down to and including the FORTRAN statement with an assigned number of 3 to be executed for 15 consecutive times.

The third FORTRAN statement, READ (5,101) X,Y, gives each cycle of the looping a value for an X and a value for a Y to be read into the computer from an IBM card. The X and Y values are punched into the IBM card according to the FORMAT statement with an assigned number of 101.

The fourth FORTRAN statement, $Z = X + Y$, causes a sum for an X and a Y to be computed.

The fifth FORTRAN statement, WRITE (6,103) N,X,Y,Z, causes values of an N, an X, a Y, a Z to be printed out according to the FORMAT statement with an assigned number of 103.

The sixth FORTRAN statement, 3 CONTINUE, is an executable statement which is often referred to as a dummy statement. This statement provided the programmer with a means for inserting a number into the FORTRAN program without the compiler generating any instructions.

For further discussions on the remaining FORTRAN statements of the program, the comments on these types of FORTRAN statements as given in the FORTRAN program of Example 10-1 should be reviewed.

Example 10-3. Polynomial Evaluation

Write a FORTRAN program for computing the value of a formation volume factor at 1,000 psia if the formation volume factor is a function of pressure as given by the following equation:

Commentary

Figure 10-6 indicates that the solution is straightforward. There is no looping and only one value is computed. The form of the polynomial is simplified to facilitate computations.

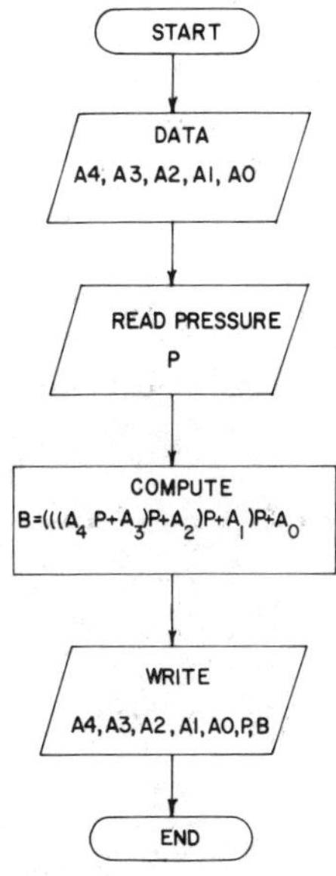

Figure 10-6. Flow diagram of the FORTRAN program for computing the value of a formation volume factor.

```
C       STEPHEN OWENS        FORMATION-VOLUME CALCULATION
C
C       PLACE SPECIFIC DATA IN STORAGE REGISTER
        DATA A4,A3,A2,A1,A0/1.1E-13,1.5E-10,1.9E-7,1.0E-3,1.05/
        READ(5,101)P
    101 FORMAT(F12.1)
C       SIMPLIFIED FORM OF POLYNOMIAL FOR COMPUTATION
        B=((((A4*P+A3)*P+A2)*P+A1)*P+A0
        WRITE (6,102) A4,A3,A2,A1,A0,P,B
    102 FORMAT (1X,5E12.4,2F12.6)
        CALL EXIT
        END

0.1100E-12  0.1500E-09  0.1900E-06  0.1000E-02  0.1050E 01 1000.000000  2.499998
```

Figure 10-7. Printout of a FORTRAN program for computing the value of a formation volume factor.

The printout of the FORTRAN program as given in Figure 10-7 illustrates the use of a DATA statement. The use of real numbers in exponential specification and computations when the specifications of real numbers are mixed (Ew.d exponential and Fw.d decimal).

The first FORTRAN statement, DATA A4,A3,A2,A1,A0/1.1E-13,1.5E-10,1.9E-7,1.0E-3,1.05/, was used so that numerical values could be assigned to the variables prior to the execution of the program. It was necessary that the coefficients A4, A3, A2, A1, A0 be subscripted in the statement. The arithmetic specifications for these variables follow the first slash, and a slash is placed after the arithmetic specifications. It is essential that the variables and the arithmetic specifications be separated by the slash. The variables and arithmetic specifications are separated from each other by commas. The FORTRAN format for 1.1×10^{-13} is 1.1E − 13. The E indicates that the arithmetic value is written in exponential form, the minus sign indicates that the exponent power is negative, and the 13 indicates the exponent power.

Example 10-4. Division by Iteration

Write a FORTRAN program using the iteration method given in the Appendix for determining the quotient for 504 divided by 24 with D_i greater than 0.9999. Have the computer print out all data.

Commentary

In writing the FORTRAN program, N_i, D_i and $1 + d_i$ were represented respectively by *AN*, *D* and *DP*; also N_{i+1} and D_{i+1} were represented by *ANP* and *AD*.

Figure 10-8 indicates that the object program, compiled from the FORTRAN program, will cause the computer to operate as follows during the computation of the quotient:

1. The headings for the data are printed out at the beginning.
2. The values for *AN*, *C*, *D*, *T*, *W*, are read from a card and stored in the computer.
3. The value of *I* is given an initial value of zero.
4. The value of *I* is advanced by 1.
5. The complement of *D* (DP) plus 1 is computed.
6. The initial values of *I*, *AN*, *D*, *DP*, are printed out.
7. The remainder for 1.00000000 minus D_1 is computed.
8. The remainder is tested against 0.0001; if the remainder is less than 0.0001, the execution of the program ceases; but if the remainder is greater than 0.0001, new values for *I*, *AN*, *D*, *DP*, are computed.

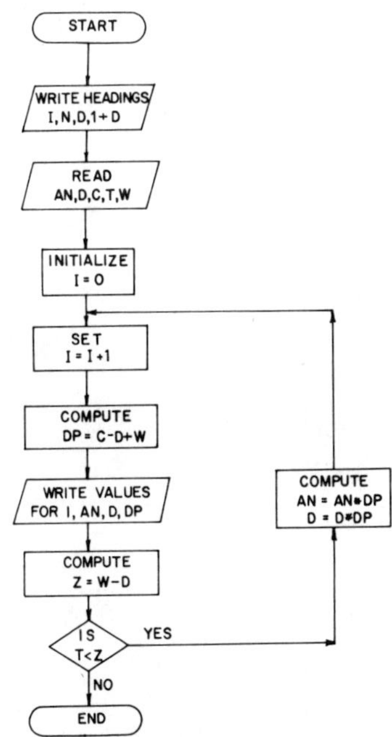

Figure 10-8. Flow diagram of a FORTRAN program for determining the quotient of two numbers by the iteration method.

9. The new values of *AN* and *D* are caused to occupy the original storage spaces for the initial *AN* and *D*.

10. And the cycle of looping continues (computing and printing out new values for *I, AN, D, DP*) until 1.00000000 − *D* is less than 0.0001; then the execution of the object program ceases.

Only one data card was required for the FORTRAN program. On the data card were the initial values of *AN, D, C, T, W*; these values were respectively 5.04000000, 0.24000000, 0.99999999, 0.0001, 1.00000000.

```
C      STEPHEN OWENS        DIVISION BY ITERATION
C
C      PRINT HEADINGS FOR DATA
       WRITE (6,101)
C      READ INITIAL DATA INTO COMPUTER
       READ (5,102) AN,D,C,T,W
C      INITIALIZE NUMBER FOR SET OF DATA
       I=0
       GO TO 5
C      COMPUTE  N(I+1) AND D(I+1)
     4 AN=AN*DP
       D=D*DP
     5 I=I+1
C      COMPUTE COMPLIMENT PLUS ONE FOR D(I)
       DP=C-D+W
C      PRINT OUT SET OF DATA
       WRITE (6,103) I,AN,D,DP
C      COMPUTE VALUE FOR TOLERANCE CHECK
       Z=W-D
C      CHECK TOLERANCE
       IF (T.LE.Z) GO TO 4
       GO TO 500
   101 FORMAT (3X,'I',6X,'N',10X,'D',9X,'1+D'//)
   102 FORMAT (5F12.9)
   103 FORMAT(1X,I4,3F12.8)
C      CEASE EXECUTION
   500 CALL EXIT
       END
```

```
   I        N           D          1+D


   1  5.03999996  0.23999995  1.75999928
   2  8.87039566  0.42239970  1.57759953
   3 13.99393177  0.66637754  1.33362198
   4 18.66261292  0.88869572  1.11130333
   5 20.73982239  0.98761046  1.01238918
   6 20.99676514  0.99984610  1.00015354
   7 20.99998474  0.99999958  1.00000000
```

Figure 10-9. Printout of a FORTRAN program for computing the quotient of two numbers.

Example 10-5. Computation of Exponential Integral Table

1. Write a FORTRAN program for computing exponential values, $Ei(-x)$, for values of x from 0.00 to 0.209 in increments of 0.001. Have the computer to print out the data in table form.
2. Repeat number 1 for values of x from 0.1 to 2.9 in increments of 0.01.

Commentary

The flow diagram of the FORTRAN program for computing the first table of exponential values ($X = 0.000$ to 0.209) has three LOOPS: (1) one LOOP causes the value of the exponential integral $Ei(-X)$ to be computed for the first value of X in each row, (2) another LOOP causes the values of the exponential integral $Ei(-X)$ to be computed for the remaining values of X in each row, and (3) the remaining LOOP controls the beginnings of the computations for each row of integral values. Since the value of $Ei(X)$ for $X = 0.0$ is infinity, the FORTRAN program causes the value of $Ei(0.0)$ to be printed out as 0.0.

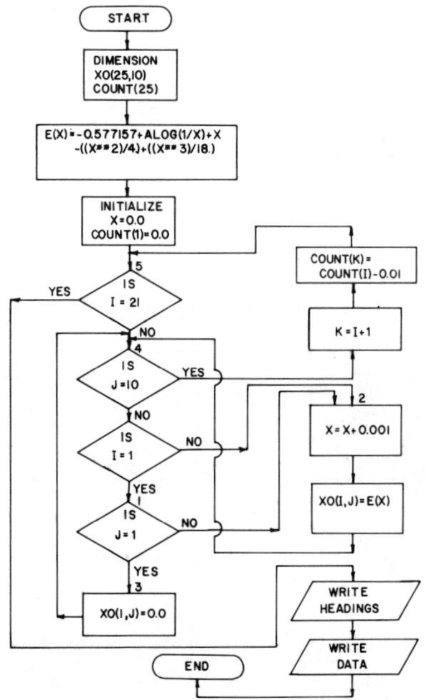

Figure 10-10. Flow diagram of a FORTRAN program for computing exponential integral values.

The key FORTRAN statements for LOOP (1) are listed statements Nos. 4, 15, and 16 (see Figure 10-11). The key FORTRAN statements for LOOP (2) are listed statements Numbers 6, 10, and 11. And the key FORTRAN statement for LOOP (3) is listed statement No. 5.

Because the flow diagram for computing the second table of exponential integral values ($X = 0.1$ to 2.9) is very similar to the flow diagram for computing the first table of exponential integral values, this flow diagram is omitted. However, the computer program for the second table of values has an additional logic FORTRAN statement and an additional algebraic FORTRAN statement.

The FORTRAN statement for computing the second table of exponential integral values (see Figure 10-12) contains two algebraic FORTRAN statements for the computation of exponential integral values. One of the FORTRAN statements is for (X) values less than one and the other FORTRAN statement is for (X) values greater than one. The listed

```
           C        STEPHEN OWENS        EXPONENTIAL INTERGRAL FUNCTION
           C        RANGE  .00 LT X LT .209
           C
           C        RESERVE 250 STORAGE REGISTERS FOR EI(X) VALUES  AND 25 STORAGE
           C        REGISTERS FOR INITIAL RAW VALUES OF (X)
0001                DIMENSION XO(25,10),COUNT(25)
           C        ALGEBRAIC EXPRESSION FOR COMPUTING EI(X)
0002                E(X)=-0.577157+ALOG(1/X)+X-((X**2)/4.)+((X**3)/18.)
           C        INITIALIZE X=0
0003                X=0.0
           C        INITIALIZE COLUMN COUNTER EQUAL TO 0
0004                COUNT(1)=0.0
           C        COMPUTE 21 ROWS OF THE EXPONENTIAL INTERGRAL
0005                DO 5 I=1,21
           C        COMPUTE 10 COLUMNS OF THE EXPONENTIAL INTERGRAL
           C        EXAMPLE OF NESTED DO-LOOPS
0006                DO 4 J=1,10
           C        CHECK IF SUBSCRIPT I IS 1
0007                IF (I.EQ.1) GO TO 1
           C        IF I=1 CHECK IF J=1, OTHERWISE GO TO 2
0008                GO TO 2
           C        IF I=1 AND J=1, GO TO 3
0009              1 IF (J.EQ.1) GO TO 3
           C        INCREMENT X BY .001
0010              2 X=X+.001
           C        COMPUTE EI(X) VALUES
0011                XO(I,J)=E(X)
           C        GO TO STATEMENT NO. 4
0012                GO TO 4
           C        GIVE EI(1,1) A VALUE OF 0.0
0013              3 XO(I,J)=0.0
           C        RETURN AND CALCULATE ANOTHER VALUE OF EI(X) IN THE SAME ROW
0014              4 CONTINUE
           C        ADVANCE K BY 1
0015                K=I+1
           C        COMPUTE THE INTIAL X VALUE FOR THE NEXT ROW
0016                COUNT(K)= COUNT(I)+.01
           C        RETURN AND BEGIN CALCULATION OF THE NEXT ROW
0017              5 CONTINUE
```

Figure 10-11. Printout of a FORTRAN program for computing exponential values; x = 0.00 to 0.209.

FORTRAN statements which cause the computer to determine which algebraic FORTRAN statements to use are Numbers 12 through 16.

Example 10-6. Computation of Y-Factor Values

1. Write a FORTRAN computer program for computing *Y*-Factor values, for values of *x* from 0.00 to 0.209 in increments of 0.001. Have the computer print out the data in table form.
2. Repeat Number 1 for values of *x* from 0.1 to 2.9 in increments of 0.01.

Commentary

The FORTRAN Examples 10-5 and 10-6 differ only in the algebraic FORTRAN statements (compare Figure 10-11 to Figure 10-13 and Figure 10-12 to Figure 10-14).

```
              C       PRINT OUT HEADINGS FOR THE EI(X) TABLE
0018                  WRITE (6,99)
              C       PRINT OUT THE EI(X) VALUES IN TABLE FORM
0019                  DO 20 I=1,21
0020                  WRITE (6,101) COUNT(I),(XO(I,J),J=1,10)
0021               20 CONTINUE
0022                  GO TO 500
0023               99 FORMAT(3X,'X',6X,'0',6X,'1',6X,'2',6X,'3',6X,'4',6X,'5',6X,'6',6X,
                     1'7',6X,'8',6X,'9'//)
0024              101 FORMAT(11F7.3)
0025              500 CALL EXIT
0026                  END
```

X	0	1	2	3	4	5	6	7	8	9
0.0	0.0	6.332	5.639	5.235	4.948	4.726	4.545	4.392	4.259	4.142
0.010	4.038	3.944	3.858	3.779	3.705	3.637	3.574	3.514	3.458	3.405
0.020	3.355	3.307	3.261	3.218	3.176	3.137	3.098	3.062	3.026	2.992
0.030	2.959	2.927	2.897	2.867	2.838	2.810	2.783	2.756	2.731	2.706
0.040	2.681	2.658	2.634	2.612	2.590	2.568	2.547	2.527	2.507	2.487
0.050	2.468	2.449	2.431	2.413	2.395	2.378	2.360	2.344	2.327	2.311
0.060	2.295	2.280	2.265	2.249	2.235	2.220	2.206	2.192	2.178	2.164
0.070	2.151	2.138	2.125	2.112	2.099	2.087	2.074	2.062	2.050	2.039
0.080	2.027	2.016	2.004	1.993	1.982	1.971	1.960	1.950	1.939	1.929
0.090	1.919	1.909	1.899	1.889	1.879	1.870	1.860	1.851	1.841	1.832
0.100	1.823	1.814	1.805	1.796	1.788	1.779	1.770	1.762	1.754	1.745
0.110	1.737	1.729	1.721	1.713	1.705	1.697	1.690	1.682	1.675	1.667
0.120	1.660	1.652	1.645	1.638	1.631	1.623	1.616	1.609	1.603	1.596
0.130	1.589	1.582	1.576	1.569	1.562	1.556	1.549	1.543	1.537	1.530
0.140	1.524	1.518	1.512	1.506	1.500	1.494	1.488	1.482	1.476	1.470
0.150	1.465	1.459	1.453	1.448	1.442	1.436	1.431	1.425	1.420	1.415
0.160	1.409	1.404	1.399	1.393	1.388	1.383	1.378	1.373	1.368	1.363
0.170	1.358	1.353	1.348	1.343	1.338	1.333	1.329	1.324	1.319	1.315
0.180	1.310	1.305	1.301	1.296	1.292	1.287	1.283	1.278	1.274	1.269
0.190	1.265	1.261	1.256	1.252	1.248	1.244	1.239	1.235	1.231	1.227
0.200	1.223	1.219	1.215	1.211	1.207	1.203	1.199	1.195	1.191	1.187

Figure 10-11. Continued.

```
C        STEPHEN OWENS        EXPONENTIAL INTERGRAL FUNCTION
C        RANGE IS FOR X FROM .1 TO 2.9
C
         DIMENSION XO(25,10),COUNT(25)
C        ALGEBRAIC EXPRESSION FOR COMPUTING EI(X) FOR VALUES OF X LT 1
         E(X)=-0.577157+ALOG(1/X)+X-((X**2)/4.)+((X**3)/18.)
        1-((X**4)/96.)+((X**5)/600.)-((X**6)/4320.)+((X**7)/35280.)
C        ALGEBRAIC EXPRESSION FOR COMPUTING EI(X) FOR VALUES OF X GT 1
         F(X)=-0.577157-ALOG(X)+X-((X**2)/4.)+((X**3)/18.)
        1-((X**4)/96.)+((X**5)/600.)-((X**6)/4320.)+((X**7)/35280.)
C
         X=0.0
         COUNT(1)=0.0
         DO 7 I=1,21
         DO 6 J=1,10
         IF (I.EQ.1) GO TO 1
         GO TO 2
       1 IF (J.EQ.1) GO TO 5
       2 X=X+.01
C        IF X GT 1 CALCULATE EI(X) USING THE EXPRESSION F(X)
         IF (X.GE.1.0) GO TO 3
         XO(I,J)=E(X)
         GO TO 6
       3 XO(I,J)=F(X)
         GO TO 6
       5 XO(I,J)=0.0
       6 CONTINUE
         K=I+1
         COUNT(K)=COUNT(I)+.1
       7 CONTINUE
         WRITE (6,99)
         DO 20 I=1,21
         WRITE (6,101) COUNT(I),(XO(I,J),J=1,10)
      20 CONTINUE
         GO TO 500
      99 FORMAT(2X,'X',8X,'0',7X,'1',7X,'2',7X,'3',7X,'4',7X,'5',7X,'6',7X,
        1'7',7X,'8',7X,'9'//)
     101 FORMAT(11F8.4)
     500 CALL EXIT
         END
```

X	0	1	2	3	4	5	6	7	8	9
0.0	0.0	4.0380	3.3548	2.9592	2.6813	2.4680	2.2954	2.1509	2.0270	1.9188
0.1000	1.8230	1.7372	1.6596	1.5890	1.5242	1.4645	1.4092	1.3578	1.3099	1.2649
0.2000	1.2227	1.1830	1.1454	1.1099	1.0763	1.0443	1.0139	0.9850	0.9574	0.9310
0.3000	0.9057	0.8816	0.8584	0.8362	0.8148	0.7943	0.7745	0.7555	0.7372	0.7195
0.4000	0.7024	0.6860	0.6701	0.6547	0.6398	0.6254	0.6114	0.5979	0.5848	0.5721
0.5000	0.5598	0.5479	0.5363	0.5250	0.5141	0.5034	0.4931	0.4830	0.4732	0.4637
0.6000	0.4544	0.4454	0.4366	0.4281	0.4197	0.4116	0.4036	0.3959	0.3884	0.3810
0.7000	0.3738	0.3668	0.3600	0.3533	0.3468	0.3404	0.3342	0.3281	0.3221	0.3163
0.8000	0.3107	0.3051	0.2997	0.2944	0.2892	0.2841	0.2791	0.2742	0.2695	0.2648
0.9000	0.2602	0.2558	0.2514	0.2471	0.2429	0.2388	0.2348	0.2308	0.2270	0.2232
1.0000	0.2194	0.2158	0.2122	0.2087	0.2053	0.2019	0.1986	0.1954	0.1922	0.1891
1.1000	0.1861	0.1831	0.1801	0.1772	0.1744	0.1716	0.1689	0.1662	0.1636	0.1610
1.2000	0.1585	0.1560	0.1536	0.1512	0.1488	0.1465	0.1442	0.1420	0.1398	0.1377
1.3000	0.1355	0.1335	0.1314	0.1294	0.1274	0.1255	0.1236	0.1217	0.1199	0.1181
1.4000	0.1163	0.1146	0.1129	0.1112	0.1095	0.1079	0.1063	0.1047	0.1032	0.1016
1.5000	0.1002	0.0987	0.0972	0.0958	0.0944	0.0930	0.0917	0.0904	0.0890	0.0878
1.6000	0.0865	0.0852	0.0840	0.0828	0.0816	0.0805	0.0793	0.0782	0.0771	0.0760
1.7000	0.0749	0.0738	0.0728	0.0718	0.0708	0.0698	0.0688	0.0679	0.0669	0.0660
1.8000	0.0651	0.0642	0.0633	0.0624	0.0616	0.0607	0.0599	0.0591	0.0583	0.0575
1.9000	0.0567	0.0559	0.0552	0.0545	0.0537	0.0530	0.0523	0.0516	0.0509	0.0503
2.0000	0.0496	0.0490	0.0483	0.0477	0.0471	0.0465	0.0459	0.0453	0.0448	0.0442

Figure 10-12. Printout of a FORTRAN program for computing exponential integral values; $x = 0.1$ to 2.9.

```
C       STEPHEN OWENS         Y FACTOR (EXPONENTIAL INTERGRAL)
C      RANGE    .00 LT XLT .209
C
C
        DIMENSION XO(25,10),COUNT(25)
C
        E(X)=.577157-ALOG(1./X)-X+((X**2)/4.)-((X**3)/18.)+(1./(X*EXP(X)))
C
        X=0.0
        COUNT(1)=0.0
        DO 5 I=1,21
        DO 4 J=1,10
        IF (I.EQ.1) GO TO 1
        GO TO 2
      1 IF (J.EQ.1) GO TO 3
      2 X=X+.001
        XO(I,J)=E(X)
        GO TO 4
      3 XO(I,J)=0.0
      4 CONTINUE
        K=I+1
        COUNT(K)= COUNT(I)+.01
      5 CONTINUE
        WRITE (6,99)
        DO 20 I=1,21
        WRITE (6,101) COUNT(I),(XO(I,J),J=1,10)
     20 CONTINUE
        GO TO 500
     99 FORMAT(3X,'X',6X,'0',6X,'1',6X,'2',6X,'3',6X,'4',6X,'5',6X,'6',6X,
       1'7',6X,'8',6X,'9'//)
    101 FORMAT(11F7.3)
    500 CALL EXIT
        END
```

```
   X      0      1      2      3      4      5      6      7      8      9


0.0      0.0  992.669493.361327.100244.054194.277161.125137.469119.745105.973
0.010 94.967 85.971 78.482 72.151 66.730 62.037 57.934 54.318 51.106 48.236
0.020 45.655 43.323 41.204 39.272 37.502 35.876 34.376 32.989 31.702 30.505
0.030 29.389 28.346 27.369 26.453 25.591 24.779 24.013 23.289 22.604 21.955
0.040 21.338 20.753 20.196 19.665 19.159 18.676 18.214 17.773 17.350 16.945
0.050 16.557 16.184 15.826 15.481 15.150 14.831 14.524 14.228 13.943 13.667
0.060 13.401 13.144 12.895 12.654 12.422 12.196 11.978 11.766 11.561 11.362
0.070 11.169 10.982 10.799 10.622 10.450 10.283 10.121  9.962  9.808  9.658
0.080  9.512  9.370  9.231  9.096  8.964  8.835  8.709  8.587  8.467  8.350
0.090  8.236  8.125  8.016  7.909  7.805  7.703  7.603  7.506  7.410  7.317
0.100  7.225  7.136  7.048  6.962  6.878  6.796  6.715  6.636  6.558  6.482
0.110  6.407  6.334  6.262  6.191  6.122  6.054  5.987  5.921  5.857  5.794
0.120  5.731  5.670  5.610  5.551  5.494  5.437  5.381  5.325  5.271  5.218
0.130  5.166  5.114  5.063  5.014  4.964  4.916  4.869  4.822  4.776  4.730
0.140  4.686  4.642  4.598  4.555  4.513  4.472  4.431  4.391  4.351  4.312
0.150  4.274  4.236  4.198  4.161  4.125  4.089  4.054  4.019  3.984  3.950
0.160  3.917  3.884  3.851  3.819  3.787  3.756  3.725  3.694  3.664  3.634
0.170  3.605  3.576  3.547  3.519  3.491  3.463  3.436  3.409  3.383  3.357
0.180  3.331  3.305  3.280  3.255  3.230  3.205  3.181  3.157  3.134  3.111
0.190  3.088  3.065  3.042  3.020  2.998  2.976  2.955  2.933  2.912  2.892
0.200  2.871  2.851  2.830  2.811  2.791  2.771  2.752  2.733  2.714  2.695
```

Figure 10-13. Printout of a FORTRAN program for computing *y*-factor values; $x = 0.00$ to 0.209.

```
C      STEPHEN OWENS        Y-FACTOR      (EXPONENTIAL INTERGRAL)
C      RANGE      .1   TO     2.9
C
C
       DIMENSION XO(25,10),COUNT(25)
C
       E(X)=.577157-ALOG(1./X)-X+((X**2)/4.)-((X**3)/18.)+(1./(EXP(X)*X))
      1+((X**4)/96.)-((X**5)/600.)+((X**6)/4320.)-((X**7)/35280.)
       F(X)= 0.577157+ALOG(X)-X+((X**2)/4.)-((X**3)/18.)+(1./(EXP(X)*X))
      1+((X**4)/96.)-((X**5)/600.)+((X**6)/4320.)-((X**7)/35280.)
C
       X=0.0
       COUNT(1)=0.0
       DO 7 I=1,21
       DO 6 J=1,10
       IF (I.EQ.1) GO TO 1
       GO TO 2
     1 IF (J.EQ.1) GO TO 5
     2 X=X+.01
       IF (X.GE.1.0) GO TO 3
       XO(I,J)=E(X)
       GO TO 6
     3 XO(I,J)=F(X)
       GO TO 6
     5 XO(I,J)=0.0
     6 CONTINUE
       K=I+1
       COUNT(K)=COUNT(I)+.1
     7 CONTINUE
       WRITE (6,99)
       DO 20 I=1,21
       WRITE (6,101) COUNT(I),(XO(I,J),J=1,10)
    20 CONTINUE
       GO TO 500
    99 FORMAT(2X,'X',8X,'0',7X,'1',7X,'2',7X,'3',7X,'4',7X,'5',7X,'6',7X,
      1'7',7X,'8',7X,'9'//)
   101 FORMAT(11F8.4)
   500 CALL EXIT
       END
```

X	0	1	2	3	4	5	6	7	8	9
0.0	0.0	94.9670	45.6552	29.3890	21.3384	16.5566	13.4007	11.1690	9.5120	8.2360
0.1000	7.2254	6.4068	5.7314	5.1656	4.6855	4.2735	3.9167	3.6049	3.3305	3.0875
0.2000	2.8709	2.6770	2.5024	2.3446	2.2013	2.0709	1.9516	1.8423	1.7419	1.6492
0.3000	1.5637	1.4844	1.4108	1.3424	1.2786	1.2191	1.1635	1.1113	1.0625	1.0165
0.4000	0.9734	0.9327	0.8943	0.8581	0.8239	0.7916	0.7609	0.7319	0.7043	0.6781
0.5000	0.6532	0.6296	0.6070	0.5856	0.5651	0.5456	0.5269	0.5091	0.4921	0.4758
0.6000	0.4602	0.4453	0.4310	0.4173	0.4042	0.3916	0.3795	0.3678	0.3567	0.3459
0.7000	0.3356	0.3256	0.3161	0.3069	0.2980	0.2894	0.2812	0.2732	0.2656	0.2582
0.8000	0.2510	0.2441	0.2374	0.2310	0.2248	0.2188	0.2129	0.2073	0.2019	0.1966
0.9000	0.1915	0.1866	0.1818	0.1771	0.1727	0.1683	0.1641	0.1600	0.1560	0.1522
1.0000	0.1484	0.1448	0.1413	0.1379	0.1346	0.1313	0.1282	0.1252	0.1222	0.1193
1.1000	0.1166	0.1138	0.1112	0.1086	0.1061	0.1037	0.1014	0.0991	0.0968	0.0946
1.2000	0.0925	0.0905	0.0884	0.0865	0.0846	0.0827	0.0809	0.0791	0.0774	0.0757
1.3000	0.0741	0.0725	0.0710	0.0694	0.0680	0.0665	0.0651	0.0637	0.0624	0.0611
1.4000	0.0598	0.0586	0.0574	0.0562	0.0550	0.0539	0.0528	0.0517	0.0506	0.0496
1.5000	0.0486	0.0476	0.0467	0.0457	0.0448	0.0439	0.0430	0.0422	0.0413	0.0405
1.6000	0.0397	0.0389	0.0382	0.0374	0.0367	0.0359	0.0352	0.0345	0.0339	0.0332
1.7000	0.0326	0.0319	0.0313	0.0307	0.0301	0.0295	0.0290	0.0284	0.0278	0.0273
1.8000	0.0268	0.0263	0.0257	0.0253	0.0248	0.0243	0.0238	0.0234	0.0229	0.0225
1.9000	0.0220	0.0216	0.0212	0.0208	0.0203	0.0199	0.0196	0.0192	0.0188	0.0184
2.0000	0.0180	0.0177	0.0173	0.0170	0.0166	0.0163	0.0160	0.0156	0.0153	0.0150

Figure 10-14. Printout of a FORTRAN program for computing *y*-factor values; $x = 0.1$ to 2.9.

Example 10-7. Build-Up Density

Write a FORTRAN program for numerically computing values for a build-up density curve in a saturated sand-packed pipe. Scale the diffusivity equation so that the variables are dimensionless and the build-up density curve can be used for solving an infinite number of build-up density problems (see Example 8-1).

Commentary

The handling of a partial differential equation is frequently simplified by transforming the partial differential equation into another partial differential equation which is in terms of dimensionless variables and is free of constants. Usually a solution to the dimensionless variable partial differential equation can be used for solving a large number of problems containing specific numerical values. To illustrate the use of a single solution of a dimensionless variable partial differential equation in obtaining a multiple of solutions, the build-up density problem was selected. The diffusivity equation and boundary conditions for the build-up density problem are:

$$\frac{\partial \rho}{\partial t} = \eta \frac{\partial^2 \rho}{\partial x^2} \qquad \text{10-1}$$

$$(0, 0) = \rho_o \qquad \text{10-2a}$$

$$(L, t) = \rho_e \qquad \text{10-2b}$$

$$(x, 0) = \rho_o + \frac{x}{L}(\rho_e - \rho_o) \qquad \text{10-2c}$$

Since the problem concerns only the increment of density $(\rho_e - \rho_o)$, the above equation may be written as

$$\frac{\partial (\rho - \rho_o)}{\partial t} = \eta \frac{\partial^2 (\rho - \rho_o)}{\partial x^2} \qquad \text{10-3}$$

$$(0, 0) = 0 \qquad \text{10-4a}$$

$$(L, t) = \Delta\rho_e \qquad \text{10-4b}$$

where $\Delta\rho_e = \rho_e - \rho_o$

$$(x, 0) = x/L(\Delta\rho_e) \qquad \text{10-4c}$$

Table 10-1
Selected Build-Up Density Values from Printout of a FORTRAN Program

Dimensionless Time t	x/L = 0.05	$\overline{\Delta\rho} = \Delta\rho/\Delta\rho_e$ x/L = 0.25	x/L = 0.50	x/L = 0.75
1	0.05	0.253	0.500	0.750
10	0.154	0.261	0.500	0.750
20	0.206	0.287	0.503	0.750
30	0.245	0.314	0.509	0.751
40	0.279	0.339	0.518	0.752
60	0.335	0.385	0.541	0.759
80	0.383	0.426	0.566	0.769
120	0.463	0.498	0.616	0.793
160	0.529	0.560	0.662	0.817
200	0.587	0.614	0.703	0.844
400	0.785	0.798	0.845	0.916
800	0.941	0.945	0.958	0.978
1200	0.984	0.985	0.988	0.994
1600	0.995	0.996	0.997	0.998
2000	0.998	0.999	0.999	1.000
2320	0.999	0.999	1.000	1.000

In transforming the diffusivity equation, the following substitutions were used

$$d\,(\rho - \rho_o) = d\bar{\rho}$$

$$x = \beta\bar{x}$$
$$t = \gamma\bar{t}$$

With the substitutions, the diffusivity equation becomes

$$\frac{\partial\,(\rho - \rho_o)}{\partial x} = \frac{\alpha}{B}\frac{\partial\bar{\rho}}{\partial\bar{x}}, \quad \frac{\partial^2\,(\rho - \rho_o)}{\partial x^2} = \frac{\alpha}{B^2}\frac{\partial^2\bar{\rho}}{\partial\bar{x}^2}$$

$$\frac{\partial\,(\rho - \rho_o)}{\partial t} = \frac{\alpha}{\gamma}\frac{\partial\bar{\rho}}{\partial\bar{t}}$$

The substitution of the dimensionless partial derivatives into Equation **10-3** gives

$$\frac{\alpha}{\gamma}\frac{\partial\bar{\rho}}{\partial\bar{t}} = \frac{\eta\alpha}{\beta^2}\frac{\partial^2\bar{\rho}}{\partial\bar{x}^2} \qquad \textbf{10-5}$$

and if

$$\gamma = \frac{\beta^2}{\eta} \qquad \textbf{10-6}$$

Equation **10-5** becomes

$$\frac{\partial \bar{\rho}}{\partial \bar{t}} = \frac{\partial^2 \bar{\rho}}{\partial \bar{x}^2} \qquad \textbf{10-7}$$

In the process of transforming the differential equation, the units α, β, γ, were taken to be identical as the units of the variables ρ, x, t. The variables, $\bar{\rho}$, $\bar{x}$, $\bar{t}$ are then dimensionless. Arbitrarily taking $\beta = L$ and $\gamma = t/\bar{t}$ and substituting these values into Equation **10-6** gives

$$t = \frac{\bar{t}\phi\mu c L^2}{k} \qquad \textbf{10-8}$$

and

$$\bar{t} = \frac{kt}{\phi\mu c L^2}$$

because $\eta = k/\phi\mu c$.

If α is assigned the value of 1.00, the boundary conditions of Equation **10-7** are

$$\rho(0,0) = 0 \qquad \textbf{10-10}$$

$$\rho(1,\bar{t}) = 1 \qquad \textbf{10-11}$$

$$\rho(\bar{x},0) = x/L \qquad \textbf{10-12}$$

The boundary conditions in Equations **10-10, 10-11,** and **10-12** indicate $\bar{x}$ has values for 0-1.0; $\bar{\rho}$ has the values from zero to 1.0 at $\bar{t}$ equal infinity. If Equation **10-7** is solved numerically and $\bar{\rho}$ versus $\bar{t}$ is plotted for various constant values of $\bar{x}$, a set of curves is obtained which can be used for solving a large number of problems with various combinations of ϕ, c, L, μ values.

A flow diagram for writing a FORTRAN program is given in Figure 10-15. The flow diagram indicates that a compiled machine language pro-

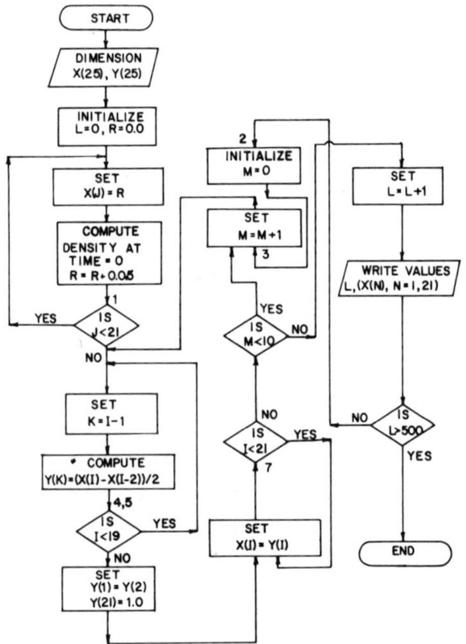

Figure 10-15. Flow diagram of a FORTRAN program for computing build-up density of an oil-saturated, sand-packed pipe.

gram from the FORTRAN program will cause the computer to calculate the $\bar{\rho}$ values at the mesh points in the following manner. First, the computer calculates the $\bar{\rho}$ values at the mesh points along the $\bar{x}$ axis at time $\bar{t} = \bar{t}$. Next, a second set of $\bar{\rho}$ values are calculated for the mesh points at $\bar{t} = \Delta\bar{t}$. Following the computation of the second set of $\bar{\rho}$ values, the second set of $\bar{\rho}$ is caused to be stored in the same storage registers as the first set of $\bar{\rho}$ values. When the second set of values is stored in the same storage registers as the first set of values, the first set of values is erased prior to the storing of the second set of values. After the restoring of the second set of $\bar{\rho}$ values, a third set of $\bar{\rho}$ values is computed. The cycles of computing and restoring of new sets of dimensionless density values are repeated for 500 cycles. Every tenth set of $\bar{\rho}$ is printed out by the computer. In order that a $\bar{\rho}$ value is at $\bar{x} = 0.0$ a value of $\bar{\rho}$ at $\bar{x} = 0$ is set equal to the corresponding $\bar{\rho}$ value at $\bar{x} = 0.05$. At the same time the computer sets $\bar{x}_i = \bar{x}_2$, $\bar{x}_{21}$ was set equal to 1.0.

The fourteenth FORTRAN statement, $Y(K) = (X(I) + X(I+2))/2$ (Figure 10-16), illustrates the use of subscripted variables in FORTRAN statements.

```
C     STEPHEN OWENS
C     BUILD UP DENSITY IN A SAND PACKED PIPE
C     RESERVE 25 STORAGE REGISTERS FOR BOTH X AND Y
      DIMENSION X(25),Y(25)
C     INITIALIZE L, THE NUMBER OF ROWS OF OUTPUT
      L=0
C     INITIALIZE R, THE DENSITY VALUES
      R=0.0
C     COMPUTE INITAL DENSITY POINTS FOR THE MESH
      DO 1 J=1,21
      X(J)=R
      R=R+.05
    1 CONTINUE
C     PRINT HEADINGS
      WRITE (6,98)
      WRITE (6,99)
      GO TO 4
    2 M=0
    3 M=M+1
C     COMPUTE 19 VALUES FOR Y
    4 DO 5 I=1,19
      K=I+1
      Y(K)=(X(I)+X(I+2))/2.
    5 CONTINUE
C     SET Y(1)=Y(2) AND Y(21)=1
      Y(1)=Y(2)
      Y(21)=1.0
C     REPLACE X(I) VALUES IN STORAGE WITH CORESPONDING Y(I) VALUES
      DO 7 I=1,21
      X(I)=Y(I)
    7 CONTINUE
C     DETERMINE WHETHER 10 SETS OF MESH NUMBERS HAVE BEEN COMPUTED
      IF (M.LT.10) GO TO 3
      L=L+1
      NOMER=L*10
C     PRINT OUT THE MESH VALUES FOR DBAR AT .05,.25,.50, AND .75
C     FOR EVER VALUE OF TBAR EQUAL TO 10
      WRITE (6,101) NOMER,X(2),X(6),X(11),X(16)
C     STOP EXECUTION AFTER 2500 SETS OF MESH VALUES HAVE BEEN COMPUTED
      IF (L.GE.250) GO TO 500
      GO TO 2
   98 FORMAT (1X,'DIMENSIONLESS',22X,'DIMENSIONLESS DENSITY')
   99 FORMAT (5X,'TIME',12X,'DBAR=.05',8X,'DBAR=.25',8X,'DBAR=.50',8X,
     1'DBAR=.75'//)
  101 FORMAT (4X,I5,5X,4E16.5)
  500 CALL EXIT
      END
```

Figure 10-16. Printout of a FORTRAN program for computing build-up density of an oil-saturated, sand-packed pipe.

The sixteenth FORTRAN statement, $Y(1) = Y(2)$, causes for a row, the value of ρ at X(1) to be replaced by the value of Y(2) and the nineteenth statement causes the value of X(1) in the storage register to be replaced by the value of Y(1) in the storage register.

The seventeenth FORTRAN statement, $Y(21) = 1.0$, causes the $\bar{\rho}$ value at $X(21)$ to always have a value of 1.0.

The seventh FORTRAN statement, $R = R + 0.05$ is the key statement in a DO Loop which assigns the initial $\bar{\rho}$ values at the mesh points when $\bar{t} = 0$.

Since the computer printed out more than 1,000 $\bar{\rho}$ values, only a few selected $\bar{\rho}$ values are given (see Table 10-1). The data in the table are plotted in Figure 10-17. The value of the constant associated with $\bar{t}$ was obtained as

$$\bar{t} = (\Delta \bar{x})^2/2 \text{ or } \bar{t} = (0.05)^2/2 = 1.25 \times 10^{-3}$$

An example follows which illustrates the use of Figure 10-17 for determining the transient build-up density in an oil-saturated, sand-packed pipe.

Example 10-7a

If the density of oil at $x = 0.0$ cms and time zero is 0.700 and the density of oil is 0.900 at $x = 10000$ cms and time zero in an oil-saturated sand-packed pipe, determine with the aid of Figure 10-17 the time for the density of the oil at $x = 0$ to become 0.800. Also, determine the density for the same time at $x = 7500$ cms. Additional data are: $\phi = 0.20$, $\mu = 1.00$ cp, $k = 1.0$ darcy, $c = 5 \times 10^{-4}$ atm.

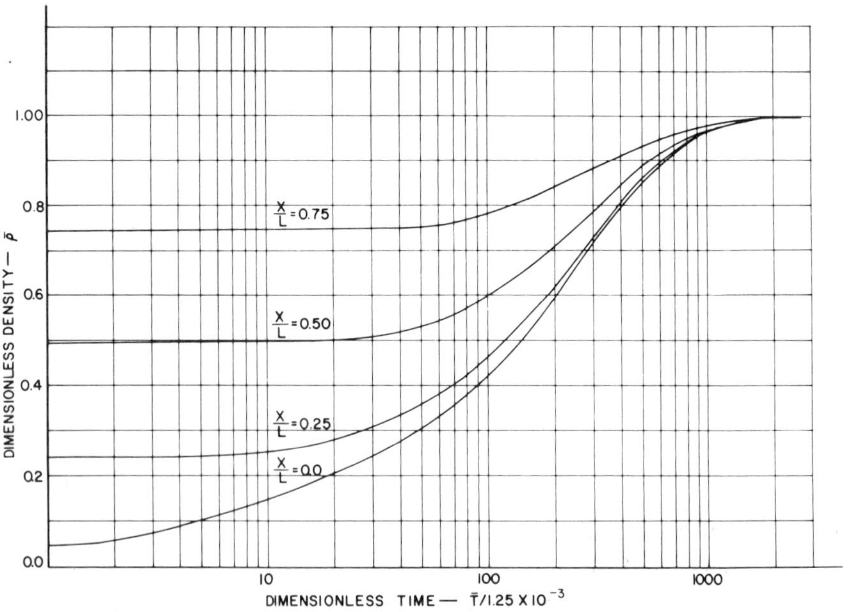

Figure 10-17. Build-up density curve; dimensionless time and dimensionless density increments.

Solution

The incremental maximum density is

$$\Delta \rho_e = 0.900 - 0.700 = 0.200$$

and

$$\bar{\rho} = (0.900 - 0.800)/0.200 = 0.500$$

From Figure 10-18, on the $x/L = 0.00$ curve at $\bar{t} = (1.25 \times 10^{-3})$ (148); therefore

$$t = \bar{t}\phi\mu c L^2/k$$
$$= 1.25 \times 10^{-3} \times 148 \times 0.2 \times 1.0 \times 5.0 \times 10^{-4} \times (10000)^2/1.0$$
$$= 1850 \text{ seconds}$$

And reading along $x/L = 0.75$ curve at $\bar{t} = (1.25 \times 10^{-3})$ (148) the dimensionless density is 0.81 or

$$\rho\bar{x} = 7500 = (0.81 \times 0.2) + 0.7 = 0.862.$$

From Figure 10-16, on the $x/L = 0.00$ curve at $t = (1.25 \times 10^{-3})$ (148); therefore,

Example 10-8. Pressure Distributions

Using the iteration method of solving difference equations, write a FORTRAN program for computing values of pressures at the interior points of a square reservoir if the values for the exterior pressures (in atmospheres) are given as in Table 10-2.

Assume the iteration cycles continue until there is a difference of less than 0.001 between two successive iteration values for the center mesh point (note Example 4-11).

Table 10-2
Mesh Point Values of Pressure

o 00	o1.3125	o5.2500	o11.8125	o21.0000
o 00	o	o	o	o19.2500
o 00	o	o	o	o17.5000
o 00	o	o	o	o15.7500
o 00	o0.875	o3.5000	o7.8750	o14.0000

Commentary

The flow diagram for Example **10-8** is given in Figure 10-18, and the printout from the FORTRAN program is given in Figure 10-19. The mesh points were given consecutive numbers by assigning numbers on each row from left to right. By this system the extreme left mesh point of the first row was assigned No. 1; the extreme right mesh point of the bottom row was assigned No. 25; and the center mesh point was assigned No. 13. To start the iteration process a value of 1.0 was given to each interior mesh point; these initial assigned values eliminated the calculations of the initial values for the interior mesh points.

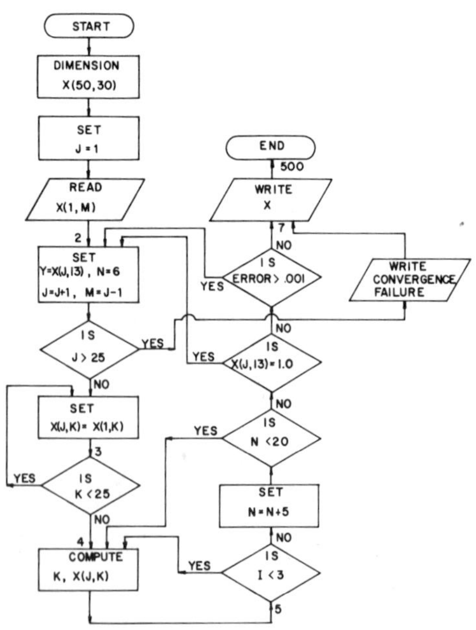

Figure 10-18. Flow diagram of FORTRAN program for computing interior values of pressure when boundary values are given.

```
C       STEPHEN OWENS      ITERATION OF LAPLACE DIFFERENCE EQUATION
C
C
        DIMENSION X(50,30)
        J=1
C       INPUT OF DATA
        READ (5,100) (X(1,M),M=1,25)
      2 Y=X(J,13)
        N=6
        J=J+1
        M=J-1
C       LIMITING CRITERIA
        IF (J.GT.25) GO TO 22
        DO 3 K=1,25
        X(J,K)=X(1,K)
      3 CONTINUE
C       CALCULATION OF INTERNAL VALUES
      4 DO 5 I=1,3
        K=I+N
        X(J,K)=(X(M,K-1)+X(M,K+1)+X(M,K-5)+X(M,K+5))/4.
      5 CONTINUE
        N=N+5
        IF (N.LT.20) GO TO 4
C       CONVERGENCE CRITERIA
        IF (X(J,13).EQ.1.0) GO TO 2
        IF (ABS(X(J,13)-Y).GT..001) GO TO 2
C       OUTPUT OF CALCULATED MATRICES
      7 DO 10 I=1,J
        WRITE (6,101)I,(X(1,K),K=1,6),(X(I,K),K=7,9),X(1,10),X(1,11),(X(I,
       1K),K=12,14),X(1,15),X(1,16),(X(I,K),K=17,19),(X(1,K),K=20,25)
     10 CONTINUE
        GO TO 500
     22 WRITE (6,103)
        GO TO 7
    100 FORMAT (5F10.4)
    101 FORMAT (1X,I3,2X,'ARRAY'//1X,5F12.6/1X,5F12.6/1X,5F12.6/1X,5F12.6
       1/1X,5F12.6//)
    103 FORMAT(1X,'ITERATION FAILED TO CONVERGE IN  25 CYCLES')
    500 CALL EXIT
        END
```

```
  1  ARRAY

   0.0        1.312500    5.250000   11.812500   21.000000
   0.0        1.000000    1.000000    1.000000   19.250000
   0.0        1.000000    1.000000    1.000000   17.500000
   0.0        1.000000    1.000000    1.000000   15.750000
   0.0        0.875000    3.500000    7.875000   14.000000

  2  ARRAY

   0.0        1.312500    5.250000   11.812500   21.000000
   0.0        0.828125    2.062500    8.265625   19.250000
   0.0        0.750000    1.000000    5.125000   17.500000
   0.0        0.718750    1.625000    6.406250   15.750000
   0.0        0.875000    3.500000    7.875000   14.000000

  3  ARRAY

   0.0        1.312500    5.250000   11.812500   21.000000
   0.0        1.031250    3.835938    9.562500   19.250000
   0.0        0.636719    2.390625    8.292969   17.500000
   0.0        0.812500    2.906250    7.593750   15.750000
   0.0        0.875000    3.500000    7.875000   14.000000

  4  ARRAY

   0.0        1.312500    5.250000   11.812500   21.000000
   0.0        1.446289    4.558594   10.797852   19.250000
   0.0        1.058594    3.917969    9.261719   17.500000
   0.0        1.104492    3.574219    8.706055   15.750000
   0.0        0.875000    3.500000    7.875000   14.000000
```

Figure 10-19. Printout of a FORTRAN program for computing interior pressure values from boundary values.

```
5  ARRAY

 0.0        1.312500     5.250000    11.812500    21.000000
 0.0        1.732422     5.353027    11.220703    19.250000
 0.0        1.617188     4.613281    10.230469    17.500000
 0.0        1.376953     4.307129     9.115234    15.750000
 0.0        0.875000     3.500000     7.875000    14.000000

6  ARRAY

 0.0        1.312500     5.250000    11.812500    21.000000
 0.0        2.070679     5.704102    11.661499    19.250000
 0.0        1.930664     5.376953    10.612305    17.500000
 0.0        1.699829     4.651367     9.540649    15.750000
 0.0        0.875000     3.500000     7.875000    14.000000

7  ARRAY

 0.0        1.312500     5.250000    11.812500    21.000000
 0.0        2.236816     6.089783    11.844727    19.250000
 0.0        2.286865     5.724609    11.019775    17.500000
 0.0        1.864258     5.029358     9.722168    15.750000
 0.0        0.875000     3.500000     7.875000    14.000000

8  ARRAY

 0.0        1.312500     5.250000    11.812500    21.000000
 0.0        2.422287     6.264038    12.043015    19.250000
 0.0        2.456421     6.106445    11.197876    17.500000
 0.0        2.047806     5.202759     9.918533    15.750000
 0.0        0.875000     3.500000     7.875000    14.000000

9  ARRAY

 0.0        1.312500     5.250000    11.812500    21.000000
 0.0        2.508240     6.455437    12.131104    19.250000
 0.0        2.644135     6.280273    11.391998    17.500000
 0.0        2.133545     5.393196    10.006409    15.750000
 0.0        0.875000     3.500000     7.875000    14.000000

10  ARRAY

 0.0        1.312500     5.250000    11.812500    21.000000
 0.0        2.603018     6.542404    12.227482    19.250000
 0.0        2.730515     6.471188    11.479446    17.500000
 0.0        2.228083     5.480057    10.102547    15.750000
 0.0        0.875000     3.500000     7.875000    14.000000

11  ARRAY

 0.0        1.312500     5.250000    11.812500    21.000000
 0.0        2.646355     6.637917    12.271088    19.250000
 0.0        2.825572     6.558105    11.575298    17.500000
 0.0        2.271393     5.575451    10.146126    15.750000
 0.0        0.875000     3.500000     7.875000    14.000000

12  ARRAY

 0.0        1.312500     5.250000    11.812500    21.000000
 0.0        2.693996     6.681385    12.318924    19.250000
 0.0        2.868963     6.653553    11.618828    17.500000
 0.0        2.319005     5.618904    10.193932    15.750000
 0.0        0.875000     3.500000     7.875000    14.000000
```

Figure 10-19. Continued.

```
13   ARRAY

     0.0         1.312500     5.250000    11.812500    21.000000
     0.0         2.715712     6.729115    12.340675    19.250000
     0.0         2.916638     6.697018    11.666599    17.500000
     0.0         2.340716     5.666622    10.215679    15.750000
     0.0         0.875000     3.500000     7.875000    14.000000


14   ARRAY

     0.0         1.312500     5.250000    11.812500    21.000000
     0.0         2.739563     6.750847    12.364552    19.250000
     0.0         2.938361     6.744740    11.688335    17.500000
     0.0         2.364565     5.688351    10.239552    15.750000
     0.0         0.875000     3.500000     7.875000    14.000000


15   ARRAY

     0.0         1.312500     5.250000    11.812500    21.000000
     0.0         2.750426     6.774712    12.375416    19.250000
     0.0         2.962216     6.766468    11.712208    17.500000
     0.0         2.375427     5.712212    10.250416    15.750000
     0.0         0.875000     3.500000     7.875000    14.000000


16   ARRAY

     0.0         1.312500     5.250000    11.812500    21.000000
     0.0         2.762357     6.785576    12.387352    19.250000
     0.0         2.973080     6.790333    11.723072    17.500000
     0.0         2.387357     5.723076    10.262352    15.750000
     0.0         0.875000     3.500000     7.875000    14.000000


17   ARRAY

     0.0         1.312500     5.250000    11.812500    21.000000
     0.0         2.767789     6.797508    12.392784    19.250000
     0.0         2.985011     6.801197    11.735004    17.500000
     0.0         2.392789     5.735008    10.267784    15.750000
     0.0         0.875000     3.500000     7.875000    14.000000


18   ARRAY

     0.0         1.312500     5.250000    11.812500    21.000000
     0.0         2.773754     6.802940    12.398750    19.250000
     0.0         2.990443     6.813129    11.740437    17.500000
     0.0         2.398754     5.740440    10.273750    15.750000
     0.0         0.875000     3.500000     7.875000    14.000000


19   ARRAY

     0.0         1.312500     5.250000    11.812500    21.000000
     0.0         2.776470     6.808907    12.401466    19.250000
     0.0         2.996409     6.818562    11.746403    17.500000
     0.0         2.401470     5.746407    10.276466    15.750000
     0.0         0.875000     3.500000     7.875000    14.000000


20   ARRAY

     0.0         1.312500     5.250000    11.812500    21.000000
     0.0         2.779453     6.811623    12.404449    19.250000
     0.0         2.999125     6.824528    11.749119    17.500000
     0.0         2.404453     5.749123    10.279449    15.750000
     0.0         0.875000     3.500000     7.875000    14.000000
```

(figure continued on next page)

Figure 10-19. Continued.

```
21  ARRAY

  0.0        1.312500    5.250000    11.812500    21.000000
  0.0        2.780811    6.814606    12.405807    19.250000
  0.0        3.002109    6.827244    11.752106    17.500000
  0.0        2.405811    5.752106    10.280807    15.750000
  0.0        0.875000    3.500000     7.875000    14.000000

22  ARRAY

  0.0        1.312500    5.250000    11.812500    21.000000
  0.0        2.782303    6.815964    12.407303    19.250000
  0.0        3.003467    6.830231    11.753464    17.500000
  0.0        2.407303    5.753464    10.282303    15.750000
  0.0        0.875000    3.500000     7.875000    14.000000

23  ARRAY

  0.0        1.312500    5.250000    11.812500    21.000000
  0.0        2.782982    6.817459    12.407982    19.250000
  0.0        3.004959    6.831589    11.754959    17.500000
  0.0        2.407982    5.754959    10.282982    15.750000
  0.0        0.875000    3.500000     7.875000    14.000000

24  ARRAY

  0.0        1.312500    5.250000    11.812500    21.000000
  0.0        2.783730    6.818138    12.408730    19.250000
  0.0        3.005638    6.833084    11.755634    17.500000
  0.0        2.408730    5.755638    10.283730    15.750000
  0.0        0.875000    3.500000     7.875000    14.000000

25  ARRAY

  0.0        1.312500    5.250000    11.812500    21.000000
  0.0        2.784069    6.818886    12.409065    19.250000
  0.0        3.006386    6.833759    11.756382    17.500000
  0.0        2.409069    5.756386    10.284065    15.750000
  0.0        0.875000    3.500000     7.875000    14.000000
```

Figure 10-19. Continued.

The subscript M in the FORTRAN program initially refers to the location of the numbers (not potential values) assigned to the mesh points for the purpose of reading in data. Initially the subscript refers to the number of the iteration cycle. The computer program sets a limit of 25 iteration cycles. FORTRAN statement No. 4 causes the computer to store in a storage register the value at mesh point No. 13 which has just been computed; this value is used in the IF Statement (No. 19) of the next iteration cycle. FORTRAN statements Numbers 5, 6, 7 are used for initializing the subscripts of the array prior to an iteration cycle for computing new values for the interior mesh points. A value of $N = 6$ is stored in the computer because there is a difference of 6 between the first mesh point of the first row and the second mesh point (first iteration point) of the second row. The number 5 is also important because there is a difference of 5 between the last iteration mesh point on the second row and the first iteration mesh point on the third row. There also is a difference of 5 between the last iteration point on the third row and the first iteration point on the fourth row.

FORTRAN statement No. 8 causes the computer to check the number of the iteration cycle; if the number of the cycle is 25, the phrase "Iteration Failed to Converge in 25 Cycles" is printed out, and the computer is instructed to print out all data and cease operation. The DO LOOP, FORTRAN statement Nos. 9, 10, 11 are used for assigning subscripts to the array for computed values at the interior mesh points.

The DO LOOP, FORTRAN statement Nos. 12, 13, 14 are used for the computation of new values for a row of interior mesh points. FORTRAN statement No. 16 adjusts the subscripts of the array for computing a set of new values for the next row of interior mesh points.

By assigning initial values of one to the interior mesh points, it is obvious that the value of mesh point No. 13 will not change during the first iteration cycle because all surrounding points have values of one. In order for the program to continue, an IF Statement is required. This IF Statement is required only for the first iteration cycle. For iteration cycles other than the first to continue, another IF Statement (No. 17) is required.

Following the completion of the iteration cycles, the computed values at the mesh points are printed out in an array form as instructed by FORTRAN statement No. 21 and in accordance with FORMAT statement No. 27.

By coincidence 25 iteration cycles were required before the difference between two consecutive values of mesh point No. 13 was less than 0.001.

Note. This FORTRAN computer program was written so that only the immediate previously computed values at the interior mesh points were used during an iteration cycle. This caused more iteration cycles before there was the required convergence at mesh point No. 13. Also this FORTRAN program illustrates the changing of symbolic notation for subscripts of an array. It is the numerical value of the symbolic notation which is effective and not necessarily the symbol.

Example 10-9. Flow Across a Lease Line

For a homogeneous, constant thickness, circular oil reservoir with infinite radius, write a FORTRAN program for computing the flow of oil across a 1,200-foot section of the lease-line for the following data (see Example **9-10**). The lease line parallels the line $x = 800$ ft; number of wells $= 3$; the x, y coordinates of the wells are (600,200), (200, 1000), (1200,1200); the x, y coordinates for the pressure points are (790,1200), (810,1200), (790,1000), (810,1000), (790,800), (810,800), (790,600), (810,600), (790,400),

(810,400), (790,200), (810,200); $p_e = 1525$ psia; $h = 10$ ft; $q_1 = 200$ bbls/day; $q_2 = 300$ bbls/day; $q_3 = 150$ bbls/day; $t_1 = 400$ hrs; $t_2 = 350$ hrs; $t_3 = 375$ hrs; $948.4\ \phi\mu B/k = 0.000033$; $1.127\ k/\mu B = 0.01442$; $70.6\ \mu B/hk = 0.01442$.

Commentary

The flow diagram for this example is given in Figure 10-20, and the printout of the FORTRAN program is given in Figure 10-21. The symbols used in the FORTRAN program are:

$T(I)$ = time
$Q(I)$ = production rate bbls/day
NW = the number of wells
PN = reservoir pressure at the boundary
DL = distance between two corresponding pressure points
$D = 1.127k/\mu B$
$C = 948.4\ c\mu\phi/k$
$B = 70.6\mu B/kh$

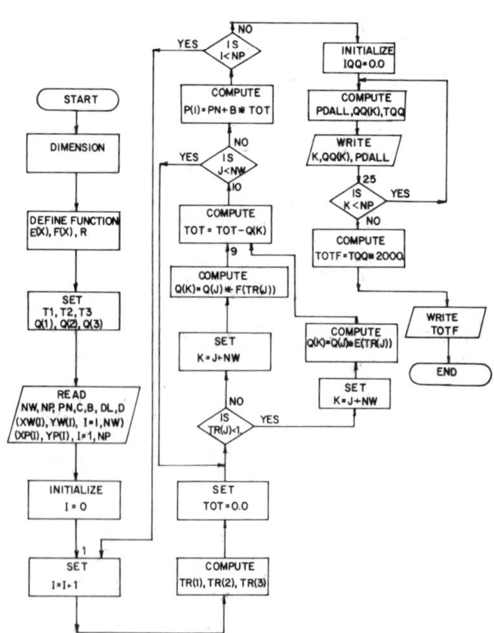

Figure 10-20. Flow diagram of a FORTRAN program for computing the flow of oil across a lease line.

```
C       STEPHEN OWENS   FLOW ACCROSS LEASE LINE
C
        DIMENSION XW(40),YW(40),XP(40),YP(40),Q(80),TR(40),QQ(50),P(80)
C       FUNCTION STATEMENTS
        F(X)=-0.577157+ALOG(1/X)+X-((X**2)/4.)+((X**3)/18.)
       1-((X**4)/96.)+((X**5)/600.)-((X**6)/4320.)+((X**7)/35280.)
        E(X)=-0.577157-ALOG(X)+X-((X**2)/4.)+((X**3)/18.)
       1-((X**4)/96.)+((X**5)/600.)-((X**6)/4320.)+((X**7)/35280.)
        R(X1,Y1,X2,Y2)=SQRT((X2-X1)**2+(Y2-Y1)**2)
C       INPUT OF DATA
        DATA T1,T2,T3,Q(1),Q(2),Q(3)/400.,350.,375.,200.,300.,150./
        READ (5,100) NW,NP,PN,C,B,DL,D
        READ (5,101)(XW(I),YW(I),I=1,NW)
        READ (5,102) (XP(I),YP(I),I=1,NP)
        I=0
C       CALCULATIOM OF DISTANCES AND
C      INDIVIDUAL FLOW RATES
      1 I=I+1
        TR(1)=((R(XP(I),YP(I),XW(1),YW(1))**2)*C)/T1
        TR(2)=((R(XP(I),YP(I),XW(2),YW(2))**2)*C)/T2
        TR(3)=((R(XP(I),YP(I),XW(3),YW(3))**2)*C)/T3
        TOT=0.0
        DO 10 J=1,NW
        IF (TR(J).LT.1.) GO TO 5
        K=J+NW
        Q(K)=Q(J)*F(TR(J))
        GO TO 9
      5 K=J+NW
        Q(K)=Q(J)*E(TR(J))
      9 TOT=TOT+Q(K)
     10 CONTINUE
        P(I)=PN+B*(TOT)
        IF (I.LE.NP) GO TO 1
C       CALCULATION OF  FLOWS ACCROSS LINE
C      AND INDIVIDUAL PRESSURE DROPS
        TQQ=0.0
        DO 25 K=1,NP,2
        PDALL=P(K)-P(K+1)
        QQ(K)=D*PDALL/DL
        TQQ=TQQ+QQ(K)
        WRITE (6,105) K,QQ(K),PDALL
     25 CONTINUE
C       CALCULATION OF TOTAL FLOW
        TOTF=TQQ*200.*10.
        WRITE (6,106) TOTF
        GO TO 500
    100 FORMAT (2I4,F10.5,F12.7,3F10.5)
    101 FORMAT(2F10.2)
    102 FORMAT(2F10.2)
    105 FORMAT(1X,I3,2X,'PRESSURE POINT   ',F20.8,2X,'BBLS/DAY/SQFT '/
       124X,F20.8,2X,'PRESSURE DROP ACCROSS L.L.'/)
    106 FORMAT (1X,'OVERALL FLOW RATE ACCROSS LEASE LINE =',F12.3,'BBLS/D
       1AY ')
    500 CALL EXIT
        END
```

```
  1  PRESSURE POINT              0.00157050  BBLS/DAY/SQFT
                                  2.17822266  PRESSURE DROP ACCROSS L.L.

  3  PRESSURE POINT              0.00388048  BBLS/DAY/SQFT
                                  5.38208008  PRESSURE DROP ACCROSS L.L.

  5  PRESSURE POINT              0.00552737  BBLS/DAY/SQFT
                                  7.66625977  PRESSURE DROP ACCROSS L.L.

  7  PRESSURE POINT              0.00661116  BBLS/DAY/SQFT
                                  9.16943359  PRESSURE DROP ACCROSS L.L.

  9  PRESSURE POINT              0.01049076  BBLS/DAY/SQFT
                                 14.55029297  PRESSURE DROP ACCROSS L.L.

 11  PRESSURE POINT              0.01771255  BBLS/DAY/SQFT
                                 24.56665039  PRESSURE DROP ACCROSS L.L.

OVERALL FLOW RATE ACCROSS LEASE LINE =       91.586BBLS/D AY
```

Figure 10-21. Printout of a FORTRAN program for computing the flow of oil across a lease line.

FORTRAN statement Numbers 3, 4, 5 as indicated on the printout from the computer Statement Functions are for computing the values of exponential integrals and distances from a well to a pressure point. FORTRAN Statement No. 3 is the Statement Function for computing the value of an exponential integral when the value of x is less that one and FORTRAN statement No. 4 is the Statement Function for computing the value of an exponential integral when x is greater than one.

FORTRAN statement No. 5 causes values of production time for each of the three wells to be placed in storage registers and values of production rates for each of the three wells. FORTRAN statement No. 7 causes the x,y coordinate values of the wells to be read and processed. FORTRAN statement No. 8 causes the x,y coordinate values of pressure points to be read and processed. One card with an x,y coordinate set of values for a pressure point will be read and processed at a time.

FORTRAN statement Numbers 11, 12, 13 cause, for given pressure points, the values of $R^2C/T(I)$ to be computed for all three wells. FORTRAN statement No. 16 checks to determine which Statement Function should be used in computing values of the exponential function where $X = R^2C/T(J)$. At the same time an exponential integral is computed, the integral value is multiplied by the appropriate production rate (see FORTRAN statements Numbers 18 and 21). FORTRAN statement No. 22 causes the summing of the products, $Q(N)Ei[-X(N))]$, at each pressure point, and FORTRAN statement No. 24 causes the completion of the calculations for pressure at the pressure points.

FORTRAN statement No. 28 causes, for a number of pressure locations along the lease line, the calculation of the pressure drops across a distance of 20 feet. FORTRAN statement No. 29 causes the calculation of the flow rates across an area of one square foot for the pressure locations along the lease line. And FORTRAN statement No. 30 causes all the individual unit flow rates to be summed. It was assumed that each unit flow rate was average for a 200-foot section of the lease line. Since the thickness of the reservoir was 10 feet, the total flow across the lease line is the sum of the unit flow rates times an area 200 feet by 10 feet. The calculation of the total flow rate across the lease line was caused by FORTRAN statement No. 33.

Example 10-10. Least Square Curve Fit

Write a FORTRAN program for determining the constants of an equation of the form $k_g/k_o = a\, e^{bS_o}$ for the data in Table 10-3.

Table 10-3
Relative Permeability Data

Oil Saturation Percent (S_o)	Relative Permeability k_g/k_o
47	10.000000
50	5.000000
55	1.600000
60	0.589532
65	0.202184
70	0.069340
75	0.023791
80	0.008156
85	0.002797
90	0.000959

Commentary

In Chapter 5 Equations **5-28** and **5-29** were derived for computing constants in the exponential equation by the least squares method. The equations are:

$$\Sigma \ln k_g/k_o = n \ln a + b\,\Sigma S_o \quad \textbf{10-13}$$

and

$$\Sigma S_o \ln k_g/k_o = \Sigma S_o \ln a + b\,\Sigma S_o^2 \quad \textbf{10-14}$$

A direct method for solving for the constants is developed as follows. If it is assumed that $X = \ln a$, $Y = b$, $C = \Sigma \ln k_g/k_o$, $A = n$, $B = \Sigma S_o$, $F = \Sigma S_o \ln k_g/k_o$ and $E = \Sigma S_o^2$, then Equations **10-13** and **10-14** can be written as

$$C = AX + BY \quad \textbf{10-15}$$
$$F = BX + EY \quad \textbf{10-16}$$

Solving Equation **10-15** for X gives $X = C/A - BY/A$ and the substitution of the value X into Equation **10-16** followed by solving for Y gives

$$Y = (F - BC/A)/(E - B^2A) \quad \textbf{10-17}$$

Solving Equation **10-15** for Y gives $Y = C/B - AX/B$ and the substitution of the value of Y into Equation **10-16** gives

$$X = (F - EC/B)/(B - EA/B) \tag{10-18}$$

The flow diagram for the FORTRAN program is given in Figure 10-22, and the printout of the program from the computer is given in Figure 10-23. FORTRAN statement Numbers 5 through 10 cause the computer to compute the values of the constants, B,C,E,F for the matrix. FORTRAN statement Numbers 11 and 12 cause the computer to compute the values of the constants, X or $\log a$ and Y or b for the exponential equation. FORTRAN statement Number 14 causes the computer to solve for the value of a.

FORTRAN statement Numbers 17, 18, 19 cause the computer to print out the phrase "From a Least Squares Fit of the Data the Coefficients Are" and the values of $\log X$, X, and Y.

FORTRAN statement Numbers 21 through 33 cause the computer to use the exponential equation with the computed values of the constants and compute values of k_g/k_o. Then, the computed values of k_g/k_o are compared with the values of k_g/k_o which were fed into the computer.

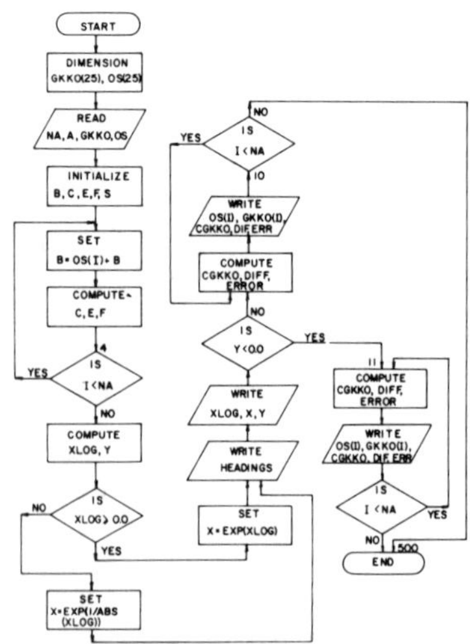

Figure 10-22. Flow diagram of a FORTRAN program for computing the constants of an exponential equation by the least-squares method.

Example 10-11. Depletion-Drive Calculations by Means of Schilthuis' Equation

Using the following data* which describe reservoir characteristics and the relative permeability-fluid saturation equation from Example **10-10,** write a FORTRAN program for computing values of cumulative production versus pressure and gas oil ratio versus pressure for the reservoir. In writing the FORTRAN program use Schilthuis' equation for the material balance.

Commentary

The form of the Schilthuis material balance equation is

$$N = \frac{N_p\,[B + \alpha\,(R_p - R_s)]}{\alpha(R_{so} - R_s) - (B_o - B)} \qquad \textbf{10-19}$$

*Pirson, Sylvan J., *Oil Reservoir Engineering,* 2nd Ed., p. 498ff, McGraw-Hill Book Company, Inc., New York (1958).

Table 10-4
Basic Reservoir Data

Pressure p	Formation Volume Factor B	Solution Gas Sg cu ft/cu ft	Viscosity Ratio $\mu o/\mu g$	Alpha $\frac{p_a T_f Z}{p T_a}$
2000	1.2690	88	48	0.00590
1850	1.2540	83	51	0.00665
1700	1.2390	78	56	0.00730
1550	1.2240	73	63	0.00815
1400	1.2100	68	65	0.09100
1250	1.1940	63	69	0.01030
1100	1.1800	58	74	0.01190
950	1.1640	53	78	0.01400
800	1.1490	48	81	0.01700
650	1.1340	43	88	0.02100
500	1.1190	38	100	0.02800
350	1.1000	32	115	0.04100
200	1.0750	24	125	0.07300
100	1'0520	16	150	0.14800
14.4	1.0000	00	198	1.03800

*Courtesy of McGraw-Hill Publishing Co.

```
C      STEPHEN OWENS
C      LEAST SQUARES CURVE FIT FOR KG/KO
C
       DIMENSION GKKO(25),OS(25)
       READ (5,99) NA,A
       READ (5,100)(GKKO(I),OS(I),I=1,NA)
       DATA B,C,E,F,S/0.0,0.0,0.0,0.0,0.0/
C      COMPUTE CONSTANTS FOR TWO SIMULTANEOUS EQUATIONS
       DO 4 I=1,NA
       B=OS(I)+B
       C=ALOG(GKKO(I))+C
       E=(OS(I)**2)+E
       F=(OS(I)*ALOG(GKKO(I)))+F
     4 CONTINUE
C      SOLVE FOR VALUES OF X AND Y IN THE TWO EQUATIONS
       XLOG=(F-(E*C/B))/(B-(E*A/B))
       Y=(F-(B*C/A))/(E-((B**2)/A))
       IF (XLOG) 7,5,5
     5 X=EXP(XLOG)
       GO TO 8
     7 X=EXP(1/(ABS(XLOG)))
C      WRITE CONSTANT VALUES
     8 WRITE (6,101)
       WRITE (6,102) XLOG,X,Y
       WRITE (6,103)
       IF (Y.LT.0.0) GO TO 11
C      EVALUATE FITTED EXPRESSION
       DO 10 I=1,NA
       CGKKO=X*EXP(Y*OS(I))
       DIFF=GKKO(I)-CGKKO
       ERROR=(DIFF/GKKO(I))*100.
       WRITE (6,104)OS(I),GKKO(I),CGKKO,DIFF,ERROR
    10 CONTINUE
       GO TO 500
    11 DO 12 I=1,NA
       CGKKO=X*(1./(EXP(ABS(Y*OS(I)))))
       DIFF=GKKO(I)-CGKKO
       ERROR=(DIFF/GKKO(I))*100.
       WRITE (6,104)OS(I),GKKO(I),CGKKO,DIFF,ERROR
    12 CONTINUE
       GO TO 500
    99 FORMAT (I5,F10.2)
   100 FORMAT (2F10.5)
   101 FORMAT (1X,' FROM A LEAST SQUARES FIT OF THE DATA THE COEFFICIENTS
      1 ARE '/)
   102 FORMAT (1X,'LOG(X)=',E16.6/1X,'    X=',E16.6/1X,'    Y=',E16.6//)
   103 FORMAT(1X,3X,'OIL SATURATION',2X,'    KG/KO',5X,'  KG/KO CALC.'
      1,6X,'DIFFERENCE',4X,'ERROR'//)
   104 FORMAT (1X,4E16.6,F8.2)
   500 CALL EXIT
       END
```

```
 FROM A LEAST SQUARES FIT OF THE DATA THE COEFFICIENTS ARE

LOG(X)=     0.123134E 02
   X=    0.222656E 06
   Y=   -0.214030E 02
```

OIL SATURATION	KG/KO	KG/KO CALC.	DIFFERENCE	ERROR
0.470000E 00	0.100000E 02	0.952544E 01	0.474562E 00	4.75
0.500000E 00	0.500000E 01	0.501220E 01	-0.121965E-01	-0.24
0.550000E 00	0.165000E 01	0.171897E 01	-0.689688E-01	-4.18
0.600000E 00	0.600000E 00	0.589532E 00	0.104675E-01	1.74
0.650000E 00	0.200000E 00	0.202184E 00	-0.218415E-02	-1.09
0.700000E 00	0.670000E-01	0.693403E-01	-0.234038E-02	-3.49
0.750000E 00	0.240000E-01	0.237807E-01	0.219271E-03	0.91
0.800000E 00	0.790000E-02	0.815582E-02	-0.255816E-03	-3.24
0.850000E 00	0.280000E-02	0.279711E-02	0.288989E-05	0.10
0.900000E 00	0.100000E-02	0.959279E-03	0.407209E-04	4.07

Figure 10-23. Printout of a FORTRAN program for computing the constants of an exponential equation by the least-squares method.

and the form of the gas oil ratio equation is

$$R = R_s + B\frac{k_{rg}}{k_{ro}}\frac{\mu_o}{\mu_g}\frac{1}{\alpha} \qquad \textbf{10-20}$$

The expanded forms of Equations **10-19** and **10-20** for computing incremental gas production are:

$$G_p = NR_p = N[(R_{so} - R_s) - 1/\alpha\,(B_o - B)] - N_p\,(B/\alpha - R_s) \qquad \textbf{10-21}$$

$$G_p = \int_0^N RdN \qquad \textbf{10-22}$$

$$G_{p2} - G_{p1} = N[(R_{s1} - R_{s2}) - (B_o - B_2)/\alpha_2 + (B_o - B_1)/\alpha_1] - N_{p2}(B_2/\alpha_2 - R_{s2}) + N_{p1}(B_1/\alpha_1 - R_{s1}) \qquad \textbf{10-23}$$

$$S_o = (1 - N_p)\,B/B_o \qquad \textbf{10-24}$$

and

$$G_{p2} - G_{p1} = \frac{R_1 + R_2}{2}(N_{p2} - N_{p1}) \qquad \textbf{10-25}$$

where

B = formation volume factor, cu ft/cu ft
k_{rg} = relative permeability of gas
k_{ro} = relative permeability of oil
N = number of units of stock-tank oil, cu ft, originally in reservoir
N_p = number of units of stock-tank oil, cu ft, produced up to a given time
R = instantaneous gas-oil ratio (standard cu ft of gas per cu ft of stock-tank oil)
R_p = net average cumulative gas-oil ratio (standard cu ft gas per cu ft stock-tank-oil)
R_s = solubility of gas in oil on a unit per unit basis (standard cu ft of gas per cu ft of stock-tank oil)
S_o = oil saturation
α = reservoir volume of one unit of gas at standard conditions of temperature and pressure
$\alpha = 14.7\,T_f Z/520$
μ_g = viscosity of gas, centipoise
μ_o = viscosity of oil, centipoise.

To facilitate the writing of a FORTRAN program the following substitutions were used: $GK = k_{rg}/k_{ro}$; $R = GOR$; $U = \mu_o/\mu_g$; $A = \alpha$; $SG = R_s$; $N = 1$; and $PN = N_p$; and using these substitutions, the equations for the FORTRAN program are Equations **10-26** to **10-30.**

$$GR(K) = (SG(I) - SG(K) - ((B(I) - B(K))/A(K))) \\ - (PN(K) * ((B(K)/A(K)) - SG(K))) \qquad \textbf{10-26}$$

where $K = I + 1$

$$GR(K) = (SG(I) - SG(K) - ((B(1) - B(K))/A(K) + ((B(1) \\ - B(I))/A(I))) - (PN(K) * ((B(K)/A(K) - SG(K))) \\ + (PN(I) * ((B(I)/A(I) - SG(I))) \qquad \textbf{10-27}$$

$$GK(J) = X/(EXP(Y * SO(J)) \qquad \textbf{10-28}$$

$$SO(J) = ((1 - PN(J)) * B(J))/B(1) \qquad \textbf{10-29}$$

and

$$GL(K) - (SG(K) + ((B(K) * U(K) * X/(EXP(Y * B(K) * ((1 \\ - PN(K))/B(1))))))/A(K)) + SG(I) + ((B(I) * U(I) \\ + X/(EXP(Y * B(I) * ((1 - PN(I))/B(1))))))/A(I))) \\ * ((PN(K) = PN(I))/2) \qquad \textbf{10-30}$$

The flow diagram for the FORTRAN program for depletion-drive calculations by means of the Schilthuis equation is given in Figure 10-24 and the printout from the computer is given in Figure 10-25. The printout in Figure 10-25 is from a Xerox Sigma VI; this printout is different from the printouts from the IBM 360. The Xerox Sigma VI assigns numbers to every line, whereas the IBM 360 only assigns numbers to FORTRAN statements.

The FORTRAN statement DIMENSION causes the computer to reserve 25 storage registers for each set of data and computed data. The first READ statement causes the computer to read into the computer the constants for the exponential equations which were computed in Example **10-10.** The second READ statement causes the computer to take the data necessary for computing the values of cumulative oil production and gas-oil ratio. The symbol NUM of the READ statement refers to the 15 values in each set of data. The DATA statement causes the computer to take in the initial values of the computed data; these values are all equal to zero. The symbol

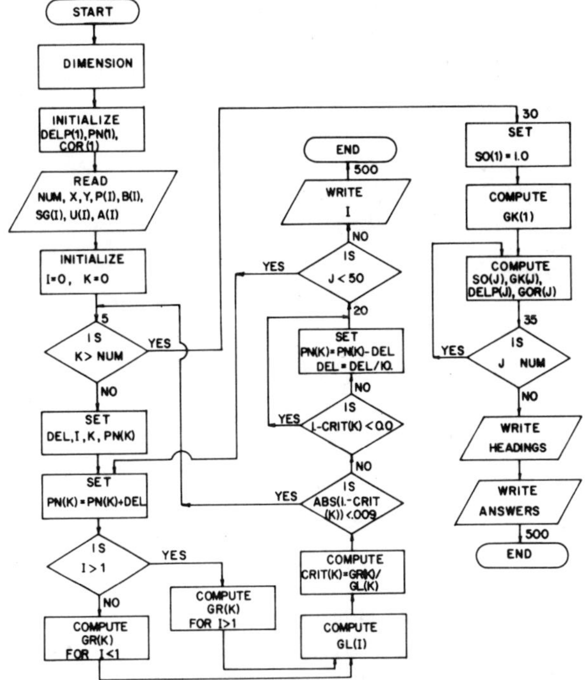

Figure 10-24. Flow diagram of a FORTRAN program for depletion-drive calculations by means of Schilthuis' equation.

CRIT(I) refers to the ratio of the incremental gas production computed by the two different methods (material balance and gas-oil-ratio formulas, $GR(K)/GL(K)$).

The first FORTRAN IF statement (No. 5), checks to determine whether values of cumulative oil production have been computed for all of the values of pressure which were fed into the computer. If a total of 15 values of cumulative oil production have been computed, the computer will compute values of oil saturation, relative permeability ratios (k_{rg}/k_{ro}), incremental cumulative oil produced, and gas-oil ratio (note FORTRAN statements under the comment "Calculation of oil saturation, *KG/KO,* delta oil prod, and GOR"). If a total of 15 values of cumulative oil production has not been computed, the computer will proceed to compute values of cumulative oil production.

The calculation of the first value for the cumulative oil production requires a different modified material balance equation than the calcula-

```
C         JOHN M. DANIEL
C        OIL PRODUCTION BY DEPLETION DRIVE
         DIMENSION P(25),B(25),SG(25),U(25),A(25),DELP(25),PN(25),GOR(25),
        1GR(25),GL(25),CRIT(25),SO(25),GK(25)
         DATA DELP(1),PN(1),GR(1),GL(1),CRIT(1),GOR(1)/0.0,0.0,0.0,0.0,0.0,
        10.0/
         READ (5,99) NUM,X,Y
         READ (5,100) (P(I),B(I),SG(I),U(I),A(I),I=1,NUM)
         I=0
         K=0
       5 IF (K.GE.NUM) GO TO 30
         DEL=.01
         I=I+1
         K=I+1
         PN(K)=PN(I)
         DO  20 J=1,50
C        INCREMENTING OIL SATURATION
         PN(K)=PN(K)+DEL
         IF (I.GT.1) GO TO 6
C        EXPRESSION FOR THE RIGHT SIDE FOR N=2
         GR(K)=(SG(I)-SG(K)-((B(1)-B(K))/A(K)))-(PN(K)*
        1((B(K)/A(K))-SG(K)))
         GO TO 7
C        EXPRESSION FOR THE RIGHT SIDE FOR N GT 2
       6 GR(K)=(SG(I)-SG(K)-((B(1)-B(K))/A(K))+((B(1)-B(I))/A(I)))-(PN(K)*
        1((B(K)/A(K))-SG(K)))+(PN(I)*((B(I)/A(I))-SG(I)))
C        EXPRESSION FOR THE LEFT SIDE FOR ALL VALUES OF N
       7 GL(K)=(SG(K)+((B(K)*U(K)*X/(EXP(Y*B(K)*((1.-PN(K))/B(1)))))/A(K))
        1+SG(I)+((B(I)*U(I)*X/(EXP(Y*B(I)*((1.-PN(I))/B(1)))))/A(I)))*((
        2PN(K)-PN(I))/2.)
C        CONVERGENCE TEST -- GR/GL
         CRIT(K)=GR(K)/GL(K)
         IF (ABS(1.-CRIT(K)).LT..009) GO TO 5
         IF ((1.-CRIT(K)).LT.0.0) GO TO 20
C        CHANGING THE INCREMENTAL INCREASE OF N BY FACTORS OF1/10
         PN(K)=PN(K)-DEL
         DEL=DEL/10.
      20 CONTINUE
         WRITE (6,101) I
         GO TO 500
C        CALCULATION OF OIL SATURATION, KG/KO, DELTA OIL PROD, AND GOR.
      30 SO(1)=1.0
         GK(1)=X/(EXP(Y))
         DO 35 J=2,NUM
         SO(J)=((1.-PN(J))*B(J))/B(1)
         GK(J)=X/(EXP(Y*SO(J)))
         DELP(J)=PN(J)-PN(J-1)
         GOR(J)=GL(J)/DELP(J)
      35 CONTINUE
         WRITE (6,107)
```

Figure 10-25. Printout of the FORTRAN program for depletion-drive calculations by means of Schilthuis' equation.

tions for the other values of cumulative oil production. The second FORTRAN IF statement causes the computer to use the correct modified material balance equation. FORTRAN statement No. 7 is used for calculating values of cumulative oil production other than the first.

A new value of cumulative oil production for an incremental decrease of pressure is computed by an iteration process:

(1) to a previously computed value of cumulative oil production ($PN(I)$), an increment of 0.01 is added;

```
      WRITE (6,108)
      WRITE (6,102)
      WRITE (6,103) (P(I),SO(I),GK(I),PN(I),DELP(I),GR(I),GL(I),CRIT(I),
     1GOR(I),I=1,NUM)
      WRITE (6,104)
      WRITE (6,105)
      WRITE (6,106) (P(I),B(I),SG(I),U(I),A(I),I=1,NUM)
      GO TO 500
   99 FORMAT(I5,2F10.5)
  100 FORMAT (5F10.5)
  101 FORMAT (1X,'ITERATION FAILED TO CONVERGE ON THE',I3,'ATTEMPT')
  102 FORMAT (3X,'PRESSURE',5X,'OIL SAT',6X,'KG/KO',7X,'N(I)',5X,'N(I+1)
     1-N(I)',5X,'GR',10X,'GL',8X,'GR/GL',7X,'GOR(I)'/)
  103 FORMAT (1X,9E12.5)
  104 FORMAT ('0',' THE DATA USED IS AS FOLLOWS'//)
  105 FORMAT (3X,'PRESSURE',6X,'BETA',7X,'SG(I)',7X,'UO/UG',7X,'ALPHA'/)
  106 FORMAT (1X,5E12.5)
  107 FORMAT (1X//2X,'OIL PRODUCTION BY DEPLETION DRIVE')
  108 FORMAT (1X,'==================================='//)
  500 CALL EXIT
      END
```

PRESSURE	OIL SAT	KG/KO	N(I)	N(I+1)-N(I)	GR	GL	GR/GL	GOR(I)
.20000E 04	.10000E 01	.11283E-03	.00000E 00	.00000E 00	.00000E 00	.00000E 00	.00000E 00	.00000E 00
.18500E 04	.97415E 00	.19622E-03	.14200E-01	.14200E-01	.12432E 01	.12353E 01	.10076E 01	.87026E 02
.17000E 04	.94385E 00	.37531E-03	.33300E-01	.19100E-01	.15908E 01	.15895E 01	.10007E 01	.83227E 02
.15500E 04	.90763E 00	.81474E-03	.59000E-01	.25700E-01	.20887E 01	.20852E 01	.10017E 01	.51138E 02
.14000E 04	.86664E 00	.19590E-02	.91100E-01	.32100E-01	.26734E 01	.26585E 01	.10056E 01	.82820E 02
.12500E 04	.82131E 00	.51688E-02	.12710E 00	.36000E-01	.33941E 01	.34069E 01	.99623E 00	.34637E 02
.11000E 04	.77728E 00	.13265E-01	.16410E 00	.37000E-01	.47747E 01	.48040E 01	.99390E 00	.12984E 03
.95000E 03	.73986E 00	.29545E-01	.19340E 00	.29300E-01	.59036E 01	.58590E 01	.10076E 01	.19997E 03
.80000E 03	.70932E 00	.56801E-01	.21660E 00	.23200E-01	.70290E 01	.70013E 01	.10038E 01	.30178E 03
.65000E 03	.68487E 00	.95860E-01	.23360E 00	.17000E-01	.73035E 01	.72886E 01	.10020E 01	.42874E 03
.50000E 03	.66329E 00	.15213E 00	.24780E 00	.14200E-01	.81543E 01	.81259E 01	.10035E 01	.57226E 03
.35000E 03	.64111E 00	.24459E 00	.26040E 00	.12600E-01	.90684E 01	.90252E 01	.10048E 01	.71631E 03
.20000E 03	.61527E 00	.42520E 00	.27370E 00	.13300E-01	.10656E 02	.10595E 02	.10057E 01	.79666E 03
.10000E 03	.59241E 00	.69360E 00	.28540E 00	.11700E-01	.91909E 01	.91336E 01	.10057E 01	.78111E 03
.14400E 02	.54326E 00	.19855E 01	.31060E 00	.25200E-01	.14370E 02	.14292E 02	.10055E 01	.56713E 03

THE DATA USED IS AS FOLLOWS

PRESSURE	BETA	SG(I)	UO/UG	ALPHA
.20000E 04	.12690E 01	.88000E 02	.48000E 02	.59000E-02
.18500E 04	.12540E 01	.83000E 02	.51000E 02	.66500E-02
.17000E 04	.12390E 01	.78000E 02	.56000E 02	.73000E-02
.15500E 04	.12240E 01	.73000E 02	.63000E 02	.81500E-02
.14000E 04	.12100E 01	.68000E 02	.65000E 02	.91000E-02
.12500E 04	.11940E 01	.63000E 02	.69000E 02	.10300E-01
.11000E 04	.11800E 01	.58000E 02	.74000E 02	.11900E-01
.95000E 03	.11640E 01	.53000E 02	.78000E 02	.14000E-01
.80000E 03	.11490E 01	.48000E 02	.81000E 02	.17000E-01
.65000E 03	.11340E 01	.43000E 02	.88000E 02	.21000E-01
.50000E 03	.11190E 01	.38000E 02	.10000E 03	.28000E-01
.35000E 03	.11000E 01	.32000E 02	.11500E 03	.41000E-01
.20000E 03	.10750E 01	.24000E 02	.12500E 03	.73000E-01
.10000E 03	.10520E 01	.16000E 02	.15000E 03	.14800E 00
.14400E 02	.10000E 01	.00000E 00	.19800E 03	.10380E 01

EXIT

Figure 10-25. Continued.

(2) for this new value of cumulative oil production values of oil saturation *SO*(*J*), *KG*/*KO*, *GR*(*K*), are computed;

(3) then a ratio *GR*(*K*)/*GL*(*K*) is computed and tested to determine whether the ratio is less than 1.0 by a difference less than 0.009;

(4) if the difference is less than 0.009 the computer continues to calculate a new value of cumulative oil production for a lower pressure;

(5) if the difference is greater than 0.009 a new increment of 0.01 is added to the previously tested value of *PN*(*I*) and the cycle is repeated;

(6) however, if the ratio *GR*(*K*)/*GL*(*K*) is greater than 1.0 the increment of *PN* is reduced by one-tenth and added to the previously tested

value of $PN(I)$ and the cycle is repeated until a value of $(1.0 - GR(K)/GL(K))$ is obtained which is less than 0.009;

(7) in the event an acceptable value for the ratio $GR(K)/GL(K)$ is not obtained with the 0.001 increment of PN, the computer will cause the increment to be further reduced by a factor of ten (0.0001).

In preliminary calculations, it was found that $GL(K)$ was less than $GR(K)$ for the first few increments of PN. The FORTRAN statements used for checking the values of the ratios of $GR(K)/GL(K)$ and causing the computer to either decrease the increment or compute a value of cumulative oil production for a lower pressure are given under the heading "Convergence Test—*GR/GL*." FORTRAN statement, $DEL = DEL/10$ actually causes the increment of PN to be decreased by a factor of ten. If after 13 cycles of adding increments of cumulative oil production the ratio $GR(K)/GL(K)$ is either greater than 1.0 or its difference from 1.0 is greater than 0.009, the FORTRAN statement WRITE (6,101)*I*, causes the phrase "Iteration Failed to Converge on the '13' Attempt" to be printed out and the computer ceases the processing of the computer program.

FORTRAN statement No. 107 causes the phrase "Oil Production by Depletion Drive" to be printed above the data on the print-out of the FORTRAN computer program and the FORTRAN statement No. 108 causes the previous heading to be underlined.

Table 10-5
Comparison of High-Speed Electronic Digital Computer Results with Conventional Computed Results

Pressure psia	Cumulative Fraction Oil Produced Daniel	Pirson	Gas-Oil-Ratio—cuft/cuft Daniel	Pirson
2000	0.0000	0.0000	000.00	000.0
1850	0.0142	0.0142	87.03	85.5
1700	0.0330	0.0340	83.23	80.5
1550	0.0590	0.0600	81.14	75.5
1400	0.0911	0.0930	82.82	78.8
1250	0.1271	0.1300	94.64	94.7
1100	0.1641	0.1660	127.84	130.0
950	0.1934	0.1960	199.97	202.5
800	0.2166	0.2180	301.78	326.0
650	0.2336	0.2330	428.74	466.0
500	0.2478	0.2470	572.26	589.0
350	0.2604	0.2600	716.31	713.0
200	0.2737	0.2735	796.66	796.0
100	0.2854	0.2860	781.11	752.0
14.4	0.3106	0.3110	567.13	536.0

Exercises

1. Write a FORTRAN program for computing the sum of products,

   $n(n-1)$ from $n = 2$ to $n = 20$.

2. Write a FORTRAN program for using the reciprocal method of division for determining the quotient $912/38 = ?$.
3. Write a FORTRAN program for converting numbers in the decimal system into numbers in the octal system.
4. Write a FORTRAN program for converting numbers in the binary system into numbers in the decimal system.

5-12. Write FORTRAN programs for solving with the aid of electronic digital com-

13. Write a FORTRAN program for least squares fitting of a polynomial to a set of data $(B = a + bp + cp^2)$ and then determine the constants for the polynomial using the curve in Figure 10-26 (B vs. p).
14. Write a FORTRAN program for computing the reservoir performance of a depletion-drive field which have the characteristics as represented by Figures 10-26 through 10-30. Use the Schilthuis material balance equation.
15. Consult a reference text and repeat Exercise 10-14 by using Muskat's equation.
16. Rewrite the material balance equations as given in Example 10-12 to include a gas cap.

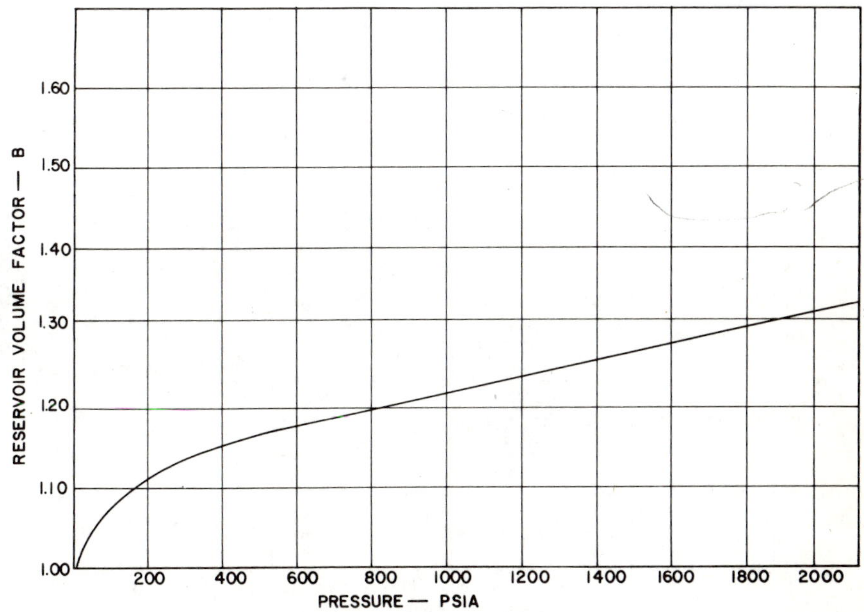

Figure 10-26. Reservoir volume factor.

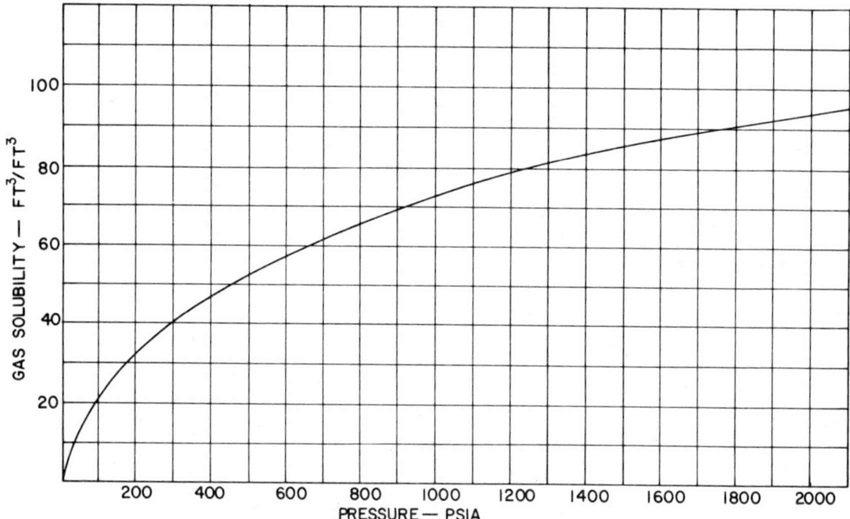

Figure 10-27. Gas solubility.

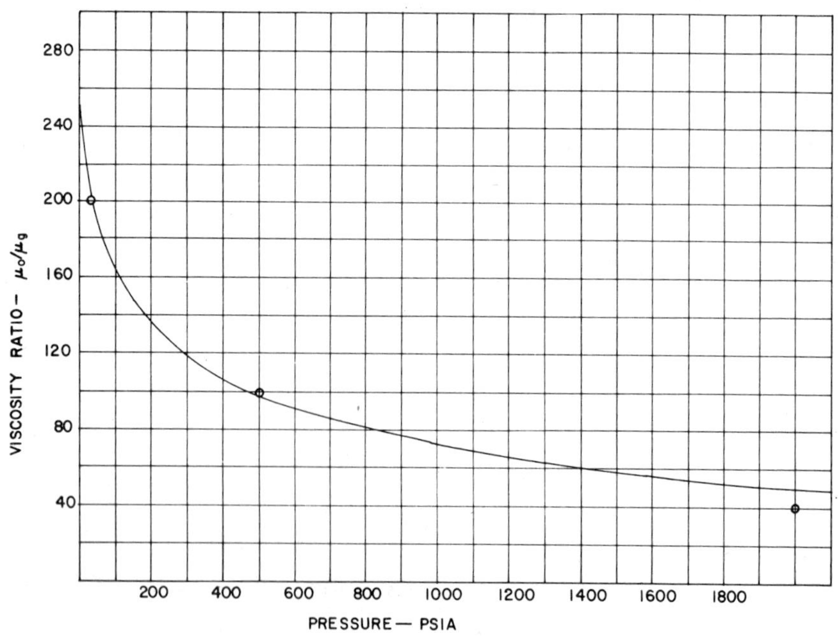

Figure 10-28. Viscosity ratio.

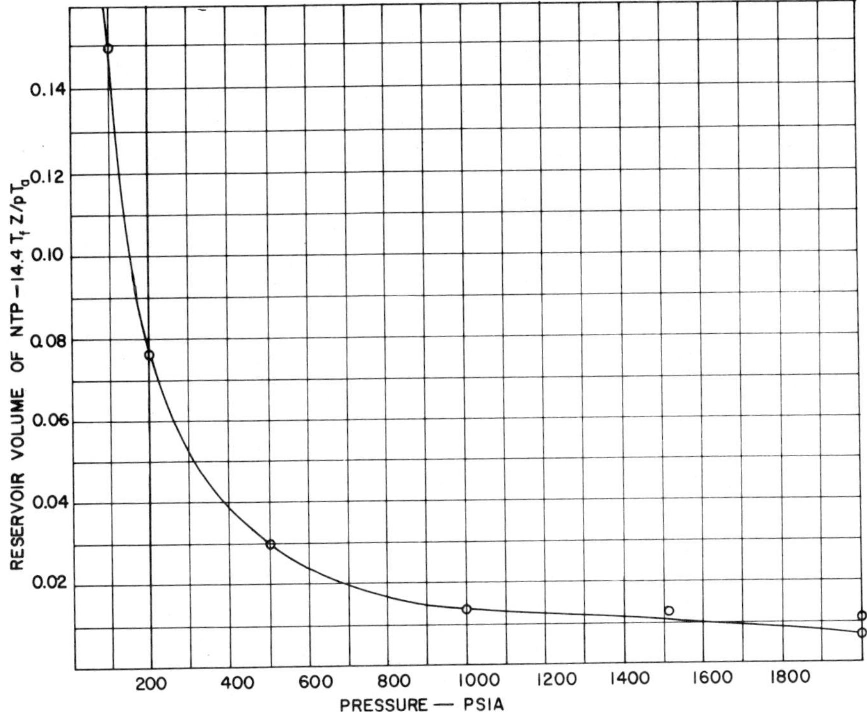

Figure 10-29. Reservoir volume of 1.0 NTP cubic foot of gas.

17. Assuming the volume of the gas cap is 0.2 that of the volume for the oil volume, repeat Exercise 10-14 to include this volume.
18. The differential equation for the flow of fluid through a circular tube is

$$\frac{dV_z}{dr} = -\left(\frac{P_o - P_1}{2uL}\right)r$$

In terms of dimensionless variables, determine the expressions for V_z, $<V_z>$, Q and R for the boundary conditions
at $r = 0$ the momentum flux is not infinite
and at $r = R$ (radius of tube) $V_s = 0$
and where

$$<V_z> = \frac{\int_0^{2\pi}\int_0^R V_z\, r dr d\theta}{\int_0^{2\pi}\int_0^R r dr d\theta}$$

and Q = Area x $<V_s>$

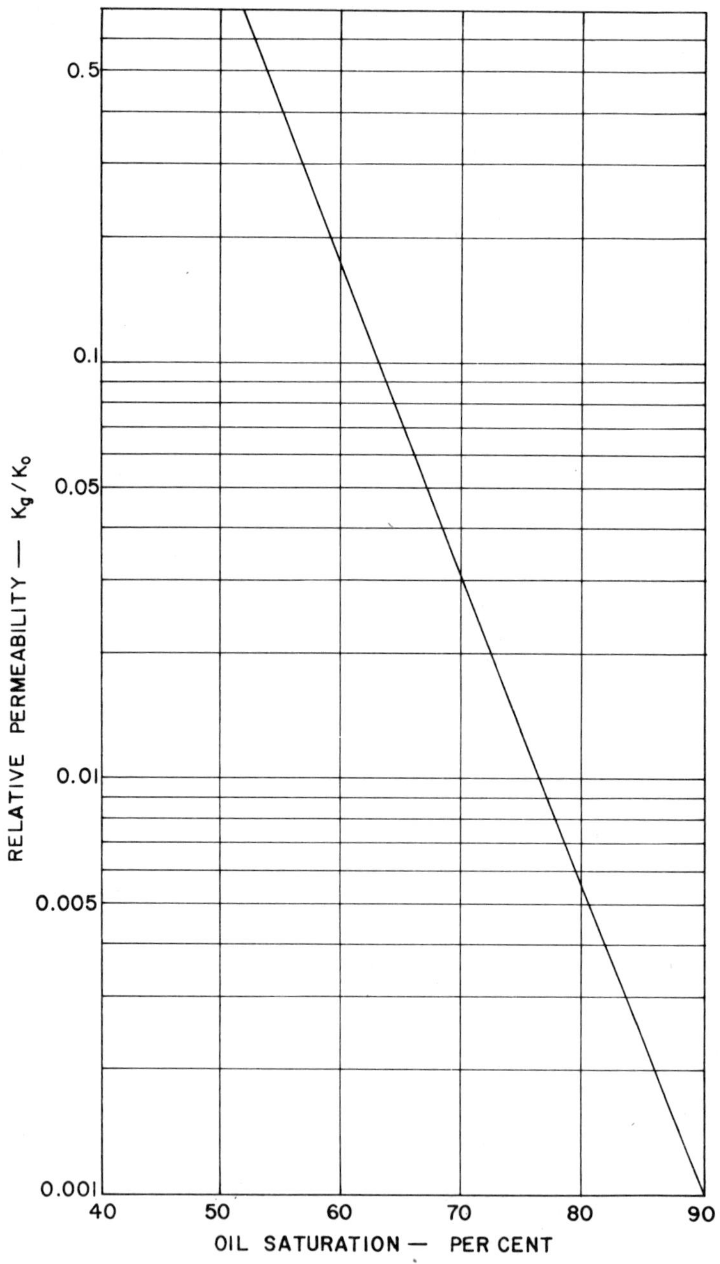

Figure 10-30. Relative permeability ratio as a function of oil saturation.

Suggested Reading

1. Bird, R. B., Stewart, W. E., and Lightfoot, E. N., *Transport Phenomena,* New York: John Wiley and Sons Inc., 1962, p. 42ff, p. 107ff.
2. Carnahan, B., Luther, H. A., and Wilkes, J. O., *Applied Numerical Methods,* New York: John Wiley and Sons Inc., 1969.
3. Craft, D. C. and Hawkins, M. F., *Applied Petroleum Reservoir Engineering,* Englewood Cliffs, N. J.: Prentice-Hall, Inc., 1959, p. 147ff.
4. Dimitry, D. L. and Mott, Jr., T. H., *Introduction to FORTRAN IV Programming,* New York: Holt, Rinehart and Winston, 1966.
5. Langhaar, H. L., *Dimensional Analysis and Theory of Models,* New York: John Wiley and Sons, Inc., 1951, p. 13ff.
6. Lipka, J., *Graphical and Mechanical Computations,* New York: John Wiley and Sons Inc., 1918, p. 120ff.
7. Murrill, P. W., and Smith, C. L., *FORTRAN Programming for Engineers and Scientists,* Scranton: International Textbook Co., 1968.
8. Muskat, M. and Wyckoff, R. D., *The Flow of Homogeneous Fluids Through Porous Media,* Ann Arbor: J. Edwards, Inc., 1946.
9. Pirson, Sylvain J., *Oil Reservoir Engineering, 2nd ed.,* New York: McGraw-Hill Book Co., Inc., 1958, p. 473ff.
10. Richards, R. K., *Arithmetic Operations in Digital Computers,* Princeton: D. Van Nostrand Co., Inc., 1955, p. 275ff.
11. Scarborough, J. B., *Numerical Mathematical Analysis, 2nd ed.,* Baltimore: The Johns Hopkins University Press, 1950, p. 312ff.

11 BASIC Programs and Examples

All but one of the FORTRAN programs given in Chapter 10 have been rewritten in BASIC language and are given in the examples which follow. Two of the programs are written for, and run on, a minicomputer, which uses a Teletype (Model 33) for the input/output device. A total of 16 BASIC programs are discussed in 16 examples. The last program is written for and will run on a mainframe computer. An Apple IIe was used as a terminal to the mainframe computer. In connection with the last example, an illustration on how to use a terminal to the mainframe is given.

In general, the introductory comments in Chapter 10 about FORTRAN programs and the selection thereof apply for BASIC programs. Also, the discussions of BASIC programs here parallel the discussions of FORTRAN programs in Chapter 10.

Examples 13 and 15 contain BASIC programs for examples given in other petroleum engineering textbooks. In writing the BASIC programs for the examples, it was assumed the reader is familiar with the texts from which data were taken. A BASIC program for solving algebraic simultaneous equations is given in Example 16.

Computer algebraic expressions are used throughout this chapter, instead of the mathematician's algebraic equations. For instance, the algebraic equation

$$\frac{x^2 + y^2}{2}$$

is written as

(X^2 + Y^2)/2

Word processing is simplified by using this method.

Programmed BASIC Examples

As the BASIC examples are studied, Chapter 3 should be referred to if needed. Also, the details in which the BASIC statements are written should be noted with care, that is, commas, parentheses, semicolons, periods, etc. Each has a meaning and must be placed in an exact manner. Flow diagrams have been omitted; however, some of the flow diagrams given in Chapter 10 also apply to the examples in Chapter 11.

If not specified, the programs for the examples are written in Applesoft BASIC and run on an Apple IIe microcomputer.

Example 11-1. Sums of Number Sets I

Write a BASIC program for computing the sums of 15 sets of two numbers such as

$$Z(I) = X(I) + Y(I) \qquad I = 1 \text{ to } 15$$

using counter looping, that is, using READ, IF/THEN, and GOTO statements. Compare the results with those of Example 10-1. The printed computed values from the program in Figure 11-1 are identical to those of Example 10-1.

The REM statements were inserted in the BASIC program for remarks about given sections of the program. The remarks were not printed out in the results. Only information (other than variables) enclosed in quotes in PRINT statements is printed out in the results; statement 220 is an example of what was used for printing out the headings for the columns of numbers.

The DATA statements 60 through 190 were used to input the data of the number sets; a DATA statement was used for each set. Statement 197 was used for initializing the counter.

```
]LIST

10  REM  COMPUTE AND PRINT Z(I)=X
     (I)+Y(I)
20  REM  FOLLOWING DATA LISTS ARE
     IN THE FORM:
30  REM         DATA  X(I),Y(I)
40  REM   AND ARE IN ORDER I = 1
    TO N
50  REM   THE TWO NUNBER SETS ARE

60  DATA  521.41,-666.66
70  DATA  425.67, 222.22
80  DATA  333.32, 333.33
90  DATA  222.22, 424.67
100  DATA  333.33, 523.46
110  DATA  444.44,-643.21
120  DATA  555.55, 543.21
130  DATA  411.11, 748.72
140  DATA  123.45, 343.45
150  DATA  678.90, 678.88
160  DATA  167.54, 424.21
170  DATA  432.31, 667.78
175  DATA  667.42, -24.64
180  DATA  777.71, 334.21
190  DATA  111.11, 897.47
195  REM  INITIALIZE COUNTER
197  LET I = 0
198  REM   SKIP LINES BEFORE PRIN
     TING RESULTS
200  PRINT : PRINT : PRINT
220  PRINT " I"; TAB( 12)"X"; TAB(
     22)"Y"; TAB( 32)"Z"
230  READ X,Y
235  LET Z = X + Y
240  LET I = I + 1
250  PRINT I; TAB( 10)X; TAB( 20)
     Y; TAB( 30)Z
260  IF I = 15 THEN 300
270  GOTO 230
300  END

]RUN

 I         X          Y          Z
1        521.41     -666.66    -145.25
2        425.67     222.22     647.89
3        333.32     333.33     666.65
4        222.22     424.67     646.89
5        333.33     523.46     856.79
6        444.44     -643.21    -198.77
7        555.55     543.21     1098.76
8        411.11     748.72     1159.83
9        123.45     343.45     466.9
10       678.9      678.88     1357.78
11       167.54     424.21     591.75
12       432.31     667.78     1100.09
13       667.42     -24.64     642.78
14       777.71     334.21     1111.92
15       111.11     897.47     1008.58

]PR#0
]RUN
```

Figure 11-1. Printout of a BASIC program for computing the sums of number sets (counter looping).

A PRINT statement with no information to follow causes a blank line for the next printout. In statement 200 the three PRINT's separated by colons cause three blank lines to appear in the printout preceding the actual printing of anything.

In statement 220 the numbers enclosed in parentheses following the word TAB designate the numbers of the columns on the row for printing out the letters enclosed in quotes; the letters in the quotes are the headings (I,X,Y,Z) for the columns of the table.

A loop is formed by statements 230 through 270. Statement 230 causes the values for a number set to be stored in the working memory of the computer, and statement 235 causes the computer to add these two numbers and place the sum in the memory of the computer at another memory location.

Statement 240 causes a 1 to be added to I, and I to be replaced by I + 1 and stored in the memory location of the preceding I value.

Statement 250 causes the values of I,X,Y,Z to be printed immediately following the increase of the I value by 1 and the computation of a Z value.

Statement 260 checks to determine if I has reached a value of 15. If the answer to the IF statement is yes, the computer is transferred to statement 300 (an END statement) and the execution of the program ceases. If the answer is no, the execution of the program continues to GOTO statement 270, which returns the computer to READ statement 230, where a new set of values are stored in the working memory of the computer and the process of computing another sum, etc. is continued. The process of looping continues until I becomes equal to 15, then the execution of the program ceases.

Example 11-2. Sum of Number Sets II

Write a BASIC program for a PDP-8 minicomputer to compute the sums of 15 sets of two numbers such as

$$Z(I) = X(I) + Y(I) \qquad I = 1 \text{ to } 15$$

Use FOR I = 1 TO 15 and NEXT I statements in the program. Compare the results with those of Example 10-2. The listing and printout of the program is given in Figure 11-2.

The values of the sums for the sets of two numbers are identical to those obtained in Example 10-2.

The looping for calculating the 15 sums begins with statement 300 and ends with statement 340. Statement 300 starts the looping by setting the value of I = 1. Statement 340 increases the value of I by 1 and compares it with the upper limit specified by statement 300. If the value of I is less than

```
]LIST

10  REM   PDP MINICOMPUTER PROGRA
     M; RUN ON BOTH MINI- AND MIC
     RO- COMPUTERS
100  REM  COMPUTE AND PRINT Z(I)
     = X(I) + Y(I)
110  REM  FOLLOWING DATA LISTS AR
     E IN THE FORM:
120  REM     DATA X(I),Y(I)
130  REM  AND ARE IN ORDER I = 1
     TO N
201  DATA  521.41, -666.66
202  DATA  425.67,  222.22
203  DATA  333.33,  333.33
204  DATA  222.22,  424.67
205  DATA  333.33,  523.46
206  DATA  444.44, -643.21
207  DATA  555.55,  543.21
208  DATA  411.11,  748.72
209  DATA  123.45,  343.45
210  DATA  678.90,  678.88
211  DATA  167.54,  424.21
212  DATA  432.31,  667.78
213  DATA  667.42,  -24.64
214  DATA  777.71,  334.21
215  DATA  111.11,  897.47
240  PRINT : PRINT : PRINT
250  PRINT " X"," Y"," Z"
300  FOR I = 1 TO 15
310  READ X,Y
320  LET Z = X + Y
330  PRINT X,Y,Z
340  NEXT I
350  PRINT : PRINT : PRINT
360  STOP
370  END

]RUN
 X                Y                Z
521.41           -666.66          -145.25
425.67           222.22           647.89
333.33           333.33           666.66
222.22           424.67           646.89
333.33           523.46           856.79
444.44           -643.21          -198.77
555.55           543.21           1098.76
411.11           748.72           1159.83
123.45           343.45           466.9
678.9            678.88           1357.78
167.54           424.21           591.75
432.31           667.78           1100.09
667.42           -24.64           642.78
777.71           334.21           1111.92
111.11           897.47           1008.58


BREAK IN 360
]

]PR#0
]RUN
```

Figure 11-2. Printout of a BASIC program for sums of number sets (FOR/NEXT looping).

the upper limit, the execution of the program is returned to statement 300. If the value of I is greater than the upper limit, the execution of the program proceeds to statement 350. Similar explanations for the other statements were given in the discussion for Example 11-1.

Example 11-3. Polynomial Evaluation

Write a BASIC program for a PDP-8 minicomputer to compute the value of a formation volume factor which is a function of pressure. The functional relationship is given by the equation

$$B = 1.05 + 0.001\ P + 0.00000019\ P\hat{\ }2 + 0.00000000015\ P\hat{\ }3 + .00000000000011\ P\hat{\ }4$$

Compute the FVF value at 1000 psia.

The listing and the printout of the program are given in Figure 11-3. Identical values are obtained in both Example 10-3 and Example 11-3. Both examples use engineering notation for the input of the constants; however, the printout in Example 11-3 is not in engineering notation because the computer found the answer to be a two-digit number.

This program illustrates a simple way of rearranging a polynomial equation for simplifying its evaluation. In addition, it shows a direct way to input data by means of LET statements. LET statements can be omitted yet the program will still be correct and give the same answer for B.

Example 11-4. Frontal Advance Calculations and Plots

If the constants are a = 12, b = 600 for the permeability ratio exponential equation, calculate Fw and dFw/dSw for the Sw values 0.2, 0.3, 0.4, 0.5, 0.6, 0.7, 0.8, 0.9 and have the computer plot the calculated values versus Sw. Take 0.6 as the viscosity ratio U. Use the TAB feature for plotting the results.

```
]PR#0
]LIST

5  REM  A PDP MINICOMPUTER PROGRA
     M
6  REM  RUN ON AN APPLE IIe MICRO
     COMPUTER
7  REM  THE PDP MINICOMPUTER USES
      A VERTICAL ARROW FOR EXPONE
     NTIATION
10  REM   COMPUTE AND PRINT THE S
     OLUTION OF THE POLYNOMIAL
15  REM  B=A4*P^4 + A3*P^3 +A A2*
     P^2 + A1*P + A0
20  LET A0 = 1.05
25  LET A1 = 1.0E - 03
30  LET A2 = 1.9E - 7
35  LET A3 = 1.5E - 10
40  LET A4 = 1.1E - 13
45  LET P = 1.0E + 03
50  LET B = (((A4 * P + A3) * P +
     A2) * P + A1) * P + A0
55  PRINT : PRINT : PRINT : PRINT

60  PRINT "   THE VALUE OF THE PO
     LYNOMIAL EVALUATED AT"P;" IS
      ";B
65  PRINT : PRINT : PRINT : PRINT
     : PRINT
69  STOP
70  END

]RUN


   THE VALUE OF THE POLYNOMIAL EVALUATED AT1000 IS 2.5




BREAK IN 69
]PR#0
]RUN
```

Figure 11-3. Printout of a BASIC program for evaluating a fourth-order polynomial.

The listing of the program, the calculated values, and the plots of the calculated values are given in Figure 11-4. Eight values of water saturation Sw designated by S(I) are to be read into the computer. Statement 10 is used for reserving eight spaces in memory for these values.

During the execution of the program, eight values of FS(I) and F(I) are calculated and stored in the memory of the computer. These values are com-

```
]LIST

10  DIM S(8)
15  REM   WATER FRACTION VALUES
20  DATA  .2,.3,.4,.5,.6,.7,.8,.9

25  REM   VISCOSITY RATIO
30 U = .6
32  REM   CONSTANTS FOR EQUATIONS

35 B = 12
38 A = 600
50  FOR I = 1 TO 8
60  READ S(I)
65  REM   CALCULATION OF FRACTION
     AL DERIVATIVES
70  LET FS(I) = ((U * B * A) /  EXP
     (B * S(I))) / ((1 + (U * A) /
     EXP (B * S(I))) ^ 2)
75  REM   CALCULATION OF FRACTION
     AL FLOW VALUES FOR WATER
77  LET F(I) = (1) / (1 + (U * A)
     /  EXP (B * S(I)))
80  PRINT F(I),FS(I)
90  NEXT I
95  FOR I = 1 TO 7
97  REM  MULTIPLYING FRACTIONAL V
     ALUES BY 60 AND TAKING INTEG
     ER VALUES
100  LET G(I) =  INT ((F(I) * 600
      + .5) / 10)
110  REM   MULTIPLYING DERIVATIVE
      VALUES BY 15 AND TAKING INT
      EGER VALUES
120  LET GS(I) =  INT ((FS(I) * 1
     50 + .5) / 10)
121  PRINT G(I),GS(I)
130  NEXT I
132  REM  PLOTTING FRACTIONAL FLO
     W VALUES
135  PRINT "--------------------
     ----------------------------
     ---------------"
140  FOR I = 1 TO 7
150  PRINT S(I); TAB( G(I));"*"
152  PRINT
154  PRINT
160  NEXT I
162  REM  PLOTTING DERIVATIVE VAL
     UES
165  PRINT "--------------------
     ----------------------------
     ---------------"
```

Figure 11-4. Frontal advance values and curves.

```
170  FOR I = 1 TO 7
180  PRINT S(I); TAB( GS(I));"+"
182  PRINT
184  PRINT
190  NEXT I
200  END

]RUN
.0297102097      .345930158
.0922803767      1.0051765
.252352624       2.26404933
.528443251       2.99029178
.788164362       2.0035356
.925110255       .831375246
.976198004       .278825539
.992709717       .0868456138
1                5
5                15
15               34
31               44
47               30
55               12
58               4
-----------------------------------------------------------------
.2*


.3  *


.4                    *


.5                                   *


.6                                                *


.7                                                        *


.8                                                           *
-------------------------------------------------------------
.2  +


.3            +


.4                           +


.5                                    +


.6                         +


.7          +


.8 +

]PR#0
]RUN
```

Figure 11-4. Continued.

puted during looping between statement 50 and statement 90. Statement 50 causes the subscript numbers to be advanced by one for each time around the loop. For each new value of S (I), statement 70 causes a new value to be computed and statement 77 causes a new value of F(I) to be computed. Statement 80 causes the printouts for the computed values. Statement 90 checks to determine when the value of the subscripts becomes greater than eight.

When I becomes greater than eight, the control of the computer is transferred to statement 95. Another loop is formed between statement 95 and statement 130. During the execution of this loop, the integer values of F(I) are taken, multiplied by 60, and stored in the memory of the computer. Also, during the execution of the loop, the integer values of FS(I) are taken, multiplied by 5, and stored in the memory of the computer. The F(I) and FS(I) values are plotted horizontally on the printout, and the maximum value for a TAB number is 40, which must be an integer. The multiplication number scales the printouts so the integer values do not become greater than 40.

Statement 135 causes a horizontal line to be printed on the line preceding the printing of the S(I) and F (I) values by statement 150; on the plot the coordinate points are designated by *. Statements 152 and 154 cause two lines to be skipped before printing another set of values. Statements 140 and 160 are the beginning and ending statements for the plotting loop.

In a similar manner the plot of the curve S(I) versus FS(I) is controlled by the loop formed between statement 170 and statement 190; the coordinate points are designated by the character +.

Example 11-5. Division by Iteration

Write a BASIC program using the iteration method given in Example A-27 for determining the quotient of 504 divided by 24 with Di greater than 0.9999.

The program and its printout are given in Figure 11-5. The value of 504/24 is 20.9999995, which is 21 when rounded to the sixth decimal place. This value agrees with the value obtained in Example 10-4.

Statement 5 causes the headings for the table to be printed. Statement 10 is a DATA statement containing the values to be READ into the computer by statement 30. The loop for calculating the different trial values of the constants begins with statement 50 and ends with statement 130.

```
]LIST

2  REM  SPACE BETWEEN PROGRAM AND
      HEADINGS AND (5) SPACING OF
      HEADINGS
3  PRINT
5  PRINT "I"; TAB( 5)"AN"; TAB( 1
     7)"D"; TAB( 29)"DP"
7  REM  DATA VALUES; DECIMAL POIN
     T SHIFTED FOR DIVIDEND AND D
     IVISOR SO THE DIVISOR WOULD
     BEGIN AT THE DECIMAL MARKING

8  PRINT
10  DATA  5.04000, .24000000, .99
     9999999, .000001, 1.0000000
15  REM  INITIALIZE NUMBER FOR SE
     T OF DATA
20  LET I = 0
25  REM  READ INITIAL SET OF DATA
      INTO COMPUTER
30  READ AN,D,C,T,W
40  GOTO 70
45  REM  COMPUTE AN(I+1) AND D(I+
     1)
50 AN = AN * DP
60 D = D * DP
70 I = I + 1
75  REM  COMPUTE COMPLIMENT PLUS
     1 FOR D(I)
80 DP = C - D + W
90  REM   SPACING BETWEEN LIST OF
      PROGRAM AND RESULTS
105  REM   USE OF TAB FOR SPACING
      RESULT VALUES
110  PRINT I; TAB( 4)AN; TAB( 16)
     D; TAB( 28)DP
115  REM   COMPUTE VALUE FOR TOLE
     RANCE CHECK
120 Z = W - D
125  REM  TOLERANCE CHECK
130  IF T < Z THEN 50
140  GOTO 150
150  END

]RUN

I  AN            D            DP

1  5.04          .24          1.76
2  8.8704        .4224        1.5776
3  13.993943     .66637824    1.33362176
4  18.6626269    .888696521   1.11130348
5  20.7398423    .987611535   1.01238846
6  20.996777     .999846525   1.00015347
7  20.9999995    .999999976   1.00000002

]PR#0
]RUN
```

Figure 11-5. Printout of a BASIC program for computing the quotient of two numbers.

Example 11-6. Least-Square Curve Fit

Write a BASIC program for determining the constants for the equation Kg/Ko = a exp(bSo) from the data in Table 11-1.

The technique for determining the constants is discussed in Example 10-10. The listing of the program and a printout of the results are given in Figure 11-6.

The considerable differences between the experimental values of Kg/Ko and the calculated values are due to poor experimental data of which all points did not fall on a straight line of a semilog plot; however, the calculated values do represent points on a semilog plot that create the best-fit curve for the experimental data. Fluid saturation data and permeability ratio data are stored in the memory of the computer by means of DIM, DATA, and READ statements. The constants B, C, E, and F are computed by the FOR/NEXT loop and bounded by statements 130 and 190.

Using the calculated constants, the value for X is computed by statement 200 and the value for Y is computed by statement 210. From Example 10-10, X = log a, a = exp(X), and Y = B.

Computed values of Kg/Ko are obtained by the loop bounded by statements 240 and 255. Also, this loop controls the printing out of the values for fluid saturation and permeability.

Table 11-1
Fluid Saturation and Permeability Ratio Values

Fluid Saturation (Percent So)	Permeability Ratio (Kg/Ko)
0.50	4.000
0.55	2.100
0.60	0.850
0.65	0.400
0.70	0.160
0.75	0.060
0.80	0.016
0.86	0.010

```
]PR#0
]LIST

20  DIM S(8),K(8)
25  REM   OIL SATURATION FRACTION
     AL VALUES
30  DATA  .5,.55,.6,.65,.7,.75,.8
     ,.86
35  REM  GAS-OIL PERMEABILITY RAT
     IOS
40  DATA  4,2.1,.85,.4,.16,.06,.0
     16,.01
50  REM  STORING INPUT DATA IN ME
     MORY OF THE COMPUTER
60  FOR I = 1 TO 8: READ S(I)
80  NEXT I
90  FOR I = 1 TO 8: READ K(I)
110  NEXT I
115  REM   CALCULATION OF CONSTAN
     TS FOR SIMULTANEOUS EQUATION
     S
120 A = 8
130  FOR I = 1 TO 8
140  LET B = B + S(I)
150  LET C = C +  LOG (K(I))
170  LET E = E + S(I) ^ 2
180  LET F = F + S(I) *  LOG (K(I
     ))
190  NEXT I
195  REM   SOLVING FOR LOG A IN T
     HE EXPONENTIAL EQUATION
200  LET X = (F - (E * C) / B) /
     (B - (E * A) / B)
205  REM   SOLVING FOR B OF THE E
     XPONENTIAL EQUATION
210  LET Y = (F - (B * C) / A) /
     (E - (B ^ 2) / A)
215  REM   SOLVING FOR A OF THE E
     XPONENTIAL EQUATION
220  LET Z =  EXP (X)
225  PRINT "LOG A =";X
227  PRINT " A  =";Z
230  PRINT " B  =";Y
232  REM   HEADINGS FOR TABLE PRI
     NT OUT
234  PRINT
235  PRINT "S(I)    K(I),ABS   K(I
     ),CAL"
236  PRINT
240  LET YA =  ABS (Y)
243  FOR I = 1 TO 8
244  REM   TESTING THE EXPONENTIA
     L EQUATION
245  LET M(I) = Z /  EXP (YA * S(
     I))
250  PRINT S(I); TAB( 9)K(I); TAB(
     20)M(I)
255  NEXT I
500  END
```

(figure continued on next page)

Figure 11-6. Printout of a BASIC program for computing the constants of an exponential equation by the least-squares method.

```
]RUN
LOG A =10.3558945
 A  =31441.8302
 B  =-17.5940969

S(I)    K(I),ABS    K(I),CAL

.5      4           4.7533329
.55     2.1         1.97218328
.6      .85         .818269404
.65     .4          .339504359
.7      .16         .140862177
.75     .06         .0584444717
.8      .016        .0242489243
.86     .01         8.4378408E-03

]PR#0
]RUN
```

Figure 11-6. Continued.

Example 11-7. Computation of Exponential Integral Table

Write a BASIC program for computing exponential values, Ei(−x), for values of x from 0.000 to 0.209. Printout the results in tabular form.

The listing of the program and the printout of the results are given in Figure 11-7. Double subscripted variables are used in the BASIC program. The use of double subscripted variables provides a convenient way of forming a loop within a loop.

The dimension statement 20 causes the computer to reserve $25 \times 15 = 375$ spaces in memory for the computed integral values; this is enough spaces for a table with 25 rows and 15 columns. During the execution of the program, only $10 \times 21 = 210$ exponential values are computed.

Statement 190 generated the x values used in statement 90 for computing the different exponential integral values. Statement 50 caused the headings for the columns in the table to be printed. This illustrates the use of a PRINT statement using numbers in quotes with spaces between the numbers. There were five spaces between x and 0, and six spaces between each of the numbers.

Statement 100 was used for rounding the computed exponential values to three decimal positions. And statement 15 added a one in the fourth decimal position to each of the computed values; this was done to have uniformity in the number of positions to the right of the decimal. With all computed values the same length, it was easy to apply the space technique given for statement 50 to control the printout of the computed exponential values; this printout feature enabled a neat table of values to be formed. Of course,

```
]PR#0


]LIST

20  DIM X(25,15)
30 M = 0
40 Z = 0
45  REM   HEADINGS FOR COLUMNS IN
      TABLE
50  PRINT  TAB( 2)"X      0       1
           2       3      4       5
           6       7      8       9

55  PRINT
60  FOR I = 1 TO 21
70  FOR J = 1 TO 10
80  IF M = 0 THEN 120
85  REM   COMPUTING EXPONENTIAL V
     ALUES
90  LET X(I,J) = ( - .577157 +  LOG
     (1 / M) + M - M ^ 2 / 4 + M ^
     3 / 18 - M ^ 4 / 96)
95  REM   ROUNDING OFF COMPUTED V
     ALUES TO THREE DECIMAL POINT
     S
100 X(I,J) =  INT (X(I,J) * 1000 +
     .5) / 1000
102  REM   ADDING .0001 TO EACH C
     OMPUTED VALUE FOR CONTROLLIN
     G THE PRINT OUTS
105  LET X(I,J) = X(I,J) + .0001
110  GOTO 140
115  REM   SETTING A VALUE FOR X(
     1,1)
120  LET X(1,1) = 9.9996
125  REM   COMPUTING THE REQUIRED
      VALUES OF M TO BE USED FOR
     COMPUTING EXPONENTAL VALUES
140  LET M = M + .001
150  NEXT J
160  NEXT I
170  FOR I = 1 TO 21
175  REM   FORMAT FOR PRINTING OU
     T THE EXPONENTIAL INTEGRAL V
     ALUES
180  PRINT Z; TAB( 5)X(I,1);" "X(
     I,2);" "X(I,3);" "X(I,4);" "
     X(I,5);" "X(I,6);" "X(I,7);"
      "X(I,8);" "X(I,9);" "X(I,10
     )
185  REM   RECALCULATING THE M VA
     LUES FOR PRINT OUTS
190 Z = Z + .01
200  NEXT I
300  END
```

(figure continued on next page)

Figure 11-7. Printout of a BASIC program for computing exponential integral values: $x = 0.000$ to 0.209.

```
]RUN
 X      0      1      2      3      4      5      6      7      8      9

0    9.9996 6.3321 5.6391 5.2351 4.9481 4.7261 4.5451 4.3921 4.2591 4.1421
.01 4.0381 3.9441 3.8581 3.7791 3.7051 3.6371 3.5741 3.5141 3.4581 3.4051
.02 3.3551 3.3071 3.2611 3.2181 3.1761 3.1371 3.0981 3.0621 3.0261 2.9921
.03 2.9591 2.9271 2.8971 2.8671 2.8381 2.8101 2.7831 2.7561 2.7311 2.7061
.04 2.6811 2.6581 2.6341 2.6121 2.5901 2.5681 2.5471 2.5271 2.5071 2.4871
.05 2.4681 2.4491 2.4311 2.4131 2.3951 2.3781 2.3601 2.3441 2.3271 2.3111
.06 2.2951 2.2801 2.2651 2.2491 2.2351 2.2201 2.2061 2.1921 2.1781 2.1641
.07 2.1511 2.1381 2.1251 2.1121 2.0991 2.0871 2.0741 2.0621 2.0501 2.0391
.08 2.0271 2.0161 2.0041 1.9931 1.9821 1.9711 1.9601 1.9501 1.9391 1.9291
.09 1.9191 1.9091 1.8991 1.8891 1.8791 1.8701 1.8601 1.8511 1.8411 1.8321
.1   1.8231 1.8141 1.8051 1.7961 1.7881 1.7791 1.7701 1.7621 1.7541 1.7451
.11 1.7371 1.7291 1.7211 1.7131 1.7051 1.6971 1.6901 1.6821 1.6751 1.6671
.12 1.6601 1.6521 1.6451 1.6381 1.6311 1.6231 1.6161 1.6091 1.6031 1.5961
.13 1.5891 1.5821 1.5761 1.5691 1.5621 1.5561 1.5491 1.5431 1.5371 1.5301
.14 1.5241 1.5181 1.5121 1.5061 1.5001 1.4941 1.4881 1.4821 1.4761 1.4701
.15 1.4651 1.4591 1.4531 1.4481 1.4421 1.4361 1.4311 1.4251 1.4201 1.4151
.16 1.4091 1.4041 1.3991 1.3931 1.3881 1.3831 1.3781 1.3731 1.3681 1.3631
.17 1.3581 1.3531 1.3481 1.3431 1.3381 1.3331 1.3291 1.3241 1.3191 1.3151
.18 1.3101 1.3051 1.3011 1.2961 1.2921 1.2871 1.2831 1.2781 1.2741 1.2691
.19 1.2651 1.2611 1.2561 1.2521 1.2481 1.2431 1.2391 1.2351 1.2311 1.2271
.2   1.2231 1.2191 1.2151 1.2111 1.2071 1.2031 1.1991 1.1951 1.1911 1.1871

]PR#0
]RUN
```

Figure 11-7. Continued.

with this feature the fourth decimal position does nct have any value; it must be disregarded. However, there is a three-place accuracy. All 80 columns on the page were utilized for printing out the table of values.

A loop within a loop is created with the use of the double subscripted variables X(I,J) and statements 60 and 70. In statements 60 and 70 and statements 150 and 160, the I subscript represents the row number and the J subscript represents the column number. The loop formed by statements 70 and 150 (FOR J/NEXT J) causes a set of values for a row to be computed, and statements 60 and 160 cause the rows of values to be computed. During the execution of the program, the J subscript controls columns in a row, but the I subscript controls the number of rows. For every time around an I loop, there are 10 times around a J loop. The J loops are inside the I loops. In all there are $21 \times 10 = 210$ loops. For additional information, see Example 10-5.

Example 11-8. Computation of Y-Factor Values

Write a BASIC program for computing Y-factor values for x = 0.000 to x = 0.209 in increments of 0.001.

The printout of the BASIC program for computing Y-factor values is given in Figure 11-8. The computation of exponential values and Y-factor values require the same type of program. Since more than 80 columns were

```
]LIST

10  REM   CALCULATION OF Y-FACTOR
     S
20  DIM X(25,15)
30 M = 0
40 Z = 0
45  REM   PARTIAL HEADINGS OF TAB
     LE COLUMNS
50  PRINT  TAB( 2)"X"; TAB( 9)"0"
     ; TAB( 17)"1"; TAB( 25)"2"; TAB(
     33)"3"; TAB( 41)"4"
55  PRINT
60  FOR I = 1 TO 21
70  FOR J = 1 TO 10
80  IF M = 0 THEN 120
85  REM  COMPUTING Y-FACTOR VALUE
     S
90  LET X(I,J) = .577157 -  LOG (
     1 / M) - M + ((M ^ 2) / 4) -
     ((M ^ 3) / 18) + (1 / (M *  EXP
     (M)))
95  REM   ROUNDING OFF COMPUTED V
     ALUES TO THREE DECIMAL POINT
     S
100 X(I,J) =  INT (X(I,J) * 1000 +
     .5) / 1000
110  GOTO 140
115  REM   SETTING A VALUE FOR X(
     1,1)
120  LET X(1,1) = 999.9
125  REM   COMPUTING THE REQUIRED
      VALUES OF M FOR COMPUTING Y
     -FACTOR VALUES
140  LET M = M + .001
150  NEXT J
160  NEXT I
170  FOR I = 1 TO 21
175  REM   PARTIAL FORMAT FOR PRI
     NTING OUT THE Y-FACTOR VALUE
     S
180  PRINT Z; TAB( 6)X(I,1); TAB(
     14)X(I,2); TAB( 22)X(I,3); TAB(
     30)X(I,4); TAB( 38)X(I,5)
185  REM   RECALCULATING THE M VA
     LUES FOR PRINT OUTS
190 Z = Z + .01
200  NEXT I
210  PRINT : PRINT : PRINT
212  REM   REMAINING HEADINGS OF
     TABLE COLUMNS
214  PRINT  TAB( 6)"5"; TAB( 14)"
     6"; TAB( 22)"7"; TAB( 30)"8"
     ; TAB( 38)"9"
215  PRINT
220  FOR I = 1 TO 21
230  PRINT  TAB( 3)X(I,6); TAB( 1
     1)X(I,7); TAB( 19)X(I,8); TAB(
     27)X(I,9); TAB( 35)X(I,10)
240  NEXT I
300  END
```

(figure continued on next page)

Figure 11-8. Printout of a BASIC program for computing y-factor values: $x = 0.000$ to 0.209.

]RUN

X	0	1	2	3	4	5	6	7	8	9
0	999.9	992.669	493.362	327.1	244.054	194.276	161.125	137.469	119.745	105.973
.01	94.967	85.971	78.482	72.151	66.73	62.037	57.934	54.318	51.106	48.236
.02	45.655	43.323	41.204	39.272	37.502	35.876	34.376	32.989	31.702	30.505
.03	29.389	28.346	27.369	26.453	25.591	24.779	24.013	23.289	22.604	21.955
.04	21.338	20.753	20.196	19.665	19.159	18.676	18.214	17.773	17.35	16.945
.05	16.557	16.184	15.826	15.481	15.15	14.831	14.524	14.228	13.943	13.667
.06	13.401	13.144	12.895	12.654	12.422	12.196	11.978	11.766	11.561	11.362
.07	11.169	10.982	10.799	10.622	10.45	10.283	10.121	9.962	9.808	9.658
.08	9.512	9.37	9.231	9.096	8.964	8.835	8.709	8.587	8.467	8.35
.09	8.236	8.124	8.015	7.909	7.805	7.703	7.603	7.506	7.41	7.317
.1	7.225	7.136	7.048	6.962	6.878	6.796	6.715	6.635	6.558	6.48200001
.11	6.407	6.333	6.261	6.191	6.122	6.054	5.987	5.921	5.857	5.794
.12	5.731	5.67	5.61	5.551	5.493	5.436	5.38	5.325	5.271	5.218
.13	5.166	5.114	5.063	5.013	4.964	4.916	4.869	4.822	4.776	4.73
.14	4.685	4.641	4.598	4.555	4.513	4.472	4.431	4.391	4.351	4.312
.15	4.274	4.236	4.198	4.161	4.125	4.089	4.053	4.019	3.984	3.95
.16	3.917	3.884	3.851	3.819	3.787	3.756	3.725	3.694	3.664	3.634
.17	3.605	3.576	3.547	3.519	3.491	3.463	3.436	3.409	3.383	3.356
.18	3.331	3.305	3.28	3.255	3.23	3.205	3.181	3.157	3.134	3.111
.19	3.087	3.065	3.042	3.02	2.998	2.976	2.955	2.933	2.912	2.892
.2	2.871	2.851	2.83	2.811	2.791	2.771	2.752	2.733	2.714	2.695

]PR#0
]RUN

Figure 11-8. Continued.

required for the table of Y-factors, the left half of the table was printed out first; then the right half was printed out. By placing the halves side by side, a complete table can be viewed or copied.

Example 11-9. Build-Up Density

Write a BASIC program for numerically computing values for build-up density curves in a saturated sand-packed pipe. Scale the diffusivity equation so the variables are dimensionless, and the build-up density curves can be used for solving large numbers of density problems. For more details about the problem, see Examples 8-1 and 10-7. The density calculation program and the results are given in Figure 11-9.

Due to round-off errors, there were errors in the calculations of X(11) values prior to the fiftieth increment of time and in X(16) values prior to the one-hundredth increment of time. The X(11) values should be constant at 0.50 until the fiftieth increment of time, and the X(16) values should be constant at 0.75 until the one-hundredth increment of time.

Beginning programmers often have the tendency to write statement 90 as LET Y(I) = (X(I−1) + X(I+1))/2; this will not work. Instead, a statement such as statement 80 must be added to go with statement 90.

With the built-in interpreter in the microcomputer, 32 minutes were required to run the build-up density program, but with the aid of a compiler, the time for the execution of the program was reduced to eight minutes.

```
]NEW

]PR#0
]LIST

8  REM    RESERVING SPACE FOR X AN
      D Y DENSITY VALUES
10  DIM X(25),Y(25)
20 R = 0.0
30  FOR J = 1 TO 19
40  LET X(J) = R
50 R = R + .05
60  NEXT J
62  REM    HEADINGS FOR PRINT OUTS

64  PRINT "  M      X(1)    X(6)
      X(11)  X(16)"
65  PRINT
70 M = 0
75  FOR I = 1 TO 19
80  LET K = I + 1
```

(figure continued on next page)

Figure 11-9. Printout of a BASIC program for computing build-up densities of an oil-saturated, sand-packed pipe.

```
90  LET Y(K) = (X(I) + X(I + 2)) /
    2
110  NEXT I
115  REM   INITIALIZING Y(1) AND
     Y(21)
120  LET Y(1) = Y(2)
130  LET Y(21) = 1.0
140  FOR I = 1 TO 21
145  REM   REPLACING CALCULATED V
     ALUES FOR INITIAL VALUES OF
     DENSITY
150  LET X(I) = Y(I)
151  REM   ROUNDING OFF DECIMAL P
     OINTS
152  LET X(I) = X(I) * 10000 + .5

154  LET X(I) =  INT (X(I))
156  LET X(I) = X(I) / 10000
170  NEXT I
180 M = M + 1
185  REM   CONTROLLING PRINT OUTS

190  IF M < 10 THEN 210
191  IF M = 50 THEN 210
192  IF M = 100 THEN 210
193  IF M = 500 THEN 210
194  IF M = 1000 THEN 210
195  IF M = 1500 THEN 210
196  IF M = 1501 THEN 230
198  REM   END OF LOOP- INSTRUCTI
     NG RETURN TO BEGINNING
200  GOTO 75
205  REM   PRINT OUT SCHEME
210  PRINT M; TAB( 8)X(1); TAB( 1
     5)X(6); TAB( 22)X(11); TAB(
     29)X(16)
220  GOTO 75
230  END

]RUN
 M      X(1)    X(6)    X(11)   X(16)

1       .05     .25     .5      .75
2       .075    .25     .5      .75
3       .0875   .25     .5      .75
4       .1      .25     .5      .6907
5       .1094   .25     .5      .7188
6       .1188   .2516   .5      .6759
7       .1266   .2524   .5      .711
8       .1345   .2548   .5      .6759
9       .1413   .2561   .4982   .711
50      .3037   .3545   .5207   .7391
100     .4182   .4559   .5866   .7745
500     .8479   .8578   .8907   .9408
1000    .9756   .9776   .9831   .9913
1500    1       1       1       1

]PR#0
]RUN
```

Figure 11-9. Continued.

Example 11-10. Pressure Distributions

Using the iteration method of solving difference equations, write a BASIC program for computing values of pressure for the problem given in Example 10-8 (also note Example 4-11). The listing and the printout of the results are given in Figure 11-10.

Statements 20 through 60 give the pressure values for the first iteration cycle. Statements 90 and 360 are the beginning and ending statements of a loop which contains six small loops. The six loops are delimited by statements 100 through 130, statements 140 through 170, statements 180 through 210, statements 220 through 240, statements 250 through 270, and statements 280 through 300. The three smaller loops are used for controlling the calculation of the three rows of interior values. A small subroutine is used for the actual calculation of the interior values and rounding the computed values to three significant decimal digits. The subroutine is bounded by statements 500 and 520. See Example 11-9 for explanations of statements 110, 150, and 190.

```
]LIST

8  REM  DIFFERENCE SOLUTION OF A
     LAPLACE EQUATION
10  DIM X(30),Y(30)
15  REM   BOUNDARY VALUES WITH IN
     TERIOR VALUES SET EQUAL TO O
     NE
20  DATA  0,1.31,5.25,11.81,21.0
30  DATA  0,1.00,1.00,1.00,19.25
40  DATA  0,1.00,1.00,1.00,17.50
50  DATA  0,1.00,1.00,1.00,15.75
60  DATA  0,.875,3.50,7.88,14.00
65 N = 1
70  FOR I = 1 TO 25: READ X(I)
80  NEXT I
85  REM  CALCULATING NEW VALUES F
     OR INTERIOR MESH POINTS
90  FOR J = 1 TO 25
100  FOR I = 1 TO 3
110  LET K = I + 6
120  GOSUB 500
130  NEXT I
140  FOR I = 1 TO 3
150  LET K = I + 11
160  GOSUB 500
170  NEXT I
180  FOR I = 1 TO 3
190  LET K = I + 16
200  GOSUB 500
210  NEXT I
220  FOR I = 7 TO 9
230  LET X(I) = Y(I)
240  NEXT I
250  FOR I = 12 TO 14
260  LET X(I) = Y(I)
270  NEXT I
280  FOR I = 17 TO 19
285  REM   INITIALIZING VALUES FO
     R NEXT SET OF CALCULATIONS
290  LET X(I) = Y(I)
300  NEXT I
302  REM   NUMBERING ARRAYS
305  PRINT N;"         ARRAY"
307  REM   FORMAT FOR PRINTING OU
     T THE ARRAYS
310  PRINT  TAB( 3)X(1); TAB( 6)X
     (2); TAB( 14)X(3); TAB( 22)X
     (4); TAB( 30)X(5)
320  PRINT  TAB( 3)X(6); TAB( 6)X
     (7); TAB( 14)X(8); TAB( 22)X
     (9); TAB( 30)X(10)
330  PRINT  TAB( 3)X(11); TAB( 6)
     X(12); TAB( 14)X(13); TAB( 2
     2)X(14); TAB( 30)X(15)
340  PRINT  TAB( 3)X(16); TAB( 6)
     X(17); TAB( 14)X(18); TAB( 2
     2)X(19); TAB( 30)X(20)
350  PRINT  TAB( 3)X(21); TAB( 6)
     X(22); TAB( 14)X(23); TAB( 2
     2)X(24); TAB( 30)X(25)
351  REM   INCREASING COUNTER IN
     STEPS OF ONE
352  LET N = N + 1
355  PRINT : PRINT
```

(figure continued on next page)

Figure 11-10. Printout of a BASIC program for computing interior pressure values from boundary values.

```
360   NEXT J
400   END
450   REM   SUBROUTINE FOR CALCULA
      TING INTERIOR VALUES AND ROU
      NDING OFF THE VALUES
500   LET Y(K) = (X(K - 1) + X(K +
      1) + X(K - 5) + X(K + 5)) /
      4
510   LET Y(K) =  INT (Y(K) * 1000
       + .5) / 1000
520   RETURN
```

```
]RUN
1       ARRAY
  0  1.31     5.25     11.81    21
  0  .828     2.063    8.265    19.25
  0  .75      1        5.125    17.5
  0  .719     1.625    6.408    15.75
  0  .875     3.5      7.88     14

2       ARRAY
  0  1.31     5.25     11.81    21
  0  1.031    3.836    9.562    19.25
  0  .637     2.391    8.293    17.5
  0  .813     2.907    7.595    15.75
  0  .875     3.5      7.88     14

3       ARRAY
  0  1.31     5.25     11.81    21
  0  1.446    4.559    10.797   19.25
  0  1.059    3.918    9.262    17.5
  0  1.105    3.575    8.708    15.75
  0  .875     3.5      7.88     14

4       ARRAY
  0  1.31     5.25     11.81    21
  0  1.732    5.353    11.22    19.25
  0  1.617    4.614    10.231   17.5
  0  1.377    4.308    9.117    15.75
  0  .875     3.5      7.88     14

5       ARRAY
  0  1.31     5.25     11.81    21
  0  2.07     5.704    11.661   19.25
  0  1.931    5.377    10.613   17.5
  0  1.7      4.652    9.542    15.75
  0  .875     3.5      7.88     14

6       ARRAY
  0  1.31     5.25     11.81    21
  0  2.236    6.09     11.844   19.25
  0  2.287    5.725    11.02    17.5
  0  1.865    5.03     9.724    15.75
  0  .875     3.5      7.88     14

7       ARRAY
  0  1.31     5.25     11.81    21
  0  2.422    6.264    12.043   19.25
  0  2.457    6.107    11.198   17.5
  0  2.048    5.204    9.92     15.75
  0  .875     3.5      7.88     14

8       ARRAY
  0  1.31     5.25     11.81    21
  0  2.508    6.456    12.131   19.25
  0  2.644    6.281    11.393   17.5
  0  2.134    5.394    10.008   15.75
  0  .875     3.5      7.88     14

9       ARRAY
  0  1.31     5.25     11.81    21
  0  2.603    6.543    12.227   19.25
  0  2.731    6.472    11.48    17.5
  0  2.228    5.481    10.104   15.75
  0  .875     3.5      7.88     14

10      ARRAY
  0  1.31     5.25     11.81    21
  0  2.646    6.638    12.271   19.25
  0  2.826    6.559    11.576   17.5
  0  2.272    5.576    10.148   15.75
  0  .875     3.5      7.88     14

11      ARRAY
  0  1.31     5.25     11.81    21
  0  2.694    6.682    12.319   19.25
  0  2.869    6.654    11.62    17.5
  0  2.319    5.62     10.195   15.75
  0  .875     3.5      7.88     14

12      ARRAY
  0  1.31     5.25     11.81    21
  0  2.715    6.729    12.341   19.25
  0  2.917    6.698    11.667   17.5
  0  2.341    5.667    10.217   15.75
  0  .875     3.5      7.88     14

13      ARRAY
  0  1.31     5.25     11.81    21
  0  2.739    6.751    12.364   19.25
  0  2.939    6.745    11.689   17.5
  0  2.365    5.689    10.241   15.75
  0  .875     3.5      7.88     14

14      ARRAY
  0  1.31     5.25     11.81    21
  0  2.75     6.775    12.375   19.25
  0  2.962    6.767    11.713   17.5
  0  2.376    5.713    10.252   15.75
  0  .875     3.5      7.88     14
```

Figure 11-10. Continued.

```
15       ARRAY
  0  1.31     5.25     11.81    21
  0  2.762    6.786    12.387   19.25
  0  2.973    6.791    11.724   17.5
  0  2.388    5.724    10.264   15.75
  0  .875     3.5      7.88     14

16       ARRAY
  0  1.31     5.25     11.81    21
  0  2.767    6.798    12.393   19.25
  0  2.985    6.802    11.736   17.5
  0  2.393    5.736    10.27    15.75
  0  .875     3.5      7.88     14

17       ARRAY
  0  1.31     5.25     11.81    21
  0  2.773    6.803    12.399   19.25
  0  2.99     6.814    11.741   17.5
  0  2.399    5.741    10.276   15.75
  0  .875     3.5      7.88     14

18       ARRAY
  0  1.31     5.25     11.81    21
  0  2.776    6.809    12.401   19.25
  0  2.997    6.819    11.747   17.5
  0  2.402    5.747    10.278   15.75
  0  .875     3.5      7.88     14

19       ARRAY
  0  1.31     5.25     11.81    21
  0  2.779    6.812    12.404   19.25
  0  2.999    6.825    11.75    17.5
  0  2.405    5.75     10.281   15.75
  0  .875     3.5      7.88     14

20       ARRAY
  0  1.31     5.25     11.81    21
  0  2.78     6.815    12.406   19.25
  0  3.002    6.828    11.753   17.5
  0  2.406    5.753    10.283   15.75
  0  .875     3.5      7.88     14

21       ARRAY
  0  1.31     5.25     11.81    21
  0  2.782    6.816    12.407   19.25
  0  3.004    6.831    11.754   17.5
  0  2.408    5.754    10.284   15.75
  0  .875     3.5      7.88     14

22       ARRAY
  0  1.31     5.25     11.81    21
  0  2.783    6.818    12.408   19.25
  0  3.005    6.832    11.756   17.5
  0  2.408    5.756    10.285   15.75
  0  .875     3.5      7.88     14

23       ARRAY
  0  1.31     5.25     11.81    21
  0  2.783    6.818    12.409   19.25
  0  3.006    6.834    11.756   17.5
  0  2.409    5.756    10.285   15.75
  0  .875     3.5      7.88     14

24       ARRAY
  0  1.31     5.25     11.81    21
  0  2.784    6.819    12.409   19.25
  0  3.007    6.834    11.757   17.5
  0  2.409    5.757    10.285   15.75
  0  .875     3.5      7.88     14

25       ARRAY
  0  1.31     5.25     11.81    21
  0  2.784    6.819    12.409   19.25
  0  3.007    6.835    11.757   17.5
  0  2.41     5.757    10.286   15.75
  0  .875     3.5      7.88     14

]PR#0
]RUN
```

Figure 11-10. Continued.

Example 11-11. Distance to a Fault

Write a BASIC program for determining the distance to a fault for the problem given in Example 9-10. The BASIC program for computing the distance to a fault is given in Figure 11-11. The discrepancy between the results in Examples 9-10 and 11-11 is due to the methods of obtaining values of the exponential integral: one set of values was read from curves, and one set of values was computed directly by the computer.

Statement 100 sets trial values for the radius of the image well from the nonproducing well. The accuracy of the computer value is less than 10 feet.

```
]NEW

]PR#0
]LIST

3  REM  DISTANCE TO A FAULT
5  REM   FLOW RATE, VISCOSITY, PE
     RMEABILITY, COMPRESSIBILITY,
      POROSITY, SHUT IN PRESSURE,
      FORMATION VOLUME FACTOR, NO
     N FLOWING PRESSURE, DISTANCE

10  DATA  200,3,200,30,.000015,.1
     5,1017,1.3,987,500
20  READ Q,U,K,H,C,O,PN,B,PW,X
30  LET QK = (70.6 * U * B * Q) /
     (K * H)
40  LET TD = (948.4 * C * U * O *
     (500 ^ 2)) / (200 * 100)
45  REM   EXPONENTIAL INTEGRAL OF
      THE FLOWING WELL
50  LET EP =  - .577162 +  LOG (1
     / TD) - TD + TD ^ 2 / 4 - T
     D ^ 3 / 18 + TD ^ 4 / 96
55  REM   PRESSURE DROP DUE TO FL
     OWING WELL
60  LET Z = PN - PW - (QK * EP)
70  LET R = 500
75  REM   EXPONENTIAL INTEGRAL OF
      THE IMAGE WELL
80  LET D2 = (948.4 * C * U * O *
     (R ^ 2)) / (200 * 100)
90  LET P2 =  - .577152 +  LOG (1
     / D2) - D2 + D2 ^ 2 / 4 - D
     2 ^ 3 / 18 + D2 ^ 4 / 96
95  REM   INCREMENTAL IMAGE WELL
     RADIUS
100  LET R = R + 10
105  REM  CHECK TO DETERMINE THE
     CORRECT RADIUS OF THE IMAGE
     WELL
110  LET W = Z - (QK * P2)
120  IF W > 0 THEN 140
130  GOTO 80
140  LET D = .5 *  SQR ((R ^ 2) -
     (X ^ 2))
150  PRINT "THE DISTANCE TO THE F
     AULT IS   ";D;" FT"
151  PRINT
152  PRINT "THE RADIUS OF THE IMA
     GE WELL IS   ";R;" FT"
200  END

]RUN
THE DISTANCE TO THE FAULT IS   191.637679 FT

THE RADIUS OF THE IMAGE WELL IS   630 FT

]PR#0
]runRUN
```

Figure 11-11. Printout of a BASIC program for computing the distance to a fault.

Example 11-12. Flow Across a Lease Line

Write a computer program for the flow across a lease line in a three-well field. With wells located at coordinate points (−600,800), (400,1000), and (−200,000) and the lease line taken as the y-axis, determine the pressures for each well at coordinate points (−10,1000), (−10,800), (−10,600), (−10,400), (−10,200), (−10,00) and (10,1000), (10,800), (10,600), (10,400), (10,200), (10,00). With the pressures, determine the pressure drops for the 20-foot distances at the y-coordinate points and, from pressure drops for the 20-foot distances at the y-coordinate points and, from these pressure drops, compute the total flow across the lease line from $y = 0$ to $y = 1000$ for a thickness of 10 feet. Other data: pe = 1525 psia; q1 = 200 bbls/day; q2 = 300 bbls/day; q3 = 150 bbls/day; t1 = 400 hrs; t2 = 350 hrs; t3 = 375 hrs; 948.4 ouB/k = 0.00033; 1.127 k/uB = 0.01442; 70.6 uB/kh = 0.01442. For additional information see Examples 10-9 and 9-10.

Between statements 40 and 150, there are five loops which are used for reading data into the computer and calculating the 36 radii for the 12 pressure points—3 radii for each point.

The values for substituting into the integral equation are computed by the loop bounded by statements 160 and 200—a total of 36 values. The exponential values are computed by the loop bounded by statements 205 and 240. Statement 260 computes the total flow across 1000 linear feet of lease line—from $y = 0$ to $y = 1000$.

```
]LIST

10   REM  FLOW ACROSS A LEASE LINE

20   DIM RA(6),RB(6),RC(6),ZA(6),Z
      B(6),ZC(6)
25   DIM KA(6),KB(6),KC(6),LA(6),L
      B(6),LC(6)
30   DIM EA(6),EB(6),EC(6),FA(6),F
      B(6),FC(6)
33   REM    CALCULATION OF RADII TO
       MESH POINTS
35   DATA  200,0,200,400,600,800
40   FOR I = 1 TO 6
45   READ YA(I)
50   NEXT I
60   FOR I = 1 TO 6
65   LET RA(I) =  SQR (YA(I) ^ 2 +
      590 ^ 2)
70   NEXT I
75   DATA  1000,800,600,400,200,0
80   FOR I = 1 TO 6
85   READ YB(I)
90   LET RB(I) =  SQR (YB(I) ^ 2 +
      190 ^ 2)
95   NEXT I
100  DATA  0,200,400,600,800,1000

105  FOR I = 1 TO 6
110  READ YC(I)
120  LET RC(I) =  SQR (YC(I) ^ 2 +
     410 ^ 2)
125  NEXT I
130  FOR I = 1 TO 6
135  LET ZA(I) =  SQR (YA(I) ^ 2 +
     610 ^ 2)
140  LET ZB(I) =  SQR (YB(I) ^ 2 +
     210 ^ 2)
145  LET ZC(I) =  SQR (YC(I) ^ 2 +
     390 ^ 2)
150  NEXT I
155  REM    CALCULATION OF THE VAL
     UES FOR THE EXPONENTIAL EQUA
     TIONS
```

(figure continued on next page)

Figure 11-12. Printout of a BASIC program for computing the flow of oil across a lease line.

```
160  FOR I = 1 TO 6
170  LET KB(I) = (.000033 * RB(I)
      ^ 2) / 350
175  LET KA(I) = (.000033 * RA(I)
      ^ 2) / 400
180  LET KC(I) = (.000033 * RC(I)
      ^ 2) / 375
185  LET LB(I) = (.000033 * ZB(I)
      ^ 2) / 350
190  LET LA(I) = (.000033 * ZA(I)
      ^ 2) / 400
195  LET LC(I) = (.000033 * ZC(I)
      ^ 2) / 375
200  NEXT I
203  REM   CALCULATION OF EXPONEN
     TIAL VALUES FOR THE RADII

205  FOR I = 1 TO 6
210  LET EA(I) =  - .577152 +  LOG
     (1 / KA(I)) + KA(I) - KA(I) ^
     2 / 4 + KA(I) ^ 3 / 18 - KA(
     I) ^ 4 / 96
215  LET FA(I) =  - .577152 +  LOG
     (1 / LA(I)) + LA(I) - LA(I) ^
     2 / 4 + LA(I) ^ 3 / 18 - LA(
     I) ^ 4 / 96
220  LET EB(I) =  - .577152 +  LOG
     (1 / KB(I)) + KB(I) - KB(I) ^
     2 / 4 + KB(I) ^ 3 / 18 - KB(
     I) ^ 4 / 96
225  LET EC(I) =  - .577152 +  LOG
     (1 / KC(I)) + KC(I) - KC(I) ^
     2 / 4 + KC(I) ^ 3 / 18 - KC(
     I) ^ 4 / 96
230  LET FB(I) =  - .577152 +  LOG
     (1 / LB(I)) + LB(I) - LB(I) ^
     2 / 4 + LB(I) ^ 3 / 18 - LB(
     I) ^ 4 / 96
235  LET FC(I) =  - .577152 +  LOG
     (1 / LC(I)) + LC(I) - LC(I) ^
     2 / 4 + LC(I) ^ 3 / 18 - LC(
     I) ^ 4 / 96
240  NEXT I
243  REM   CALCULATION OF PRESSUR
     E DROPS AT THE CHOSEN REFERE
     NCE POINTS
245  FOR I = 1 TO 6
250  LET D(I) = 2.884 * (FA(I) -
     EA(I)) + 4.32 * (FB(I) - EB(
     I)) + 2.16 * (FC(I) - EC(I))

255  NEXT I
260  LET Q = (.01442 * 1000 * 10 /
     20) * (D(1) + D(2) + D(3) +
     D(4) + D(5) + D(6))
275  PRINT "A NEGATIVE Q MEANS FL
     OW TO THE LEFT"
277  PRINT
278  PRINT Q;"   BBLS/DAY"
400  END

]RUN
A NEGATIVE Q MEANS FLOW TO THE LEFT

-13.0171205   BBLS/DAY

]PR#0
]RUN
```

Figure 11-12. Continued.

Example 11-13. Depletion Drive

Using data from Cole*, Example 4-1, and the constants for the exponential curve of Example 11-6, write a BASIC program for computing values of cumulative production and gas/oil ratio versus pressure for the reservoir. In writing the program use Schilthuis' material balance equation. For additional information, see Example 10-11.

The printout of the BASIC program for the depletion-drive calculations is given in Figure 11-13. The data for the program is given in statements 20 to 80. Statement 10 indicates the assignment of variables for the data. The constants for the exponential curve are given in statements 120 to 130.

There is a large loop between statements 144 and 270, and a small loop within the large loop bounded by statements 150 and 225. During the execution of the small loop, the incremental values of differential gas production are calculated and tested for the appropriate values of Np. Statement

*Cole, F. W., *Reservoir Engineering Manual,* Second Edition, Houston: Gulf Pub. Co., 1969, p. 188ff.

```
]PR#0
]LIST

3  REM   SCHILTHUIS MATERIAL BALA
     NCE
5  PRINT  TAB( 1)"PRESSURE"; TAB(
     15)"NP"; TAB( 29)"RP"
6  PRINT
7  PRINT  TAB( 1)"2100"; TAB( 15)
     "0"; TAB( 29)"1340"
8  REM   DATA: PRESSURE; SOLUTION
      GAS; FVF; GASFVF; 2-PHASE F
     VF; VISCOSITY RATIO
10  DIM P(7),RS(7),B(7),BG(7),BT(
     7),U(7)
20  DATA 2100,1340,1.480,.001283,
     1.480,34.1
30  DATA 1800,1280,1.468,.001518,
     1.559,38.3
40  DATA 1500,1150,1.440,.001853,
     1.792,42.4
50  DATA 1200,985,1.399,.002365,2
     .239,48.8
60  DATA 1000,860,1.360,.002885,2
     .745,53.6
70  DATA 700,662,1.287,.004250,4.
     167,62.5
80  DATA 400,465,1.202,.007680,7.
     922,79.0
90  FOR I = 1 TO 7
100  READ P(I),RS(I),B(I),BG(I),B
     T(I),U(I)
110  NEXT I
112  REM   INITIALIZING CONSTANTS

115 SR = RS(1)
120  LET KA = 31441
130  LET KB = 17.594
135 NP = 0
140  LET X = 0
142  LET BM = 0
144  FOR I = 2 TO 7
148  REM   INCREMENTING TRIAL NP
     VALUES
150  LET NP = NP + .0001
155  REM  CALCULATING INCREMENTAL
      GAS BY SCHILTHUIS EQUATION
160 GM(I) = ((BT(I) - 1.480) - NP
      * (BT(I) - 1340 * BG(I))) /
     BG(I) - BM
165  REM   CALCULATING INCREMENTA
     L GAS BY G-O-R EQUATIONS
170  LET SO = ((1 - NP) * B(2) /
     B(1)) * .85
180  LET K = KA /  EXP (KB * SO)
190  LET R(I) = K * U(I) * B(I) /
     BG(I) + RS(I)
200  LET GG(I) = ((R(I) + SR) / 2
      * (NP - X))
210  LET M(I) = GM(I) / GG(I)
220  LET Z = M(I) - 1
223  REM  TESTING TO END CYCLE
224  IF  ABS (Z) < .005 THEN 230
225  GOTO 150
```

(figure continued on next page)

Figure 11-13. Printout of a BASIC program for depletion-drive calculations using Schilthuis' equation.

```
230   PRINT  TAB( 1)P(I); TAB( 15)
      NP; TAB( 29)R(I)
235   REM  CALCULATING INCREMENTAL
       GAS VALUES FOR THE NEXT CYC
      LE
240   LET X = NP
250   LET BM = ((BT(I) - 1.48) - X
       * (BT(I) - 1340 * BG(I))) /
      BG(I)
260   LET SR = R(I)
270   NEXT I
280   END

]RUN
PRESSURE           NP                 RP

2100               0                  1340
1800               .0379              2018.13854
1500               .0969000028        2725.52678
1200               .149000005         3974.6222
1000               .177900007         4877.41332
700                .215300008         5902.77397
400                .24940001          6142.72211

]PR#0
]RUN
```

Figure 11-13. Continued.

150 sets the trial values of Np. Statement 250 computes the values of NpRp to be used in statement 160 for a new value of Np for the next pressure value. Statement 260 sets the correct value of gas/oil ratio to be used in the differential gas production equation for calculating the next Np value. In other words statements 240, 250, and 260 are used for calculating or setting initial values to be used for computing a new Np value.

Example 11-14. Hydrocarbon Mixture Separation

A hydrocarbon system has the following composition:

Component	Mole Fraction
CH4	0.15
C2H6	0.05
C3H8	0.25
C4H10	0.20
C5H12	0.25
C6H14	0.10

Write a BASIC computer program for determining the composition of the liquid and the vapor if separation is conducted at 200 psia and 100°F. Assume nonideal behavior.

The BASIC program for computing the liquid and vapor compositions in a one-stage separator is given in Figure 11-14. The liquid composition of the feed liquid and the K-values are given in DATA statements 30 to 80.

The loop bounded by statements 120 and 155 computes for a given value of NL the liquid component fractions in the separator. Then statements 157 and 160 check to determine if the sum of the components is not equal to one within a set limit. If the sum of the X-values is not equal to one, then statement 167 returns the execution of the program to statement 115 where a new trial value of NL is set and the looping continues until a correct NL is obtained. Statement 202 causes the computed values of X and Y to be printed.

```
]PR#0
]LIST


10   REM  FLASH VAPORIZATION OF HY
       DROCARBON MIXTURE
12   REM   SEPARATOR TEMPERATURE =
        100F; SEPARATOR PRESSURE =
       200 PSIA
14   REM   INPUT LIQUID COMPOSITIO
       NS AND K-VALUES--C(1) THROUG
       H C(6)
16   PRINT "I"; TAB( 4)"Z(I)"; TAB(
       12)"X(I)"; TAB( 26)"Y(I)"
17   PRINT
20   DIM Z(6),K(6)
30   DATA  .15,14.1
40   DATA  .05,2.78
50   DATA  .25,.97
60   DATA  .20,.35
70   DATA  .25,.116
80   DATA  .10,.041
86 NL = 0
88   REM   READING INTO THE COMPUT
       ER SEPARATOR INPUT DATA
90   FOR I = 1 TO 6
100   READ Z(I),K(I)
110   NEXT I
112   REM   SETTING NEW TRIAL VALU
       ES FOR NL
115 NL = NL + .01
116   REM   INITIALIZING X-VALUES
       FOR NEW CYCLE
117 W = 0
120   FOR I = 1 TO 6
130   REM   EQUATION FOR CALCULATI
       NG SEPARATOR LIQUID MOLE FRA
       CTIONS
140   LET X(I) = Z(I) / ((NL * (1 -
       K(I))) + K(I))
150   LET W = W + X(I)
155   NEXT I
156   REM   TESTING TO DETERMINE I
       F THE SUMMATION OF THE X'S E
       QUALS ZERO
157   LET M =  ABS (W - 1)
160   IF M < .01 THEN 200
167   GOTO 115
200   FOR I = 1 TO 6
205   REM  CALCULATION OF Y VALUES

210   LET Y(I) = K(I) * X(I)
220   PRINT I; TAB( 4)Z(I); TAB( 1
       0)X(I); TAB( 24)Y(I)
230   NEXT I
231   REM  CALCULATION OF NV VALUE
       S
232   LET NV = 1 - NL
234   PRINT
235   PRINT "   NL = ";NL
236   PRINT
237   PRINT "   NV = ";NV
238   PRINT
239   PRINT "   SUMMATION OF X'S =
       ";W
500   END

]RUN
I  Z(I)     X(I)            Y(I)

1  .15    .0361969113    .510376449
2  .05    .0350336323    .0973934978
3  .25    .251813054     .244258662
4  .2     .236966825     .0829383886
5  .25    .317323314     .0368095045
6  .1     .129897121     5.32578198E-03

  NL = .76

  NV = .24

  SUMMATION OF X'S = 1.00723086

]PR#0
]RUN
```

Figure 11-14. Printout of a BASIC program for computing vapor-liquid compositions in a separator.

Example 11-15. Combination Casing String Design—Tree Program

Write a BASIC program for computing the following values during the design of a combination casing string:

1. Reservoir pressure.
2. Minimum internal pressure.
3. Least collapse pressure.
4. Collapse pressure with load.
5. Casing joint strength.
6. Yield load.

Assume all weights of API 7-in. casing in grades J-55 and N-80 are available. Use INPUT statements in the program for determining selections and to input data from the keyboard. Test the BASIC program by computing the first collapse pressure with load for Example 2-13 by Craft, Holder, and Graves.*

The printout of the BASIC program for the design of a combination casing string is given in Figure 11-15. The steps used in running the casing design problem are:

1. The word RUN is typed and RETURN is pressed. A listing of casing grades appears on the CRT with the question: WHAT IS THE NUMBER OF THE CASING GRADE?

2. The number 7 is typed and RETURN is pressed. The unit weight of the casing, the cross-sectional area of the casing, and the number of the grade appears on the CRT. Also, the choices for the next calculations appear with the phrase: ENTER THE NUMBER OF YOUR CHOICE.

3. The number 4 is typed, RETURN is pressed, and the phrase PIPE SECTION WEIGHT = appears on the CRT.

4. The number 12180 is typed, RETURN is pressed, and the phrase COLLAPSE PRESSURE WITH LOAD = 528.77602 appears on the CRT. Also, the calculation choices reappear with the phrase: ENTER THE NUMBER OF YOUR CHOICE.

5. Finally, a zero is typed and the cursor starts to blink—the execution of the program is over.

---

*Craft, B. C., Holder, W. R., and Graves, Jr., E. D., *Well Design: Drilling and Production*, Englewood Cliffs, NJ, Prentice-Hall, 1962.

```
]LIST

10   REM  CASING DESIGN CALCULATIO
       NS
20   REM  BY M. A. NOBLES 9/20/83
90   REM   DIMENSIONS FOR 7-IN CAS
       ING PARAMETERS
100  DIM W(11),PI(11),PE(11),K(11
       ),F(11),Y(11),A(11)
101  REM   CASING DATA
102  DATA  20,3740,2500,747000,25
       4000,55000,4.198
104  DATA  23,4360,3290,865000,30
       0000,55000,5.105
106  DATA  26,4980,4060,981000,34
       5000,55000,5.998
108  DATA  23,4360,3290,865000,34
       4000,55000,5.105
110  DATA  26,4980,4060,981000,39
       5000,55000,5.998
112  DATA  23,6340,4300,1132000,4
       00000,80000,5.105
114  DATA  26,7240,5320,1283000,4
       60000,80000,5.998
116  DATA  29,8160,6370,1436000,5
       20000,80000,6.899
118  DATA  32,9060,7400,1584000,5
       78000,80000,7.766
120  DATA  35,9960,8420,1729000,6
       35000,80000,8.622
122  DATA  38,10800,9080,1863000,
       688000,80000,9.408
128  REM   READING THE CASING DAT
       A INTO THE MEMORY OF THE COM
       PUTER
130  FOR I = 1 TO 11
135  READ W(I),PI(I),PE(I),K(I),F
       (I),Y(I),A(I)
140  NEXT I
150  REM  LISTING THE TYPES OF CA
       SING
160  PRINT "      1  J-20-FJS  "
162  PRINT "      2  J-23-FJS  "
164  PRINT "      3  J-26-FJS  "
165  PRINT "      4  J-23-FJL  "
166  PRINT "      5  J-26-FJL  "
168  PRINT "      6  N-23-FJL  "
170  PRINT "      7  N-26-FJL  "
172  PRINT "      8  N-29-FJL  "
174  PRINT "      9  N-32-FJL  "
175  PRINT "     10  N-35-FJL  "
176  PRINT "     11  N-38-FJL  "
180  PRINT
183  REM   THE NUMBER OF THE CASI
       NG IS ALSO THE SUBSCRIPT NUM
       BER
185  INPUT "  WHAT IS THE NUMBER
       OF THE CASING GRADE?  ";L
186  PRINT
188  REM   SETTING THE SUBSCRIPT
       NUMBER FOR A ROUND OF CALCUL
       ATIONS
190  LET I = L
192  PRINT W(I),A(I),L
193  PRINT
200  PRINT "  CALCULATION CHOICES
       "
202  PRINT
205  PRINT "  0  NO CHOICE  "
210  PRINT "  1  RESERVOIR PRESSU
       RE "
215  PRINT "  2  MINIMUM INTERNAL
        PRESSURE  "
220  PRINT "  3  LEAST COLLAPSE P
       RESSURE "
222  PRINT "  4  COLLAPSE PRESSUR
       E WITH LOAD "
225  PRINT "  5  CASING JOINT STR
       ENGTH "
230  PRINT "  6  YIELD LOAD  "
231  PRINT
239  PRINT
240  INPUT " ENTER THE NUMBER OF
       YOUR CHOICE  ";X
241  PRINT
245  REM   THE NUMBER CONTROLS TH
       E TYPE OF CALCULATION
246  REM   AN IF STATEMENT SENDS
       THE COMPUTER TO THE APPROPRI
       ATE SECTION OF THE PROGRAM F
       OR CALCULATIONS
250  IF X = 0 THEN 1000
252  IF X = 1 THEN 300
254  IF X = 2 THEN 325
256  IF X = 3 THEN 350
258  IF X = 4 THEN 400
260  IF X = 5 THEN 430
264  IF X = 6 THEN 480
290  REM   REM  INPUT STATEMENTS
       ARE USED FOR SUPPLYING INFOR
       MATION FROM THE KEYBOARD
300  INPUT "  WELL DEPTH = ";WL
302  INPUT " DP/DL    ";DR
305  PRINT
310  LET PWS = WL * DR
315  PRINT " RESERVOIR PRESSURE =
        ";PWS
323  GOTO 200
325  INPUT " RESERVOIR PRESSURE =
        ";PWS
326  INPUT " DESIGN FACTOR  = ";N
       1
327  LET P1 = PWS * N1
328  PRINT "  MINIMUM INTERNAL YI
       ELD = ";P1
329  GOTO 200
330  INPUT "  DESIGN FACTOR  = "N
       2
332  INPUT "  MUD DENSITY    = "D
       E
334  INPUT "  WELL DEPTH     = "W
       L
336  LET P3 = .052 * N2 * D * WL
338  PRINT "  COLLAPSE PRESSURE N
       O LOAD  ";P3
348  GOTO 200
```

(figure continued on next page)

Figure 11-15. Printout of a BASIC program for the design of a combination casing string.

```
350  INPUT "  DESIGN FACTOR  =";N
     2
352  INPUT "  MUD DENSITY    =";D

354  INPUT "  WELL DEPTH     =";W
     L
356  LET P3 = .052 * N2 * D * WL
358  PRINT "  COLLAPSE PRESSURE N
     O LOAD =";P3
398  GOTO 200
400  INPUT "PIPE SECTION WEIGHT =
     ";TW
405  LET P4 = (PE(I) / K(I)) * ( SQR
     (K(I) ^ 2 - 3 * (TW ^ 2)) -
     TW)
410  PRINT "COLLAPSE PRESSURE WIT
     H LOAD = ; "P4
428  GOTO 200
430  INPUT "  DESIGN FACTOR J-S
     = ";JS
435  LET WMAX = F(I) / JS
440  PRINT "  MAX JOINT STRENGTH
     ="WMAX
445  GOTO 200
480  INPUT " DESIGN FACTOR Y-S =
     ";YS
485  LET WYMAX = Y(I) * A(I) / YS

490  PRINT " YIELD STRENGTH  = ";
     WYMAX
528  GOTO 200
1000  END
```

```
]RUN
     1  J-20-FJS
     2  J-23-FJS
     3  J-26-FJS
     4  J-23-FJL
     5  J-26-FJL
     6  N-23-FJL
     7  N-26-FJL
     8  N-29-FJL
     9  N-32-FJL
    10  N-35-FJL
    11  N-38-FJL

  WHAT IS THE NUMBER OF THE CASING GRADE?  7

26              5.998            7

  CALCULATION CHOICES

  0  NO CHOICE
  1  RESERVOIR PRESSURE
  2  MINIMUM INTERNAL PRESSURE
  3  LEAST COLLAPSE PRESSURE
  4  COLLAPSE PRESSURE WITH LOAD
  5  CASING JOINT STRENGTH
  6  YIELD LOAD

 ENTER THE NUMBER OF YOUR CHOICE  4

PIPE SECTION WEIGHT =1213080
COLLAPSE PRESSURE WITH LOAD = ; 5268.77602
  CALCULATION CHOICES

  0  NO CHOICE
  1  RESERVOIR PRESSURE
  2  MINIMUM INTERNAL PRESSURE
  3  LEAST COLLAPSE PRESSURE
  4  COLLAPSE PRESSURE WITH LOAD
  5  CASING JOINT STRENGTH
  6  YIELD LOAD

 ENTER THE NUMBER OF YOUR CHOICE  0


]PR#0
]RUN
```

Figure 11-15. Continued.

Following the calculations, the execution of the program returns to the list of calculation choices and another type of selection becomes available. This can be repeated as many times as one chooses—provided the grade of casing has not been changed. Briefly, the steps in the design are:

1. A table of values describing 7-in. API casing in grades J-55 and N-80 are read into the computer by statements 100 through 140.
2. Each grade of casing is given a number and programmed to appear on the CRT—statements 160 to 176.
3. An INPUT statement is used for the selection of a casing grade by its number.
4. Each type of calculation is given a number and programmed to appear on the CRT—statements 205-230.
5. An INPUT statement is used for selecting the types of calculation by a number.
6. IF/THEN statements are used to test for the number of the calculation type and for sending the execution of the computer to the section of the program for performing the selected calculation.
7. If there are not enough data for the calculation, the data values are asked for by means of INPUT statements.
8. As soon as all information is in the memory of the computer, the calculation is performed and printed.
9. At the end of each subprogram for performing the selected calculation, there is a GOTO statement which returns the execution to the "calculation choice" section.

Example 11-16. Algebraic Simultaneous Equations Solutions

Write a BASIC mainframe computer program, using MATRIX statements for determining the unknown values in the three equations:

$$
\begin{aligned}
3X1 + 7X2 + 3X3 &= 12 \\
9X1 + 43X2 + 21X3 &= 31 \\
2X1 + 8X2 + 19X3 &= 32
\end{aligned}
$$

The coefficients of the X's are written in matrix form

$$
\begin{matrix} 3 & 7 & 32 \\ 9 & 43 & 21 \\ 2 & 8 & 19 \end{matrix} \quad = \quad \text{matrix A}
$$

The constants of the equation to the right of the equal signs are written in matrix form

$$\begin{matrix} 12 \\ 31 \\ 32 \end{matrix} = \text{matrix B}$$

Mathematicians have found the values of the unknowns (X's) are equal to the inversion of matrix A times matrix B. Therefore, MAT X = MAT(inversion A) * MAT B or the values for the unknowns in the three equations.

The BASIC program printout for solving the simultaneous equations, which was performed with the VAX 11-780, is given in Figure 11-16. The procedure used is:

1. Using an Apple IIe fitted with an SSM Modemcard (SSM Microcomputer Products Inc., San Jose, California), contact is made through a telephone line by dialing from the keyboard of the microcomputer.

```
]IN#2
]
VAX/VMS AT YOUR SERVICE!


USERNAME: MAN9506
PASSWORD:


          WELCOME TO VAX/VMS VERSION V3.4 ON THE TTU ACADEMIC VAX SYSTEM.
                    15-NOV-1983 13:10:17.71, ON LINE TTH0:

TO UNDERSTAND A PROGRAM YOU MUST BECOME BOTH THE MACHINE AND THE PROGRAM
 -- ALAN PERLIS


$ EDIT MIMEQ.BAS

INPUT FILE DOES NOT EXIST
[EOB]

*INSERT

                    10 DIM A(3,3)

                    20 DIM B(3,1)
```

Figure 11-16. Printout (from a mainframe) of a BASIC program for solving algebraic simultaneous equations.

```
              30 DIM C(3,3)

              40 DIM X(3,1)

              50 MAT READ A

              60 DATA 3,7,32,9,43,21,2,8,19

              70 MAT READ B

              80 DATA 12,31,32

              90 MAT INV   C = INV(A)

              100 MAT X = C*B

              110 MAT PRINT X

              120 END

              ^Z

[EOB]

*EXIT

USER:[MAN9506]MIMEQ.BAS;1 12 LINES

$ TYPE MIMEQ.BAS;1

10 DIM A(3,3)
20 DIM B(3,1)
30 DIM C(3,3)
40 DIM X(3,1)
50 MAT READ A
60 DATA 3,7,32,9,43,21,2,8,19
70 MAT READ B
80 DATA 12,31,32
90 MAT C = INV(A)
100 MAT X = C*B
110 MAT PRINT X
120 END

$ EXECUTE MIMEQ.BAS;1

$ BASIC/CHECK MIMEQ.BAS;1
$ LINK  MIMEQ
$ RUN MIMEQ
-46.5218
 9.11913
 2.74161

$ BYE

  MAN9506      LOGGED OUT AT 15-NOV-1983 13:19:53.15




]RUN
```

Figure 11-16. Continued.

2. When the mainframe receives information from the microcomputer, the words "VAX/VMS AT YOUR SERVICE" are displayed on the CRT.
3. The words USER NAME appear on the CRT.
4. To this the reply is: MAN9506.
5. The mainframe asks PASSWORD:
6. The user replies.
7. Following the appearance of a few sayings, a dollar sign appears; this indicates the computer is ready for the user.
8. The user responds by typing EDIT MINEQ.BAS
9. The computer checks and finds the user does not have a file in its memory by that name; then the computer displays "INPUT FILE DOES NOT EXIST"
10. Next, the user types INSERT, followed by typing the BASIC program.
11. After the END statement, the user presses the CONTROL key and the Z key at the same time.
12. The user types EXIT to get out of the file.
13. The computer then displays USER:[MAN9506]MINEQ.BAS;1 12LINES.
14. To list the program, the user types TYPEMINEQ.BAS;1, and the computer lists the program. To RUN the program, the user types EXECUTE MINEQ.BAS;1.
15. The computer checks the program for errors and finds no errors; then it is LINKED and RUN.
16. The values of X's are printed.
17. To cease operations, the word BYE is typed and the call is terminated.

The values for the X's check those obtained by using a program by Jules H. Gilder (see Suggested Reading). This program contains approximately 80 statements and uses the Gauss-Jordan elimination technique.

The MAT statements are not included in all versions of BASIC nor is there a standard method for representing the single-column matrix. The VAX-780 requires the DIM statement for the B matrix to be written as DIM B(3,1); some computers will accept DIM B(3) and some will accept B(3,0)

A machine language program, MATRIX][, is available for extending the capabilities of Applesoft to matrix manipulations; the software contains a program for simultaneous linear equation solutions and a program for least-squares polynomial curve fit.

BASIC Program Development

In the development of a BASIC program a small section of the overall program is written and tested with a PRINT statement. The PRINT statement is then deleted and another section is written, tested, and so on until the whole program is completed. The following procedure is used in the development of programs which have a considerable amount of input data:

1. All data are typed using DATA statements.
2. With the aid of DIM statements, major variables are assigned and dimensioned.
3. The data are stored in the memory of the computer with READ statements.
4. With the aid of a PRINT statement, this section is tested to determine if the data have been stored correctly.
5. If the data have been stored correctly, the PRINT statement is deleted.
6. Additional constants and initializing data are added to the program.
7. Key algebraic-like expressions are written for determining single-variable values.
8. A PRINT statement is added to test the accuracy of the algebraic expressions.
9. The program to this point is RUN.
10. When the two sections become correct, the PRINT statement is deleted.
11. An observation is made for looping to compute a series of values for subscripted variables. The questions are: Is a single loop sufficient? Will it take a loop in a loop or two series loops? Etc.
12. Finally, methods are determined for displaying the printed results and rounding to significant figures.

Exercises

1-17. Repeat Exercises 1 through 17 of Chapter 10 in BASIC.

18. Write a BASIC program using a TAB feature for plotting y = a*exp (b*x); note the maximum value for y is 40.
19. Substituting the data from Example 10-10 into the BASIC program of Example 11-6, determine the constants for the exponential equation. Compare the two sets of constants which were obtained by the two methods.
20. Substituting the data from Example 10-11 into the BASIC program of Example 11-13, use constants obtained in Exercise 10-10 in the program. Compare the results obtained by using two different languages.
21. With the aid of Example 11-16, write a program for determining the constants in a third-order polynomial equation.
22. With the aid of Example 11-15, write a BASIC tree program for sucker rod design.

Suggested Reading

1. Burcik, E. J., *Properties of Petroleum Reservoir Fluids,* New York: John Wiley and Sons, 1957.
2. Craft, B. C. and Hawkins, M. F., *Petroleum Reservoir Engineering,* Englewood Cliffs: Prentice-Hall Inc., 1959.
3. Craft, B. C., Holder, W. R., and Graves, Jr., E. D., *Well Design: Drilling and Production,* Englewood Cliffs: Prentice-Hall, Inc., 1962.
4. Cole, F. W., *Reservoir Engineering Manual,* Second Edition, Houston: Gulf Publishing Company, 1969.
5. Gilder, J. H., *BASIC Computer Programs in Science and Engineering,* Rochelle Park: Hayden Book Company, 1980.
6. Gottfried, B. S., *Theory and Problems of Programming with Microcomputers, 2nd Ed.,* New York: McGraw-Hill Book Company, Schaum Outline Series, 1982.
7. Hennefeld, J., *Using BASIC—An Introduction to Computer Programming,* Boston: Prindle, Weber and Schmidt, 1981.
8. Kenney, P., *Minicomputers, Low-Cost Computer Power for Management,* New York: Amacom, 1981.
9. Murrill, P. W. and Smith, C. L., *BASIC Programming,* Scranton: Intext Educational Publishers, 1971.
10. *Programming Languages PDP Family,* Maynard: Digital Equipment Corporation, 1970.
11. *Vax-11 BASIC Language Reference Manual,* Digital Equipment Corporation, 1980.

Appendix A
Number Systems

As civilization developed, counting became more extensive, and the counting process was systematized. This systematization was achieved by arranging numbers in basic groups, the sizes of which were largely determined by the "matching" process. For instance, the fingers furnished a convenient matching device. Simply described, the counting process consisted of a number, *b* selected as a base and assigning names to the numbers 1, 2, 3, . . ., b. Names for larger numbers were usually assigned somewhat as combinations of the names of the numbers already selected. Other names for a base are radix and scale. Bases have been selected for the number of fingers on one hand, the number of fingers on both hands, the number of fingers on both hands and numbers of toes on both feet; these bases are 5, 10 and 20 respectively.

Number systems with bases of 5, 10, 20, and 60 have been used. The abacists (Oriental) used a number system with a base of 5 (the quinary number system); the algorists (Arabic) used a number system with a base of 10 (the decimal number system); the Mayan Indians used a number system with a base of 20 (vigesimal number system); and the ancient Babylonians used a number system with a base of 60 (sexagesimal number system). Historical records indicate that other bases have also been used. Values in the Mayan Indian number system were represented by shells, dots, and dashes. Perhaps the dash represented a stick, and the dot represented a pebble. Figure A-1 is a simple grouping scheme for the Mayan Indians.

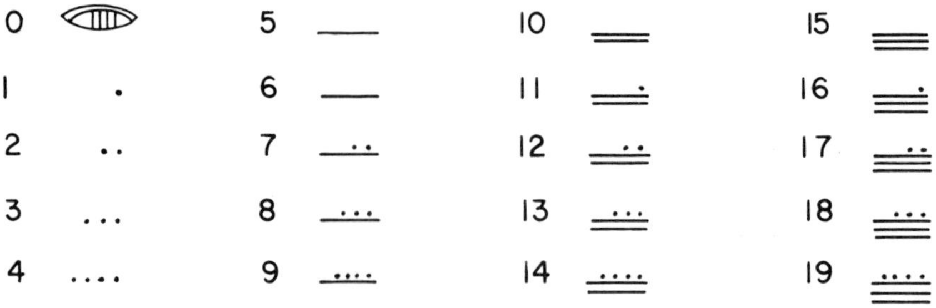

Figure A-1. Mayan Indian numeral system.

Prior to the sixteenth century, a 400-year conflict existed in Europe between the algorists and the abacists. By 1500 A.D. the algorists had won, and by 1600 A.D., the abacists were almost forgotten.

Advances in science, engineering, and everyday technology have placed new and increasing demands upon scientists, mathematicians, and engineers to solve problems requiring large numbers of calculations. As a result, engineers and scientists were compelled to develop computers. Since the decimal number system is not suited for computers, suitable number systems were developed. Most computers essentially use either the binary number system or a modification of that system; among the number systems now used are the binary, octal, and hexadecimal. The binary is the simplest of all number systems, as it can be represented by 0 and 1. In a machine, for instance, it can be represented by a charge and no-charge section of a core, a light on and a light off, a location on a punched card either with a hole, or without a hole, or a magnetized section of a magnetic core magnetized in one direction or magnetized in the other. The digit 0 is assigned to one of the conditions, and 1 is assigned to the other.

Number systems are classified as ciphered and positional. A ciphered number system for a base b can be designated as $1, 2, \ldots, (b-1); b, 2b \ldots (b-1)b; b^2, 2b^2, \ldots, (b-1)b^2; b^3$ and so on. The Greek numeral system is of the ciphered type and can be traced back to 450 B.C. The base for this system is 10; it has 27 characters—all 24 letters of the Greek alphabet and symbols for the obsolete digamma, koppa, and sampi.

The positional number systems are the most common. Our own decimal system is an example. The general formula is as follows:

$$N = a_n b^n + a_{n-1} b^{n-1}, \ldots, a_1 b^1 + a_0 b^0 + a_{-1} b^{-1} + a_{-2} b^{-2}, \ldots, a_{-k} b^{-k}$$

where b represents the base of the number system, N is any number, a, n, and k are constants. The decimal point for the number is between a_0b^0 and $a_{-1}b^{-1}$. The remainder of this chapter is devoted to positional number systems. For most number systems, each of the constants a takes on values of 0 through (b − 1).

Positional Number Representation

The relations between the binary numbers, octal numbers, decimal numbers, and hexadecimal numbers are shown in Table A-1.

Letters of the alphabet have been used for the six additional terms in the hexadecimal number system. The letters f,g,j,k,q,w, and K,S,N,J,F,L have been used; the first set of letters were used with the Royal McBee computer, and the second set of letters were used with the Illiac computer at The University of Illinois. Now it appears the first six letters of the alphabet A,B,C,D,E,F are the ones which microcomputers are using.

Table A-1
Relations Between Binary, Octal, Decimal, and Hexadecimal Numbers

Decimal	Binary	Octal	Sexidecimal or Hexadecimal
0	0	0	0
1	1	1	1
2	10	2	2
3	11	3	3
4	100	4	4
5	101	5	5
6	110	6	6
7	111	7	7
8	1000	10	8
9	1001	11	9
10	1010	12	A
11	1011	13	B
12	1100	14	C
13	1101	15	D
14	1110	16	E
15	1111	17	F
16	10000	20	10
17	10001	21	11
18	10010	22	12
19	10011	23	13
20	10100	24	14

To designate a number in a given system, the number is enclosed in parentheses, and the base is written as a subscript to the right of the parentheses. Examples of numbers for various number systems are:

$$(1101.101)_2, \ (6402.146)_8, \ (7340.685)_{10}, \ (\text{A0B}.15)_{16}$$

These four numbers have, respectively, the bases 2, 8, 10 and 16; and the numbers represent binary, octal, decimal, and hexadecimal number systems. These systems are compared with the decimal system in Examples A-1, A-2, A-3 and A-4.

Example A-1

The binary number $(1101.101)_2$ means

$$
\begin{aligned}
&(1 \times 2^3) &&= (8)_{10} \\
+ &(1 \times 2^2) &&= (4)_{10} \\
+ &(0 \times 2^1) &&= (0)_{10} \\
+ &(1 \times 2^0) &&= (1)_{10} \\
+ &(1 \times 2^{-1}) &&= (0.5)_{10} \\
+ &(0 \times 2^{-2}) &&= (0)_{10} \\
+ &(1 \times 2^{-3}) &&= \underline{(0.125)_{10}} \\
&\text{Total} && \quad (13.625)_{10}
\end{aligned}
$$

or $(1101.101)_2 = (13.625)_{10}$

Example A-2

The octal number $(6402.146)_8$ means

$$
\begin{aligned}
&(6 \times 8^3) &&= (3072)_{10} \\
+ &(4 \times 8^2) &&= (256)_{10} \\
+ &(0 \times 8^1) &&= (0)_{10} \\
+ &(2 \times 8^0) &&= (2)_{10} \\
+ &(1 \times 8^{-1}) &&= (0.125)_{10} \\
+ &(4 \times 8^{-2}) &&= (0.0625)_{10} \\
+ &(6 \times 8^{-3}) &&= \underline{(0.01171875)_{10}} \\
&\text{Total} && \quad (3330.19921875)_{10}
\end{aligned}
$$

or $(6402.146)_8 = (3330.19921875)_{10}$

Example A-3

The decimal number $(7340.685)_{10}$ means

$$
\begin{array}{rll}
 & (7 \times 10^{3}) & = (7000)_{10} \\
+ & (3 \times 10^{2}) & = (300)_{10} \\
+ & (4 \times 10^{1}) & = (40)_{10} \\
+ & (0 \times 10^{0}) & = (00)_{10} \\
+ & (6 \times 10^{-1}) & = (0.6)_{10} \\
+ & (8 \times 10^{-2}) & = (0.08)_{10} \\
+ & (5 \times 10^{-3}) & = (0.005)_{10} \\
\hline
 & \text{Total} & (7340.685)_{10}
\end{array}
$$

Example A-4

The hexadecimal number $(A0B.15)_{16}$ means

$$
\begin{array}{rll}
 & (10 \times 16^{2}) & = (2560)_{10} \\
+ & (0 \times 16^{1}) & = (0)_{10} \\
+ & (11 \times 16^{0}) & = (11)_{10} \\
+ & (1 \times 16^{-1}) & = (0.0625)_{10} \\
+ & (5 \times 16^{-2}) & = (0.01953125)_{10} \\
\hline
 & \text{Total} & (2571.08203125)_{10}
\end{array}
$$

Arithmetic in Positional Number Systems

The four basic arithmetic operations (addition, subtraction, multiplication, and division) are similar for all positional number systems. In other words, these operations are similar for binary, octal, decimal, or hexadecimal bases. Examples of arithmetic problems in the binary, octal, and hexadecimal number systems follow. Arithmetic tables for each of the number systems are given as they are needed for the arithmetic operations.

Remarks concerning addition in the decimal number system are in preparation for the discussions of addition by these three number systems. The rules for arithmetical operations apply in the same manner for all bases of the positional number systems. Symbols for the numbers begin with 1 and continue to base minus one $(b - 1)$. The symbol for the base is 10. The next number is written 11; the process continues until the number twice the base minus one (2b − 1) and the following number is 2 followed by a zero. This process applies for all number systems. Table A-2 gives the addition of two numbers in the decimal number system.

The use of Table A-2 for the addition of two numbers is illustrated by the problem

```
   11 1
  67964
+ 56427
-------
 124391
```

In solving the problem, the sum of 4 and 7 is found by locating the element in the table which is in Column 4, Row 7. The number is $\overline{1}$; then a 1 is written below the line in the extreme right-hand position (units position) of the problem, and a 1 is carried to the next column to the left. To add the next column, the number in the table is found which is in Column 6, Row 2; the number is 8. Now the number in the table is found which is in Column 1, Row 8; the number is 9. To complete the addition for this column, the number 9 is written below the line in the second column from the right. Values placed below the line for each of the other three columns are found in a similar manner as those

Table A-2
Addition Table for Base 10

Addend							Augend			
	0	1	2	3	4	5	6	7	8	9
0	0	1	2	3	4	5	6	7	8	9
1	1	2	3	4	5	6	7	8	9	$\overline{0}$*
2	2	3	4	5	6	7	8	9	$\overline{0}$	$\overline{1}$
3	3	4	5	6	7	8	9	$\overline{0}$	$\overline{1}$	$\overline{2}$
4	4	5	6	7	8	9	$\overline{0}$	$\overline{1}$	$\overline{2}$	$\overline{3}$
5	5	6	7	8	9	$\overline{0}$	$\overline{1}$	$\overline{2}$	$\overline{3}$	$\overline{4}$
6	6	7	8	9	$\overline{0}$	$\overline{1}$	$\overline{2}$	$\overline{3}$	$\overline{4}$	$\overline{5}$
7	7	8	9	$\overline{0}$	$\overline{1}$	$\overline{2}$	$\overline{3}$	$\overline{4}$	$\overline{5}$	$\overline{6}$
8	8	9	$\overline{0}$	$\overline{1}$	$\overline{2}$	$\overline{3}$	$\overline{4}$	$\overline{5}$	$\overline{6}$	$\overline{7}$
9	9	$\overline{0}$	$\overline{1}$	$\overline{2}$	$\overline{3}$	$\overline{4}$	$\overline{5}$	$\overline{6}$	$\overline{7}$	$\overline{8}$

*The bar above a number means that a 1 is carried or 1 is added to the next significant position.

for the two extreme right columns. For the extreme left-hand column of the two numbers which were to be added, the sum of 6, 5, and 1, according to the table, is $\overline{2}$. For this column, a 2 was written below the line and a 1 was carried; since there were no numbers to which to add this number, a 1 was written below the line of the problem to the left of the 2.

An addition table for adding two numbers in the binary system is given in Table A-3.

Example A-5 illustrates addition in the binary system.

Example A-5

Binary		Decimal Equivalent
1111		1
11001	(a)	25
+ 1111		+ 15
101000		40
111		13
1101	(b)	+ 11
+ 1011		24
11000		

The table for addition in the binary system is simple because the values in the table are either 1 or 0. A third row was added to the table because, frequently due to a carry, the addition of three 1's would be necessary. The sum of $(1 + 1 + 1)_2 = (11)_2$. For the second binary example, the value for the two numbers in the extreme right-hand column is found

Table A-3
Addition Table for Base 2

Addend	Augend 0	Augend 1
0	0	1
1	1	$\overline{0}$*
0	$\overline{0}$	$\overline{1}$

*The bar above a number means that a 1 is carried or 1 is added to the next significant position.

from the binary table corresponding to the value in the second row and the second column. This value is $\bar{0}$. Since there is a carry, 1 is placed above the first row in the second column from the right, and a 0 is placed in the first column below the line. To obtain the value for the second column from the right, the table is used and the sum of $1 + 1$ is again found in the second row in the second column of the table. The 0 is placed below the line and a 1 is placed in the third column from the right. For the value in the third column from the right, 1 and 1 is $\bar{0}$. This time, a 0 is placed below the line under the third column from the right, and a 1 is placed in the fourth column from the right above the first row. For the fourth column from the right, a value from the binary table for $1 + 1 + 1$ is obtained. This value is obtained by looking up the value of $1 + 1$ in the table which is $\bar{0}$. Then a 1 is added to the $\bar{0}$ and the table indicates that the value of $1 + \bar{0} = \bar{1}$. A 1 is placed in the fourth column from the right of the column beneath the line. Since the fourth column to the left is the extreme left column for the data of the problem, a 1 is placed below the line in the position of a fifth column from the right. The answer to the second problem is $(11000)_2 = (24)_{10}$.

Table A-4 is a table for the addition of two numbers in the octal number system.

Problems of addition in the octal number system are given in Example A-6.

Table A-4
Addition Table for Base 8

Addend	Augend 0	1	2	3	4	5	6	7
0	0	1	2	3	4	5	6	7
1	1	2	3	4	5	6	7	$\bar{0}$*
2	2	3	4	5	6	7	$\bar{0}$	$\bar{1}$
3	3	4	5	6	7	$\bar{0}$	$\bar{1}$	$\bar{2}$
4	4	5	6	7	$\bar{0}$	$\bar{1}$	$\bar{2}$	$\bar{3}$
5	5	6	7	$\bar{0}$	$\bar{1}$	$\bar{2}$	$\bar{3}$	$\bar{4}$
6	6	7	$\bar{0}$	$\bar{1}$	$\bar{2}$	$\bar{3}$	$\bar{4}$	$\bar{5}$
7	7	$\bar{0}$	$\bar{1}$	$\bar{2}$	$\bar{3}$	$\bar{4}$	$\bar{5}$	$\bar{6}$

*The bar above a number means that a 1 is carried or a 1 is added to the next significant position.

Example A-6

Octal		Decimal
1		11
4673	(a)	2491
+ 5063		+ 2611
11756		5102
111		1 1
3267	(b)	1719
+ 6674		+ 3516
12163		5235

The procedure followed in the example problems for finding the sum of two octal numbers is very similar to the procedure for finding the sum of two binary numbers as well as the sum of two decimal numbers. Referring to the second octal example of addition, the sums for each column were found in Table A-4. For each column, the table indicated a carry; therefore, the number found in the table for the sum in a column was placed below the line in the respective column, and a 1 was added to the next column to the right. This procedure was repeated for all columns except the extreme left-hand column. For this column, the number 2 from the addition table was placed below the line, and a 1 was placed under the line in the position of the next column to the left.

A check for the second example of the octal addition is given in Example A-7.

Example A-7

$(3267)_8$ is equivalent to

$$
\begin{array}{rl}
3 \times 8^3 = & (1536)_{10} \\
+ 2 \times 8^2 = & + (\ 128)_{10} \\
+ 6 \times 8^1 = & + (\ \ 48)_{10} \\
+ 7 \times 8^0 = & + (\ \ \ 7)_{10} \\
\hline
\text{Total} & (1719)_{10}
\end{array}
$$

$(6674)_8$ is equivalent to

$$
\begin{array}{rl}
6 \times 8^3 = & (3072)_{10} \\
+\,6 \times 8^2 = + & (\ \ 384)_{10} \\
+\,7 \times 8^1 = + & (\ \ \ \ 56)_{10} \\
+\,4 \times 1^0 = + & (\ \ \ \ \ \ 4)_{10} \\
\text{Total} & \overline{(3516)_{10}}
\end{array}
$$

The sum of the two octal numbers, which is $(12163)_8$, is equivalent to

$$
\begin{array}{rl}
1 \times 8^4 = & (4096)_{10} \\
+\,2 \times 8^3 = & (1024)_{10} \\
+\,1 \times 8^2 = & (\ \ \ \ 64)_{10} \\
+\,6 \times 8\ \ = & (\ \ \ \ 48)_{10} \\
+\,1 \times 3\ \ = & (\ \ \ \ \ \ 3)_{10} \\
\text{Total} & \overline{(5235)_{10}}
\end{array}
$$

Table A-5 shows the addition of two numbers in the hexadecimal system.

Table A-5
Addition Table for Base 16

								Augend								
Addend	**0**	**1**	**2**	**3**	**4**	**5**	**6**	**7**	**8**	**9**	**A**	**B**	**C**	**D**	**E**	**F**
0	0	1	2	3	4	5	6	7	8	9	A	B	C	D	E	F
1	1	2	3	4	5	6	7	8	9	A	B	C	D	E	F	$\overline{0}$
2	2	3	4	5	6	7	8	9	A	B	C	D	E	F	$\overline{0}$	$\overline{1}$
3	3	4	5	6	7	8	9	A	B	C	D	E	F	$\overline{0}$	$\overline{1}$	$\overline{2}$
4	4	5	6	7	8	9	A	B	C	D	E	F	$\overline{0}$	$\overline{1}$	$\overline{2}$	$\overline{3}$
5	5	6	7	8	9	A	B	C	D	E	F	$\overline{0}$	$\overline{1}$	$\overline{2}$	$\overline{3}$	$\overline{4}$
6	6	7	8	9	A	B	C	D	E	F	$\overline{0}$	$\overline{1}$	$\overline{2}$	$\overline{3}$	$\overline{4}$	$\overline{5}$
7	7	8	9	A	B	C	D	E	F	$\overline{0}$	$\overline{1}$	$\overline{2}$	$\overline{3}$	$\overline{4}$	$\overline{5}$	$\overline{6}$
8	8	9	A	B	C	D	E	F	$\overline{0}$	$\overline{1}$	$\overline{2}$	$\overline{3}$	$\overline{4}$	$\overline{5}$	$\overline{6}$	$\overline{7}$
9	9	A	B	C	D	E	F	$\overline{0}$	$\overline{1}$	$\overline{2}$	$\overline{3}$	$\overline{4}$	$\overline{5}$	$\overline{6}$	$\overline{7}$	$\overline{8}$
A	A	B	C	D	E	F	$\overline{0}$	$\overline{1}$	$\overline{2}$	$\overline{3}$	$\overline{4}$	$\overline{5}$	$\overline{6}$	$\overline{7}$	$\overline{8}$	$\overline{9}$
B	B	C	D	E	F	$\overline{0}$	$\overline{1}$	$\overline{2}$	$\overline{3}$	$\overline{4}$	$\overline{5}$	$\overline{6}$	$\overline{7}$	$\overline{8}$	$\overline{9}$	$\overline{A}$
C	C	D	E	F	$\overline{0}$	$\overline{1}$	$\overline{2}$	$\overline{3}$	$\overline{4}$	$\overline{5}$	$\overline{6}$	$\overline{7}$	$\overline{8}$	$\overline{9}$	$\overline{A}$	$\overline{B}$
D	D	E	F	$\overline{0}$	$\overline{1}$	$\overline{2}$	$\overline{3}$	$\overline{4}$	$\overline{5}$	$\overline{6}$	$\overline{7}$	$\overline{8}$	$\overline{9}$	$\overline{A}$	$\overline{B}$	$\overline{C}$
E	E	F	$\overline{0}$	$\overline{1}$	$\overline{2}$	$\overline{3}$	$\overline{4}$	$\overline{5}$	$\overline{6}$	$\overline{7}$	$\overline{8}$	$\overline{9}$	$\overline{A}$	$\overline{B}$	$\overline{C}$	$\overline{D}$
F	F	$\overline{0}$	$\overline{1}$	$\overline{2}$	$\overline{3}$	$\overline{4}$	$\overline{5}$	$\overline{6}$	$\overline{7}$	$\overline{8}$	$\overline{9}$	$\overline{A}$	$\overline{B}$	$\overline{C}$	$\overline{D}$	$\overline{E}$

*The bar above a number means that 1 is carried or a 1 is added to the next significant figure.

At this stage the reader is familiar with the arithmetical procedure to follow for the addition of two numbers, by the use of an addition table, for all bases of positional numbers. Therefore, the details for adding two hexadecimal numbers are omitted. An example of an addition problem in the hexadecimal system is given in Example A-8.

Example A-8

Hexadecimal	Decimal Equivalent
2 4 B E	9 4 0 6
+ 9 7 7 F	+ 3 8 7 8 3
B C 3 D	4 8 1 8 9

A check for the solution of the problem in the hexadecimal system is given in Example A-9.

Example A-9

$(24BE)_{16}$ is equivalent to

$$\begin{array}{rrl} & 2 \times 16^3 = 2 \times 4{,}096 = & (8192)_{10} \\ + & 4 \times 16^2 = 4 \times 256 = & (1024)_{10} \\ + & 11 \times 16^1 = 11 \times 16 = & (\ 176)_{10} \\ + & 14 \times 16^0 = 14 \times 1 = & (\ 14)_{10} \\ \hline & \text{Total} & (9406)_{10} \end{array}$$

$(977F)_{16}$ is equivalent to

$$\begin{array}{rrl} & 9 \times 16^3 = 9 \times 4{,}096 = & (36864)_{10} \\ + & 7 \times 16^2 = 7 \times 256 = & (\ 1792)_{10} \\ + & 7 \times 16^1 = 7 \times 16 + & (\ 112)_{10} \\ + & 15 \times 16^0 = 15 \times 1 = & (\ 15)_{10} \\ \hline & \text{Total} & (38783)_{10} \end{array}$$

$(BC3D)_{16}$ is equivalent to

$$
\begin{aligned}
11 \times 16^3 &= 11 \times 4096 = (45056)_{10} \\
12 \times 16^2 &= 12 \times 256 = (3072)_{10} \\
3 \times 16^1 &= 3 \times 16 = (48)_{10} \\
13 \times 16^0 &= 13 \times 1 = (13)_{10} \\
\hline
\text{Total} &\quad (48189)_{10}
\end{aligned}
$$

Similarly as for the arithmetical operation of addition, the discussion for the arithmetical operation of subtraction begins with numbers in the decimal system. The subtraction procedure is the same arithmetical operation in all positional number systems. Students learn to subtract and carry out two other arithmetical operations in the decimal system by means of tables during their elementary school training. The student memorizes the arithmetic tables for addition, subtraction and multiplication.

To subtract with the use of tables requires additional expense in the construction of computers with a device for borrowing; therefore, most electronic computers are constructed so that the differences of two numbers are obtained by adding complements of the subtrahend to the minuend. Subtraction by using tables and adding complements are illustrated by examples for decimal, binary, octal, and hexadecimal number systems. Table A-6 is a subtraction table for the decimal system.

Problems for the subtraction of decimal numbers by a subtraction table are given in Example A-10.

Table A-6
Subtraction Table for Base 10

	Minuend									
Subtrahend	**0**	**1**	**2**	**3**	**4**	**5**	**6**	**7**	**8**	**9**
0	0	1	2	3	4	5	6	7	8	9
1	$\underline{9}$	0	1	2	3	4	5	6	7	8
2	$\underline{8}$	$\underline{9}$	0	1	2	3	4	5	6	7
3	$\underline{7}$	$\underline{8}$	$\underline{9}$	0	1	2	3	4	5	6
4	$\underline{6}$	$\underline{7}$	$\underline{8}$	$\underline{9}$	0	1	2	3	4	5
5	$\underline{5}$*	$\underline{6}$	$\underline{7}$	$\underline{8}$	$\underline{9}$	0	1	2	3	4
6	$\underline{4}$	$\underline{5}$	$\underline{6}$	$\underline{7}$	$\underline{8}$	$\underline{9}$	0	1	2	3
7	$\underline{3}$	$\underline{4}$	$\underline{5}$	$\underline{6}$	$\underline{7}$	$\underline{8}$	$\underline{9}$	0	1	2
8	$\underline{2}$	$\underline{3}$	$\underline{4}$	$\underline{5}$	$\underline{6}$	$\underline{7}$	$\underline{8}$	$\underline{9}$	0	1
9	$\underline{1}$	$\underline{2}$	$\underline{3}$	$\underline{4}$	$\underline{5}$	$\underline{6}$	$\underline{7}$	$\underline{8}$	$\underline{9}$	0

* A bar below a number means that a 1 was taken from the adjacent digit to the left of the minuend.

Example A-10

(a)	(b)
$\begin{array}{r} ^{4\;5\quad 8} \\ 5\;6\;4\;9\;2 \\ -\;3\;7\;6\;4\;5 \\ \hline 1\;8\;8\;4\;7 \end{array}$	$\begin{array}{r} ^{7\;6} \\ 6\;9\;8\;7\;4 \\ -\;4\;5\;7\;9\;9 \\ \hline 2\;4\;0\;7\;5 \end{array}$

Referring to the first example, the difference in the first column from the right was found from Table A-6 to be $\underline{7}$; the bar below the 7 means that 7 is to be placed below the line in this column and that 1 has been taken from the 9 in the ten's column of the minuend. The ten's digit of the minuend now becomes 8; therefore, the 9 was crossed out and an 8 was written above the 9. The difference for the ten's column $(8 - 4)$ was found from the table to be 4; the 4 was placed in the second column below the line. In a similar manner differences to be placed below the line for the hundreds, thousands and ten-thousands columns are found from the table.

The two example problems are solved by complement method as shown in Example A-11.

Example A-11

(a) $56492 - 37645 = 56492 + (100{,}000 - 37645) - 100{,}000 =$
$56492 + 62355 - 100{,}000 = 18847$

(b) $69874 - 45799 = 69874 + (100{,}000 - 45799)$
$- 100{,}000 = 69874 + 54201 - 100{,}000 = 24075$

or

(a) $56492 - 37645 = 56492 + (99{,}999 - 37645) - 100{,}000 + 1 = 56492$
$+ 62354 + 1 - 100{,}000 = 18847$

(b) $69874 - 45799 = 69874 + (99{,}999 - 45799) + 1 - 100{,}000 = 69874$
$+ 54200 + 1 - 100{,}000 = 24075$

The first complements method is referred to as tens complement; and the second method is referred to as nines complement. Subtraction by the nines is perhaps the easiest, because it is easier to subtract from 9 than from 0. Subtraction by the nines complement can be thought of as adding the minuend to the nines complement of the subtrahend and then removing the 1 from the extreme left position of the sum and adding it to the units position of the sum as shown in Example A-12.

Example A-12

(a)

```
  5 6 4 9 2                        5 6 4 9 2
- 3 7 6 4 5   is equivalent to   + 6 2 3 5 4
  ---------                      -----------
                                 1 1 8 8 4 6
                                         + 1
                                 -----------
                                   1 8 8 4 7
```

and

(b)

```
  6 9 8 7 4                        6 9 8 7 4
- 4 5 7 9 9   is equivalent to   + 5 4 2 0 0
  ---------                      -----------
                                 1 2 4 0 7 4
                                         + 1
                                 -----------
                                   2 4 0 7 5
```

The subtraction table for binary numbers is Table A-7.

Example A-13 shows subtraction for binary numbers.

Table A-7
Subtraction Table for Base 2

	Minuend	
Subtrahend	0	1
0	0	1
1	1* (with bar below)	0

* A bar below the number means that a 1 was taken from the ajacent digit to the left of the minuend.

Example A-13

Binary		Decimal Equivalent
	(a)	
1 0 1 0 1		2 1
− 1 1 0 1		− 1 3
1 0 0 0		8
	(b)	
1 1 0		
1 0 0 1 0		1 8
− 1 1 1 1		− 1 5
0 0 1 1		3

In Example A-13a it was unnecessary to take a 1 from the digit to the left; the differences for each column were taken from Table A-7. The solution for the second example was somewhat more complex. The difference for the units column was found to be a $\underline{1}$; a 1 was placed below the line in the units column, and a 1 was taken from the minuend, and this left a 0. From the subtraction table the difference was found to be $\underline{1}$, and a 1 was placed below the line in the second column (twos).

Since a 1 was taken from the third column (fours column) and the digit in this column was 0, a 1 was taken from the fourth column (eights column). And since the digit in this column was 0, a 1 was taken from the last column (sixteens column); the digits of the minuend now become 0 1 1 0 for the twos, fours, eights and sixteens columns. Finally, the differences for these columns were obtained from the subtraction table.

The two binary examples are solved respectively by the twos and ones complements as shown in Example A-14.

Example A-14

Twos complement solutions:

(a) $10101 - 1101 = 10101 + (10{,}000 - 1101) - 10{,}000 = 10101 + 11 - 10{,}000 = 1000$

(b) $10010 - 1111 = 10010 + (10{,}000 - 1111) - 10{,}000 = 10010 + 0001 - 10{,}000 = 0011$

Ones complement solutions:

(a) 10101 − 1101 = 10101 + (1111 − 1101) + 1 − 10,000 = 10101 + 10 + 1 − 10,000 = 1000
(b) 10010 − 1111 = 10010 + (1111 − 1111) + 1 − 10,000 = 10010 + 0 + 1 − 10,000 = 00011

The subtraction table for octal numbers is Table A-8.

Subtraction of octal numbers is illustrated in Example A-15.

Example A-15

Octal		Decimal Equivalent
	(a)	
5 6 3 1		3
6 7 4 2 1		2 8 4 3 3
− 4 7 5 6 4		− 2 0 3 4 0
1 7 6 3 5		8 0 9 3
	(b)	
6 5 2		
4 7 6 3 1		2 0 3 7 7
− 2 6 7 7 4		− 1 1 7 7 2
2 0 6 3 5		8 6 0 5

Table A-8
Subtraction Table for Base 8

Subtrahend	Minuend 0	1	2	3	4	5	6	7
0	0	1	2	3	4	5	6	7
1	$\underline{7}$*	0	1	2	3	4	5	6
2	$\underline{6}$	$\underline{7}$	0	1	2	3	4	5
3	$\underline{5}$	$\underline{6}$	$\underline{7}$	0	1	2	3	4
4	$\underline{4}$	$\underline{5}$	$\underline{6}$	$\underline{7}$	0	1	2	3
5	$\underline{3}$	$\underline{4}$	$\underline{5}$	$\underline{6}$	$\underline{7}$	0	1	2
6	$\underline{2}$	$\underline{3}$	$\underline{4}$	$\underline{5}$	$\underline{6}$	$\underline{7}$	0	1
7	$\underline{1}$	$\underline{2}$	$\underline{3}$	$\underline{4}$	$\underline{5}$	$\underline{6}$	$\underline{7}$	0

* A bar below a number means that a 1 was taken from the adjacent digit to the left of the minuend

Referring to Example A-15a, the remainders for the columns were taken from the octal subtraction table. As noted for the first column, 4 from 1 gave 5 which meant that 1 was taken from the eights column; the 2 in the minuend was changed to a 1. The difference for the second column (1 − 6) was found in the table to be 3; this meant that 1 was taken from the 4 in sixty-fours column. A 3 was placed below the line in the eights column, and a 5 was placed below the line in the units column. For the sixty-fours column, the difference (3 − 5) was found to be 6; and a 1 was taken from the 7 in the 512s column. This left a 6 in the minuend for this column. The remainder (6 − 7) in the 512s column was found in the table to be 7; and a 1 was taken from the 6 in the 4096s column. The difference (5 − 4) for the last column was found from the table to be 1. Finally, 6, 7 and 1 were placed respectively below the line in the positions as indicated in the example.

The two examples are solved in Example A-16 by the 8's and 7's complements method.

Example A-16

Eights complement solution:

(a) $69721 - 47564 = 67421 + (100{,}000 - 47564) - 100{,}000 = 67421 + 30214 - 100{,}000 = 17635$

(b) $47631 - 26774 = 47631 + (100{,}000 - 26774) - 100{,}000 = 47631 + 51004 - 100{,}000 = 20635$

Sevens complement solution:

(a) $67421 - 47564 = 67421 + (77{,}777 - 47564) + 1 - 100{,}000 = 67421 + 30213 + 1 - 100{,}000 = 17635$

(b) $47631 - 26774 = 47631 - (77{,}777 - 26774) + 1 - 100{,}000 = 47631 + 51003 + 1 - 100{,}000 = 20635$

The subtraction table for hexadecimal numbers is given in Table A-9.

Table A-9
Subtraction Table for Base 16

	Minuend															
Subtrahend	0	1	2	3	4	5	6	7	8	9	A	B	C	D	E	F
0	0	1	2	3	4	5	6	7	8	9	A	B	C	D	E	F
1	$\underline{F}$*	0	1	2	3	4	5	6	7	8	9	A	B	C	D	E
2	$\underline{E}$	$\underline{F}$	0	1	2	3	4	5	6	7	8	9	A	B	C	D
3	$\underline{D}$	$\underline{E}$	$\underline{F}$	0	1	2	3	4	5	6	7	8	9	A	B	C
4	$\underline{C}$	$\underline{D}$	$\underline{E}$	$\underline{F}$	0	1	2	3	4	5	6	7	8	9	A	B
5	$\underline{B}$	$\underline{C}$	$\underline{D}$	$\underline{E}$	$\underline{F}$	0	1	2	3	4	5	6	7	8	9	A
6	$\underline{A}$	$\underline{B}$	$\underline{C}$	$\underline{D}$	$\underline{E}$	$\underline{F}$	0	1	2	3	4	5	6	7	8	9
7	$\underline{9}$	$\underline{A}$	$\underline{B}$	$\underline{C}$	$\underline{D}$	$\underline{E}$	$\underline{F}$	0	1	2	3	4	5	6	7	8
8	$\underline{8}$	$\underline{9}$	$\underline{A}$	$\underline{B}$	$\underline{C}$	$\underline{D}$	$\underline{E}$	$\underline{F}$	0	1	2	3	4	5	6	7
9	$\underline{7}$	$\underline{8}$	$\underline{9}$	$\underline{A}$	$\underline{B}$	$\underline{C}$	$\underline{D}$	$\underline{E}$	$\underline{F}$	0	1	2	3	4	5	6
A	$\underline{6}$	$\underline{7}$	$\underline{8}$	$\underline{9}$	$\underline{A}$	$\underline{B}$	$\underline{C}$	$\underline{D}$	$\underline{E}$	$\underline{F}$	0	1	2	3	4	5
B	5	$\underline{6}$	$\underline{7}$	$\underline{8}$	$\underline{9}$	$\underline{A}$	$\underline{B}$	$\underline{C}$	$\underline{D}$	$\underline{E}$	$\underline{F}$	0	1	2	3	4
C	$\underline{4}$	$\underline{5}$	$\underline{6}$	$\underline{7}$	$\underline{8}$	$\underline{9}$	$\underline{A}$	$\underline{B}$	$\underline{C}$	$\underline{D}$	$\underline{E}$	$\underline{F}$	0	1	2	3
D	$\underline{3}$	$\underline{4}$	$\underline{5}$	$\underline{6}$	$\underline{7}$	$\underline{8}$	$\underline{9}$	$\underline{A}$	$\underline{B}$	$\underline{C}$	$\underline{D}$	$\underline{E}$	$\underline{F}$	0	1	2
E	$\underline{2}$	$\underline{3}$	$\underline{4}$	$\underline{5}$	$\underline{6}$	$\underline{7}$	$\underline{8}$	$\underline{9}$	$\underline{A}$	$\underline{B}$	$\underline{C}$	$\underline{D}$	$\underline{E}$	$\underline{F}$	0	1
F	$\underline{1}$	$\underline{2}$	$\underline{3}$	$\underline{4}$	$\underline{5}$	$\underline{6}$	$\underline{7}$	$\underline{8}$	$\underline{9}$	$\underline{A}$	$\underline{B}$	$\underline{C}$	$\underline{D}$	$\underline{E}$	$\underline{F}$	0

*A bar below a number means that a 1 was taken from the adjacent digit to the left of the minuend.

Examples of hexadecimal numbers are given in Example A-17.

Example A-17

	Hexadecimal	Decimal Equivalent
(a)	E A B	3 7 5 5
	− 4 6 C	− 1 1 3 2
	A 3 F	2 6 2 3
(b)	7 6 4 3 1	4 8 4, 4 0 1
	− 4 A B C E	− 3 0 6, 1 2 6
	2 8 8 6 3	1 7 8, 2 7 5

Referring to the first example, the remainder for the units column was obtained from Table A-9. The difference for this column was F which means that 1 was taken from the sixteens column and an F was placed below the line. A line was crossed through the A, and the middle number of the minuend became 9. From the subtraction table, the difference $(9 - 6)$ of the sixteens column was 3. Therefore, a 3 was placed below the line in the middle column. The difference for the left-hand column $(E - 4)$ was found from the table to be A; therefore, A was placed below the line in the extreme left-hand column.

The two examples are solved in Example A-18 by the 10's and F's complements.

Example A-18

10's complement:

(a) $EAB - 46C = EAB + (1000 - 46C) - 1000 = EAB + B94 - 1000$
$= ABF$

(b) $76431 - 4ABCE = 76431 + (100{,}000 - 4ABCE) - 100{,}000$
$= 76431 + B5432 - 100{,}000 = 2B863$

F's complement:

(a) $EAB - 46C = EAB + (FFF - 46C) + 1$
$- FFF = EAB + B93 + 1 - 1000 = A3F$

(b) $76431 - 4ABCE = 76431 + (FFFFF - 4ABCE) + 1$
$- 100000 = 76431 + B5431 + 1 - 100000 = 2B863$

The procedure for the arithmetical operation of multiplication begins with the aid of a multiplication table. The procedure for the multiplication operation is the same for all positional number systems. A review of the multiplication operation in the decimal system will serve as preparation for multiplication in other number systems. The multiplication table for the decimal system is given in Table A-10.

Table A-10
Multiplication Table for Base 10

Multiplier	Multiplicand 0	1	2	3	4	5	6	7	8	9
0	0	0	0	0	0	0	0	0	0	0
1	0	1	2	3	4	5	6	7	8	9
2	0	2	4	6	8	10	12	14	16	18
3	0	3	6	9	12	15	18	21	24	27
4	0	4	8	12	16	20	24	28	32	36
5	0	5	10	15	20	25	30	35	40	45
6	0	6	12	18	24	30	36	42	48	54
7	0	7	14	21	28	35	42	49	56	63
8	0	8	16	24	32	40	48	56	64	72
9	0	9	18	27	36	45	54	63	72	81

Example A-19 illustrates multiplication in the decimal system.

Example A-19

(a)

```
  452
 5674
   67
------
39718
34044
------
380158
```

(b)

```
  6432
    98
------
 51456
57888
------
630336
```

Referring to Table A-10, the procedure for obtaining the product of the two numbers in the first example is described. The product of 7 x 4 was found in the table to be 28; the 8 was placed below the first line in the units position, and the 2 was written above the multiplicand in the tens position. Next, 7 x 7 was found in the table to be 49; the 2 was added to the 49 which gave 51; and the 1 was placed below the line in the tens position. The 5 was written above the 6 of the multiplicand in the hundreds position. From the table 7 x 6 was found to be 42, and 42 + 5 equals 47. The 7 was placed below the first line in the hundreds position, and the 4 was placed above the 5 of the multiplicand in the thousands position. From the table, 7 x 5 was found to be 35; the 4 above the 5 was added to the 34 which gave 39, and the 39 was placed below the first line in the ten-thousands and thousands positions.

Using the same procedure for the multiplication of the multiplicand by the 7 in the units position of the multiplier, the multiplier was multiplied by the 6 in the tens position of the multiplier. The product of the multiplicand and the 6 of the multiplier was placed in the second row below the first line and shifted one position to the left, and this procedure was continued for the other numbers in the multiplicand. Finally, the two rows (two products) between the two lines were added. The sum of the two rows was placed below the second line, and this is the product of two numbers or final answer for the example.

The IBM 1620 uses a table for the multiplication of two numbers. Desk calculators and some electronic computers multiply by successive addition. By successive addition, the first example could have been solved as shown in Example A-20.

Example A-20

```
    5674
  x   67
  ------
    5674
  + 5674
  + 5674
  + 5674
  + 5674
  + 5674
  + 5674
 + 5674
 + 5674
 + 5674
 + 5674
 + 5674
 + 5674
 -------
  380158
```

Table A-11 is a multiplication table for the binary system.

Table A-11
Multiplication Table for Base 2

Multiplier	Multiplicand 0	Multiplicand 1
0	0	0
1	0	1

Problems of binary multiplication are given in Example A-21.

Example A-21

	Binary	Decimal Equivalent
	11011	27
(a)	101	5
	11011	135
	11011	
	10,000,111	
(b)	101011	43
	111	7
	101011	301
	101011	
	101011	
	100101101	

It is readily apparent that a multiplication table was not needed for the multiplication of binary numbers. The multiplication could have been performed by successive addition. While examining each digit of the multiplier, if the digit were 0, there was no adding; but if the digit were 1, the multiplicand was added once in the appropriate postion. The multiplication of binary numbers is very simple.

A multiplication table for the octal system is Table 1-12.

Table A-12 is a multiplication table for the octal system.

Table A-12
Multiplication Table for Base 8

	Multiplicand							
Multiplier	0	1	2	3	4	5	6	7
0	0	0	0	0	0	0	0	0
1	0	1	2	3	4	5	6	7
2	0	2	4	6	10	12	14	16
3	0	3	6	11	14	17	22	25
4	0	4	10	14	20	24	30	34
5	0	5	12	17	24	31	36	43
6	0	6	14	22	30	36	44	52
7	0	7	16	25	34	43	52	61

Example A-22 illustrates multiplication of octal numbers.

Example A-22

	Octal	Decimal Equivalent
(a)	123	
	7256	3758
	x 64	x 52
	35270	7516
	54024	18790
	575530	195416
(b)	4357	2287
	x 75	x 61
	26253	2287
	37211	13722
	420363	139507

The procedure for the multiplication of the first octal example is described. The multiplicand $(7256)_8$ was first multiplied by $(4)_8$ with the aid of Table A-12. From the table $(6 \times 4)_8$ equals $(30)_8$, the 0 was placed below the line in the units position, and the 3 was carried to the next left column (eights position). For the next digit of the multiplicand, $(5 \times 4)_8$ equals $(24)_8$, and with the 3 carried forward $(24)_8 + (3)_8$ equals $(27)_8$; the 7 was placed below the line, and the 2 was carried. The product of the next digit, sixty-fours position, of the multiplicand is $(2 \times 4)_8 = 10$ and with the 2 carried forward gave $(12)_8$. The 2 was placed below the line, and the 1 was carried forward. For the last digit, 512s position, the product is $(7 \times 4)_8 = (34)_8$; and, with the 1 carried forward, it became $(35)_8$. The 35 is placed below the first line in the two extreme left positions.

Multiplication of octal numbers by successive addition is shown in Example A-23.

Example A-23

Octal	Decimal Equivalent
564	372
x 23	x 19
564	372
564	372
564	372
564	372
564	372
15634	372
	372
	372
	372
	372
	7068

Table A-13 is for the multiplication of hexadecimal numbers.

Hexadecimal multiplication is shown in Example A-24.

Table A-13
Multiplication Table for Base 16

	Multiplicand															
Multiplier	**0**	**1**	**2**	**3**	**4**	**5**	**6**	**7**	**8**	**9**	**A**	**B**	**C**	**D**	**E**	**F**
0	0	0	0	0	0	0	0	0	0	0	0	0	0	0	0	0
1	0	1	2	3	4	5	6	7	8	9	A	B	C	D	E	F
2	0	2	4	6	8	A	C	E	10	12	14	16	18	1A	1C	1E
3	0	3	6	9	C	F	12	15	18	1B	1E	21	24	27	2A	2D
4	0	4	8	C	10	14	18	1C	20	24	28	2C	30	34	38	3C
5	0	5	A	F	14	19	1E	23	28	2D	32	27	3C	41	46	4B
6	0	6	C	12	18	1E	24	2A	30	36	3C	42	48	4E	54	5A
7	0	7	E	15	1C	23	2A	31	38	3F	46	4D	54	5B	62	69
8	0	8	10	18	20	28	30	38	40	48	50	58	60	68	70	78
9	0	9	12	1B	24	2D	36	3F	48	51	5A	63	6C	75	7E	87
A	0	A	14	1E	28	32	3C	46	50	5A	64	6E	78	82	8C	96
B	0	B	16	21	2C	37	42	4D	58	63	6E	79	84	8F	9A	A5
C	0	C	18	24	30	3C	48	54	60	6C	78	84	90	9C	A8	B4
D	0	D	1A	27	34	41	4E	5B	68	75	82	8F	9C	A9	B6	C3
E	0	E	1C	2A	38	46	54	62	70	7E	8C	9A	AB	B6	C4	D2
F	0	F	1E	2D	3C	4B	5A	69	78	87	96	A5	B4	C3	D2	E1

Example A-24

Hexadecimal	Decimal Equivalent
(a)	
	43973
ABC5	150
x 96	00000
4069E	219865
609ED	43973
64A56E	6595950
(b)	
	35172
8964	239
EF	316548
80CDC	105516
78378	70344
80445C	8406108

For the first example, the multiplication of $(ABC5)_{16}$ by $(6)_{16}$ is explained. By the multiplication $(6 \times 5)_{16} = (1E)_{16}$, E was written below the line in the units position, and 1 was carried to the sixteens position. For the sixteens digit $(C \times 6)_{16} = (48)_{16}$; and by adding the 1, it became 49. The 9 was placed below the line, and the 4 was carried to the next left position. The product for the 256s position $(B \times 6)_{16}$ was found from the table to be $(42)_{16}$. When the 4 was added, it became $(46)_{16}$; the 6 was drawn down, and the 4 was carried to the next left position. The product for the 4096s digit $(A \times 6)_{16}$ was found from the table to be $(3C)_{16}$, and it became $(40)_{16}$ when the 4 was added—this is the value for the two extreme left digits of the products. Likewise, the product $(ABC5 \times 9)_{16}$ was determined in a similar manner by using the multiplication table; then the product was placed under the first product and shifted one position to the left. Finally, to obtain the answer for the example problem, the two products were added.

Since multiplication of two hexadecimal numbers by successive addition is rather lengthy, examples of this method have been omitted.

The arithmetic operations of division are more complicated than the operations of addition, subtraction or multiplication. Division cannot be performed by using a table because each new divisor would require a new table with computers. Division is most commonly performed by a sequence of successive subtractions, shifts and tallies. Some of the large electronic computers do not have circuitry for performing divisions. The basic IBM 1620 computer unit does not have circuitry for performing division; however, separate attachments for automatic division are available. Division can be performed by multiplication of the reciprocal of the divisor and the dividend. Iteration processes are used for finding reciprocals. Division can also be performed by an iterative process of multiplication.

For a review, two examples of division of decimal numbers are Examples A-25 and A-26. The divisions are performed, first, by longhand, and, second, by successive subtraction, next by determining the reciprocal of the divisor followed by a multiplication by the dividend and finally by an iteration process. Division by longhand of decimal numbers is shown in Example A-25.

Example A-25

(a)

```
        54
243  13122
     1215
     -----
       972
       972
       ---
```

(b)

```
         63
675   42525
      4050
      -----
       2025
       2025
       ----
```

Division of the examples by successive subtractions is illustrated in Example A-26.

Example A-26

	(a)				(b)		
	13122				42525		
	− 243	1			− 675	1	
	10692				35775		
	− 243	1			− 675	1	
	8262				29025		
	− 243	1			− 675	1	
	5832				22275		
	− 243	1			− 675	1	
	3402				15525		
	− 243	1			− 675	1	
Shift	972				8775		
	− 243		1		− 675	1	
	729				2025		
	− 243		1	Shift	− 675		1
	486				1350		
	− 243		1		− 675		1
	243				675		
	− 243		1		− 675		1
Tally			54	Tally			63

Equations for calculating reciprocals of numbers are as follows:

$$b_{M+1} = b_M(2 - xb_M) \qquad \text{A-1}$$

$$b_{M+1} = b_M 3(1 - xb_M) + (xb_M)^2 \qquad \text{A-2}$$

where x is the number for which the reciprocal is being obtained, and b_M is the successive approximation to the reciprocal. The first approximation b_0 may be chosen in an arbitrary manner, but it must be greater than 0 and less than $2/x$. If b_0 is not chosen within this range, the series will not converge.

Equation **A-1** is a second-order equation, and Equation **A-2** is a third-order equation. This means that the number of correct digits in the approximations is doubled and tripled, respectively, upon each application of the equations. The reciprocals of the divisor for the first example problems are determined by the procedure which follows.

A first approximation of b_0 must be less than 2.000/243 and greater than 0; the examination of $1.00/250 = 0.004$ yields an acceptable value for for b_0. Using this value, 0.004, a value for b_1 is calculated with the aid of the equation. This process is repeated until two successive values of *b's* are identical for the required number of significant figures. At this stage, the process is terminated and the last value is taken as the reciprocal. The reciprocal for 243 is calculated as follows:

$$b_1 = 0.004[2 - (243 \times 0.004)] = 0.004112$$
$$b_2 = 0.004112[2 - (243 \times 0.004112)] = 0.0041152238$$
$$b_3 = 0.0041152238[2 - (243 \times 0.0041152238)] = 0.0041152263$$
$$b_4 = 0.00411522633\,[2 - (243 \times 0.00411522633)] = 0.0041152267$$

To obtain the quotient, the dividend is multiplied by the reciprocal as follows:

$$13122 \times 0.0041152267 = 54.000004$$

which agrees very closely with 54; this value was obtained by the other two methods.

The iteration method applies the equation

$$\frac{N_{i+1}}{D_{i+1}} = \frac{(2 - D_i)\, N_i}{(2 - D_i)\, D_i} \qquad \textbf{A-3}$$

where N_0 is the dividend, and D_0 is the divisor for obtaining the quotient. In applying Equation **A-3,** D_0 is made to be less than 1 and equal to or greater than 0.1. After repeated calculations, D_i approaches 1 and N_i approaches the quotient because the ratio between N_i and D_i is not changed when both the numerator and denominator are multiplied by the factor, $2 - D_i$. If d_i is taken as equivalent to $1 - D_i$, then $D_i = 1 - d$ and $2 - D_i = 1 + d$. An approximate value for d_i is determined by taking the nines complement of the highest order non-nine digit of D_i. The procedure for determining the quotient for the first decimal example is illustrated in Example A-27.

Example A-27

i	N_i	D_i	$1 + d_i$
0	13.122	.243	1.7
1	22.307	.4131	1.5
2	33.4605	.61965	1.3
3	43.4865	.80554	1.1
4	47.848515	.88609	1.1
5	52.633366	.974699	1.02
6	53.686033	.9941930	1.005
7	53.954463	.9991640	1.0008
8	53.997626	.9999633	1.00003
9	53.99925	.999993	

13122/243 = 53.99925 approximately or 243 × 53.9985958 = 13121.6587648

The procedures for finding the quotients of binary numbers are similar to those employed for finding quotients of decimal numbers. Perhaps the application of successive subtractions, shifting, and tallying is the simplest method of determining quotients of binary numbers. Longhand division of binary numbers is given in Example A-28.

Example A-28

	Binary	Decimal Equivalent
	1001	9
(a)	1011) 1101110 remainder	11) 108
	1011	99
	10100	9 remainder
	1011	
	1001 remainder	
	1101	13
(b)	1010) 10001010	10) 138
	1010	10
	1110	38
	1010	30
	10010	8 remainder
	1010	
	1000 remainder	

The solutions of the two binary examples by successive subtractions, shifting and tallying are shown in Example A-29.

Example A-29

(a)

	1101100	
−	1011	1
	101	
Shift −	1011	0
	1010	
Shift −	1011	0
	10100	
Shift −	1011	1
	1001	
Talley		1001

(b)

	10001010	
−	1010	1
	111010	
Shift —	1010	1
	10010	
	1010	0
	10010	
Shift —	1010	1
	1000	
Tally		1101

From observations of the two examples of division by successive subtractions, etc., it is easily seen that the process of division can be carried out according to the following rules:

1. Place the divisor under the dividend in the extreme left-hand position.
2. Subtract the divisor from the dividend.
3. If the difference is positive, place a tally in the extreme left-hand position of the tally column. If the difference is negative, place a 0 in the extreme left-hand position of the tally column and add the divisior back to the difference.
4. Shift the divisor one position to the right and subtract the divisor from the difference or difference plus the divisor—whichever is appropriate.
5. If the difference is positive, place a 1 in the next position to the right in the tally column; if the difference is negative, place a 0 in the next position to the right in the tally column and add the divisor to the difference.
6. Repeat steps 4 and 5 until the last units position of the divisor is under the units position of the dividend. If there is a difference after the subtraction, this is the remainder for the problem.
7. If during the subtractions the difference became 0, this would indicate that the process of successive subtraction was completed, and it would be

necessary to add sufficient zeros to the right-hand columns in the tally columns.

8. Finally, the answer is obtained by writing down the sums of the columns in the tally row in the appropriate positions.

Equations **A-1** and **A-2** can also be applied for obtaining reciprocals of divisors in the binary system. For the binary system, these equations are written as follows:

$$b_{n+1} = b_n(10 - xb_n) \quad \textbf{A-1a}$$
$$b_{n+1} = b_n11(1 - xb_n) + (xb_n)^2 \quad \textbf{A-2a}$$

The procedure for applying Equation **A-1a** for determining the quotient for the primary problem 1001/11 = 11 is illustrated in Example A-30.

Example A-30

Let $b_0 = 0.01$
then

$$\begin{aligned}
b_1 &= 0.01\ (10 - 11 \times 0.01) \\
&= 0.01 \times 1.01 = 0.0101 \\
b_2 &= 0.0101\ (10 - 11 \times 0.0101) \\
&= 0.0101 \times 1.0001 = 0.01010101 \\
b_3 &= 0.01010101\ (10 - 11 \times 0.01010101) \\
&= 0.01010101 \times 1.00000001 \\
&= 0.0101010101010101
\end{aligned}$$

To obtain the quotient, the dividend is multiplied by the reciprocal as

$$1001 \times 0.0101010101010101 = 10.1111111111111111.$$

The true quotient and the quotient obtained by the reciprocal method differ by 1 in the sixteenth place to the right of the decimal point. Throughout this text the decimal point will be used as a reference point for separating digits which are either equal to one, greater than one or less than one. The digits which give values equal to or greater than one are placed to the left of the decimal point. The digits which give values less than one are placed to the right of the decimal point.

The procedure for applying Equation **A-2a,** which is an iteration method for determining quotients, is illustrated with the last binary problem in Example A-31.

Example A-31

i	N	D	$1 + d$
1	10.01	.11	1.001
2	10.10001	.11011	1.001
3	10.11011	.11110	1.00001
4	10.1110111	.1111101	1.000001
5	10.1111101	.1111110	1.0000001
6	10.111111111110		

By carrying the iteration method for six steps, a quotient was obtained which checks the true quotient to 1 in the twelfth place to the right of the decimal point.

The application of Equations **A-1a** and **A-2a** indicates that quotients for binary numbers can also be obtained by iteration processes and that special circuitry for division in electronic computers is not always necessary.

The processes for division in the decimal system can also be applied for division in the octal system. Example A-32 shows octal longhand division.

Example A-32

(a)

```
       13
465  6507
     465
     ----
     1637
     1637
     ----
     0000
```

(b)

```
        23
672  20316
     1564
     -----
      2456
      2456
     -----
      0000
```

The two octal examples of division are solved in Example A-33 by successive subtractions, shiftings, and tallying.

Example A-33

(a)		(b)	
6507		20316	
− 465	1	− 672	1
1637		1137	
− 465	1	− 672	1
1152		2456	
− 465	1	− 672	1
465		1564	
− 465	1	− 672	1
000		672	
Tally sum	13	− 672	1
		000	
		Tally sum	23

The reciprocal of the divisor $(456)_8$ of the first octal example is determined by the Equation **A-1** as shown in Example A-34.

Example A-34

$$
\begin{aligned}
\text{Let } b_0 &= 0.0015 \\
\text{then } b_1 &= 0.0015\,[2-(465\times 0.0015)] \\
&= 0.0015\times 1.0117 = 0.00152003 \\
b_2 &= 0.00152003\,[2-(465\times 0.00152003)] \\
&= 0.00152003\times 1.0001414 = 0.00152027 \\
b_3 &= 0.00152027\,[2-(465\times 0.00152027)] \\
&= 0.00152027\times 1.00000075 = 0.0015027 \\
b_4 &= 0.00152030\,[2-(465\times 0.00152030)] \\
&= 0.00152030\times 1.00000 = 0.00152030
\end{aligned}
$$

The quotient for the first octal example is found by multiplying the dividend $(6507)_8$ by the reciprocal of the dividend as:

$$6507\times 0.0015203 = 13.00000525$$

The first octal example of division is solved by the application of Equation **A-2** as shown in Example A-35.

Example A-35

i	Ni	Di	$1 + di$
1	6.507	0.465	1.3
2	11.1015	0.6507	1.1
3	12.2117	0.73577	1.04
4	12.72237	0.773667	1.004
5	12.77610	0.777646	1.0001
6	12.77737	0.777746	1.00003
7	12.777777	0.777775	1.000002
8	13.000001	0.777777	

Quotients of the octal example obtained by the application of Equations **A-1** and **A-2** agree for seven digits with the quotients which were obtained by standard methods; the differences beyond the seven digits are due to round off errors.

Examples of longhand division of hexadecimal numbers are given in Example A-36.

Example A-36

(a)

```
       52
AB / 36C6
     357
    ----
     156
     156
    ----
     000
```

(b)

```
       61
5E / 239E
     234
    ----
      5E
      5E
    ----
      00
```

Solutions of the two hexadecimal examples by successive subtractions, shifting, and tallying are illustrated in Example A-37.

Example A-37

(a)		(b)	
36C6		239E	
− AB	1	− 5E	1
2C1		1DB	
− AB	1	− 5E	1
216		17D	
− AB	1	− 5E	1
16B		11F	
− AB	1	− 5E	1
C0		C1	
− AB	1	− 5E	1
156		63	
− 5E	1	− 5E	1
AB		5E	
− AB	1	− AB	1
00		00	
Tally sum	52	Tally sum	61

The calculations of the reciprocal for the divisor $(AB)_{16}$ of the first hexadecimal example of division are calculated by Equation **A-1** as shown in Example A-38.

Example A-38

$$\text{Let } b_0 = 0.01$$
$$\text{then } b_1 = 0.01\ (2 - AB \times 0.010)$$
$$= 0.01 \times 1.55 = 0.0155$$
$$b_2 = 0.0155\ (2 - AB \times 0.0155)$$
$$= 0.0155 \times 1.1C39 = 0.017A97$$
$$b_3 = 0.016A97\ (2 - AB \times 0.017A97)$$
$$= 0.017A97 \times 1.031D23 = 0.017F31BD$$
$$b_4 = 0.017F32\ (2 - AB \times 0.017F32) = 0.01732\ (2.0 - \mathbf{FFF666})$$
$$= 0.017F32 \times 1.00099A = 0.017F405F4614$$
$$b_5 = 0.017F406\ (2 - AB \times 0.017F406) = 0.017F406\ (2.0 - 1.0000002)$$
$$= 0.017F406 \times 1.000000 = 0.17F406$$

To obtain the quotient for the hexadecimal example, the dividend $(36C6)_{16}$ is multiplied by the reciprocal of the divisor as:

$$36C6 \times 0.017F406 = 52.00001A4$$

The first example for the hexadecimal division is solved by the use of Equation **A-2** as shown in Example A-39.

Example A-39

i	N_i	D_i	$1+d_i$
1	36.C6	0.AB	1.5
2	47.E3E	0.E07	1.1
3	4C.621E	0.EE77	1.1
4	51.283FE	0.FD5E7	1.02
5	51.CA906FC	0.FF593	1.00A
6	51.FBE9A	0.FFF8C	1.0007

It is noted that the quotient obtained by the reciprocal method was accurate to 1 in the fifth place to the right of the decimal point, and the quotient obtained by the method using the iteration equation was accurate to 3 in the second place to the right of the decimal point. The quotients would have been more accurate in the calculations if more significant digits had been used. Since calculations for hexadecimal numbers are rather tedious when carried out with pencil and paper, only sufficient digits and steps were used to prove that the two equations apply for hexadecimal numbers.

In applying Equation **A-2** to the binary, octal, and hexadecimal numbers, complements of 1, 7, and F were employed respectively for determining the values of d_i.

Conversion of One Positional Number System to Another Positional Number System

In the section on positional number representation, example procedures for converting binary numbers to decimal numbers, octal numbers to decimal numbers and hexadecimal numbers to decimal numbers are given. This section deals with the conversion of decimal numbers to binary, octal and hexadecimal numbers. For numbers which contain decimal fractions, the digits to the left of the decimal point and the digits to the right of the decimal point require slightly different procedures for their conversions. Procedures for the conversion of the digits to the left of the decimal point (whole numbers) will be given first. Example A-40 illustrates a procedure for converting a decimal number to a binary number.

Example A-40

$$
\begin{aligned}
(33)_{10} &= \frac{33}{2} \times 2 = (16 + \tfrac{1}{2})2 = 16 \times 2^1 + 1 \times 2^0 \\
&= \left(\frac{16}{2} \times 2\right) 2^1 + 1 \times 2^0 = \left(8 + \frac{0}{2}\right) 2^2 + 1 \times 2^0 \\
&= 8 \times 2^2 + 0 \times 2^1 + 1 \times 2^0 \\
&= \left(\frac{8}{2} \times 2\right) 2^2 + 0 \times 2 + 1 \times 2^0 \\
&= \left(4 + \frac{0}{2}\right) 2^3 + 0 \times 2^2 + 0 \times 2^1 + 1 \times 2^0 \\
&= 4 \times 2^3 + 0 \times 2^2 + 0 \times 2^1 + 1 \times 2^0 \\
&= \left(\frac{4}{2} \times 2\right) 2^3 + 0 \times 2^2 + 0 \times 2^1 + 1 \times 2^0 \\
&= \left(2 + \frac{0}{2}\right) 2^4 + 0 \times 2^3 + 0 \times 2^2 + 0 \times 2^1 + 1 \times 2^0 \\
&= 2 \times 2^4 + 0 \times 2^3 + 0 \times 2^2 + 0 \times 2^1 + 1 \times 2^0 \\
&= \left(\frac{2}{2} \times 2\right) 2^4 + 0 \times 2^3 + 0 \times 2^2 + 0 \times 2^1 + 1 \times 2^0 \\
&= \left(1 + \frac{0}{2}\right) 2^5 + 0 \times 2^3 + 0 \times 2^2 + 0 \times 2^1 + 1 \times 2^0 \\
&= 1 \times 2^5 + 0 \times 2^4 + 0 \times 2^3 + 0 \times 2^2 + 0 \times 2^1 + 1 \times 2^0
\end{aligned}
$$

which written as a binary number becomes

$$(33)_{10} = (100001)_2$$

It is readily seen that the procedure for converting the decimal number to the binary number can be simplified as shown in Example A-41.

Example A-41

$$
\begin{array}{rl}
 & 16 \\
2 & \overline{33} \\
 & \underline{32} \\
 & 1 \text{ remainder} \qquad \text{first digit } 1
\end{array}
$$

$$
\begin{array}{r}
8 \\
2\,\overline{)16} \\
16 \\
\hline
0
\end{array}
$$

0 remainder first two digits 01

$$
\begin{array}{r}
4 \\
2\,\overline{)8} \\
8 \\
\hline
0
\end{array}
$$

0 remainder first three digits 001

$$
\begin{array}{r}
2 \\
2\,\overline{)4} \\
4 \\
\hline
0
\end{array}
$$

0 remainder first four digits 0001

$$
\begin{array}{r}
1 \\
2\,\overline{)2} \\
2 \\
\hline
0
\end{array}
$$

0 remainder first five digits 00001

$$
\begin{array}{r}
0 \\
2\,\overline{)1} \\
0 \\
\hline
1
\end{array}
$$

1 remainder total digits 100001

The remainder after each division by 2 furnishes the respective digit of the binary number. Each remainder is placed respectively to the left of the existing digits.

Decimal numbers are sometimes converted to binary numbers with the aid of tables of positive powers of 2. Table A-14 lists the values of 2 raised to various whole number powers. By using Table A-14 the decimal number $(33)_{10}$ is converted to its binary equivalence by the procedure:

1. From Table A-14, $(33)_{10}$ is between 2^5 and 2^6, therefore the highest power of 2 in $(33)_{10}$ is the fifth which is $(32)_{10}$.
2. The value of the highest power of 2 which is $(32)_{10}$ is subtracted from $(33)_{10}$; this leaves a remainder of 1.

3. From the table 2^0 or $(1)_{10}$ is the highest power of 2 which will go into 1 and $2^0 - 1 = 0$.

4. From Steps 2 and 3 it is self-evident that $(33)_{10}$ is the sum of $1 + 2^5$ and 1×2^0; the terms 2^4, 2^3, 2^2, and 2^1 are not present; therefore, the coefficient for each of these terms is 0.

5. The decimal number $(33)_{10}$ is equivalent to

$$(1 \times 2^5) + (0 \times 2^4) + (0 \times 2^3) + (0 \times 2^2) + (0 \times 2^1) + (1 \times 2^0)$$

and this can be written as the binary number $(100001)_2$.

Example A-42 illustrates a procedure for converting decimal numbers to octal numbers.

Example A-42

$$(2377)_{10} = \left(\frac{2{,}377}{8} \times 8\right) = \left(297 + \frac{1}{8}\right) 8^1$$

$$= 297 \times 8^1 + 1 \times 8^0$$

$$= \left(\frac{297}{8} \times 8\right) 8^1 + 1 \times 8^0$$

$$= \left(37 + \frac{1}{8}\right) 8^2 + 1 \times 8^0$$

$$= 37 \times 8^2 + 1 \times 8^1 + 1 \times 8^0$$

$$= \left(\frac{37}{8} \times 8\right) 8^2 + 1 \times 8^1 + 1 \times 8^0$$

$$= \left(4 + \frac{5}{8}\right) 8^3 + 1 \times 8^1 + 1 \times 8^0$$

$$= 4 \times 8^3 + 5 \times 8^2 + 1 \times 8^1 + 1 \times 8^0$$

or in octal notation

$$(2{,}377)_{10} = (4{,}511)_8$$

Table A-14
Values of 2 Raised to Positive Powers

n	2^n
0	1
1	2
2	4
3	8
4	16
5	32
6	64
7	128
8	256
9	512
10	1 024
11	2 048
12	4 096
13	8 192
14	16 384
15	32 768
16	65 536
17	131 072
18	262 144
19	524 288
20	1 048 576
21	2 097 152
22	4 194 304
23	8 388 608
24	16 777 216
25	33 554 432
26	67 108 864
27	134 217 728
28	268 435 456
29	536 870 912
30	1 073 741 824
31	2 147 483 648
32	4 294 967 296
33	8 589 934 592
34	17 179 869 184
35	34 359 738 368
36	68 719 476 736
37	137 438 953 472
38	274 877 906 944
39	549 755 813 888

This procedure is simplified in Example A-43.

Example A-43

$$
\begin{array}{r l}
 & 297 \\
8 & \overline{)2,377} \\
 & 2,376 \\
\hline
 & 1 \text{ remainder} \qquad \text{First digit 1}
\end{array}
$$

$$
\begin{array}{r l}
 & 37 \\
8 & \overline{)297} \\
 & 296 \\
\hline
 & 1 \text{ remainder} \qquad \text{First two digits 11}
\end{array}
$$

$$
\begin{array}{r l}
 & 4 \\
8 & \overline{)37} \\
 & 32 \\
\hline
 & 5 \text{ remainder} \qquad \text{First three digits 511}
\end{array}
$$

$$
\begin{array}{r l}
 & 0 \\
8 & \overline{)4} \\
 & 0 \\
\hline
 & 4 \text{ remainder} \qquad \text{Total digits 4511}
\end{array}
$$

$$(2,377)_{10} = (4,511)_8$$

Table A-15 is a listing of the values of 8 raised to various positive whole number powers. By the use of Table A-15, $(2377)_{10}$ is converted to its octal equivalence by the following procedure:

1. The table indicates that 3 is the highest power of 8 in $(2,377)_{10}$.
2. The decimal number $(2,377)_{10}$ is divided by 8^3 or 512.

This gives 4 with a remainder of 329.

3. The remainder 329 is greater than 8^2 and less than 8^3, and the remainder is divided by 8^2 which is 64. This gives 5 with a remainder of 9.
4. The remainder is greater than 8^1 and less than 8^2, therefore the remainder is divided by 8. The result is 1 with a remainder of 1.
5. Summarizing steps 2, 3 and 4, $(2,377) = (4 \times 8^3) + (5 \times 8^2) + (1 \times 8^1) + (1 \times 8^0)$ and when written in octal notation, $(2,377)_{10} = (4,511)_8$.

Table A-15
Values of 8 Raised to Positive Powers

n	8^n
0	1
1	8
2	64
3	512
4	4 096
5	32 768
6	262 144
7	2 097 152
8	16 777 216
9	134 217 728
10	1 073 741 824
11	8 589 934 592

The decimal number $(6893)_{10}$ is converted to the hexadecimal number as illustrated in Example A-44.

Example A-44

$$
\begin{aligned}
6(6{,}893)_{10} &= \left(\frac{6{,}893}{16} \times 16\right) = \left(430 + \frac{13}{16}\right) 16^1 \\
&= 430 \times 16^1 + 13 \times 16^0 \\
&= \left(\frac{430}{16} \times 16\right) 16^1 + 13 \times 16^0 \\
&= \left(26 + \frac{14}{16}\right) 16^2 + 13 \times 16^0 \\
&= \left(1 + \frac{10}{16}\right) 16^3 + 14 \times 16^1 + 13 \times 16^0 \\
&= 1 \times 16^3 + 10 \times 16^2 + 14 \times 16^1 + 13 \times 16^0
\end{aligned}
$$

In hexadecimal

13 = D, 14 = E, and 10 = A

Therefore, in hexadecimal notation,

$(6893)_{10} = (1AED)_{16}$

The procedure for converting the decimal number to a hexadecimal number is simplified in Example A-45.

Example A-45

```
        430
16 | 6,893
      6,880
      -----
         13 remainder     First digit D

         26
16 |    430
        416
        ---
         14 remainder     First two digits ED

          1
16 |     26
         16
         --
         10 remainder     First three digits AED

          0
16 |      1
          0
          -
          1 remainder     Total digits (1AED)16
```

Table A-16 is a listing of the values of 16 raised to positive whole number powers. By using this table, $(6893)_{10}$ is converted to its octal equivalence by the following procedure:

1. Table A-16 indicates that 16 cubed is the highest power in $(6893)_{10}$.
2. The value of 16 cubed, which is 4096, is subtracted from 6893; the difference is 2797.
3. The difference 2797 is a multiple of 16 squared; therefore, 2797 is divided by 16 squared which is 256. The division gives 10 with a remainder of 237.
4. The remainder 237 is now divided by 16 raised to the first power; the result is 14 with a remainder of 13.
5. Summarizing steps 2, 3 and 4, $(6893)_{10} = 1 \times 16^3 + 1 \times 16^2 + 14 \times 16^1 + 13 \times 16^0 = (1AED)_{16}$.

Table A-16
Values of 16 Raised to Positive Powers

n	16^n
0	1
1	16
2	256
3	4 096
4	65 536
5	1 048 576
6	16 777 216
7	268 435 456
8	4 294 967 296
9	68 719 476 736

The procedures for converting decimal fractions to fractions in binary, octal and hexadecimal systems are somewhat similar to the procedures for converting decimal whole numbers to these systems. In the algebraic procedure multiplication by the respective base is performed instead of division by the respective base; however, the procedures which make use of tables are identical both for whole numbers and fractions, except that tables for whole negative powers are used for converting decimal fractions to other systems.

The algebraic procedure for converting a decimal fraction to a binary fraction is illustrated in Example A-46.

Example A-46

$$(0.65625)_{10} = \frac{(0.65625 \times 2)}{2} = (1.3125)\,2^{-1}$$

$$= (1.0 + 0.3125)\ 2^{-1} = 1 \times 2^{-1} + (0.3125)\ 2^{-1}$$

$$= 1 \times 2^{-1} + \frac{(0.3125 \times 2)}{2}\,2^{-1} = 1 \times 2^{-1} = (0.6250)\,2^{-2}$$

$$= 1 \times 2^{-1} + (0 + 0.625)\,2^{-2} = 1 \times 2^{-1} + 0 \times 2^{-2} + (0.625)\,2^{-2}$$

$$= 1 \times 2^{-1} + 0 \times 2^{-2} + \frac{(0.625 \times 2)}{2}\,2^{-2} = 1 \times 2^{-1} + 0 \times 2^{-2} + (1.250)\,2^{-3}$$

$$= 1 \times 2^{-1} + 0 \times 2^{-2} + (1 + 0.25)\,2^{-3}$$

$$= 1 \times 2^{-1} + 0 \times 2^{-2} + 1 \times 2^{-3} + (0.25)\,2^{-3}$$

$$= 1 \times 2^{-1} + 0 \times 2^{-2} + 1 \times 2^{-3} + \frac{(0.25 \times 2)}{2} 2^{-3}$$

$$= 1 \times 2^{-1} + 0 \times 2^{-2} + 1 \times 2^{-3} + (0.50) 2^{-4}$$

$$= 1 \times 2^{-1} + 0 \times 2^{-2} + 1 \times 2^{-3} + (0 + 0.5) 2^{-4}$$

$$= 1 \times 2^{-1} + 0 \times 2^{-2} + 1 \times 2^{-3} + 0 \times 2^{-4} + (0.5) 2^{-4}$$

$$= 1 \times 2^{-1} + 0 \times 2^{-2} + 1 \times 2^{-3} + 0 \times 2^{-4} + \frac{(0.5 \times 2)}{2} 2^{-4}$$

$$= 1 \times 2^{-1} + 0 \times 2^{-2} + 1 \times 2^{-3} + 0 \times 2^{-4} + 1 \times 2^{-5}$$

and when written in binary form is $(0.10101)_2$.

The algebraic example for the conversion of the decimal fraction to the binary fraction can be simplified as shown in Example A-47.

Example A-47

$$\begin{array}{r} 0.65625 \\ \times \quad 2 \\ \hline 1.31250 \end{array}$$ First digit 0.1

$$\begin{array}{r} 0.31250 \\ \times \quad 2 \\ \hline 0.6250 \end{array}$$ First two digits 0.10

$$\begin{array}{r} 0.6250 \\ \times \quad 2 \\ \hline 1.2500 \end{array}$$ First three digits 0.101

$$\begin{array}{r} 0.25 \\ \times \quad 2 \\ \hline 0.50 \end{array}$$ First four digits 0.1010

$$\begin{array}{r} 0.50 \\ \times \quad 2 \\ \hline 1.00 \end{array}$$ Total digits $(0.10101)_2$

Table A-17
Values of 2 Raised to Negative Powers

n	2^{-n}
0	1.0
1	0.5
2	0.25
3	0.125
4	0.062 5
5	0.031 25
6	0.015 625
7	0.007 812 5
8	0.003 906 25
9	0.001 953 125
10	0.000 976 562 5
11	0.000 488 281 25
12	0.000 244 140 625
13	0.000 122 070 312 5
14	0.000 061 035 156 25
15	0.000 030 517 578 125
16	0.000 015 258 789 062 5
17	0.000 007 629 394 531 25
18	0.000 003 814 697 265 625
19	0.000 001 907 348 632 812 5
20	0.000 000 953 674 316 406 25
21	0.000 000 476 837 158 203 125
22	0.000 000 238 418 579 101 562 5
23	0.000 000 119 209 289 550 781 25
24	0.000 000 059 604 644 775 390 625
25	0.000 000 029 802 322 387 695 312 5
26	0.000 000 014 901 161 193 847 656 25
27	0.000 000 007 450 580 596 923 828 125
28	0.000 000 003 725 290 298 461 914 062 5
29	0.000 000 001 862 645 149 230 957 031 25
30	0.000 000 000 931 322 574 615 478 515 625
31	0.000 000 000 465 661 287 307 739 257 812 5
32	0.000 000 000 232 830 643 653 869 628 906 25
33	0.000 000 000 116 415 321 826 934 814 453 125
34	0.000 000 000 058 207 660 913 467 407 226 562 5

The values of 2 raised to negative powers are given in Table A-17. The decimal fraction $(0.65625)_{10}$ is converted to the binary fraction as:

1. The decimal fraction $(0.65625)_{10}$ is greater than 2^{-1}; therefore, the value of 2^{-1}, which is 0.50000, is subtracted from the decimal fraction. The difference is 0.15625.
2. This difference is less than 2^{-2} but greater than 2^{-3}; this means that the decimal fraction is composed of a 0×2^{-2}) term. The value of 2^{-3}, which is 0.125, is subtracted from the difference of step 1; this time a difference of 0.03125 is obtained.

3. Reference to the table now indicates that 2^{-4} is greater than 0.03125 and that $2^{-5} = 0.03125$. The original decimal fraction contains a (0×2^{-4}) term and a (1×2^{-5}) term.
4. Summarizing the steps, the decimal $(0.65625)_{10} = 1 \times 2^{-1} + 0 \times 2^{-2} + 1 \times 2^{-3} + 0 \times 2^{-4} + 1 \times 2^{-5} = (0.10101)_2$.

The algebraic procedure for converting a decimal fraction to an octal fraction is illustrated in Example A-48.

Example A-48

$$
\begin{aligned}
(0.466{,}796{,}875)_{10} &= \frac{(0.466{,}796{,}875 \times 8)}{8} \\
&= (3.734375)8^{-1} = (3 + 0.734375)8^{-1} \\
&= 3 \times 8^{-1} + \frac{(0.734375 \times 8)}{8}8^{-1} \\
&= 3 \times 8^{-1} + (5 + 0.875)8^{-2} \\
&= 3 \times 8^{-1} + 5 \times 8^{-2} + (0.875)8^{-2} \\
&= 3 \times 8^{-1} + 5 \times 8^{-2} + \frac{(0.875 \times 8)}{8}8^{-2} \\
&= 3 \times 8^{-1} + 5 \times 8^{-2} + (7 + 0)8^{-3} \\
&= 3 \times 8^{-1} + 5 \times 8^{-2} + 7 \times 8^{-3}
\end{aligned}
$$

when written in octal form, $(0.466{,}796{,}875)_{10} = (0.357)_8$.

The algebraic procedure for converting a decimal fraction to the octal fraction can be further simplified as shown in Example A-49.

Example A-49

$$
\begin{array}{r}
0.466{,}796{,}875 \\
\times \qquad\qquad 8 \\
\hline
3.734375000
\end{array}
\qquad \text{First digit } 0.3
$$

$$\begin{array}{r} 0.734375 \\ \times \quad\quad 8 \\ \hline 5.875000 \end{array}$$

First two digits 0.35

$$\begin{array}{r} 0.875 \\ \times \quad 8 \\ \hline 7.000 \end{array}$$

$$(0.466{,}796{,}875)_{10} = (0.357)_8$$

Values of 8 raised to negative powers are listed in Table A-18. The use of this table for converting decimal fractions to octal fractions is illustrated by the following procedure for the decimal fraction $(0.466796875)_{10}$:

1. The decimal fraction is divided by 8^{-1}, which is 0.125; this gives a quotient of 3 with a remainder of 0.091796875.

2. The remainder (0.091796875) is now divided by 8^{-2} which from the table is 0.015625; a quotient of 5 with a remainder of 0.013671875 is obtained.

3. The last remainder is divided by 8^{-3}; a quotient of 7 is obtained.

4. Summarizing the three steps, $(0.466796875)_{10} = 3 \times 8^{-1} + 5 \times 8^{-2} + 7 \times 8^{-3} = (0.357)_8$.

The algebraic procedure for converting a decimal fraction to a hexadecimal fraction is illustrated in Example A-50.

Example A-50

$$(0.646240274)_{10} = \frac{(0.646240274 \times 16)}{16}$$

$$= (10 + 0.339844384)\ 16^{-1}$$
$$= (10 \times 16^{-1}) + \frac{(0.339844384 \times 16)}{16}\ 16^{-1}$$
$$= 10 \times 16^{-1} + (5 + 0.437510144)\ 16^{-2}$$
$$= 10 \times 16^{-1} + 5 \times 16^{-2} + \frac{(0.437510144 \times 16)}{16}\ 16^{-2}$$
$$= 10 \times 16^{-1} + 5 \times 16^{-2} + (7 + 0.000162304)\ 16^{-3}$$
$$= 10 \times 16^{-1} + 5 \times 16^{-2} + 7 \times 16^{-3} + 0.000162304 \times 16^{-3}$$

By rounding off and writing in hexadecimal form, it becomes

$$(\text{O.A57})_{16} \text{ for } (0.646240274)_{10}$$

Table A-18
Values of 8 Raised to Negative Powers

n	8^{-n}
0	1.0
1	0.125
2	0.015 625
3	0.001 953 125
4	0.000 244 140 625
5	0.000 030 517 578 1
6	0.000 003 814 697 26
7	0.000 000 476 837 158
8	0.000 000 059 604 644 7
9	0.000 000 007 450 580 59
10	0.000 000 000 931 322 574
11	0.000 000 000 116 415 321

The algebraic example for the conversion of the decimal fraction to the hexadecimal fraction can be simplified as shown in Example A-51.

Example A-51

$$\begin{array}{r} 0.646240274 \\ \times \quad\quad 16 \\ \hline 10.339844384 \end{array}$$ First digit 0.A

$$\begin{array}{r} 0.339844384 \\ \times \quad\quad 16 \\ \hline 5.437510144 \end{array}$$ First two digits 0.A5

$$\begin{array}{r} 0.437510144 \\ \times \quad\quad 16 \\ \hline 7.000162304 \end{array}$$ Total digits (0.A57)

The values of the negative powers of 16 are listed in Table A-19. By the use of this table, the decimal fraction $(0.646240274)_{10}$ is converted to the hexadecimal fraction as:

1. The value of 16^{-1} is found from the table to be 0.0625; the decimal fraction (0.646240274) is divided by 16^{-1} which gives a quotient of A plus a remainder of 0.021240274.

Table A-19
Values of 16 Raised to Negative Powers

n	16^{-n}
0	1.0
1	0.062 5
2	0.003 906 25
3	0.000 244 140 625
4	0.000 015 258 789 1
5	0.000 000 953 674 316
6	0.000 000 059 604 644 8
7	0.000 000 003 725 290 30
8	0.000 000 000 232 830 643
9	0.000 000 000 014 551 915 2

2. The remainder is divided by 16^{-2} which is 0.00390625; this gives a quotient of 5 with a remainder of 0.001709024.

3. The remainder from step 2 is divided by 16^{-3} which is 0.000244140625; this gives a quotient of 7.

4. Summarizing steps 1, 2 and 3, $(0.646240274)_{10} = A \times 16^{-1} + 5 \times 16^{-2} + 7 \times 16^{-3} = (O.A57)_{16}$.

Coding of Digits

The IBM 1620 operates as a decimal machine, but it is coded in binary. The arithmetic and control circuits operate in groups of four bits; in other words, four binary bits are required to represent one decimal digit. The decimal digits are represented by the binary bits as given in Table A-20.

The value of a decimal digit is the sum of the bits in the 8, 4, 2 and 1 columns.

The IBM 650 is coded in a biquinary system; each decimal digit is represented by one of two binary variables and one of five quinary variables. The binary variables are assigned values of 5 and 0; the five quinary variables are assigned values of 4, 3, 2, 1, 0. The coding for the IBM 650 is given in Table A-21.

The LGP-30 (Royal McBee) computer is strictly a binary computer. Approximately 3⅓ binary bits are required for representing one decimal digit. To represent 9 decimal digits, approximately 30 binary bits are required. For this computer, 30 binary bits are reserved for each decimal number; one of the bits designates the sign and the other bit designates the spacer.

Table A-20
Binary Coding of Decimal Digits

Decimal Digit	Binary Bits 8	4	2	1
0	0	0	0	0
1	0	0	0	1
2	0	0	1	0
3	0	0	1	1
4	0	1	0	0
5	0	1	0	1
6	0	1	1	0
7	0	1	1	1
8	1	0	0	0
9	1	0	0	1

Table A-21
Biquinary Coding

Decimal	Binary		Quinary				
	5	0	4	3	2	1	0
0	0	1	0	0	0	0	1
1	0	1	0	0	0	1	0
2	0	1	0	0	1	0	0
3	0	1	0	1	0	0	0
4	0	1	1	0	0	0	0
5	1	0	0	0	0	0	1
6	1	0	0	0	0	1	0
7	1	0	0	0	1	0	0
8	1	0	0	1	0	0	0
9	1	0	1	0	0	0	0

Table A-22
Hexadecimal Coding of Binary Bits

Hexadecimal Numbers	Binary Bits			
0	0	0	0	0
1	0	0	0	1
2	0	0	1	0
3	0	0	1	1
4	0	1	0	0
5	0	1	0	1
6	0	1	1	0
7	0	1	1	1
8	1	0	0	0
9	1	0	0	1
A	1	0	1	0
B	1	0	1	1
C	1	1	0	0
D	1	1	0	1
E	1	1	1	0
F	1	1	1	1

As an example of the binary number (32 bits including spacer and sign left)

Working directly with the large number of binary bits (32) is rather cumbersome. The LGP-30 is designed such that binary bits can be worked in groups of four. The use of hexadecimal numbers permits this grouping. The coding for binary bits in terms of hexadecimal digits is given in Table A-22.

5	7	1	1	7	D	3	0
0 1 0 1,	0 1 1 1,	0 0 0 1,	0 0 0 1,	0 1 1 1,	1 1 0 1,	0 0 1 1,	0 0 0 0

is written in hexadecimal as 57117D30.

Exercises

1. Perform the arithmetic operations of binary addition:

a.
```
  101010
+  11011
```

b.
```
  111011
+ 111111
```

c.
```
  101101
+ 100111
```

2. Perform the arithmetic operations of octal addition:

a.
```
  2347
+ 7651
```

b.
```
  3742
+ 4672
```

c.
```
  7777
+ 3105
```

3. Perform the arithmetic operations of hexadecimal addition:

a.
```
  ABC8
+ 179E
```

b.
```
  BF78
+ 509F
```

c.
```
  7789
+ 7651
```

4. Perform the arithmetic operations of binary subtraction by three methods (table, 2's complements and 1's complements):

a.
```
  101010
-  11011
```

b.
```
  111101
- 101011
```

c.
```
  110101
-  11010
```

5. Perform the arithmetic operations of octal subtraction by three methods (use of table, 8's complements, 7's complements):

a.
```
  5674
- 3767
```

b.
```
  6753
- 2564
```

c.
```
  4321
- 1432
```

6. Perform the arithmetic operations of hexadecimal subtraction by three methods (table, 16's complements, 15's complements):

a. ABC8
   − 179E

b. BF78
   − 509F

c. 7789
   − 6541

7. Perform the arithmetic operations of binary multiplication by two methods (multiplication table and successive additions):

a. 11011
   × 101

b. 11101
   × 1011

c. 1110101
   × 1101

8. Perform the arithmetic operations of octal multiplication by two methods (multiplication table and successive additions):

a. 7645
   × 27

b. 3265
   × 65

c. 4762
   × 43

9. Perform the arithmetic operations of hexadecimal multiplication by two methods (multiplication table and successive additions):

a. ABC8
   × 2E

b. 8F78
   × 57

c. 7789
   × D3

10. Perform the arithmetic operations of binary division by four methods (longhand, successive subtractions, reciprocals and iteration):

a. 1010 $\overline{)1110111}$

b. 1101 $\overline{)1010101}$

c. 1001 $\overline{)111111}$

11. Perform the arithmetic operations of octal division by four methods (longhand, successive subtractions, reciprocals and iteration):

a. 361 $\overline{)57063}$

b. 453 $\overline{)14621}$

c. 762 $\overline{)27645}$

12. Perform the arithmetic operations of hexadecimal division by four methods (longhand, successive subtractions, reciprocals and iteration):

a. AB $\overline{)C604}$

b. E5 $\overline{)F965}$

c. 98 $\overline{)1045}$

13. Perform the base conversions by two methods (algebraic and use of tables):

a. $(11011)_2 = (\quad)_8 = (\quad)_{10}$
b. $(5723)_8 = (\quad)_{10} = (\quad)_{16}$
c. $(94CE)_{16} = (\quad)_{10} = (\quad)_8$
d. $(5783)_{10} = (\quad)_8 = (\quad)_{16}$
e. $(29)_{10} = (\quad)_2 = (\quad)_8$
f. $(11.0101)_2 = (\quad)_{10} = (\quad)_8$
g. $(5.672)_8 = (\quad)_{10} = (\quad)_{16}$
h. $(7.B12)_{16} = (\quad)_{10} = (\quad)_8$
i. $(0.6431)_{10} = (\quad)_2 = (\quad)_8$
j. $(0.511)_{10} = (\quad)_8 = (\quad)_{16}$
k. $(0.233)_{10} = (\quad)_2 = (\quad)_{16}$
l. $(0.798)_{10} = (\quad)_8 = (\quad)_{16}$

Suggested Reading

1. Engineering Research Associates, Inc., supervised by C. B. Tompkins, and J. H. Wakelin and edited by W. W. Stifler, Jr., *High-Speed Computing Devices,* New York: McGraw-Hill Book Company, 1950.
2. Eves, H., *An Introduction to the History of Mathematics,* New York: Rinehart and Company, 1953.
3. McCormick, E. M., *Digital Computer Primer,* New York: McGraw-Hill Book Company, 1959.
4. McCracken, D. D., *Digital Computer Programming,* New York: John Wiley & Sons, 1957.
5. Mims, Forrest M., III, "Understanding Digital Computers," Catalog No. 62-2027, Fort Worth, Texas: Tandy Corporation, 1978.
6. Osborne, A., *An Introduction to Microcomputers,* Vol. 1, 2nd Ed., Berkeley, California: Osborne/McGraw-Hill, 1980.
7. Reference Manual A26-4500-2, *IBM 1620 Data Procesing Systems,* New York: International Business Machines Corporation, 1961.
8. Reference Manual F-28-8074-1, *IBM 1620 FORTRAN,* New York: International Business Machines Corporation, 1962.
9. Richards, R. K., *Arithmetic Operations in Digital Computers,* New York: D. Van Nostrand Company, 1955.
10. *Royal Precision Electronic Computer LPG-30 Programming Manual,* Port Chester, New York: Royal McBee Corporation, 1957.
11. Waite, M. and Pardee, M., *Basic Programming Primer,* 2nd Ed., Indianapolis, Indiana: Howard W. Sams and Co., Inc., 1982.
12. Wrubel, M. H., *A Primer of Programming for Digital Computers,* New York: McGraw-Hill Book Company, 1959.

Author Index

Subject Index